Polyamines
in
Biomedical Research

Polyamines in Biomedical Research

Edited by

JOSEPH M. GAUGAS

Cancer Research Campaign,
Gray Laboratory,
Northwood, Middlesex, UK

A Wiley–Interscience Publication

JOHN WILEY & SONS

Chichester · New York · Brisbane · Toronto

British Library Cataloguing in Publication Data

Polyamines in biomedical research.
1. Polyamines in the body
I. Gaugas, Joseph M
574.1'924 QP801.A48 79-40651

ISBN 0 471 27629 4

Phototypeset by Dobbie Typesetting Service, Plymouth, Devon
and printed by Page Bros. (Norwich) Ltd., Norwich.

Contributors

A. C. ALLISON — *Division of Cell Pathology, MRC Clinical Research Centre, Watford Road, Harrow, Middlesex HA1 3UJ, UK.*
Present Address:
International Laboratory for Research on Animal Diseases, Nairobi, Kenya.

A. C. ANDERSSON — *Dept. of Physiology and Biophysics, University of Lund, S-223 62 Lund, Sweden.*

G. ANDERSSON — *AB Ferrosan, Fack, S-201 10 Malmö, Sweden.*

U. BACHRACH — *Dept. of Molecular Biology, Hebrew University — Hadassah Medical School, Jerusalem, Israel.*

N. M. BARFOD — *The Institute of Cancer Research, Radiumstationen, DK-8000 Aarhus C, Denmark.*

P. BEY — *Centre de Recherche Merrell International, 16, rue d'Ankara, 67084 - Strasbourg Cedex, France.*

E. S. BONE — *Bacterial Metabolism Research Laboratory, Central Public Health Laboratory, Colindale Avenue, London NW9 5HT, UK.*
Present Address:
Tissue Cell Relationships Laboratory, Imperial Cancer Research Fund, Lincoln's Inn Fields, London WC2, UK.

R. K. BOUTWELL — *McArdle Laboratory for Cancer Research, University of Wisconsin, Medical Center, Madison, Wisconsin 53706, USA.*

A. L. BOYNTON — *Animal and Cell Physiology Group, Division of Biological Sciences, National Research Council of Canada, Ottawa, Canada K1A 0R6.*

E. S. Canellakis	*Dept. of Pharmacology, Yale University Medical School, New Haven, CT 06510, USA.*
M. Carteni'-Farina	*Dept. of Biochemistry, First Medical School University of Naples, Naples, Italy.*
P. Correa	*Depts. of Microbiology and Pathology, Louisiana State University Medical Center, New Orleans, Louisiana 70119, USA.*
M. De Rosa	*Laboratory for the Chemistry of Molecules of Biological Interest, C.N.R., Arco Felice, Naples, Italy.*
H. Desser	*Ludwig Botzmann Institute for Leukaemia Research & Haematology, Hanuschkrankenhaus, A-1140 Vienna, Austria.*
D. L. Dewey	*Cancer Research Campaign Gray Laboratory, Mount Vernon Hospital, Northwood, Middlesex HA6 2RN, UK.*
T. Eloranta	*Dept. of Biochemistry, University of Kuopio, Kuopio, Finland.*
J. Ferulga	*Division of Cell Pathology, MRC Clinical Research Centre, Watford Road, Harrow, Middlesex HA1 3UJ, UK.*
A. Gambacorta	*Laboratory for the Chemistry of Molecules of Biological Interest, C.N.R., Arco Felice, Naples, Italy.*
J. M. Gaugas	*Cancer Research Campaign Gray Laboratory, Mount Vernon Hospital, Northwood, Middlesex HA6 2RN, UK.*
J. Harada	*Dept. of Biochemistry, University of Washington, Seattle, Washington 98195, USA.*
O. Heby	*Dept. of Zoophysiology, University of Lund, Helgonävagen 3b, S-223 62 Lund, Sweden.*
J. S. Heller	*Dept. of Pharmacology, Yale University Medical School, New Haven, CT 06510, USA.*
S. Henningsson	*Dept. of Physiology & Biophysics, University of Lund, S-223 62 Lund, Sweden.*

H. HIBASAMI — *Dept. of Biochemistry, Mie University School of Medicine, Tsu, Mie 514, Japan.*

K. J. HIMMELSTEIN — *Dept. of Chemical and Petroleum Engineering, University of Kansas, Lawrence, Kansas 66054, USA.*

T. IKEDA — *Dept. of Biochemistry, Mie University School of Medicine, Tsu, Mie 514, Japan.*

H. M. KEIR — *Dept. of Biochemistry, University of Aberdeen, Marischal College, Aberdeen AB9 1AS, Scotland.*

J. KOCH-WESER — *Centre de Recherche Merrell International, 16 rue d'Ankara, 67084 - Strasbourg Cedex, France.*

P. S. MAMONT — *Centre de Recherche Merrell International, 16, rue d'Ankara, 67084 - Strasbourg Cedex, France.*

R. MÄNTYJÄRVI — *Dept. of Clinical Microbiology, University of Kuopio, Kuopio, Finland.*

P. P. MCCANN — *Centre de Recherche Merrell International, 16, rue d'Ankara, 67084 - Strasbourg Cedex, France.*

M. A. L. MELVIN — *Dept. of Biochemistry, University of Aberdeen, Marischal College, Aberdeen AB9 1AS, Scotland.*

D. M. L. MORGAN — *Division of Cell Pathology, MRC Clinical Research Centre, Watford Road, Harrow, Middlesex HA1 3UJ, UK.*

D. R. MORRIS — *Dept. of Biochemistry, University of Washington, Seattle, Washington 98195, USA.*

M. L. MURRAY — *Depts. of Microbiology & Pathology, Louisiana State University Medical Center, New Orleans, Louisiana 70119, USA.*

J. NAGAI — *Dept. of Biochemistry, Mie University School of Medicine, Tsu, Mie 514, Japan.*

T. G. O'BRIEN — *The Wistar Institute of Anatomy and Biology, 36th Street at Spruce, Philadelphia, Pennsylvania 19104, USA.*

R. L. PAJULA — *Dept. of Biochemistry, University of Kuopio, Kuopio, Finland.*

A. RAINA — *Dept. of Biochemistry, University of Kuopio, Kuopio, Finland.*

M. G. ROSENBLUM — *University of Arizona, Health Sciences Center, Tucson, Arizona 85724, USA.*
Present Address:
University of Texas, M. D. Anderson Hospital and Tumor Institute, Houston, Texas, USA.

E. ROSENGREN — *Dept. of Physiology & Biophysics, University of Lund, S-223 62 Lund, Sweden.*

N. SEILER — *Centre de Recherche Merrell International, 16, rue d'Ankara, 67084 - Strasbourg Cedex, France.*

E. STEVENS — *Dept. of Biochemistry, University of Stirling, Stirling, Scotland.*

L. STEVENS — *Dept. of Biochemistry, University of Stirling, Stirling, Scotland.*

M. TANAKA — *Dept. of Biochemistry, Mie University School of Medicine, Tsu, Mie 514, Japan.*

K. TUOMI — *Dept. of Clinical Microbiology, University of Kuopio, Kuopio, Finland.*

A. K. VERMA — *McArdle Laboratory for Cancer Research, University of Wisconsin, Medical Center, Madison, Wisconsin 53706, USA.*

P. R. WALKER — *Animal and Cell Physiology Group, Division of Biological Sciences, National Research Council of Canada, Ottawa, Canada K1A 0R6.*

J. F. WHITFIELD — *Animal and Cell Physiology Group, Division of Biological Sciences, National Research Council of Canada, Ottawa, Canada K1A 0R6.*

V. ZAPPIA — *Dept. of Biochemistry, First Medical School University of Naples, Naples, Italy.*

Contents

Introduction

Polyamines are synthesized in animal and bacterial cells. They are structurally simple cationic compounds whose trivial names (putrescine, cadaverine, spermidine and spermine) divert the attention of the casual reader from their importance in cell proliferation (Chapters 1, 2 and 5) and their potential involvement in certain tissue-degenerative pathologies.

The early history of polyamines in biomedicine has been reviewed elsewhere[1-3]. Indeed until recently polyamines had the reputation as metabolic end-products or waste materials. Cadaverine is derived from lysine, and putrescine from arginine, during animal tissue putrefaction, so they had been assumed to be of no physiological significance (this remains true only for cadaverine). Importantly, putrescine was next shown to be formed in eukaryotic cells by enzymic decarboxylation of ornithine. Mammalian cell polyamines are represented by putrescine which is the precursor for spermidine and spermine; the latter is the end-product of the biosynthetic pathway which is restricted to both G phases of the cell's growth-division cycle (Chapter 2). Though strictly a diamine, putrescine (1,4-diaminobutane) is grouped for convenience together with polyamines. Hence the new nomenclature of 'oligoamines' might gain in popularity (Chapter 11).

Celia and Herbert Tabor, S. Cohen, U. Bachrach, D. Russell and a few others, proposed that polyamines were a prerequisite for vital processes of cell proliferation. Consequently, limited pharmacological interest arose, now widening, for use as a means of control of proliferation in both cancer cells and microbial pathogens. Prokaryotic cells can produce novel polyamines, perhaps with interest for the design of analogues possessing antimicrobial activity (Chapter 16).

Particular attention is paid to studies with cell ornithine decarboxylase which is the rate-limiting enzyme in polyamine biosynthesis, and its regulation by cytoplasmic 'antizyme' (Chapters 6-9).

Dykstra and Herbst in 1965 were probably the first to demonstrate the participation of polyamines in cell proliferation mechanisms. Subsequently, the polyamine story was somewhat confused until it was realized that a cell can apparently retain a reserve pool of inert spermine for transfer to its daughters at division. Specific inhibition of polyamine biosynthesis with a variety of drugs results in proliferation arrest of exposed daughter but not of exposed parent cells (Chapters 10, 11 and 22). This suggests a requirement for activation of polyamines (Chapter 8). The precise role of polyamines in support of proliferation has yet to be fully elucidated. Their support of protein synthesis and nucleic-acid replication appears essential (Chapters 1-6, 8, 10, 22) with reservations (Chapter 5).

Polyamines are chemically stable; they are secreted by cells (Chapter 23), occur in physiological fluids as conjugates and are excreted in the urine (Chapters 25-27). Methods for polyamine measurement are evaluated (Chapter 27) since levels are currently being scrutinized as markers for disease activity in malignancy and other tissue-proliferative disorders (Chapter 26). Excretion increases in body states of elevated tissue-growth, e.g. leukaemia (where the tumour burden is great), pregnancy and liver regeneration (Chapters 25-27).

Catabolism of exogenous diamines and polyamines is due to reactivity with specific oxidative enzymes (Chapter 17 & 18). Interestingly, primary products are labile aldehydes and *in vitro* at least the oxidation system potently arrests cell proliferation (Chapters 20-22). Malfunction of catabolism in tissue pathologies, especially involving amine oxidase-rich tissues including the liver, kidney, lymphoid organs, placenta and brain is therefore implicated. Research into the participation of such aldehydes in either the aetiology or symptomatology of certain pathologies is now beginning. In fact, polyamine oxidase in human pregnancy and hepatitis sera are recent discoveries (Chapters 19 and 22). Aldehydes derived from monoamine oxidase interaction with substrates is associated with psychiatric disease and specificity of neuron toxicity accords with substrate[4]. No toxicity specificity should occur for cells affected by oxidized diamines or polyamines, however, since unlike monoamines they are ubiquitous for body tissues.

Oxidation of polyamines and polyamine-complexes have been implicated in 'chalones' (Chapters 20 and 21), and in immunoregulation (Chapter 22). Occurrence of polyamine oxidase in macrophages is inferred in the elusive killing mechanism for both foreign cells and phagocytosed pathogens (Chapter 19). Gaps inevitably occur in our knowledge of the polyamine oxidation system which is apparently distinct from mere polyamine catabolism. Unknowns include characterization and toxicity of products of labile oxidized polyamines and whether the system generates free radicals (superoxide and hydroxyl) thought to be deleterious to cells; roles for plasma membrane lipid peroxidation, mediation of inflammation and ageing. There could be a clinical need for drugs which inhibit polyamine and/or diamine oxidase.

Secondary amines from gut bacterial flora are suitable agents for nitrosation (Chapter 13). Nitroso-amines and nitroso-amides are known carcinogens and polyamines are suitable substrates for nitrosation (Chapter 14). Under defined conditions polyamines could, conversely, be involved in activity of anticarcinogens (Chapter 12).

JOSEPH M. GAUGAS

REFERENCES

1. Cohen, S. S., (1971) *Introduction to the Polyamines.* Prentice-Hall, New Jersey.
2. Bachrach, U., (1973). *Function of the Naturally Occurring Polyamines.* Academic Press, New York and London.
3. Kapeller-Adler, R., (1970) *Amine Oxidases and Methods for their Study.* Wiley, New York.
4. Cohen, G., (1978) The generation of hydroxyl radicals in biologic systems: toxological aspects. *Photochem. & Photobiol.*, **28**, pp. 669–675.

Chapter 1

Participation of Polyamines in the Proliferation of Bacterial and Animal Cells

DAVID R. MORRIS AND JOHN J. HARADA

I. INTRODUCTION

The relationship of polyamines to cell proliferation has been a subject of intense interest in recent years (see Chapters 2–10, 20–22). One of the first suggestions that polyamines might be involved in the proliferative response of cells and tissues came from the pioneering work of Dykstra and Herbst (1965) with regenerating liver. These workers demonstrated increases in tissue levels of spermidine early in the regenerative process. Since this germinal observation, the correlation between stimulation of cell growth and large increases in the rate of polyamine biosynthesis has been extended to a wide variety of systems (reviewed in Morris and Fillingame, 1974). This correlative approach was extended to its ultimate in the statistical study by Heby and coworkers (1975).

Although these correlations were highly suggestive, they clearly did not prove that increased cellular polyamine levels were essential for cell multiplication. Accelerated rates of polyamine biosynthesis, with concomitant increases in cellular polyamine levels, simply could have been non-essential by-products of the pleiotypic growth response of cells (Hershko and coworkers, 1971). It was only with the isolation of well-defined, putrescine auxotrophs of *Escherichia coli* that the growth requirement for polyamines was proved (Hirshfield and coworkers, 1970; Morris and Jorstad, 1970, 1973). Studies with these mutants have provided considerable insight into the physiology of the polyamine-deficient state (Morris, 1973) and a review of this work will comprise one major section of this chapter. Mutants have also recently revealed a polyamine requirement for growth in yeast (Whitney, Pious and Morris, 1978; Cohn, Tabor and Tabor, 1978; Whitney and Morris, 1978).

The mutational approach has not yet been extended to studies with animal systems, although it might be fruitful. Alternatively, drugs which inhibit polyamine biosynthesis have received considerable attention by workers in the animal cell field. Specific inhibitors are now available for S-adenosylmethionine decarboxylase (Corti and coworkers, 1974) and ornithine decarboxylase (Abdel-Monem, Newton and Weeks, 1974; Mamont and coworkers, 1976, 1978). Studies with these inhibitors provided the first evidence that appropriate intracellular levels of polyamines were indeed essential for optimal animal cell proliferation (Otani and coworkers, 1974; Fillingame, Jorstad and Morris, 1975; Mamont and coworkers, 1976). This work and subsequent studies relating to the role of polyamines in animal cell growth, will be reviewed in the second section of this chapter.

II. BACTERIAL MUTANTS

A. Mutant Isolation and Genetics

Attempts to isolate putrescine auxotrophs of *E. coli* were unsuccessful until it was recognized that this organism possessed multiple biosynthetic routes to this metabolite (reviewed in Morris and coworkers, 1970; Morris and Fillingame, 1974). By taking into consideration the complexities of putrescine biosynthesis, mutants were isolated initially either by direct screening (Morris and Jorstad, 1970) or by taking advantage of their arginine-sensitive phenotype (Hirshfield and coworkers, 1970). Since these early studies, Maas and his colleagues have isolated a collection of mutants defective in each of the enzymes of putrescine biosynthesis: arginine decarboxylase, agmatine ureohydrolase and ornithine decarboxylase. The genetic loci for these enzymes have been designated *speA*, *speB* and *speC*, respectively. They map between 56 and 57 min on the *E. coli* map, close to *metK*, the gene for S-adenosylmethionine synthetase (Maas, 1972; Cunningham-Rundles and Maas, 1975). Recently, using a screening procedure based on $[^{14}C]O_2$ production (Tabor, Tabor and Hafner, 1976), mutants deficient in S-adenosylmethionine decarboxylase have been isolated. This locus has been designated *speD*, and maps away from the other genes of polyamine synthesis, at about 2.7 min (Hafner, Tabor and Tabor, 1977). The physiological properties of these mutants will be of interest since they contain putrescine, but not spermidine.

B. Characteristics of Polyamine-limited Growth

To examine the physiology of the polyamine-deficient state, we have used a mutant that is partially blocked in arginine decarboxylase (*speA*) (Morris and

Jorstad, 1973). This mutant was chosen for several reasons. (1) Since the mutation is in the first step of polyamine biosynthesis, there is no accumulation of agmatine or putrescine during starvation. (2) This leaky mutant can be maintained indefinitely in exponential, polyamine-limited growth. Hence, the measurements described below were made on growing cultures, and not on cells in an undefined terminal state of polyamine starvation. (3) Because of the steady-state of growth, results were quite reproducible from experiment to experiment.

An extensive study has been made of the macromolecular composition of *E. coli* during polyamine-limited growth (Morris and Jorstad, 1973). During starvation which resulted in a three-fold reduction in growth rate, cell size, as defined microscopically and by cellular protein content, was reduced approximately two-fold (Morris and Jorstad, 1973). This is in distinct contrast to the filamentous forms described in another report (Maas, Leifer and Poindexter, 1970). In the latter study, however, the cells were harvested from agar plates and were quite possibly not in a steady-state of growth. The DNA content per cell was reduced approximately 20% in the putrescine-starved cells, but was still approximately two-fold higher than anticipated from the growth rate (discussed below). Analysis of the RNA content of polyamine-starved cells revealed that growth was not limited by either the level of transfer RNA and active ribosomes or by the degree of RNA methylation (Morris and Jorstad, 1973; Jorstad and Morris, 1974).

We have initiated a study of the specificity of the polyamine requirement of these mutants using a series of homologues of the structure:

$$H_2N(CH_2)_xNH(CH_2)_3NH_2$$

The compounds with $x = 3$ and $x = 4$ (spermidine) were equally as effective as putrescine in restoring growth. The homologues with $x = 5$ and $x = 6$ showed decreasing ability to support growth, while those with $x = 7$ and $x = 8$ were totally ineffective. These growth studies were carried out at identical intracellular concentrations of the various triamines (C. M. Jorstad and D. R. Morris, unpublished). These results suggest that the polyamine requirement in these mutants is not simply for a triamine in general, but that a definite spacial relationship between the amine groups is required. This may be related to a model for polyamine binding across the minor groove of DNA (Liquori and coworkers, 1967).

C. Protein and RNA Synthesis

During polyamine starvation, the rates of RNA and protein synthesis were decreased identically to the reduction in growth rate (Morris, 1973). To define the mechanism of inhibition of RNA and protein synthesis during polyamine starvation, we investigated the synthesis of β-galactosidase and its messenger

RNA (Morris and Hansen, 1973). The rates of addition of monomers to these two macromolecules were decreased identically to the reduction in growth rate. In the case of polypeptide chain elongation, this result was confirmed for total cellular protein by two other independent techniques (Jorstad and Morris, 1974). Since the chain elongation rates and the cellular growth rate were decreased similarly, it seemed likely that the intracellular site of polyamine action was rather intimately related to these processes. Since translation occurs simultaneously with transcription of nascent messenger RNA in bacteria, however, it is impossible to say whether polyamine starvation directly affects transcription, translation or both (Morris and Hansen, 1973; Jorstad and Morris, 1974). These reduced rates of RNA and protein chain elongation are integral parts of the model for polyamine action in bacteria which will be presented below.

D. DNA Replication

As pointed out above, polyamine-limited growth of *E. coli* results in a cellular DNA content which is approximately twice as high as expected from the growth rate. This situation could arise if the rate of replication fork movement in starved cells were reduced 50%, while maintaining a similar number of replication forks (Morris and Jorstad, 1973). This would reduce the overall rate of DNA replication by half, the amount by which cell growth is decreased. This prediction has now been confirmed experimentally (Geiger and Morris, 1978). This is an unusual response to decreased growth rate, since the velocity of growing fork movement is generally constant in *E. coli.* The rate of chromosome replication is normally adjusted to the cellular growth rate by variations in the frequency of initiation, and hence in the number of growing forks per cell (Cooper and Helmstetter, 1968). Therefore, the decreased rate of replication fork movement during polyamine limitation of growth is not simply a normal cellular regulatory mechanism responding to decreased protein and RNA synthesis, but quite likely reflects an intracellular function of the polyamines.

E. Function of Polyamines in Bacteria: A Model

The effects of polyamine-limited growth on macromolecular synthesis in *E. coli* mutants have been clearly defined. As described above, putrescine starvation results in decreased rates of monomer addition to nascent chains of protein, RNA and DNA (i.e. decreased step times or chain elongation rates). As has been discussed previously (Morris and Hansen, 1973; Geiger and Morris, 1978), these are not normal cellular responses to reduced growth rate; macromolecular elongation rates are usually invariant at such growth rates. These three effects of polyamine starvation can be accommodated in a model involving a single cellular site of polyamine action, the enzyme DNA gyrase.

DNA gyrase catalyses the ATP-dependent introduction of negative super-coils into closed circular DNA and shows a strong dependence on spermidine for activity (Gellert and coworkers, 1976a). This enzyme has been implicated in cellular DNA replication by virtue of its antibiotic sensitivity (Gellert and coworkers, 1976b). Since the enzyme quite probably controls the super-coiling of cellular DNA during replication, one would expect that depressing DNA gyrase activity as a result of cellular spermidine deficiency should have profound effects on macromolecular synthesis. Decreased intracellular gyrase activity should reduce DNA replication velocity for the enzyme seems to counteract the positive superhelical turns introduced by replication (Gellert and coworkers, 1976a). As discussed above, decreased velocity of replication fork movement indeed was found in polyamine-deficient cells (Geiger and Morris, 1978). In addition, increased torsional stress on chromosomal DNA in the polyamine-deficient state could also interfere with RNA polymerase activity, accounting for the reduced rate of RNA chain elongation seen under these conditions. As previously suggested (Morris and Hansen, 1973), inhibition of polypeptide chain growth could then be due simply to jamming of ribosomes behind the slowly moving RNA polymerase molecules. Thus, reduced DNA, RNA and protein synthesis in polyamine-deficient *E. coli* can be accounted for solely by the single intracellular action of spermidine on DNA gyrase.

To test this model, we have taken advantage of the antibiotic novobiocin, an inhibitor of DNA gyrase (Gellert and coworkers, 1976b). We have examined the physiological properties of *E. coli* growing in the presence of partially inhibitory concentrations of this drug. The phenotype was strikingly similar to that of polyamine-deficient cells. Partial inhibition of DNA gyrase by novobiocin resulted in decreased chain elongation rates for both messenger RNA and protein, as well as reduced DNA synthesis (C. M. Jorstad and D. R. Morris, unpublished). Hence these findings are consistent with the inhibition of RNA and protein synthesis, observed in polyamine-starved *E. coli* being the result of reduced DNA gyrase activity.

III. ANIMAL CELLS

This section will review primarily those studies aimed at defining the role of polyamines in the proliferation of animal cells grown *in vitro*. More general coverage of the animal cell literature may be found in other recent publications (Tabor and Tabor, 1976; Campbell and coworkers, 1978a,b).

A. Variations in the Levels of the Polyamine Biosynthetic Enzymes During the Cell Cycle

Synthesis of the polyamines in animal cells is regulated primarily through the activities of two enzymes: L-ornithine decarboxylase catalyses the only known route of putrescine biosynthesis in animal cells, and S-adenosyl-L-methionine decarboxylase provides the *n*-propylamine moiety necessary for both spermidine and spermine biosynthesis (Morris and Fillingame, 1974; Tabor and Tabor, 1976). Changes in the activities of these enzymes occurring during the cell cycle are discussed below.

The level of the polyamine biosynthetic enzymes have been observed to increase when quiescent, G_1-arrested (G_0) cells are stimulated to proliferate. In small lymphocytes, the activities of both ornithine decarboxylase and S-adenosylmethionine decarboxylase were elevated very soon after activation by mitogens. In addition, S-adenosylmethionine decarboxylase activity exhibited a second peak of activity which corresponded with the entry of these cells into S phase (Fillingame and Morris, 1973a; Fillingame, 1973; Kay and Lindsay, 1973). A similar response was observed in mouse kidney cells during lytic infections with polyoma virus (Goldstein, Heby and Marton, 1976) in which the activities of both enzymes were elevated biphasically; the first rise in activity occurred almost immediately after infection, while the second peak closely followed the induction of DNA synthesis.

Cell cycle-specific variations in the activities of the polyamine biosynthetic enzymes in continuous cell lines have been studied in populations of cells synchronized either by mitotic selection (Russell and Stambrook, 1975; Heby and coworkers, 1976; Fuller, Gerner and Russell, 1977) or by colcemid treatment (Friedman Bellantone and Canellakis, 1972). Ornithine decarboxylase activity increased biphasically, reaching activity maxima during late G_1 and again in late S/G_2. The activity of S-adenosylmethionine decarboxylase closely followed the initial rise of ornithine decarboxylase and peaked during S phase. Thus, the stimulation of polyamine biosynthesis appears to occur just prior to, or concomitant with, DNA replication in animal cells. The significance of this observation will be discussed below.

B. Involvement of Polyamines in the Cell Cycle

A major thrust of polyamine research has been directed towards defining the involvement of polyamines in the cell cycle. A number of workers have used inhibitors of polyamine biosynthesis in order to examine the effects of polyamine-deficiency on the cell cycle. When such inhibitors are used, it is mandatory to determine whether the results obtained either represent the physiological effects of polyamine-deficiency or if they are due to an unrelated pharmacological action of the inhibitor. An example of this comes from studies on protein synthesis in mitogen-activated lymphocytes. When a relatively high concentration of methylglyoxal *bis*(guanylhydrazone) (methyl-GAG), an inhibitor of S-adenosylmethionine decarboxylase (Corti and coworkers, 1974), was used to depress polyamine accumulation during mitogenesis, the rate of phenylalanine incorporation was reduced (Kay and

Pegg, 1973). In contrast another study demonstrated that a much lower concentration of methyl-GAG, which inhibited polyamine accumulation to the same extent as the higher dose, had no effect on protein synthesis (Fillingame and Morris, 1973b). In addition, methyl-GAG has been shown to cause ultrastructural changes in mitochondria *in vivo* (Pathak, Porter and Dave, 1977) and also in isolated nuclei (Brown, Nelson and Brown, 1975). In the latter instance, the alterations were certainly unrelated to polyamine function. Thus, although inhibitors such as methyl-GAG have proved to be an invaluable tool in the study of the function of polyamines, the results must be interpreted with caution.

Evidence has accumulated to indicate that one site of polyamine function involves the DNA synthetic phase of the cell cycle. Putrescine added to the culture medium stimulated the growth rate of human fibroblasts by decreasing the length of S phase (Pohjanpelto, 1975). Modulation of the growth rate in animal cells normally is mediated through changes in the length of the G_1 phase of the cell cycle (Tobey, Anderson and Peterson, 1967). This decrease in the length of S phase therefore suggests that putrescine acts specifically to increase the rate of DNA synthesis.

Methyl-GAG has been used to inhibit spermidine and spermine accumulation in mitogen-activated lymphocytes. In these polyamine-deficient cells, the rate of thymidine incorporation was inhibited (Otani and coworkers, 1974; Fillingame, Jorstad and Morris, 1975), with no alteration in either the time of entry of the cells into S phase or the percentage of cells in S phase at any time during the course of the experiment. Thus, DNA synthesis was specifically inhibited, with no effect on pre-replicative events. Use of α-methylornithine (α-MO), a competitive inhibitor of ornithine decarboxylase (Abdel-Monem, Newton and Weeks, 1974; Mamont and coworkers, 1976), in combination with methyl-GAG provided evidence that putrescine was capable of supporting DNA synthesis, though to a lesser extent than either spermidine or spermine (Morris, Jorstad and Seyfried, 1977). Considerable evidence has been amassed suggesting that the inhibition of DNA synthesis by methyl-GAG was due to reduced polyamine levels rather than a non-specific pharmacological effect of the inhibitor (Fillingame, Jorstad and Morris, 1975; Morris, Jorstad and Seyfried, 1977). This conclusion has been recently strengthened considerably by the observation that the inhibitor of ornithine decarboxylase, α-difluoromethylornithine (Mamont and coworkers, 1978), elicits a physiological response in lymphocytes identical to that found with methyl-GAG (Morris and Seyfried, in preparation).

DNA synthesis has been studied in nuclei isolated from polyamine-deficient HeLa cells and lymphocytes (Krokan and Eriksen, 1977; Knutson and Morris, 1978). Isolated nuclei exhibited the same relative inhibition of DNA synthesis as that observed in whole cells. Thus, the inhibition of thymidine incorporation observed in polyamine-deficient cells was not due to a decreased specific radioactivity of the intracellular dTTP pool, since this parameter was controlled in the *in vitro* systems. Addition of polyamines to the isolated nuclei did not restore optimal rates of DNA synthesis. Because nuclei are incapable of initiating replicons *in vitro* (Hershey, Stieber and Mueller, 1973), one interpretation of this result is that DNA synthesis was inhibited in polyamine-deficient nuclei because of a decrease in the number of active replication units. Consistent with this hypothesis is evidence suggesting that the rate of DNA chain elongation was only minimally inhibited in polyamine-deficient HeLa cells (Krokan and Eriksen, 1977).

It is apparent that polyamines are required for optimal rates of cellular DNA replication. Others have reached this same conclusion using different biological systems and in many cases, with structurally unrelated inhibitors of polyamine biosynthesis (Inoue and coworkers, 1975; Pösö and Jänne, 1976; Sunkara, Rao and Nishioka, 1977; Wiegand and Pegg, 1978). There has been a recent report, however, that thymidine uptake and phosphorylation were inhibited in methyl-GAG treated lymphocytes (Otani, Matsui and Morisawa, 1977). These workers concluded that the decreased intracellular level of radioactive dTTP was sufficient to account for the inhibition of thymidine incorporation. This conclusion conflicts with the results obtained with nuclei from polyamine-deficient cells in which exogenous dTTP was supplied to the DNA synthesizing systems (Krokan and Eriksen, 1977; Knutson and Morris, 1978). The possibility exists that depletion of the intracellular thymine nucleotide pool somehow altered the replication machinery and made the nuclei less active. This alternative seems unlikely, however. When dTTP was supplied to nuclei from amethopterin-treated cells, they were as active in DNA replication as nuclei isolated from actively replicating cells (Hershey, Stieber and Mueller, 1973). The resolution of this apparent conflict awaits further experimentation.

Examination of the cell cycle of polyamine-deficient cells has yielded results consistent with other sites of polyamine action besides DNA synthesis. In one study, methyl-GAG was used to inhibit spermidine and spermine accumulation during serum-stimulation of G_1-arrested rat embryo fibroblasts (Rupniak and Paul, 1978b). These cells were arrested in G_1 after approximately one doubling, during which DNA synthesis proceeded normally. This arrest was reversed when exogenous spermidine or spermine was supplied to the cells. The lag time between the addition of spermidine and the re-initiation of DNA synthesis was comparable to that observed during stimulation of serum-deprived, G_1-arrested cells. This observation suggested that polyamine depletion arrested these cells at the restriction point in G_1. Studies with SV40 transformed 3T3 cells, which do not G_1 arrest in response to serum or nutritional deprivations (Paul, 1973), were consistent with this conclusion. Under conditions where non-transformed 3T3 cells arrested in G_1 due to polyamine deficiency, the transformed 3T3 cells did not G_1 arrest (Rupniak and Paul, 1978a). Other workers have reported inhibition of growth and partial accumulation of cells in G_1 when rat brain tumour cells were grown in the presence of methyl-GAG (Heby and coworkers, 1977). Because this latter effect was not reversed by the addition of exogenous polyamines, it may have represented a secondary pharmacological effect of the inhibitor.

Mamont and coworkers (1976, 1978) have examined the influence of the inhibitors of ornithine decarboxylase, α-methylornithine and α-difluoromethylornithine, on the growth of rat hepatoma tissue culture (HTC) cells. These inhibitors severely depressed both putrescine and spermidine levels in the cells, but did not significantly alter spermine levels. Cell division was slowed approximately one doubling after addition of the inhibitors. If the cultures were refed regularly during the course of the experiment, the polyamine-limited cells continued to multiply exponentially, although at a much reduced growth rate as compared to untreated cultures.

We have obtained results which are consistent with those reported by Mamont and coworkers (1976, 1978). When Chinese hamster ovary (CHO) cells were grown in the presence of α-MO, the growth rate of the culture continued to decrease until, after two to three doublings, no further increase in cell number was observed (Figure 1, open symbols). Addition of putrescine to the

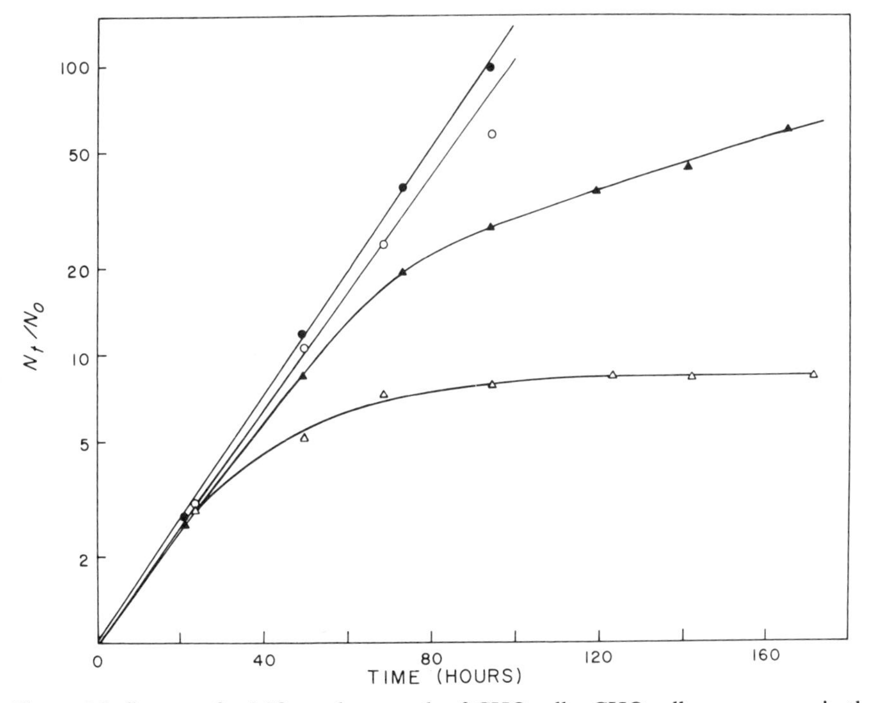

Figure 1 Influence of α-MO on the growth of CHO cells. CHO cells were grown in the presence (△,▲) or absence (○,•) of 5 mM α-MO in media containing either newborn calf serum (○,△) or horse serum (•,▲). α-MO was added at time 0. All of the cultures were refed daily. N_t/N_0 signifies the cell number at time t divided by the cell number at time 0

Table 1 Influence of α-MO on polyamine levels

Treatment	Putrescine	Polyamines (nmoles mg^{-1} protein) Spermidine	Spermine
Control	3	16	12
α-MO			
25 h	ND	1	9
48 h	ND	1	10
72 h	ND	ND	7
96 h	ND	ND	6
α-MO + putrescine	12	18	6

CHO cells were cultured in monolayers with Nutrient Mixture F-10 containing 15% newborn calf serum. During the experiment, all cultures were refed daily. At time 0, α-MO was added to a final concentration of 5 mM. At the times indicated, samples were taken for polyamine analysis. Control cultures were not treated with α-MO. Putrescine was added to the α-MO-treated cultures to a final concentration of 50 μM at 96 h. The α-MO + putrescine samples were taken 53 h after putrescine addition. The minimum limit of detection for the polyamine analyses are: putrescine, 2 nmole mg^{-1} protein; spermidine, 1 nmole mg^{-1} protein. ND signifies levels below the limit of detection.

plateau-phase cells restored growth in the cultures. Polyamine analysis revealed that α-MO effectively inhibited putrescine biosynthesis. Table 1 shows that intracellular putrescine and spermidine fell to insignificant levels, while spermine levels were less dramatically affected. Measurement of the rate of thymidine incorporation in the plateau-phase cells indicated that the rate of DNA synthesis per S phase cell was inhibited to a level (70 cpm 10^{-3} S phase cells) which was approximately 25% of the control cell value (300 cpm 10^{-3} S phase cells). Putrescine addition restored the rate of thymidine incorporation (360 cpm 10^{-3} S phase cells) to control levels. Despite this inhibition of DNA synthesis, examination of the cell cycle parameters both by flow microfluorography and by [^{3}H]thymidine autoradiography revealed no expansion of S phase relative to the other phases of the cell cycle; cells were distributed in all phases of the cell cycle similarly to the control (Table 2). Also, long term labelling of the plateau-phase cells with [^{3}H]thymidine indicated that 70% of the cells passed through S phase during a 24-h period. Thus, these results indicated that the polyamine-limited cells continued to cycle in the plateau-phase with cell death possibly balancing cell growth. This view is supported by the finding that treatment of cells with α-MO in media containing horse serum in place of calf serum resulted in continued increase in cell number at a reduced growth rate (Figure 1, closed symbols). Cells have been maintained for more than 25 generations under these conditions of polyamine-limited growth. Thus, α-MO treatment of cells in the presence of horse serum seems to prevent or reduce the cell death seen in the presence of calf serum. Whether this difference can be attributed to the presence of amine oxidase activity, which is present at high levels in calf serum but only at low levels in horse serum (Blaschko, 1962), remains to be seen.

Table 2 Effect of α-MO on the cell cycle parameters

Treatment	Cell cycle parameters			
	G_1	S	% labelled cells	G_2 + M
Control				
29 h	0.35	0.51	60	0.15
53 h	0.49	0.34	51	0.17
α-MO				
29 h	0.49	0.37	46	0.14
53 h	0.56	0.29	40	0.12
76 h	0.60	0.29	40	0.11
97 h	0.53	0.35	39	0.12
α-MO + putrescine	0.50	0.34	42	0.16

CHO cells were cultured, and α-MO (5 mM) and putrescine (50 μM) were added as described in Table 1. Samples for flow microfluorography and autoradiography were taken at the times indicated. The α-MO + putrescine sample was taken 51 h after putrescine addition. The cell cycle parameters, G_1, S and G_2 + M, were obtained by computer modelling of the data from flow microfluorometry. The per cent labelled cells was determined by ^{3}HTdR autoradiography.

C. Present Status of Polyamines in the Cell Cycle

On the basis of evidence to date, the following conclusions seem quite certain. (1) A minimum cellular level of polyamines is required for optimal rates of DNA replication. (2) An increase in the intracellular polyamine level is not a necessary prerequisite for initiation of S phase. (3) Polyamine limitation of non-transformed cells produces an arrest in G_1. Each of these conclusions is considered below.

A number of studies have indicated that polyamines are required for optimal rates of DNA replication. In lymphocytes, inhibition of polyamine accumulation during mitogenesis resulted in a reduced rate of DNA synthesis (Otani and coworkers, 1974; Fillingame, Jorstad and Morris, 1975; Morris, Jorstad and Seyfried, 1977). Thus the basal levels of polyamines in unactivated lymphocytes were sufficient to support only a depressed rate of DNA synthesis. In other cell types (HTC cells, rat embryo fibroblasts), inhibition of polyamine biosynthesis during the stimulation of growth allowed one round of DNA replication with no apparent inhibition of the rate (Mamont and coworkers, 1976; Rupniak and Paul, 1978a,b). Contrary to the situation in lymphocytes, it appears in this case that the basal levels of polyamines present in the quiescent cells were sufficient to fulfil the polyamine requirement of DNA synthesis in the first cycle after activation. Thus, a minimum level of polyamines, which in some cells is well below that observed in steady state growth, is required to support optimal rates of DNA replication. The mechanism of polyamine action in DNA replication is not understood. Polyamines may possibly serve a co-factor role in DNA synthesis, perhaps by stimulating DNA polymerase (Chiu and Sung, 1972; Stalker, Mosbaugh and Meyer, 1976) or possibly through some other interaction with the replication machinery (DNA gyrase?).

Because of the strong correlation between cell growth and polyamine biosynthesis, an attractive hypothesis has been that the large increase in polyamine levels may serve to signal subsequent events in the proliferative response. It seems extremely unlikely, however, that an increase in polyamine levels regulates the initiation of S phase. Studies with lymphocytes (Fillingame, Jorstad and Morris, 1975; Morris, Jorstad and Seyfried, 1977), HTC cells (Mamont and coworkers, 1976), and rat embryo fibroblast cells (Rupniak and Paul, 1978a,b) have demonstrated that inhibition of polyamine accumulation during activation from a quiescent state did not alter the time of entry of these cells into S phase. Therefore, while an appropriate level of polyamines seems necessary for optimal DNA replication (see above), an elevation in polyamine levels appears not to signal the initiation of S phase. Hence, the exact timing of polyamine biosynthesis in the cell cycle is probably not as important as simply reaching the required level prior to initiation of DNA replication.

Polyamine limitation of rat embryo fibroblasts arrests these cells in G_1. One interpretation of this observation is that the arrest represents the normal response of non-transformed cells to any nutrient limitation (Rupniak and Paul, 1978a,b). Non-transformed animal cells typically arrest in G_1 upon limitation of serum or of many nutrients (Baserga, 1976; Prescott, 1976). Since, as discussed above, polyamines are required for optimal movement through S phase, cells may G_1 arrest if the intracellular concentration of polyamines is not sufficient to allow completion of the cell cycle. An alternate explanation for the arrest of cells in G_1 may be that a specific polyamine-requiring process occurs in G_1 and this is a necessary event for passage into S phase. Presently, the available evidence does not eliminate either of the possibilities. Studies which demonstrate that polyamine-limited SV-3T3 cells behave differently than normal 3T3 cells would appear to support the former interpretation (Rupniak and Paul, 1978a).

The fact that polyamine-limited HTC and CHO cells continue to grow, albeit at a slower rate, demonstrates that not all cells are subject to cell cycle arrest in response to polyamine deficiency. The behaviour of these cells might be explained as a typical response of transformed cells to polyamine limitation (Rupniak and Paul, 1978a), although the transformed nature of CHO cells is questionable, since these cells will arrest in G_1 in response to isoleucine starvation (Ley and Tobey, 1970). Alternatively, the continued presence of spermine in the α-methylornithine-treated cells may fulfil the requirement for passage through the restriction point in G_1. Cell cycle studies on polyamine-limited CHO cells indicated that a 4-fold reduction in the rate of DNA synthesis per S phase cell did not result in an expansion of S phase relative to the other phases of the cell cycle (Harada and Morris, in preparation). Our working hypothesis to explain this observation is that polyamine deficie lengthens both the G_1 and S phases of the cell cycle. Presumably, the low lev of polyamines do not adequately fulfil the requirement in DNA replication, resulting in an expansion of S phase. The lengthening of G_1 may represent the response of the cells to adverse growth conditions (low polyamine levels). CHO cells have been shown to adjust their growth rate by lengthening G_1 in response to poor growth conditions (Tobey, Anderson and Peterson, 1967). An alternate interpretation of these results is that the entire cell cycle is non-specifically expanded as a result of polyamine limitation. This situation would require that cellular polyamine deficiency inhibits a cell cycle independent process. Clearly, further studies are required to resolve this issue.

Why do not mitogen-activated lymphocytes arrest in G_1? Cyclic AMP arrests cells at the restriction point in G_1 (Pardee, 1974). By this definition, the restriction point occurs approximately 15 hours after activation of lymphocytes (D. R. Morris, unpublished observations), before any appreciable accumulation of polyamines has occurred (Fillingame and Morris, 1973b). Thus, the basal levels of polyamines are sufficient for lymphocytes to pass through the restriction point and into S phase.

IV. CONCLUSION

The basic outlines of polyamine function in cell proliferation are beginning to take shape. There are, of course, many details yet to be defined. For example, is the block in G_1 simply a regulatory response of animal cells to nutrient deficiency, or is there a polyamine-requiring process necessary for progression into S phase? Are the effects of polyamine deficiency on the cell cycle restricted to G_1 and S, or are other phases affected as well? Although our knowledge is not complete, one salient point comes through: appropriate intracellular levels of polyamines are required for optimal DNA replication in both bacterial and animal cells. Given this biological frame of reference, it is now possible to meaningfully approach studies on the molecular mechanisms of polyamine action in cell-free systems. Over the past 20 years, polyamines have been shown to influence a plethora of reactions and interactions involving nucleic acids *in vitro*. In order to make physiological sense out of these results, however, it is necessary to sort out biologically relevant effects from less interesting, non-specific ionic interactions of these polycations (discussed in Morris, 1978). In approaching *in vitro* studies, the enzymatic reactions of DNA replication clearly should come under careful scrutiny. In particular, the enzyme DNA gyrase is of immediate interest. Gellert and his colleagues have described a strong spermidine requirement for DNA gyrase (Gellert and coworkers, 1976a). As discussed above, all of the physiological ramifications of polyamine deficiency in *E. coli* can be explained through a single cellular action of spermidine on this enzyme. How can this hypothesis be tested? Is activation of DNA gyrase *in vitro* simply another example of a non-specific polyamine interaction? We now have a new experimental tool to assess the biological significance of gyrase activation and of *in vitro* interactions of polyamines in general. We have synthesized a homologous series of spermidine analogues and described their specificity in restoring growth to polyamine-deficient cells (*see above*). A minimal criterion for any physiologically relevant *in vitro* reaction is that it show the same specificity with the spermidine homologues as that observed *in vivo*. This is currently being tested with DNA gyrase and, in addition, this criterion soon will be extended to a variety of other *in vitro* interactions of polyamines. Hopefully, this approach will allow us to concentrate on only a limited number of polyamine interactions and proceed with the business of defining the molecular role of polyamines in cell proliferation.

REFERENCES

Abdel-Monem, M. M., Newton, N. E., and Weeks, C. E. (1974). *J. Med. Chem.*, **17**, 447–451

Baserga, R. (1976). *Multiplication and Division in Mammalian Cells*, Marcel Dekker, Inc., New York.

Blaschko, H. (1962). The amine oxidases of mammalian blood plasma. In O. Lowenstein (Ed.), *Advances in Comparative Physiology and Biochemistry*, Vol. 1, Academic Press, New York and London, pp. 67–116.

Brown, K. B., Nelson, N. F., and Brown, D. G. (1975). *Biochem. J.*, **151**, 505–512.

Campbell, R. A., Morris, D. R., Bartos, D., Daves, G. D., and Bartos, F. (Eds) (1978a). *Advances in Polyamine Research*, Vol. 1, Raven Press, New York.

Campbell, R. A., Morris, D. R., Bartos, D., Daves, G. D., and Bartos, F. (Eds) (1978b). *Advances in Polyamine Research*, Vol. 2, Raven Press, New York.

Chiu, S.-F., and Sung, S. C. (1972). *Biochim. Biophys. Acta*, **281**, 535–542.

Cohn, M. S., Tabor, C. W., and Tabor, H. (1978). *J. Bacteriol.*, **134**, 208–213.

Cooper, S., and Helmstetter, C. E. (1968). *J. Mol. Biol.*, **31**, 519–540.

Corti, A., Dave, C., Williams-Ashman, H. G., Mihich, E., and Schenone, A. (1974). *Biochem. J.*, **139**, 351–357.

Cunningham-Rundles, S., and Maas, W. K. (1975). *J. Bacteriol.*, **124**, 791–749.

Dykstra, W. G., Jr. and Herbst, E. J. (1965). *Science*, **149**, 428–429.

Fillingame, R. H. (1973). Accumulation of polyamines during lymphocyte transformation: An analysis of its relation to nucleic acid synthesis and accumulation. Ph.D. thesis, University of Washington, Seattle.

Fillingame, R. H., and Morris, D. R. (1973a). *Biochem. Biophys. Res. Commun.*, **52**, 1020–1025.

Fillingame, R. H., and Morris, D. R. (1973b). *Biochemistry*, **12**, 4479–4487.

Fillingame, R. H., Jorstad, C. M., and Morris, D. R. (1975). *Proc. Natl. Acad. Sci. USA*, **72**, 4024–4045.

Friedman, S. J., Bellantone, R. A., and Canellakis, E. S. (1972). *Biochim. Biophys. Acta*, **261**, 188–193.

Fuller, D. J. M., Gerner, E. W., and Russell, D. H. (1977). *J. Cell. Physiol.*, **93**, 81–88.

Geiger, L. E., and Morris, D. R. (1978). *Nature, Lond*, **272**, 730–732.

Gellert, M., Mizuuchi, K., O'Dea, M. H., and Nash, H. A. (1976a). *Proc. Natl. Acad. Sci. USA*, **72**, 3872–3876.
Gellert, M., O'Dea, M. H., Itoh, T., and Tomizawa, J.-I. (1976b). *Proc. Natl. Acad. Sci. USA*, **73**, 4474–4478.
Goldstein, D. A., Heby, O., and Marton, L. J. (1976). *Proc. Natl. Acad. Sci. USA*, **73**, 4022–4026.
Hafner, E. W., Tabor, H., and Tabor, C. W. (1977). *Fed. Proc.*, **36**, 783.
Heby, O., Marton, L. J., Wilson, C. B., and Gray, J. W. (1977). *Europ. J. Cancer*, **13**, 1009–1017.
Heby, O., Gray, J. W., Lindl, P. A., Marton, L. J., and Wilson, C. B. (1976). *Biochem. Biophys. Res. Commun.*, **71**, 99–105.
Heby, O., Marton, L. J., Wilson, C. B., and Martinez, H. M. (1975). *J. Cell. Physiol.*, **86**, 511–521.
Hershey, H. V., Stieber, J. F., and Mueller, G. C. (1973). *Eur. J. Biochem.*, **34**, 383–394.
Herschko, A., Mamont, P., Shields, R., and Tomkins, G. M. (1971). *Nature (New Biol.)*, **232**, 206–211.
Hirschfield, I. N., Rosenfeld, H. J., Leifer, Z., and Maas, W. K. (1970). *J. Bacteriol.*, **101**, 725–730.
Inoue, H., Kato, Y., Takigawa, M., Adachi, K., and Takeda, Y. (1975). *J. Biochem.*, **77**, 879–893.
Jorstad, C. M., and Morris, D. R. (1974). *J. Bacteriol.*, **119**, 857–860.
Kay, J. E., and Lindsay, V. J. (1973). *Exptl. Cell Res.*, **77**, 428–436.
Kay, J. E., and Pegg, A. E. (1973). *FEBS Lett.*, **29**, 301–304.
Knutson, J. C., and Morris, D. R. (1978). *Biochim. Biophys. Acta*, **520**, 291–301.
Krokan, H., and Eriksen, A. (1977). *Eur. J. Biochem.*, **72**, 501–508.
Ley, K. D., and Tobey, R. A. (1970). *J. Cell Biol.*, **47**, 453–459.
Liquori, A. M., Constantino, L., Crescenzi, V., Elia, V., Giglio, E., Puliti, R., De Santis Savino, M., and Vitagliano, V. (1967). *J. Mol. Biol.*, **24**, 113–122.
Maas, W. K. (1972). *Molec. Gen. Genet.*, **119**, 1–9.
Maas, W. K., Leifer, Z., and Poindexter, J. (1970). *Ann. N. Y. Acad. Sci.*, **171**, 957–967.
Mamont, P. S., Böhlen, P., McCann, P. P., Bey, P., Schuber, F., and Tardif, C. (1976). *Proc. Natl. Acad. Sci. USA*, **73**, 1626–1630.
Mamont, P. S., Duchesne, M.-C., Grove, J., and Bey, P. (1978). *Biochem. Biophys. Res. Commun.*, **81**, 58–66.
Morris, D. R. (1978). Section introduction: molecular and cell biology of polyamines. In R. A. Campbell, D. R. Morris, D. Bartos, G. D. Daves and F. Bartos (Eds), *Advances in Polyamines Research*, Raven Press, New York, pp. 103–104.
Morris, D. R. (1973). RNA and protein synthesis in a polyamine-requiring mutant of *Escherichia coli*. In D. H. Russell (Ed.), *Polyamines in Normal and Neoplastic Growth*, Raven Press, New York, pp. 111–122.
Morris, D. R., and Fillingame, R. H. (1974). *Ann. Rev. Biochem.*, **43**, 303–325.
Morris, D. R., and Jorstad, C. M. (1973). *J. Bacteriol.*, **113**, 271–277.
Morris, D. R., and Jorstad, C. M. (1970). *J. Bacteriol.*, **101**, 731–737.
Morris, D. R., and Hansen, M. T. (1973). *J. Bacteriol.*, **116**, 588–592.
Morris, D. R., Jorstad, C. M., and Seyfried, C. E. (1977). *Cancer Res.*, **37**, 3169–3172.
Morris, D. R., Wu, W. H., Applebaum, D., and Koffron, K. L. (1970). *Ann. N. Y. Acad. Sci.*, **171**, 968–976.
Otani, S., Matsui, I., and Morisawa, J. (1977). *Biochim. Biophys. Acta*, **478**, 417–427.

Otani, S., Mizoguchi, Y., Matsui, I., and Morisawa, S. (1974). *Mol. Biol. Reports*, **1**, 431–436.

Pardee, A. B. (1974). *Proc. Natl. Acad. Sci. USA*, **71**, 1286–1290.

Pathak, S. N., Porter, C. W., and Dave, C. (1977). *Cancer Res.*, **37**, 2246–2250.

Paul, D. (1973). *Biochem. Biophys. Res. Commun.*, **53**, 745–753.

Pohjanpelto, P. (1975). *Biomedicine*, **23**, 350–352.

Pösö, H., and Jänne, J. (1976). *Biochem. J.*, **158**, 485–488.

Prescott, D. M. (1976). *Reproduction of Eukaryotic Cells*, Academic Press, New York and London.

Rupniak, H. T. and Paul, D. (1978a). Regulation of the cell cycle by polyamines in normal and transformed fibroblasts. In R. A. Campbell, D. R. Morris, D. Bartos, G. D. Daves and F. Bartos (Eds), *Advances in Polyamine Research*, Vol. 1, Raven Press, New York, pp. 117–126.

Rupniak, H. T., and Paul, D. (1978b). *J. Cell. Physiol.*, **94**, 161–170.

Russell, D. H., and Stambrook, P. J. (1975). *Proc. Natl. Acad. Sci. USA*, **72**, 1482–1486.

Stalker, D. M., Mosbaugh, D. W., and Meyer, R. R. (1976). *Biochemistry*, **15**, 3114–3121.

Sunkara, P. S., Rao, P. N., and Nishioka, K. (1977). *Biochem. Biophys. Res. Commun.*, **74**, 1125–1133.

Tabor, C. W., and Tabor, H. (1976). *Ann. Rev. Biochem.*, **45**, 285–306.

Tabor, H., Tabor, C. W., and Hafner, E. W. (1976). *J. Bacteriol.*, **128**, 485–486.

Tobey, R. A., Anderson, E. C., and Peterson, D. F. (1967). *J. Cell Biol.*, **35**, 53–59.

Whitney, P. A., and Morris, D. R. (1978). *J. Bacteriol*, **134**, 214–220.

Whitney, P. A., Pious, L. O., and Morris, D. R. (1978). Polyamine-deficient mutants in yeast. In *Advances in Polyamine Research* (Eds. R. A. Campbell, D. R. Morris, D. Bartos, G. D. Daves, and F. Bartos), pp. 195–199, Raven Press, New York.

Wiegand, L., and Pegg, A. E. (1978). *Biochim. Biophys. Acta*, **517**, 169–180.

Chapter 2

Polyamines and the Cell Cycle

OLLE HEBY AND GUNNAR ANDERSSON

I. INTRODUCTION

Putrescine, spermidine and spermine is a group of polyamines that plays an important physiological role. They stimulate a large number of processes of replication, transcription, and translation essential to growth and multiplication of the cell. Both prokaryotic and eukaryotic cells contain a biosynthetic apparatus that provides the cells with putrescine and spermidine. Spermine, however, only appears to be synthesized in eukaryotes. Polyamine synthesis is predominantly an intracellular event and extracellular polyamine biosynthetic activity has been found to a significant extent only in semen.

Intracellularly the polyamines are mainly non-conjugated and protonated, thus possessing high positive charge densities, which may be crucial for their interaction with the nucleic acids. The present report is concerned with the role of the polyamines in the progression of the mammalian cell through its reproductive cycle.

II. POLYAMINE BIOSYNTHESIS

Two amino acids, L-ornithine and L-methionine, are the primary precursors of the polyamines. L-Ornithine is converted to putrescine by decarboxylation catalysed by L-ornithine decarboxylase (ODC). L-Methionine is first activated to S-adenosyl-L-methionine by ATP in a reaction catalysed by ATP:L-methionine S-adenosyltransferase. S-Adenosyl-L-methionine is subjected to decarboxylation, catalysed by S-adenosyl-L-methionine decarboxylase (SAMD), and then serves as the donor of the propylamine moiety for the synthesis of spermidine and spermine. Spermidine synthase and spermine synthase catalyse the propylamine transfer reactions in which putrescine and spermidine act as the acceptor molecules, thus forming spermidine and spermine, respectively, as the end-products.

ODC is considered to be the rate-controlling enzyme in polyamine biosynthesis because its basal activity is generally lower than those of SAMD, spermidine synthase and spermine synthase. In the cell, ODC exhibits a striking inducibility and a very short half-life, two properties that may be causally related. SAMD also exhibits the latter properties but at variance with ODC its activity may be specifically stimulated by putrescine and spermidine. Contrary to ODC and SAMD, the propylamine transferases (spermidine and spermine synthase) seem to be rather stable enzymes with long biological half-lives. Hence ODC and SAMD appear to be the enzymes that control the formation of the polyamines (Chapter 6).

III. CLASSIFICATION OF CELLS

Cell populations in the adult animal can be divided with respect to their ability to synthesize DNA and to divide into three categories: (1) continuously dividing cells that repeatedly traverse the cell cycle, e.g. the epithelial cells lining the crypts of the small intestine, the stem cells of the bone marrow, the cells of the basal layer of the epidermis, exponentially growing cells in culture; (2) quiescent cells that have left the cell cycle and normally do not synthesize DNA nor divide but can be stimulated to do so by applying an appropriate stimulus, e.g. liver and kidney cells; and (3) non-dividing cells that have left the cell cycle and are destined to die without dividing again (Baserga, 1976). Cells that continuously traverse the cell cycle, and cells that have been stimulated to do so from a quiescent state, exhibit much higher activities of the enzymes in the polyamine biosynthetic pathway and contain much larger amounts of the polyamines (especially of putrescine and spermidine) than do resting cells. Furthermore, the rate of polyamine synthesis increases in parallel with the rate of cell proliferation both *in vitro* (Heby and coworkers, 1975a) and *in vivo* (Andersson and Heby, 1977). These observations indicate that increased polyamine synthesis is an event associated with the traverse of the cell cycle.

It probably remains to be seen whether non-dividing cells possess the capacity to synthesize polyamines at all, because in the category of non-dividing cells one can now safely include only cells without a nucleus, or cells that are partially enucleated, like mature red blood cells of mammals, keratinizing cells of the epidermis, polymorphonuclear leucocytes, and a few others (Baserga, 1976). Many cells previously considered to be non-dividing cells, e.g. hen erythrocytes and neurons, actually have the potential to synthesize DNA and divide if an appropriate stimulus is applied. Henceforth we will limit ourselves to discuss changes observed in the metabolism of the polyamines in cell populations stimulated to synthesize DNA and divide, and in continuously dividing populations of cells.

IV. POLYAMINES IN CELL POPULATIONS STIMULATED TO PROLIFERATE

Populations of quiescent cells that can be stimulated to synthesize DNA and divide, and that have been studied with respect to polyamine synthesis, include: (1) stationary cell cultures stimulated by nutritional changes (Heby and coworkers, 1975b; McCann and coworkers, 1975), infection with oncogenic DNA viruses (polyoma) (Goldstein, Heby and Marton, 1976), or RNA viruses (murine sarcoma) (Gazdar and coworkers, 1976); (2) lymphocytes stimulated with phytohaemagglutinin (Kay and Lindsay, 1973) or concanavalin A (Fillingame and Morris, 1973); (3) mammary tissue stimulated with insulin (Aisbitt and Barry, 1973) and/or prolactin (Oka and Perry, 1976); (4) epidermal cell cultures stimulated with phorbol esters (Yuspa and coworkers, 1976); (5) skin stimulated with epidermal growth factor (Stastny and Cohen, 1970); (6) skin stimulated with ethylphenylpropionate (Takigawa and coworkers, 1977); (7) the liver stimulated to regenerate after partial hepatectomy (Hölttä and Jänne, 1972); (8) the liver after triiodothyronine-treatment (Gaza, Short and Lieberman, 1973); (9) the liver of fasted rats after high-protein diet (Hayashi, Aramaki and Noguchi, 1972); (10) the kidney after contralateral nephrectomy (Brandt, Pierce and Fausto, 1972); (11) the isoproterenol-stimulated salivary gland (Inoue and coworkers, 1974); and (12) the oestrogen-stimulated uterus (Kaye, Icekson and Lindner, 1971). These and other examples of stimulated cell proliferation are all characterized by a lag between the application of the stimulus and the onset of DNA synthesis. Even though most quiescent cells (arrested in G0 or G1) go through S before they divide, one should keep in mind that cell populations may contain quiescent cells (arrested in G2) that proceed directly to mitosis upon stimulation (Andersson, 1977; Gelfant, 1977).

During the lag period (the G1 phase) that precedes the S phase, the cell becomes prepared to synthesize DNA. One of the molecular events that are stimulated during the G1 period, and that may be involved in the cell's preparation for DNA replication, is polyamine synthesis. A dramatic increase in the ODC activity appears to be one of the very first changes taking place after stimulation of cell proliferation in many of the systems mentioned above. In these experimental systems the ODC activity reaches a first peak in early G1 and a second peak at the time of initiation of DNA synthesis. It appears to be a general phenomenon that cells that have been stimulated to traverse the cell cycle increase their ODC activity as they go through late G1 and early S. Not all cells, however, increase their ODC activity as they go through early G1. The stimulation of ODC, occurring in early or late G1, might be a priming event that triggers all the ensuing changes in polyamine metabolism. This is suggested by the fact that the concentration of putrescine changes in parallel with the ODC activity during stimulated cell proliferation. Furthermore, putrescine activates SAMD and is then utilized as the substrate in the spermidine synthase reaction. The increase in the activity of SAMD, which is the rate-controlling enzyme in spermidine and spermine synthesis, results in a concomitant increase in the spermidine concentration. Even though the concentration of spermidine increases after activation of SAMD, that of spermine does not necessarily do so. In regenerating liver, for example, the increase in the concentration of spermidine is paralleled by a decrease in that of spermine (Heby and Lewan, 1971; Jänne, Hölttä and Guha, 1976). This might be due to the elevated concentration of putrescine, which is a competitive inhibitor for spermidine in the spermine synthase reaction. It is apparent that putrescine, in addition to being a substrate in the synthesis of the polyamines, plays an important role as a regulator. Thus putrescine is a weak competitive inhibitor of ODC, and acts as an activator (possibly allosteric) of SAMD and as a powerful competitive inhibitor of spermine synthase (Williams-Ashman and coworkers, 1972). In addition to being a competitive inhibitor of ODC, putrescine induces the synthesis of a non-competitive protein inhibitor of ODC, named antizyme (Heller, Fong and Canellakis, 1976).

The increases observed in the ODC activity after stimulation of cell proliferation appear to involve the synthesis of new enzyme molecules as revealed by experiments with inhibitors of RNA and protein synthesis and by estimates of the amount of immunoreactive ODC (Hölttä, 1975). The ultimate cause of the increased ODC activity appears to differ among the various models of stimulated cell proliferation. In some systems, in which gene activation appears to be involved, the synthesis of new mRNA for ODC is initiated promptly after stimulation. In regenerating liver the ODC messenger is probably synthesized during the first few hours, because actinomycin D or α-amanitin were able to block the stimulation of ODC (the early peak at 4 h) only when administered at the time of the operation or during the first few hours (Russell and Snyder, 1969; Kallio and coworkers, 1977a,b). The newly synthesized ODC messenger appears to be relatively stable, with an approximate half-life of 6 h, as estimated on the basis of data obtained in experiments with α-amanitin (Kallio and coworkers, 1977b). The corresponding half-life of the SAMD messenger was only about 4 h. Contrary to regenerating liver, the increase in ODC activity which results from dilution of high density suspension cultures of rat hepatoma (HTC) cells, and which reaches a maximum at about

4 h, is largely resistant to actinomycin D (Hogan, 1971). This fact suggests that the ODC messenger continues to be synthesized, or is stored, in stationary phase, high-density HTC cells and is available for more frequent translation when conditions for protein synthesis improve; in other words, that the synthesis of ODC is under post-transcriptional control.

Experiments with primary cultures of mouse kidney cells, stimulated to proliferate by infection with polyoma virus, also suggest that post-transcriptional control might be involved in the regulation of ODC activity (Heby, Goldstein and Marton, 1978b). Actinomycin D, at a dose that inhibited mRNA synthesis, was found to cause a superinduction of the ODC activity both at the time of the early and late G1 peaks. The activity of SAMD, however, was inhibited by the high dose of actinomycin D. This fact suggests that, at variance with ODC, SAMD is primarily subjected to transcriptional control in the polyoma-virus-infected cells. The mechanism of superinduction is not clear, and two opposing hypotheses have been advanced. Tomkins and coworkers (1969) suggest that actinomycin D is able to stimulate *de novo* synthesis of inducible enzymes by preventing the transcription of a repressor gene. The repressor is assumed to inhibit enzyme synthesis by combining reversibly with its mRNA to form an inactive complex, which facilitates the degradation of the mRNA. Assuming that both the repressor mRNA and the repressor exhibit turnover rates that are much greater than those of the structural gene products, inhibition of transcription by actinomycin D will result in selective elimination of the repressor and thus in increased synthesis of the enzyme. On the other hand, Reel and Kenney (1968), argue that actinomycin D causes superinduction by inhibiting the rate of degradation of the enzyme and not by stimulating *de novo* synthesis. Clark (1974) has presented data which appear to be in agreement with the latter proposal for the mechanism of superinduction. He found that actinomycin D, but not cordycepin, superinduced ODC, and that actinomycin D markedly increased the half-life of ODC. Importantly, even though the mechanism of ODC superinduction is not clear, the present data is consistent in one respect; an increased ODC activity does not always require gene activation.

V. POLYAMINES IN CONTINUOUSLY PROLIFERATING CELL POPULATIONS

In attempting to relate cellular polyamine synthesis to events in the mammalian cell cycle a variety of synchronizing techniques have been employed. Thus far, most studies have been performed on cell populations stimulated to proliferate from a quiescent state, i.e. on cell systems in which only partial synchrony was achieved. The procedures for establishing more highly synchronous cell populations generally employ environmental alteration of one kind or another to force the system into synchrony. Unfortunately the utility of these coercive methods for study of the normal relationships between intracellular events occurring during the division cycle is often diminished, because the physical or chemical stresses imposed on the cells may result in severe metabolic and temporal distortions. For example, although cells are slowed down or stopped in the S phase of their reproductive cycle by DNA synthesis inhibitors, RNA and protein synthesis may proceed unabated, resulting in so-called unbalanced growth. These cells will contain much more RNA and protein, possibly even polyamine biosynthetic enzymes and their messengers, and will be considerably larger than their normal counterparts (Rueckert and Mueller, 1960; Studzinsky and Lambert, 1969). Also, because some cells are arrested for significantly longer periods than others, the hazard of achieving an unbalanced state of growth is greater in some cells than in others. Furthermore, unbalanced growth has been found to result in a shortening of the time required for daughter cells to traverse the subsequent G1 period (Tobey, Anderson and Petersen, 1967). With regard to polyamine metabolism, it has been observed that 5-fluorodeoxyuridine, in concentrations that inhibit DNA synthesis in polyoma-virus-infected mouse kidney cells, has no effect on the synthesis and accumulation of the polyamines (Goldstein, Heby and Marton, 1976). This is merely another manifestation of unbalanced growth and further emphasizes the possible importance of such altered metabolic patterns for subsequent cell cycle traverse.

Colchicine and vinca alkaloids, mitotic poisons that have been used to obtain synchronous cell cultures for polyamine analysis (Friedman, Bellantone and Canellakis, 1972; Russell and Stambrook, 1975; Fuller, Gerner and Russell, 1977), also inhibit the activity of ODC (Chen, Heller and Canellakis, 1976). These inhibitors cause disturbances of the normal cell cycle, affecting the metabolism of cells in interphase in addition to those in mitosis; they inhibit rRNA synthesis (Wagner and Roizman, 1968), elevate RNase and DPNase activities (Erbe and coworkers, 1966) and exert effects on the citric acid and urea cycles (Johnson and coworkers, 1960). Since the polyamines play a role in DNA, RNA and protein synthesis, inhibitors interfering with these events should be avoided when attempting to relate polyamine metabolism to normal cell cycle events. Also, if unbalanced growth permits the accumulation of polyamines, one might expect a profound effect on polyamine synthesis in the subsequent cell cycle, inasmuch as putrescine and spermidine at elevated levels have been shown to markedly inhibit the activity of ODC (Jänne and Hölttä, 1974; Heller, Fong and Canellakis, 1976; McCann, Tardif and Mamont, 1977).

In order to obviate these possible objections, we have avoided methods which coerce the cells into synchrony by metabolic inhibition of one sort or another. Instead we have used a purely selective method of synchronization, in which tenuously attached mitotic cells were detached from monolayer cell cultures by shaking (Petersen, Anderson and Tobey, 1968; Terasima and

Tolmach, 1963). This technique imposes minimal stress on the system and causes no distortion of the normal pattern of DNA synthesis. According to Terasima and Tolmach (1963), the pattern of DNA synthesis and the duration of the different phases of the cell cycle in synchronous cultures are, within the limits of experimental accuracy, identical with the corresponding growth parameters found in randomly dividing cultures under the same growth conditions.

A time-sequence of cell samples was collected following plating of mitotic Chinese hamster ovary (CHO) cells, essentially all metaphases, into culture vessels. Each cell sample was analysed for its DNA distribution, ODC activity and polyamine content (Heby and coworkers, 1976a,b). Figure 1 shows the time-sequence of DNA distributions obtained by high-speed flow cytometric analysis. The DNA distribution usually showed two peaks with relative DNA contents of 1.0 and 2.0, respectively (dashed lines); the first peak represents cells with 2C DNA content (G1 phase) and the second peak represents cells with 4C DNA content (G2 and M phases). The region between the 2C and 4C peaks represents cells synthesizing DNA (S phase). The second peak may be contaminated with some G1 cells stuck together as doublets or by mitotic cells which failed to divide, but this effect is not large, as shown in the first three (2–6 h) distributions (Figure 1). These distributions show only about 3% of the cells in the 4C peak. The presence of this small fraction of cells does not interfere with the interpretation of the experiments.

After 2 h of growth, nearly all the cells had divided and the cell population almost exclusively consisted of G1 phase cells (Figure 1). From 2 through 6 h, essentially all cells possessed a 2C DNA content, i.e., they were progressing through the G1 phase of the cell cycle. No cells were engaged in DNA synthesis for about 6 h following plating. After this time, however, virtually the entire population commenced the production of DNA; 8 h after plating 40% of the cells had entered the S phase and by 10 h 70% of the cells were in the S phase, 15% were in the G1 phase and 15% had entered the G2 and M phases. At 12 h after plating the majority of the cells were progressing through the G2 and M phases of the cell cycle; 50% were in G2 and M, 35% were in S and 15% in G1. Approximately 15 h after plating, the majority of the cells had completed one cycle and possessed the DNA content of G1 phase cells. Thus, the cellular DNA content was monitored throughout one complete cell cycle traverse.

As shown in Figure 1, the cell population underwent its second division over an extended time period. This represents a desynchronization with respect to the initially collected population which was largely in metaphase and which completed mitosis shortly after plating. Thus, as the population behaved quite homogenously at the time of plating, the decay of synchrony with time must be due to variation in the generation time among the individual cells of the population. None the less, the cell cycle dispersion, which describes the standard deviation in cell cycle transit time over one complete cycle, was very low; only 1.7 h out of 14.5 h for the total cycle time.

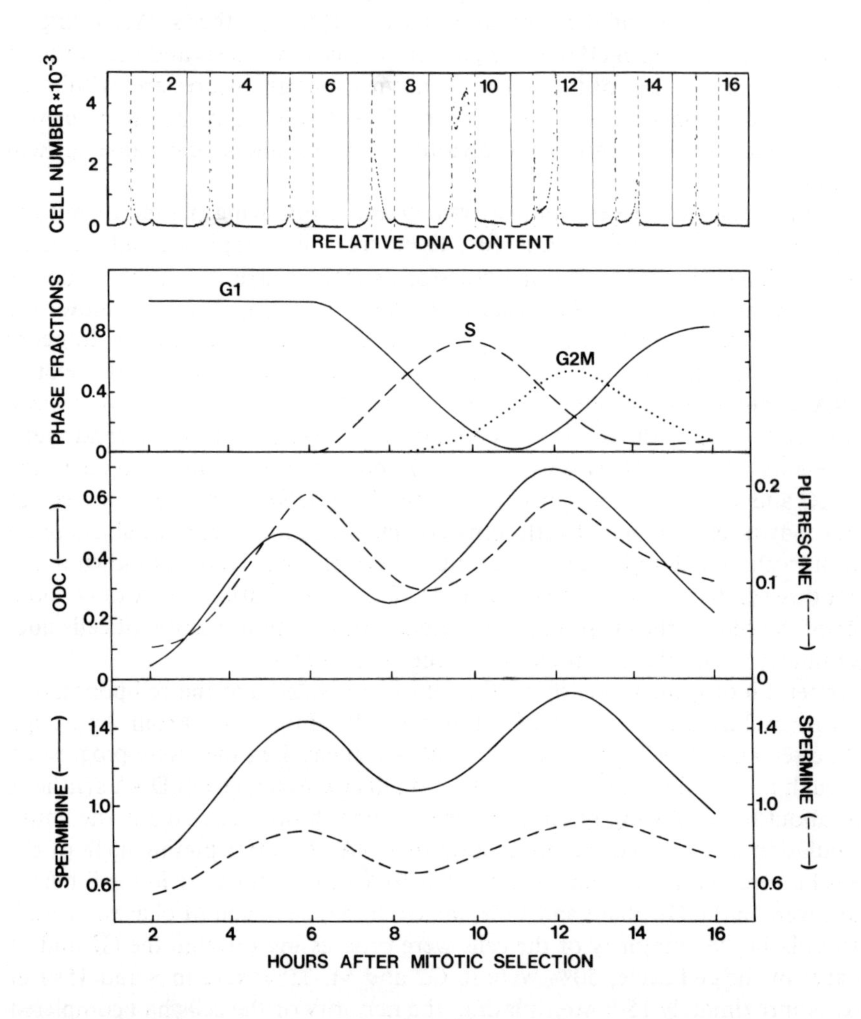

Figure 1 DNA distributions, cell cycle phase fractions, ODC activity, and cellular putrescine, spermidine and spermine content in CHO fibroblasts following synchronization by selective detachment of mitotic cells.

The cellular DNA distributions were obtained by flow cytometric analysis of CHO cells harvested at various times (2, 4, 6, 8, 10, 12, 14 and 16 h) after plating of the mitotically selected cells. To reveal estimates for the fraction of cells in each phase of the cell cycle as a function of time, the time sequence of DNA distributions from the synchronous cell populations was analysed by computer modelling. At these same times the cellular ODC activity and polyamine content were determined. ODC, nmoles of $[^{14}C]O_2/h/10^6$ cells; polyamines, nmoles/10^6 cells. Redrawn from Heby and coworkers (1976b)

By combining the results obtained by flow cytometric analysis with the changes observed in the rate of polyamine synthesis (ODC activity) and in cellular polyamine content (Figure 1), one finds that the synthesis of the polyamines was initiated in mid-G1 and that the polyamines started to accumulate towards the end of the G1 phase. The rate of synthesis reached a peak as the cells started to synthesize DNA and the highest polyamine content was observed at the beginning of the S phase. ODC activity and the levels of all three polyamines decreased significantly during mid-S, but towards the end of the S phase they increased again. Prior to cell division the polyamine biosynthetic activity and the concentration of the polyamines reached a second maximum of about the same height as the one in late G1 to early S. As expected, when the mitotic cells divided the cellular polyamine content decreased to a value about half of the mitotic level. The ODC activity decreased by more than 70% during the second division, and after the first division its activity was approximately one order of magnitude lower than in metaphase cells (Heby and coworkers, 1976a). This precipitous decrease may be due to the fact that ODC has a very short half-life and that the rate of protein synthesis declines during mitosis; it begins to fall in late prophase and drops by about 75% by late telophase (Prescott, 1976). Because dispersion develops during the division cycle, the peaks observed in ODC activity and polyamine content probably are somewhat lower than they should be if there was no dispersion. At 16 h after plating, i.e. when the cells were in the G1 phase of the second cycle, the ODC activity and the cellular polyamine levels were significantly higher than following the first cell division (Figure 1). This fact may be attributable to the dispersion developed during the traverse of the cell cycle; the 16 h sample probably included cells from the entire G1 phase and thus some of these cells are engaged in polyamine synthesis.

VI. ON THE TEMPORAL ORDERING OF POLYAMINE SYNTHESIS DURING THE CELL CYCLE

The data presented shows that in continuously dividing cells, polyamine synthesis is initiated just prior to the S phase, and that before division, the cells initiate a second pulse of polyamine synthesis. Contrary to cells stimulated to proliferate from a quiescent state, continuously dividing cells do not seem to increase their rate of polyamine synthesis nor accumulate polyamines during early G1. The fact that no early G1 peak was observed in other synchronously growing cells (Friedman, Bellantone and Canellakis, 1972; Russell and Stambrook, 1975; Hibasami and coworkers, 1977; Fuller, Gerner and Russell, 1977) suggests that early G1 polyamine synthesis is not part of the normal cell cycle of continuously dividing cells. The two pulses of polyamine synthesis that occur in late G1/early S and in late S/G2 may be part of the cell's preparation for DNA synthesis and division. This idea might also be deduced from experiments which have shown that: (1) there exists a temporal correlation between late G1 ODC activity and the time of initiation of DNA synthesis; factors that delay the S phase (ageing, 8-azaguanine- or 5-azacytidine-treatment) also delay the induction of ODC activity (Cavia and Webb, 1972; Hölltä and Jänne, 1972), and (2) the magnitude of the putrescine accumulation occurring in late G1 is related to the fraction of cells that are stimulated to synthesize DNA (Heby and coworkers, 1975b).

The early G1 polyamine biosynthetic activity observed after so many stimuli might be due to cell hypertrophy caused by humoral or medium factors and it might be completely unrelated to the cell's preparation for the traverse of the cell cycle. In support of this possibility it has been shown that stimuli which do not cause cell proliferation may still produce a rapid increase in putrescine synthesis (Schrock, Oakman and Bucher, 1970). This synthetic activity, however, tapers off rapidly and no subsequent activity follows. The early increase in ODC activity, particularly in conditions associated with a general increase in protein synthesis, might be due partly to its high rate of turnover, i.e. its rapid synthesis and degradation. The half-life of ODC, which can be as short as 5 min in some conditions (Hogan and Murden, 1974) is generally much shorter than that of any other enzyme, and thus it is synthesized more rapidly. Schimke and Doyle (1970) have pointed out that any general increase in protein synthesis will manifest itself first in those enzymes that turn over rapidly. Tabor and Tabor (1976) have estimated that if all protein synthesis were to increase by a factor of 4, the concentration of ODC (with a half-life of 10 min) would double in 10 min, while an enzyme with a turnover rate of 24 h would only increase about 1% in 10 min. In addition, a decrease in the rate of ODC degradation, as observed after the application of many growth stimuli (Hogan and Murden, 1974; Hogan, McIlhinney and Murden, 1974), may contribute to the increase in ODC activity.

Another possible explanation for the early increase in polyamine synthesis, is that induction of ODC activity is part of the change necessary for making a G0 cell enter the cell cycle. In fact, a number of events are initiated in G0 cells stimulated to proliferate, whereas these events are not expressed in continuously dividing cells (Baserga, 1976). Interestingly, the gene activation observed in stimulated G0 cells (Baserga, Costlow and Rovera, 1973) apparently involves the synthesis of ODC messenger (Kallio and coworkers, 1977a,b). There is also some evidence that is not in agreement with this explanation; the early G1 peak does not always occur in cells that are stimulated to proliferate from a quiescent state (Heby and coworkers, 1975b).

Obviously, the composition of the population of cells that is stimulated to proliferate determines the sequence of events that follow stimulation. In most cell systems the majority of the cells that have left the cell cycle and normally do not proliferate, are arrested in the G0 phase. A small but significant portion of the quiescent cells, however, may be arrested in late G1 or in G2

(Gelfant, 1977). When these cells are stimulated to enter the cell cycle, the late G1-arrested cells will start to synthesize DNA and the G2-arrested cells will start to divide only a few hours after application of the stimulus. Provided that an increased polyamine synthesis precedes the initiation of DNA synthesis and cell division, as seen in continuously dividing cells, one would expect the ODC activity to increase shortly after stimulation (within a few hours) in cells entering the cell cycle from late G1 and G2. These phenomena may also be involved in the appearance of the early peak in ODC activity observed in cells stimulated to proliferate.

Premitotic (i.e. G2 phase) stimulation of polyamine synthesis and accumulation of polyamines, may not be discernible in some experimental systems because of deteriorating cell synchrony. The low dispersion which developed in the course of the experiment described in the present report, however, allowed for an accurate determination of polyamine biosynthetic activity and cellular polyamine content during the entire traverse of the cell cycle. At variance with the results obtained with synchronously growing Don C cells (Friedman, Bellantone and Canellakis, 1972) and CHO cells (Figure 1, Heby and coworkers, 1976a,b; Hibasami and coworkers, 1977) did not observe any increase in the G2 phase of baby hamster kidney cells, synchronized by excess thymidine treatment. The authors suggested that the absence of this peak may be related to the differences in the method of synchronization and cell line employed. As previously discussed it could be a result of unbalanced growth developed during thymidine treatment. A possible role for the polyamines in chromosome condensation might be suggested when considering the fact that polyamine synthesis is stimulated prior to mitosis, and that Rao and Johnson (1971) have found that putrescine and spermine specifically promote premature chromosome condensation of interphase nuclei whereas spermidine is unique in inhibiting this event.

It is interesting to note that the activity of ODC is inhibited by colchicine and vinblastine, but not by lumicolchicine and colchiceine, two colchicine analogues which do not affect microtubular structure (Chen, Heller and Canellakis, 1976), and that there is a temporal relationship between the premitotic stimulation of ODC activity (Figure 1) and microtubule protein synthesis (Klevecz, 1975).These observations suggest that the microtubule system may be involved in the regulation of ODC activity.

VII. EFFECTS OF CELLULAR POLYAMINE DEPLETION ON THE TRAVERSE OF THE CELL CYCLE

Ample circumstantial evidence suggests that polyamine synthesis is an event associated with DNA synthesis and cell division. These observations, however, are only indirect measures of the importance of the polyamines for the cell's progression through the cell cycle and its subsequent division. The obvious means of elucidating whether polyamines are required for cell growth and multiplication, is to specifically deplete the cells of their polyamines.

Development of inhibitors of the polyamine biosynthetic enzymes, made it possible to block the increase in polyamine synthesis, normally ensuing upon stimulation of cell proliferation, and, at least partly, to deplete cells of their polyamines. The results imply that polyamines are required for DNA synthesis, i.e. for the traverse through the S phase of the cell cycle.

ODC and SAMD probably constitute the most vulnerable targets for inhibitors intended to lower the cellular polyamine content. Compounds that specifically inhibit these enzymes and that block cell proliferation are listed in Table 1. The results of most studies indicate that partial polyamine depletion causes inhibition of DNA synthesis. Thus, Fillingame, Jorstad and Morris (1975) found that cells progressed normally from G0 through G1 and into S, but that there was a prolongation of the S phase. This inhibition of DNA replication was accentuated when a combination of an ODC inhibitor and a SAMD inhibitor was used (Morris, Jorstad and Seyfried, 1977). Krokan and Eriksen (1977) attributed the decreased rate of DNA synthesis to a reduction of the number of replication units active in DNA synthesis. The rate of DNA chain elongation was found to be only slightly reduced. In contrast, Geiger and Morris (1978) have presented data indicative of a reduced rate of DNA replication fork movement in polyamine-starved *E. coli.* In consistency with both conclusions, Pohjanpelto (1975) has shown that putrescine preferentially shortens the S phase when added to human fibroblast cultures.

Studies in other experimental systems suggest that polyamine synthesis inhibitors, alone or in combination, may block entry of cells into the S phase (Boynton, Whitfield and Isaacs, 1976; Mamont and coworkers, 1976; Wiegand and Pegg, 1978). Interference with the initiation of DNA synthesis is suggested also by the observation that cells accumulate in the G1 phase of the cell cycle following methyl-GAG treatment (Heby and coworkers, 1977; Rupniak and Paul, 1978). The effect of methyl-GAG on rat brain tumour cells grown *in vitro* was assessed by perturbation of cell cycle progression by DNA distribution analysis using a high-speed flow cytometric technique (Figure 2; Heby and coworkers, 1977). The major kinetic response pattern, evidenced from serial DNA distributions, consisted of compartment shifts out of S, G2 and M into G1 (Heby and coworkers, 1977).

Under various conditions it has been observed that cells treated with polyamine synthesis inhibitors will approximately double in number before they become arrested in the G1 phase (Mamont and coworkers, 1976; Heby and coworkers, 1977; Rupniak and Paul, 1978). These results suggest that cells may have a high enough polyamine content to be able to traverse one cell cycle in the absence of polyamine synthesis. Rupniak and Paul (1978) have suggested that there may be a restriction point in G1 at which it is decided whether the intracellular level of the polyamines is sufficiently high to enable a

Table 1 Polyamine synthesis inhibitors effective in blocking cell proliferation

Inhibitor	Cell cycle phase specificity	Experimental system	References
A. ODC inhibitors			
1. DL-α-Hydrazino-δ-aminovaleric acid (HAVA)	Inhibition of DNA synthesis	Isoproterenol-stimulated parotid gland *in vivo*	Inoue and coworkers, 1975
HAVA	Inhibition of DNA synthesis	Sarcoma-180 *in vivo*	Kato and coworkers, 1976
2. *trans*-3-Dehydro-DL-ornithine	n.d.	Chick embryo muscle cells *in vitro*	Relyea and Rando, 1975
3. DL-α-methyl-ornithine (α-MO)	Inhibition of DNA synthesis; G1 phase block	HTC cells *in vitro*	Mamont and coworkers, 1976
α-MO	S/G2 phase block	Ehrlich ascites tumour *in vivo*	Heby and coworkers, 1978a
4. DL-α-Difluoro-methylornithine	n.d.	HTC, L1210 and prostatic adenoma cells *in vitro*	Mamont and coworkers, 1978
B. SAMD inhibitors			
1. Methylglyoxal-*bis* (guanyl-hydrazone) (Methyl-GAG)	Inhibition of DNA synthesis	Phytohaemagglutinin- or concanavalin A-stimulated lymphocytes *in vitro*	Kay and Pegg, 1973; Otani and coworkers, 1974; Fillingame and coworkers, 1975
Methyl-GAG	Inhibition of DNA synthesis	Isoproterenol-stimulated parotid gland *in vivo*	Inoue and coworkers, 1975
Methyl-GAG	Inhibition of DNA synthesis	WI-38, BALB/3T3 and HeLaS_3 cells *in vitro*	Boynton and coworkers, 1976; Krokan and Eriksen, 1977
Methyl-GAG	G1 phase block	Rat brain tumour cells and rat embryo fibroblasts *in vitro*	Heby and coworkers, 1977; Rupniak and Paul, 1978
Methyl-GAG	S/G2 phase block	Ehrlich ascites tumour *in vivo*	Heby and coworkers, 1978a
2. 1,1′—[(methyl-ethanediylidene) dinitrilo]-*bis*(3-aminoguanidine) (MBAG)	Inhibition of DNA synthesis	Regenerating liver *in vivo*	Wiegand and Pegg, 1978

n.d. = the effect on cell proliferation not determined in relation to the cell cycle

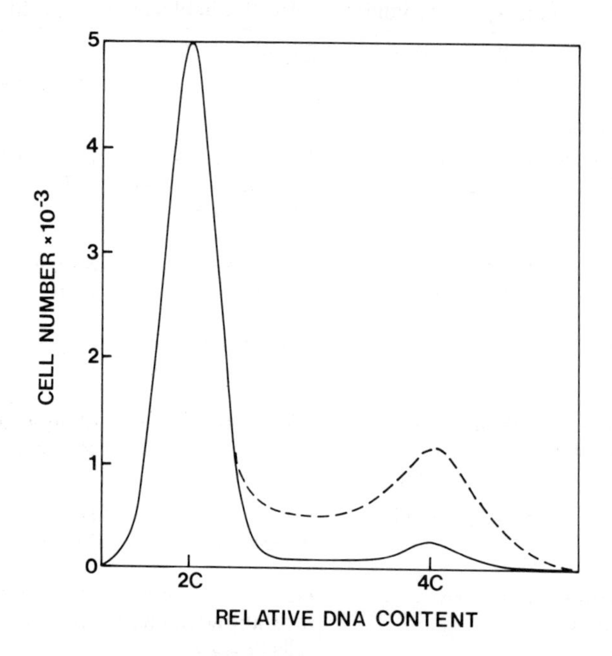

Figure 2 Flow cytometric analyses of the cellular DNA distribution in populations of rat brain tumour cells exhibiting exponential growth (dashed line) compared to populations of cells blocked in their proliferation by a 48 h methyl-GAG-treatment (solid line). This data yielded bimodal cell distributions dominated by the 2C peaks. C, the amount of DNA in a set of haploid mammalian chromosomes. The major kinetic response to methyl-GAG treatment was a block in G1 transit after about one cell cycle time (Heby and coworkers, 1977)

cell to enter a new cell cycle. This hypothesis may be extended to include a polyamine-sensitive restriction point in G2 for some tumour cells (Andersson and Heby, Chapter 4). Thus Ehrlich ascites tumour cells accumulate in the G2 phase as a result of polyamine deficiency.

If the blocking effect of polyamine synthesis inhibitors on DNA synthesis and cell proliferation is the result of intracellular polyamine deficiency, then one might expect it to be reversed by the addition of polyamines. Indeed, the addition of putrescine, spermidine or spermine has usually been found to cause an immediate resumption of DNA synthesis and cell division. The fact that normal transcription and translation continue after polyamine synthesis is blocked, suggests that the ensuing inhibition of DNA synthesis, which shows the same dose-response curve as does inhibition of polyamine synthesis (Fillingame, Jorstad and Morris, 1975), is specific. These observations indicate that DNA synthesis is more sensitive to partial polyamine depletion than are other events in the cell cycle, and that it is critically dependent on an adequate supply of polyamines.

In the cell, inhibitors like α-MO and methyl-GAG have been shown to markedly increase the amount of the enzyme which they inhibit, probably by binding to the enzyme and stabilizing it against intracellular proteolysis. In contrast to these inhibitors, there is a group of polyamine synthesis inhibitors, mainly diamines, which appear to decrease the amount of the enzyme. Thus, administration of putrescine has been found to inhibit the activity of ODC, not by simple product inhibition, but by inducing the synthesis of an ODC inhibitory protein (Heller, Fong and Canellakis, 1976; McCann, Tardif and Mamont, 1977). As a result, the ODC activity decays rapidly with an apparent half-life comparable to that observed after general inhibition of protein synthesis by cycloheximide (McCann, Tardif and Mamont, 1977; Kallio and coworkers, 1977b). Interestingly, putrescine administration during Ehrlich ascites tumour growth has been found to retard cell proliferation without markedly decreasing the cellular polyamine content (Andersson and Heby, Chapter 4). The ODC activity was completely blocked and an ODC inhibitory protein was identified. This observation might suggest that the effect of polyamine synthesis inhibitors might not be entirely attributable to a decreased intracellular polyamine content. Another diamine, 1,3-diaminopropane (DAP), has been found to inhibit DNA synthesis in regenerating liver (Kallio, Pösö and Jänne, 1977a) and in synchronized CHO cells (Sunkara, Rao and Nishioka, 1977). Inhibition of putrescine synthesis had no effect on the progression of CHO cells from G1 to S, but reduced the rate of DNA synthesis as a probable result of an effect on DNA chain elongation (Sunkara, Rao and Nishioka, 1977). DAP-treatment of synchronized S phase cells did not affect their progression to mitosis. On the basis of these data, it was concluded that putrescine is necessary for DNA synthesis but might not be essential for mitosis and cell division; thus granting late G1/early S synthesis of putrescine, but not late S/G2 synthesis (Figure 1), an important role in the traverse of the cell cycle. In view of the fact that DAP-treatment had little or no effect on the cellular spermidine and spermine content, it becomes necessary to study the effect of simultaneous depletion of putrescine, spermidine and spermine, before regarding the premitotic increase in polyamine synthesis as an extraneous event not required for the progression through mitosis. As a matter of fact, Sunkara, Seman and Rao (1978) recently showed that a 50% decrease in the cellular spermidine and spermine content was associated with a high incidence of binucleate cells. On the basis of their observations they suggested that polyamines may play a role in the regulation of nuclear division.

VIII. SUMMARY AND CONCLUSIONS

The physiological functions of the polyamines in the mammalian cell are gradually becoming disclosed. These amines play an essential role in many important metabolic processes involved in cell growth and proliferation including DNA, RNA and protein synthesis. In addition to being absolutely required for some of these processes it appears that the polyamines, by virtue of their intrinsic chemical and physical properties, serve to enhance the efficiency and fidelity of various steps in these processes, thereby contributing towards an increased rate of progression through the cell cycle.

REFERENCES

Aisbitt, R. P. G., and Barry, J. M. (1973). *Biochim. Biophys. Acta*, **320**, 610–616.

Andersson, G. (1977). *J. Cell. Physiol.*, **90**, 329–336.

Andersson, G., and Heby, O. (1977). *Cancer Res.*, **37**, 4361–4366.

Baserga, R., (1976). *Multiplication and Division in Mammalian Cells.* Dekker, New York and Basel.

Baserga, R., Costlow, M., and Rovera, G. (1973). *Fed. Proc.*, **32**, 2115–2118.

Boynton, A. L., Whitfield, J. F., and Isaacs, R. J. (1976). *J. Cell. Physiol.*, **89**, 481–488.

Brandt, J. T., Pierce, D. A., and Fausto, N. (1972). *Biochim. Biophys. Acta*, **279**, 184–193.

Cavia, E., and Webb, T. E. (1972). *Biochim. Biophys. Acta*, **262**, 546–554.

Chen, K., Heller, J., and Canellakis, E. S. (1976). *Biochem. Biophys. Res. Commun.*, **68**, 401–408.

Clark, J. L. (1974). *Biochemistry*, **13**, 4668–4674.

Erbe, W., Preiss, J., Seifert, R., and Hiltz, H. (1966). *Biochem. Biophys. Res. Commun.*, **23**, 392–397.

Fillingame, R. H., Jorstad, C. M., and Morris, D. R. (1975). *Proc. Nat. Acad. Sci. USA.*, **72**, 4042–4045.

Fillingame, R. H., and Morris, D. R. (1973). *Biochemistry*, **12**, 4479–4487.

Friedman, S. J., Bellantone, R. A., and Canellakis, E. S. (1972). *Biochim. Biophys. Acta*, **261**, 188–193.

Fuller, D. J. M., Gerner, E. W., and Russell, D. H. (1977). *J. Cell. Physiol.*, **93**, 81–88.

Gaza, D. J., Short, J., and Lieberman, I. (1973). *FEBS Lett.*, **32**, 251–253.

Gazdar, A. F., Stull, H. B., Kilton, L. J., and Bachrach, U. (1976). *Nature, Lond.*, **262**, 696–698.

Geiger, L. E., and Morris, D. R. (1978). *Nature, Lond.*, **272**, 730–732.

Gelfant, S. (1977). *Cancer Res.*, **37**, 3845–3862.

Goldstein, D. A., Heby, O., and Marton, L. J. (1976). *Proc. Nat. Acad. Sci. USA.*, **73**, 4022–4026.

Hayashi, S., Aramaki, Y., and Noguchi, T. (1972). *Biochem. Biophys. Res. Commun.*, **46**, 795–800.

Heby, O., Andersson, G., and Gray, J. W. (1978a). *Exp. Cell Res.*, **111**, 461–464.

Heby, O., Goldstein, D. A. and Marton, L. J. (1978b). Regulation of cellular polyamine synthesis during lytic infection with polyoma virus: superinduction with actinomycin D. In R. A. Campbell, (Ed.), *Advances in Polyamine Research*, Raven Press, New York, pp. 133–152.

Heby, O., Gray, J. W., Lindl, P. A., Marton, L. J., and Wilson, C. B. (1976a). *Biochem. Biophys. Res. Commun.*, **71**, 99–105.

Heby, O., and Lewan, L. (1971). *Virchows Arch. Abt. B Zellpath.*, **8**, 58–66.

Heby, O., Marton, L. J., Gray, J. W., Lindl, P. A., and Wilson, C. B. (1976b).

Polyamine metabolism in synchronously growing mammalian cells. In F. Bierring (Ed.), *Proceedings of the 9th Congress of the Nordic Society for Cell Biology*, Odense University Press, pp. 155-164.
Heby, O., Marton, L. J., Wilson, C. B., and Gray, J. W. (1977). *Europ. J. Cancer*, **13**, 1009-1017.
Heby, O., Marton, L. J., Wilson, C. B., and Martinez, H. M. (1975a). *J. Cell. Physiol.*, **86**, 511-522.
Heby, O., Marton, L. J., Zardi, L., Russell, D. H., and Baserga, R. (1975b). *Exp. Cell Res.*, **90**, 8-14.
Heller, J. S., Fong, W. F., and Canellakis, E. S. (1976). *Proc. Nat. Acad. Sci. USA.*, **73**, 1858-1862.
Hibasami, H., Tanaka, M., Nagai, J., and Ikeda, T. (1977). *Australian J. Exp. Biol. Med. Sci.*, **55**, 379-383.
Hogan, B. L. M. (1971). *Biochem. Biophys. Res. Commun.*, **45**, 301-307.
Hogan, B. L. M., McIlhinney, A., and Murden, S. (1974). *J. Cell. Physiol.*, **83**, 353-358.
Hogan, B. L. M., and Murden, S. (1974). *J. Cell. Physiol.*, **83**, 345-352.
Hölttä, E. (1975). *Biochim. Biophys. Acta*, **399**, 420-427.
Hölttä, E., and Jänne, J. (1972). *FEBS Lett.* **23**, 117-121.
Inoue, H., Kato, Y., Takigawa, M., Adachi, K., and Takeda, Y. (1975). *J. Biochem.*, **77**, 879-893.
Inoue, H., Tanioka, H., Shiba, K., Asada, A., Kato, Y., and Takeda, Y. (1974). *J. Biochem.*, **75**, 679-687.
Johnson, I. S., Wright, H. F., Svoboda, G. H., and Vlantis, J. (1960). *Cancer Res.*, **20**, 1016-1022.
Jänne, J., and Hölttä, E. (1974). *Biochem. Biophys. Res. Commun.*, **61**, 449-456.
Jänne, J., Hölttä, E., and Guha, S. K. (1976). Polyamines in mammalian liver during growth and development. In H. Popper and F. Schaffner (Ed.). *Progress in Liver Diseases.* Vol. 5, Grune and Stratton, New York, pp. 100-124.
Kallio, A., Pösö, H., Scalabrino, G., and Jänne, J. (1977b). *FEBS Lett.*, **73**, 229-234.
Kallio, A., Pösö, H., and Jänne, J. (1977a). *Biochim. Biophys. Acta*, **479**, 345-353.
Kato, Y., Inoue, H., Gohda, E., Tamada, F., and Takeda, Y. (1976). *Gann*, **67**, 569-576.
Kay, J. E., and Lindsay, V. J. (1973). *Exp. Cell Res.*, **77**, 428-436.
Kay, J. E., and Pegg, A. E. (1973). *FEBS Lett.* **29**, 301-304.
Kaye, A. M., Icekson, I., and Lindner, H. R. (1971). *Biochim. Biophys. Acta*, **252**, 150-159.
Klevecz, R. R. (1975). Molecular manifestations of the cellular clock. In J. C. Hampton (Ed.) *The Cell Cycle in Malignancy and Immunity*, NTIS, Springfield, Virginia, pp.1-19.
Krokan, H., and Eriksen, A. (1977). *Eur. J. Biochem.*, **72**, 501-508.
Mamont, P. S., Böhlen, P., McCann, P. P., Bey, P., Schuber, F., and Tardif, C. (1976). *Proc. Nat. Acad. Sci. USA.*, **73**, 1626-1630.
Mamont, P. S., Duchesne, M. C., Grove, J., and Bey, P. (1978). *Biochem. Biophys. Res. Commun.*, **81**, 58-66.
McCann, P. P., Tardif, C., and Mamont, P. S. (1977). *Biochem. Biophys. Res. Commun.*, **75**, 948-954.
McCann, P. P., Tardif, C., Mamont, P. S., and Schuber, F. (1975). *Biochem. Biophys. Res. Commun.*, **64**, 336-341.
Morris, D. R., Jorstad, C. M., and Seyfried, C. E. (1977). *Cancer Res.*, **37**, 3169-3172.
Oka, T., and Perry, J. W. (1976). *J. Biol. Chem.*, **251**, 1738-1744.
Otani, S., Mizoguchi, Y., Matsui, I., and Morisawa, S. (1974). *Mol. Biol. Reports*, **1**, 431-436.

Petersen, D. F., Anderson, E. C., and Tobey, R. A. (1968). Mitotic cells as a source of synchronized cultures. In D. M. Prescott (Ed.), *Methods in Cell Physiology*, Vol. 3, Academic Press, New York, pp. 347–370.

Pohjanpelto, P. (1975). *Biomedicine*, **23**, 350–352.

Prescott, D. M. (1976). *Reproduction of Eukaryotic Cells*. Academic Press, New York and London.

Rao, P. N., and Johnson, R. T. (1971). *J. Cell. Physiol.*, **78**, 217–224.

Reel, J. R., and Kenney, F. T. (1968). *Proc. Nat. Acad. Sci. USA.*, **61**, 200–206.

Relyea, N., and Rando, R. R. (1975). *Biochem. Biophys. Res. Commun.*, **67**, 392–402.

Rueckert, R. R., and Mueller, G. C. (1960). *Cancer Res.*, **20**, 1584–1591.

Rupniak, H. T., and Paul, D. (1978). *J. Cell. Physiol.*, **94**, 161–170.

Russell, D. H., and Snyder, S. H. (1969). *Mol. Pharmacol.*, **5**, 253–262.

Russell, D. H., and Stambrook, P. J. (1975). *Proc. Nat. Acad. Sci. USA.*, **72**, 1482–1486.

Schimke, R. T., and Doyle, D. (1970). *Ann. Rev. Biochem.*, **39**, 929–976.

Schrock, T. R., Oakman, N. J., and Bucher, N. L. R. (1970). *Biochim. Biophys. Acta*, **204**, 564–577.

Stastny, M., and Cohen, S. (1970). *Biochim. Biophys. Acta*, **204**, 578–589.

Studzinski, G. P., Lambert, W. C. (1969). *J. Cell. Physiol.*, **73**, 109–118.

Sunkara, P. S., Rao, P. N., and Nishioka, K. (1977). *Biochem. Biophys. Res. Commun.*, **74**, 1125–1133.

Sunkara, P. S., Seman, G., and Rao, P. N. (1978) *Proc. Am. Assoc. Cancer Res.*, **19**, 62.

Tabor, C. W., and Tabor, H. (1976). *Ann. Rev. Biochem.*, **45**, 285–306.

Takigawa, M., Inoue, H., Gohda, E., Asada, A., Takeda, Y., and Mori, Y. (1977). *Exp. Mol. Pathol.*, **27**, 183–196.

Terasima, T., and Tolmach, L. J. (1963). *Exp. Cell. Res.*, **30**, 344–362.

Tobey, R. A., Anderson, E. C., and Petersen, D. F. (1967). *J. Cell. Biol.*, **35**, 53–59.

Tomkins, G. M., Gelehrter, T. D., Granner, D., Martin, D., Samuels, H. H., and Thompson, E. B. (1969). *Science*, **166**, 1474–1480.

Wagner, E. K., and Roizman, B. (1968). *Science*, **162**, 569–570.

Wiegand, L., and Pegg, A. E. (1978). *Biochim. Biophys. Acta*, **517**, 169–180.

Williams-Ashman, H. G., Jänne, J., Coppoc, G. L., Geroch, M. E., and Schenone, A. (1972). New aspects of polyamine biosynthesis in eukaryotic organisms. In G. Weber (Ed.), *Advances in Enzyme Regulation, Vol. 10*, Pergamon Press, Oxford and New York, pp. 225–245.

Yuspa, S. H., Lichti, U., Ben, T., Patterson, E., Hennings, H., Slaga, T. J., Colburn, N., and Kelsey, W. (1976). *Nature, Lond.*, **262**, 402–404.

Chapter 3

Polyamines in Rapidly Growing Animal Tissues

AARNE RAINA, TERHO ELORANTA, RAIJA-LEENA PAJULA,
RAUNO MÄNTYJÄRVI AND KYLLIKKI TUOMI

I. INTRODUCTION

Numerous observations reported in recent years have indicated that polyamines may play an essential role in growth processes, which greatly increased interest in their biochemistry and biological function (Chapters 1–10, 20–22). We noticed that during embryonic development, the peak in the concentrations of polyamines coincided with that of ribonucleic acid and suggested that polyamines are in some way connected with growth and protein synthesis (Raina, 1963). Since then much additional evidence has been obtained in a number of systems characterized by rapid growth or increased protein synthesis, such as regenerating liver, hypertrophic heart or kidney, healing wound, experimental granuloma, tissues responding to anabolic hormonal stimuli or tissues actively engaged in protein synthesis, malignant tumours, and cultured cells stimulated to proliferate. More direct information at the molecular level of the involvement of polyamines in the protein-synthetic machinery or in nucleic acid synthesis has recently been provided by designing compounds which potently and fairly specifically inhibit polyamine synthesis in animal cells. Much of the new data dealing with this topic are discussed in detail in several recent reviews (Raina and Jänne, 1975; Jänne, Hölttä and Guha, 1976; Tabor and Tabor, 1976; Jänne, Pösö and Raina, 1978) and also in Chapters 2, 7, 8 and 11 of this monograph. This section is concerned with selected animal systems, and reports some new observations on the effect of inhibitors of polyamine synthesis in cultured mammalian cells.

II. TISSUE POLYAMINES DURING DEVELOPMENTAL GROWTH

A. Synthesis and Accumulation of Polyamines during Embryonic Development

Polyamine synthesis has been more extensively studied in three vertebrate systems, including chick, rat and *Xenopus* embryos. Our early studies with developing chick embryos (Raina, 1963) showed two distinct peaks in the concentrations of spermidine and spermine in whole embryos, one within the first few days and the second on the 15th-16th day of incubation. We also demonstrated the conversion of radioactive ornithine to putrescine in chick embryos *in vivo*. It was later shown by Russell and coworkers (Snyder and Russell, 1970) that chick embryo homogenates contain very high ornithine decarboxylase activity, which reaches its maximum on the fifth day of incubation and then gradually declines towards hatching. The idea that a high ornithine decarboxylase activity is characteristic for embryonic tissues was also supported by results obtained with rat and amphibian embryos (Snyder and Russell, 1970; Russell and McVicker, 1972). This high enzyme activity at least partly explains the high concentration of putrescine found in embryonic tissues (Russell and Lombardini, 1971; Russell and McVicker, 1972) in contrast to most adult tissues where the putrescine concentration is very low, only of the order of 10^{-5} M. Also in human fetal liver and brain, ornithine decarboxylase activity and the concentration of putrescine are reported to be several fold higher compared to those of mature tissues (Sturman and Gaull, 1974). On the other hand, the activity of S-adenosylmethionine decarboxylase of embryonic tissues shows much less variation during development and does not markedly differ from that found in adult tissues (Russell and Lombardini, 1971; Russell and McVicker, 1972). The activities of propylamine transferases (spermidine and spermine synthases) in embryonic tissues have not yet been systematically studied.

B. Polyamines during Postnatal Development

The first systematic study of the changes occurring in tissue polyamine concentrations during postnatal development was performed in the rat (Jänne, Raina and Siimes, 1964). The total polyamine concentration was in most rat tissues about 1-2 mM. The spermidine concentration was highest in tissues of newborn animals and then gradually decreased with increasing age. Much less age-dependent variation was seen in the tissue spermine content. A correlation was also sought between tissue polyamine and nucleic acid concentrations. Although the molar ratio of total polyamine nitrogen to RNA phosphate varied from 0.5 in the thymus to about 0.2 in skeletal muscle, it remained fairly constant in a particular tissue during ageing (Raina and Jänne, 1970). No such correlation was found between the total polyamine-N and DNA. In the developing rat brain the changes in RNA and spermidine concentrations were most closely correlated (Seiler and Lamberty, 1975).

No systematic study of the four enzymes of the polyamine biosynthetic pathway has been performed until recently. We have simultaneously assayed the activities of the polyamine-synthesizing enzymes in a number of rat tissues (Raina, Pajula and Eloranta, 1976). Our results demonstrated that the activities of spermidine and spermine synthases were much higher (in terms of specific activity) than those of ornithine decarboxylase and S-adenosylmethionine decarboxylase. In young rats aged three months, the prostate and small intestine were very rich in all four enzymes. Whereas the brain contained high activities of S-adenosylmethionine decarboxylase and spermidine and spermine synthases, ornithine decarboxylase activity was low in this tissue. The activity of spermidine synthase was remarkably high in the pancreas. Comparison of the activities of the polyamine-synthesizing enzymes to the concentrations of polyamines (Jänne, Raina and Siimes, 1964) in a particular tissue reveals that no single enzyme or even the whole pattern of enzymes alone cannot explain the actual polyamine pool which must also depend on other factors, e.g. availability of S-adenosylmethionine and other precursors, modification of enzyme activities by their products, and the rate of elimination of polyamines.

Of the four polyamine-synthesizing enzymes in animal tissues, the two decarboxylases possess an extremely short biological half-life, whereas the propylamine transferases turn over much more slowly (Raina and Jänne, 1975; Jänne, Pösö and Raina, 1978). As a great variety of 'non-specific' factors, such as the nutritional state of the animal, can markedly influence the activities of ornithine and S-adenosylmethionine decarboxylases (Eloranta, Mäenpää and Raina, 1976; Eloranta and Raina, 1977) in certain tissues, e.g. liver, comparison of these enzyme activities at different stages of development should be done with some reservation. In agreement with the developmental changes in the concentrations of putrescine and spermidine, the enzyme activities involved in their synthesis tend to decrease with increasing age (Raina, Pajula and Eloranta, 1976). As an exception, the activity of brain S-adenosylmethionine decarboxylase appears to be much lower at the time of birth compared to the adult level.

III. POLYAMINES IN TISSUE REGENERATION, REPAIR AND HYPERTROPHIC RESPONSE

A. Synthesis and Accumulation of Polyamines in the Regenerating Liver

The changes in the synthesis of polyamines in developing chick embryos (Raina, 1963) and rats (Jänne, 1964) suggested that polyamines may have an important function in the synthesis of proteins and nucleic acids, especially in rapidly growing tissues. Regenerating liver offered a useful tool to test this hypothesis. We found that partial hepatectomy of the rat caused an early stimulation of spermidine synthesis in the regenerating liver remnant, as indicated by a marked increase in the incorporation of labelled methionine and arginine into this polyamine *in vivo* (Raina, Jänne and Siimes, 1965, Jänne and Raina, 1966; Raina, Jänne and Siimes, 1966). A net accumulation of spermidine, paralleling that of ribonucleic acid was seen as early as 12–16 h after the operation (see Figure 1). Independently, similar results were reported by Dykstra and Herbst (1965). An important observation for understanding the regulatory mechanisms in the biosynthesis of polyamines was made by Jänne (1967) who demonstrated that putrescine synthesis from ornithine *in vivo* was markedly increased within the first few hours after operation, with a rapid accumulation of endogenous putrescine. That the increased production of putrescine was due to a marked stimulation of ornithine decarboxylase activity occurring soon after partial hepatectomy, was independently demonstrated by two groups (Jänne and Raina, 1968; Russell and Snyder, 1968). The activities of S-adenosylmethionine decarboxylase and spermidine and spermine synthases also increase but considerably later and to a much smaller extent than the ornithine decarboxylase activity (Hannonen, Raina and Jänne, 1972). It therefore appears that in the regenerating liver, the markedly increased ornithine decarboxylase activity with resultant accumulation of putrescine is the main reason for the accumulation of spermidine. As putrescine acts as a substrate in the spermidine synthase reaction and stimulates S-adenosylmethionine decarboxylase, but inhibits competitively spermine synthesis, the increased putrescine concentration favours spermidine synthesis (Raina and coworkers, 1970; Hannonen, Raina and Jänne, 1972; Jänne, Pösö and Raina, 1978).

The increase of ornithine decarboxylase activity is a very early event in the regenerating rat liver. It is almost maximally stimulated as early as 4 h after partial hepatectomy (Russell and Snyder, 1968; Raina and coworkers, 1970). A closer analysis of the changes in ornithine decarboxylase activity during the early period of liver regeneration has revealed that the enzyme is stimulated in two phases, the first peak occurring at 4 h after the operation, independently of the age of the animal (Hölttä and Jänne, 1972). In young animals the second peak occurred almost immediately after the first one, but was delayed by several hours in older animals. It appears that the latter peak is also influenced by the feeding schedule (Barbiroli and coworkers, 1975; Yanagi and Potter, 1977). Pituitary factors and possibly the adrenals seem to play a role in the early stimulation of ornithine decarboxylase activity normally occurring 4 h after partial hepatectomy which is delayed or diminished in hypophysectomized animals (Russell and Snyder, 1969a), and after adrenalectomy or treatment with α-adrenergic blocking agents (Thrower and Ord, 1974).

A single treatment of the rat with carbon tetrachloride, which causes acute necrosis of the liver followed by a rapid regenerative process, resulted in a marked activation of polyamine synthesis (Jänne, Hölttä and Guha, 1976). The induction of ornithine decarboxylase as well as the increase in the concentration of putrescine and spermidine occurred 12 to 24 h after the treatment, i.e. considerably later as compared to liver regenerating after partial hepatectomy.

Although the mechanisms regulating ornithine decarboxylase activity in the regenerating liver have been the subject of intensive study, they still are largely unresolved. In fact, experimental data suggest that several mechanisms are involved, but their physiological significance remains to be determined. The increase in enzyme activity in the regenerating liver appears to involve synthesis of new enzyme protein, as revealed by experiments with inhibitors of protein synthesis or by titration of immunoreactive enzyme protein (Russell and Snyder, 1969b; Hölttä, 1975). Furthermore, the increase in ornithine decarboxylase activity is abolished by actinomycin D (Russell and Snyder, 1969b) or α-amanitin (Jänne, Pösö and Raina, 1978), suggesting that the synthesis of ornithine decarboxylase is at least partly regulated at the transcriptional level. Because the administration of exogenous polyamines to partially hepatectomized rats results in a rapid decay of liver ornithine decarboxylase activity (Jänne, Pösö and Raina, 1978), it is possible that the enhanced synthesis and accumulation of endogenous putrescine and spermidine taking place in the regenerating liver represses the synthesis of ornithine decarboxylase. The involvement of cyclic AMP and cyclic AMP-dependent protein kinase(s) in the induction of ornithine decarboxylase has been suggested by some authors (Byus, Hedge and Russell, 1977; Chapter 8). The results obtained with β-adrenergic blocking agents, which prevent the increases in cyclic AMP during the early period of liver regeneration, without affecting the rise of ornithine decarboxylase activity (Thrower and Ord, 1974), contradict this hypothesis.

The stimulation of liver ornithine decarboxylase activity after partial hepatectomy is by no means a phenomenon specific for growth. In fact, liver ornithine decarboxylase is stimulated in response to a number of other stimuli, such as sham operation, hypertonic infusions and other stressful procedures (Schrock, Oakman and Bucher, 1970), administration of amino acids, fasting and refeeding, and treatment with various hormones (Eloranta and Raina, 1977; Jänne, Pösö and Raina, 1978), although usually less intensively. It has been pointed out (Tabor and Tabor, 1976) that the rapid changes in ornithine decarboxylase activity can largely be explained by the rapid turnover of this enzyme. Any general increase in protein synthesis or a decrease in the rate of degradation would therefore result in a rapid rise in the activities of those enzymes turning over rapidly. On the other hand, there is no explanation as to why the activity of liver S-adenosylmethionine decarboxylase, having a very short half-life of 20-35 min (Hannonen, Raina and Jänne, 1972; Eloranta, Mäenpää and Raina, 1976), actually decreases during the early period of liver regeneration.

B. Polyamine Synthesis in Tissue Repair

The formation of granulation tissue is a basic biological phenomenon in mammalian repair processes as a response to tissue injury. We have studied polyamine synthesis during the formation of granulation tissue induced experimentally by implanting viscose cellulose sponges subcutaneously in the rat (Raina and coworkers, 1973). The activities of the polyamine-synthesizing enzymes rose steadily during the first week, i.e. during the period of rapid cell proliferation. Furthermore, a remarkable parallelism between the accumulation of polyamines and RNA was observed. A rapid increase in ornithine decarboxylase activity and accumulation of RNA and polyamines, notably spermidine, has also been demonstrated in healing wounds in rat skin (Mizutani, Inoue and Takedi, 1974).

C. Polyamine Synthesis in Tissue Hypertrophy

Treatment of the rat with thioacetamide results in an enlargement of the liver and stimulation of RNA synthesis. As shown in Table 1, a single administration of thioacetamide caused a marked increase in liver ornithine decarboxylase activity which reached its maximum 12-24 h after the injection, whereas no significant changes were observed in the activities of S-adenosylmethionine decarboxylase or methionine adenosyltransferase. Thioacetamide treatment also increased the concentration of liver spermidine. An increase in the concentration of liver spermidine, paralleling the changes in RNA, has also been observed in the livers of mice after repeated application of perfluorovaleric acid (Seiler and Askar, 1972).

Renal hypertrophy, resulting from a prolonged treatment of castrated mice with androgens (Henningsson, Persson and Rosengren, 1978) or from unilateral nephrectomy, and cardiac hypertrophy following aortic constriction are also associated with increased synthesis of polyamines (see Jänne, Pösö and Raina, 1978), notably with stimulation of ornithine decarboxylase activity, leading to accumulation of polyamines. These conditions are characterized by an increase in tissue mass due to enhanced cell size, without an increase in the actual number of cells.

Table 1 Effect of thioacetamide on the synthesis of polyamines and S-adenosylmethionine in rat liver

Female Wistar rats weighing 110–130 g received a single intraperitoneal injection of thioacetamide, 15 mg 10 g^{-1} body weight, at the time points indicated. The animals were fasted for 12 h before analysis. The values are means of 5 animals in each group. Controls were treated with 0.9% NaCl 24 h before killing.

Time of treatment h	Polyamines μmoles g^{-1} wet wt. Spermidine	Spermine	Ornithine decarboxylase pmoles mg^{-1}protein per 30 min	S-adenosylmethionine decarboxylase	Methionine adenosyltransferase nmoles mg^{-1} protein per min
Controls	0.82	1.09	28	261	6.49
4	0.89	1.11	30	331	5.87
12	0.80	0.95	275	104	6.52
24	0.96	0.78	566	186	5.64
48	1.41	0.80	96	405	4.92
72	1.30	0.89	21	332	5.99

D. Effect of Growth-promoting Hormones on Polyamine Synthesis

The stimulation of polyamine synthesis in specific target tissues induced by the administration of a variety of hormones, including growth hormone, thyrotropin, adrenocorticotrophic hormone, gonadotrophins, and androgens and oestrogens has been discussed in detail in a recent review (Jänne, Pösö and Raina, 1978).

Since the early observation by Kostyo (1966) demonstrating that hypophysectomy results in a rapid decrease in liver spermidine, the role of growth hormone in the regulation of polyamine synthesis has been explored in a number of studies. Administration of growth hormone to intact rats caused a sharp and marked stimulation of liver ornithine decarboxylase activity, reaching a peak 3–4 h after administration (Jänne, Raina and Siimes, 1968; Raina and Hölttä, 1972), with a concomitant accumulation of putrescine. Growth hormone is able to enhance polyamine synthesis in a number of other tissues, including kidney, brain, heart and adrenals (Jänne, Pösö and Raina, 1978). The observation that a specific neutralization of endogenous growth hormone by antisera results in a decrease of ornithine decarboxylase activity in several tissues (Rao and Li, 1977) indicates that growth hormone may play a physiological role in the regulation of polyamine synthesis. It also appears that the sites affecting the synthesis of ornithine decarboxylase and the production of somatomedin reside in different portions of the growth hormone molecule (Jänne, Pösö and Raina, 1978).

In many systems characterized by increased synthesis of polyamines, e.g. in regenerating livers or livers of animals treated with thioacetamide, the activities of ornithine decarboxylase and S-adenosylmethionine decarboxylase behave differently (see Table 1). In marked contrast, the administration of androgens to castrated rats caused a rapid and remarkable stimulation of both enzymes (Pegg, Lockwood and Williams-Ashman, 1970). Also treatment of the chick with oestradiol resulted in an early stimulation of S-adenosylmethionine decarboxylase and a parallel accumulation of RNA and polyamines in the liver (Eloranta, Mäenpää and Raina, 1976). It therefore appears that, at least in some instances, S-adenosylmethionine decarboxylase might also be the rate-controlling enzyme in polyamine synthesis.

Treatment of the rat with antithyroid drugs or thyrotropin results in an increase in thyroid weight and increased synthesis and accumulation of polyamines and nucleic acids (Matsuzaki and Suzuki, 1974; Richman and coworkers, 1975). Induction of ornithine decarboxylase by thyrotropin has also been demonstrated in thyroid tissue *in vitro* (Scheinman and Burrow, 1977; Spaulding, 1977). There is some evidence to suggest that the induction of ornithine decarboxylase by thyrotropin in the thyroid is mediated by cyclic AMP (Scheinman and Burrow, 1977).

E. S-Adenosylmethionine and the Regulation of Polyamine Synthesis

Biological transmethylations and the polyamine-biosynthetic pathway share a common substrate, i.e. S-adenosylmethionine. We have therefore explored whether these two important metabolic routes have common regulatory features. Our results showed that although the activity of methionine adenosyltransferase in the rat greatly varied from tissue to tissue (Raina, Pajula and Eloranta, 1976), there were relatively small differences between various tissues in the concentration of S-adenosylmethionine, ranging from 20 to 70 nmole g^{-1} of tissue wet wt. (Eloranta, 1977). Treatment of the rat with thioacetamide or partial hepatectomy neither significantly changed the methionine adenosyltransferase activity (Table 1) nor the concentration of S-adenosylmethionine in the liver (Eloranta and Raina, 1977). The concentration of liver S-adenosylmethionine was also uninfluenced by starvation for two days, but rose somewhat on refeeding of starved animals. We can conclude that under a variety of physiological and experimental conditions the concentration of S-adenosylmethionine remains fairly constant and that changes in polyamine metabolism are not primarily connected with changes in the concentration of S-adenosylmethionine.

IV. ROLE OF POLYAMINES IN TISSUE GROWTH

A. Relationship of Polyamines to Nucleic Acids and Protein Synthesis *in vivo*

A large number of observations strongly suggest that polyamines have an important role in growth and cell proliferation, although no specific function

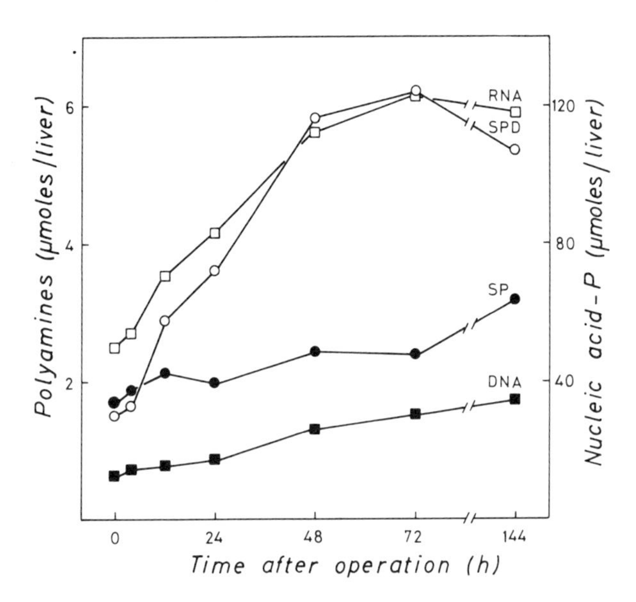

Figure 1 Accumulation of polyamines and nucleic acids in regenerating rat liver. Female Wistar rats weighing 100–120 g were partially hepatectomized 4–144 h before killing. The animals were fasted for 12 h before analysis. Each point represents the mean of three pooled preparations, each obtained from two or three livers. Symbols: ○, spermidine (SPD); ●, spermine (SP), □, ribonucleic acid; ■, deoxyribonucleic acid

has definitely been proved as yet (see also Chapters 1–5, 22). In a variety of systems characterized by rapid growth, polyamines and RNA accumulate in parallel. There is evidence to suggest that a large fraction of cellular polyamines is probably non-covalently bound to ribosomes, which may be an important site of action of polyamines (Raina and Jänne, 1975). In addition, polyamines apparently have other sites of action in the protein-synthetic machinery, e.g. in the initiation of translation.

Do polyamines play a role in the synthesis of nucleic acids? It has repeatedly been demonstrated that polyamines are able to stimulate RNA polymerase reaction *in vitro* even at an optimal concentration of divalent cation. Recent results obtained with inhibitors of polyamine synthesis have been interpreted to mean that polyamines are not indispensable for RNA synthesis, but may have an essential function in DNA synthesis in animal cells (Jänne, Pösö and Raina, 1978).

A critical test of the biological role of polyamines in animal cells would be a total depletion of cells of polyamines. Much effort has recently been put on the synthesis of compounds which would potently and specifically inhibit polyamine synthesis. (A survey of these compounds and their effects on polyamine synthesis and cell proliferation is presented in Chapters 10 and 11).

B. Effect of 1,3-diaminopropane on Polyamine and Macromolecular Synthesis in the Regenerating Liver

It was shown by Pösö and Jänne (1976a,b) that treatment of the rat after partial hepatectomy with 1,3-diaminopropane completely abolished the stimulation of ornithine decarboxylase activity. Repeated injections of diaminopropane also prevented the early increases in the concentrations of hepatic putrescine and spermidine normally occurring in response to partial hepatectomy, whereas no change was observed in the spermine concentration (Figure 2). As illustrated in Figure 2, repeated injections of diaminopropane to the rat also profoundly decreased the synthesis of DNA in the regenerating liver without any changes in the synthesis of liver RNA and protein, as measured by the incorporation of radioactive precursors into macromolecules (Pösö and Jänne, 1976b; Kallio *et al.*, 1977a).

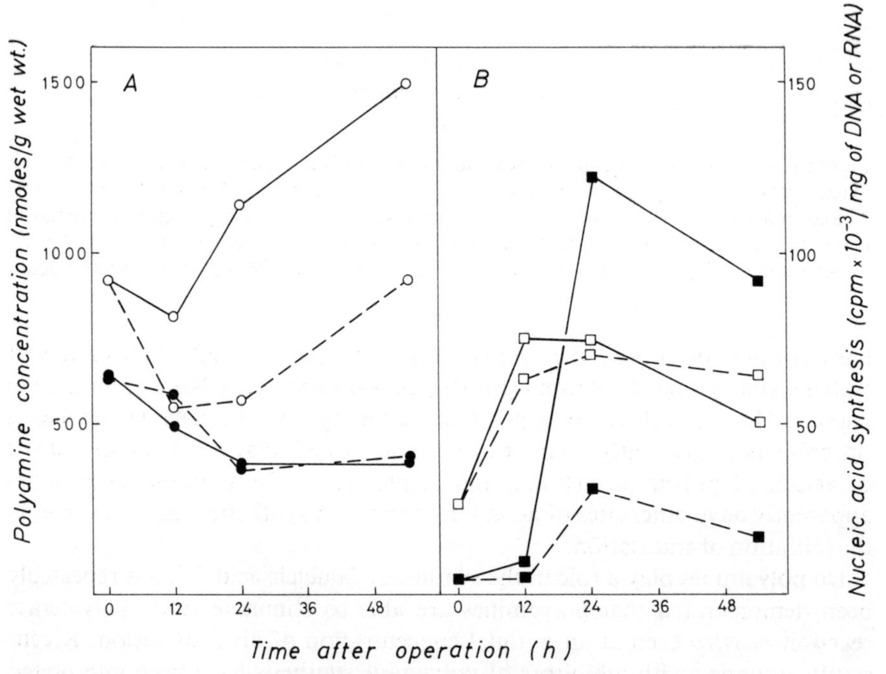

Figure 2 Effect of 1,3-diaminopropane on the accumulation of polyamines (A) and on the synthesis of RNA and DNA (B) in regenerating rat liver. The treated animals (————) received 75 μmole 100 g^{-1} body wt. of 1,3-diaminopropane intraperitoneally every 3 h starting at the time of the operation. At 2 h before killing all animals received 5 μCi of [6-^{14}C]orotic acid and 20 μCi of [^{3}H]thymidine as an intraperitoneal injection. Each point represents the mean of three or four animals. ○, spermidine; ●, spermine; □, synthesis of RNA from [^{14}C]orotic acid; ■, synthesis of DNA from [^{14}C]thymidine. (Reproduced from Pösö and Jänne, 1976b., and by permission of the Biochemical Society)

The mechanism of the inhibition of ornithine decarboxylase *in vivo* by various diamines and polyamines seems to be complex. Exogenously administered amines apparently cause a repression-type inhibition of the enzyme (Pösö and Jänne, 1976b). Evidence has been presented which suggests that the regulation of ornithine decarboxylase in the regenerating liver occurs at the transcriptional and post-transcriptional level (Kallio and coworkers, 1977b).

C. Effect of Inhibitors of Polyamine Synthesis on the Proliferation of BKT-cells

We have studied the effect of various types of inhibitors, including DL-α-hydrazino-δ-aminovaleric acid (DL-HAVA), DL-α-methylornithine, DL-α-difluoromethyl ornithine and 1,3-diaminopropane, on the synthesis of polyamines and nucleic acids in BKT-cells. DL-HAVA, a potent competitive inhibitor of ornithine decarboxylase *in vitro*, efficiently prevented the accumulation of nucleic acids and polyamines at a concentration of 1 mM (Raina and coworkers, 1978). Although DL-HAVA may have non-specific effects, such as complex formation with pyridoxal phosphate, its inhibitory effect on cell proliferation and DNA synthesis in isoproterenol-stimulated mouse salivary glands (Inoue and coworkers, 1975), in mouse sarcoma (Kato and coworkers, 1976) and in mouse skin (Takigawa and coworkers, 1977) can specifically be reversed by putrescine. The specificity of the inhibition of polyamine synthesis, however, definitely needs verification.

α-Methylornithine and α-difluoromethyl ornithine are reported to decrease the proliferation of several cell lines in culture (Chapter 10). These compounds reduce the cellular concentration of putrescine and spermidine, but have little effect on spermine. We have obtained similar results with BKT-cells, but relatively high concentrations, i.e. 2-5 mM are needed to prevent cell proliferation and the accumulation of nucleic acids (Pajula and Raina, unpublished observations).

The effect of 1,3-diaminopropane on the cellular concentrations of polyamines and the accumulation of nucleic acids in cultures of BKT-cells is shown in Figure 3. Contrary to the results obtained with ornithine analogues, diaminopropane reduced the concentration of both spermidine and spermine. The accumulation of DNA and RNA was inhibited approximately to the same extent. Further work is in progress to determine the site of the inhibitory effect of diaminopropane on nucleic acid synthesis.

D. The Value of Inhibitors of Polyamine Synthesis

It is evident that most inhibitors of polyamine synthesis presently available, efficiently reduce the cellular concentration of putrescine and/or spermidine, whilst the concentration of spermine remains unchanged or even increases.

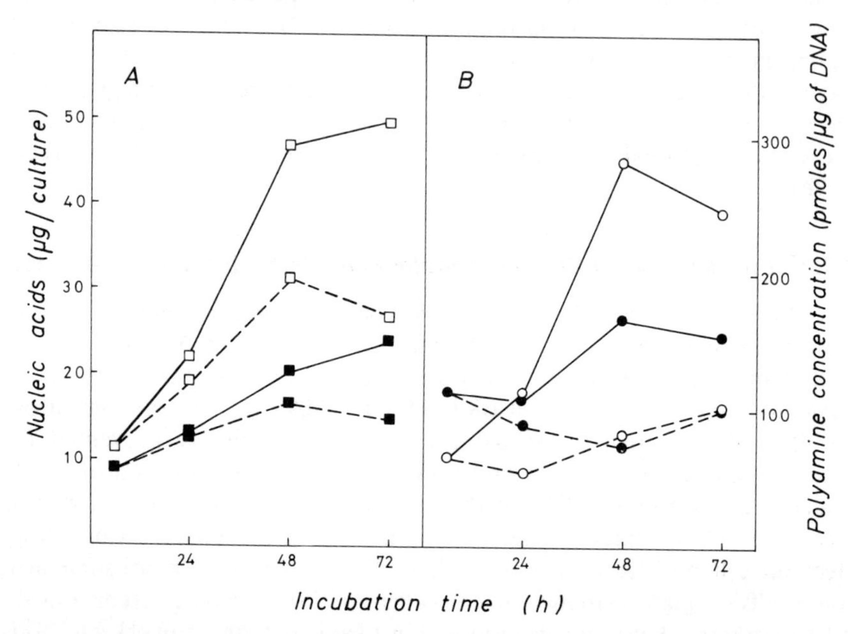

Figure 3 Effect of 1,3-diaminopropane on the accumulation of nucleic acids (A) and polyamines (B) in BKT-cells. The cells (a hamster cell line derived from BK-virus-induced tumours (Sten and coworkers, 1976)) were cultivated until confluency in the presence of enriched medium, harvested, seeded on plastic plates (diameter 10 cm), and incubated in MEM medium supplemented with 5% fetal calf serum, 0.03% glutamine, 200 μml^{-1} of penicillin. 1,3-Diaminopropane (5 mM) was added at 6 h after seeding. The growth medium was changed every 24 h. Polyamines and nucleic acids were analysed as described previously (Raina *et al.*, 1978). Symbols: ———, control; — — — —, diaminopropane; □, RNA; ■, DNA; ○, spermidine; ●, spermine

This is especially true for ornithine analogues. Some of the inhibitors have undesirable side effects, such as mitochondrial damage or complex formation with pyridoxal phosphate (see Jänne, Pösö and Raina, 1978). The reversal by exogenous polyamines of the inhibition of cell proliferation and macromolecular synthesis may speak for a specific effect caused by polyamine depletion. Other mechanisms, however, have to be considered. For example, polyamines and the inhibitor may compete for a common cellular uptake system (see Jänne, Pösö and Raina, 1978).

Depletion of the cellular polyamine pool by inhibitors of polyamine synthesis is a slow process which might seriously limit their value. The effect which is seen after a prolonged treatment with the inhibitor might or might not indicate the primary site of action of polyamines (Chapters 10 and 11). In future research special attention has to be paid to the immediate metabolic consequences at the molecular level, following the reversal of the cellular polyamine depletion by exogenous polyamines and/or removal of the inhibitor.

V. SUMMARY AND CONCLUSION

It seems evident from our investigation that polyamines have an essential function in cell proliferation. It is clear that one major approach in future research of animal cells involves the use of selective inhibitors of polyamine synthesis. The inhibitors presently available are only effective in tissue cultures, and at relatively high concentrations.

We have recently developed a rapid method for the assay of spermidine and spermine synthases (Raina, Pajula and Eloranta, 1976) and purified the spermine synthase from brain to apparent homogeneity (Pajula, Raina and Eloranta, to be published). In addition to this enzymological approach, special efforts will be made to find new types of inhibitors of polyamine synthesis, especially those acting on the propylamine transfer step.

ACKNOWLEDGEMENTS

We have received financial support from the National Research Councils for Natural and Medical Sciences (Finland), the Sigrid Jusélius Foundation and the Finnish Foundation for Cancer Research.

REFERENCES

Barbiroli, B., Moruzzi, M. S., Tadolini, B., and Monti, M. G. (1975). *J. Nutr.*, **105**, 408-412.

Byus, C. V., Hedge, G. A., and Russell, D. H. (1977). *Biochim. Biophys. Acta*, **498**, 39-45.

Dykstra, Jr., W. G., and Herbst, E. J. (1965). *Science*, **149**, 428-429.

Eloranta, T. O. (1977). *Biochem. J.*, **166**, 521-529.

Eloranta, T. O., Mäenpää, P. H., and Raina, A. M. (1976). *Biochem. J.*, **154**, 95-103.

Eloranta, T. O., and Raina, A. M. (1977). *Biochem. J.*, **168**, 179-185.

Hannonen, P., Raina, A., and Jänne, J. (1972). *Biochim. Biophys. Acta*, **273**, 84-90.

Henningsson, S., Persson, L., and Rosengren, E. (1978). *Acta Physiol. Scand.*, **102**, 385-393.

Hölttä, E. (1975). *Biochim. Biophys. Acta*, **399**, 420-427.

Hölttä, E., and Jänne, J. (1972). *FEBS Letters*, **23**, 117-121.

Inoue, H., Kato, Y., Takigawa, M., Adachi, K., and Takeda, Y. (1975). *J. Biochem.*, **77**, 879-893.

Jänne, J. (1967). *Acta Physiol. Scand.*, Suppl. **300**, 1-71.

Jänne, J., Hölttä, E., and Guha, S. K. (1976). Polyamines in mammalian liver during growth and development. In *Progress in Liver Diseases*, Vol 5, H. Popper and F. Schaffner (Eds), Grune & Stratton, Inc., New York. pp. 100-124.

Jänne, J., Pösö, H., and Raina, A. (1978). *Biochim. Biophys. Acta*, **473**, 241-293.

Jänne, J., and Raina, A. (1966). *Acta Chem. Scand*, **20**, 1174–1176.
Jänne, J., and Raina, A. (1968). *Acta Chem. Scand.*, **22**, 1349–1351.
Jänne, J., Raina, A., and Siimes, M. (1964). *Acta Physiol. Scand.*, **62**, 352–358.
Jänne, J., Raina, A., and Siimes, M. (1968). *Biochim. Biophys. Acta*, **166**, 419–426.
Kallio, A., Pösö, H., and Jänne, J. (1977a). *Biochim. Biophys. Acta*, **479**, 345–353.
Kallio, A., Pösö, H., Scalabrino, G., and Jänne, J. (1977b). *FEBS Letters*, **73**, 229–234.
Kato, Y., Inoue, H., Gohda, E., Tamada, F., and Takeda, Y. (1976). *Gann*, **67**, 569–576.
Kostyo, J. L. (1966). *Biochem. Biophys. Res. Commun.*, **23**, 150–155.
Matsuzaki, S., and Suzuki, M. (1974). *Endocrin. Japon.*, **21**, 529–537.
Mizutani, A., Inoue, H., and Takeda, Y. (1974). *Biochim. Biophys. Acta*, **338**, 183–190.
Pegg, A. E., Lockwood, D. H., and Williams-Ashman, H. G. (1970). *Biochem. J.*, **117**, 17–31.
Pösö, H., and Jänne, J. (1976a). *Biochem. Biophys. Res. Commun.*, **69**, 885–892.
Pösö, H., and Jänne, J. (1976b). *Biochem. J.*, **158**, 485–488.
Raina, A. (1963). *Acta Physiol. Scand.*, **60**, Suppl. 218, 1–81.
Raina, A., and Hölttä, E. (1972). The effect of growth hormone on the synthesis and accumulation of polyamines in mammalian tissues. In A. Pecile and E. E. Müller (Eds) *Growth and Growth Hormone* Excerpta Med. Int. Congress Series No. 244, Amsterdam. pp. 143–149.
Raina, A., and Jänne, J. (1970). *Fed. Proc.*, **29**, 1568–1574.
Raina, A., and Jänne, J. (1975). *Med. Biol.*, **53**, 121–147.
Raina, A., Jänne, J., Hannonen, P., and Hölttä, E. (1970). *Ann. N.Y. Acad. Sci.*, **171**, 697–708.
Raina, A., Jänne, J., Hannonen, P., Hölttä, E., and Ahonen, J. (1973). Polyamine-synthesizing enzymes in regenerating liver and in experimental granuloma. In D. H. Russell (Ed) *Polyamines in Normal and Neoplastic Growth*, Raven Press, New York, pp. 167–180.
Raina, A., Jänne, J., and Siimes, M. (1965). *Abstr. 2nd FEBS Meeting*, **A114**, Vienna.
Raina, A., Jänne, J., and Siimes, M. (1966). *Biochim. Biophys. Acta*, **123**, 197–201.
Raina, A., Pajula, R-L., and Eloranta, T. (1976). *FEBS Letters*, **67**, 252–255.
Raina, A., Pajula, R-L., Eloranta, T., and Tuomi, K. (1978). Synthesis of polyamines and S-adenosylmethionine in rat tissues and tumour cells. Effect of DL-α-hydrazino-δ-aminovaleric acid on cell proliferation In R. A. Campbell, D. R. Morris, D. Bartos, G. D. Daves, and F. Bartos (Eds) *Advances in Polyamine Research*, Vol. 1, Raven Press, New York, pp. 75–82.
Rao, A. J., and Li, C. H. (1977). *Arch. Biochem. Biophys.*, **180**, 169–171.
Richman, R., Park, S., Akbar, M., Yu, S., and Burke, G. (1975). *Endocrinology*, **96**, 1403–1412.
Russell, D. H., and Lombardini, J. B. (1971). *Biochim. Biophys. Acta*, **240**, 273–286.
Russell, D. H., and McVicker, T. A. (1972). *Biochim. Biophys. Acta*, **259**, 247–258.
Russell, D., and Snyder, S. H. (1968). *Proc. Natl. Acad. Sci. USA*, **60**, 1420–1427.
Russell, D. H., and Snyder, S. H. (1969a). *Endocrinology*, **84**, 223–228.
Russell, D. H., and Snyder, S. H. (1969b). *Mol. Pharmacol.*, **5**, 253–262.
Scheinman, S. J., and Burrow, G.N. (1977). *Endocrinology*, **101**, 1088–1094.
Schrock, T. R., Oakman, N. J., and Bucher, N. L. R. (1970). *Biochim. Biophys. Acta*, **204**, 564–577.
Seiler, N., and Askar, A. (1972). *Hoppe-Seyler's Z. Physiol. Chem.*, **353**, 623–633.
Seiler, N., and Lamberty, U. (1975). *J. Neurochem.*, **24**, 5–13.
Snyder, S. H., and Russell, D. H. (1970). *Fed. Proc.*, **29**, 1575–1582.

Spaulding, S. W. (1977). *Endocrinology*, **100**, 1039–1046.
Sten, M., Tolonen, A., Pitko, V. M., Nevalainen, T., and Mäntyjärvi, R. A. (1976). *Arch. Virol.*, **50**, 73–82.
Sturman, J. A., and Gaull, G. E. (1974). *Pediat. Res.*, **8**, 231–237.
Tabor, C. W., and Tabor, H. (1976). *Ann. Rev. Biochem.*, **45**, 285–306.
Takigawa, M., Inoue, H., Gohda, E., Asada, A., Takeda, Y., and Mori, Y. (1977). *Exp. Mol. Pathol.*, **27**, 183–196.
Thrower, S., and Ord, M. G. (1974). *Biochem. J.*, **144**, *361–369.*
Yanagi, S., and Potter, V. R. (1977). *Life Sci.*, **20**, 1509–1519.

Chapter 4

Polyamines in Ehrlich Ascites Tumour Growth

GUNNAR ANDERSSON AND OLLE HEBY

I. INTRODUCTION

Research in the area of cancer prevention, has revealed that the ability of a chemical to act as a tumour promoter is related to its ability to induce L-ornithine decarboxylase (ODC), the initial and rate-controlling enzyme in polyamine synthesis (Chapters 6–9). Furthermore, it has been shown that virus-transformed cells contain higher activities of ODC and greater amounts of putrescine, than their normal counterparts even though their growth rate is the same (Bachrach, Don and Wiener, 1974; Don, Wiener and Bachrach, 1975; Gazdar and coworkers, 1976).

In an attempt to study the role of the polyamines during tumour growth *in vivo* we have used an Ehrlich ascites tumour, primarily because of the technical ease of sampling and relatively high precision in the measurement of growth parameters. The kinetic behaviour of this tumour was studied in detail. An account of our findings, which provide the basis for the understanding of the role of polyamines in Ehrlich ascites tumour growth, is presented in the first section of this chapter.

II. CHARACTERISTICS OF EHRLICH ASCITES TUMOUR GROWTH

The Ehrlich ascites tumour is grown as a free-cell suspension in the peritoneal cavity of NMRI male mice (Andersson and Agrell, 1972). Growth of the

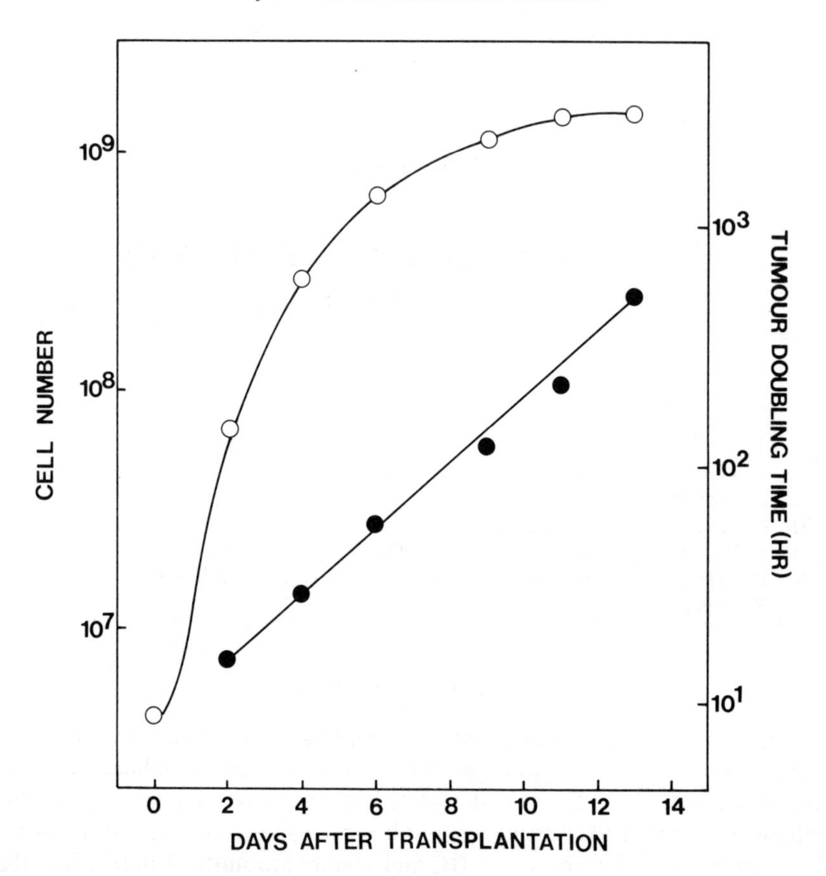

Figure 1 Growth curve (○) and temporal change in tumour doubling time (●) of the hyperdiploid Ehrlich ascites tumour. The growth curve follows a Gompertz function, according to which the time required for doubling of the tumour population increases exponentially

tumour follows a sigmoidal pattern (Figure 1). Initially, the tumour grows rapidly but with increasing tumour mass there is a continuous decrease in growth rate. Gradually the tumour reaches a plateau phase where cell production is almost entirely balanced by cell loss. Six days after transplantation of 4×10^6 cells, the tumour cells constitute more than 85% of the total number of cells in the peritoneal cavity. During plateau phase growth the number of tumour cells asymptotically approaches a figure of about 1.6×10^9 (Figure 1). This final number of cells seems to be dependent on the ploidy of the tumour inasmuch as the hyperdiploid subline reaches a cell number which is approximately twice the final number reached by the hypertetraploid subline (Hauschka and coworkers, 1957). From this observation it may be concluded that growth of the Ehrlich ascites tumour is limited by a maximum tumour mass.

Cell size distribution analysis of tumour cell populations obtained at various stages of growth has revealed that there is a shift from smaller-size-cells to larger-size-cells with increasing tumour mass. During the main part of growth, the cell size distribution displays two peaks but as the tumour approaches the plateau phase an additional peak appears (Andersson and Agrell, 1972). Concomitant with the progressive increase in the relative number of larger cells, there is an increase in the cellular DNA content (Figure 2), indicating that cells synthesize DNA without dividing, and that, with increasing tumour mass,

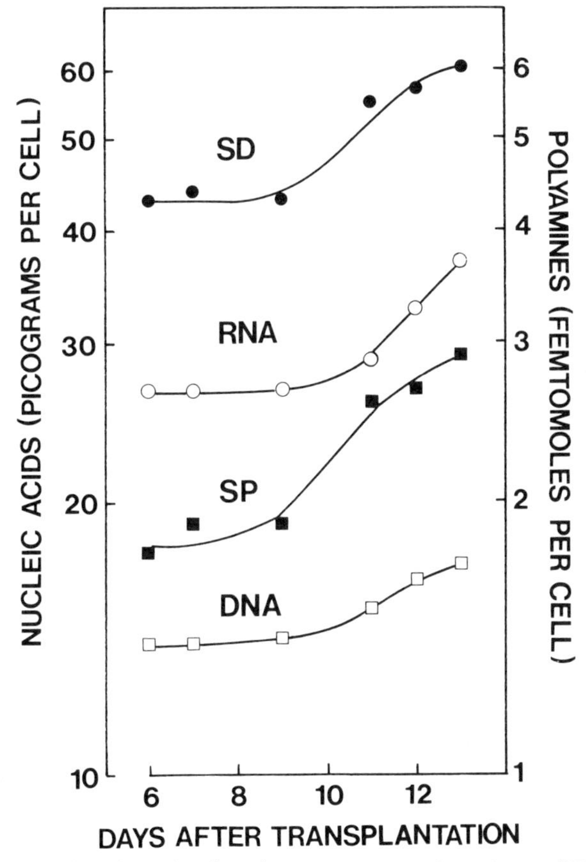

Figure 2 Cellular nucleic acid and polyamine content at various times of Ehrlich ascites tumour growth (SD = spermidine, SP = spermine)

cells slow down in a premitotic phase of the cell cycle. Analysis of the cellular DNA distribution using a microfluorometric technique has revealed that there

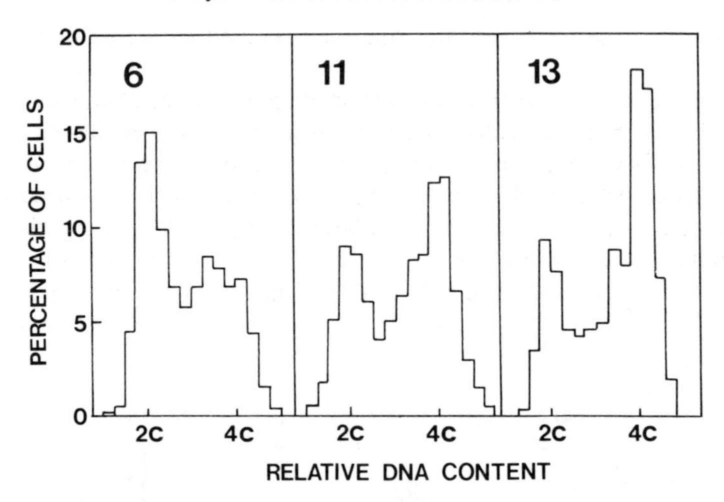

Figure 3 DNA distributions of tumour cell populations on Day 6, 11 and 13 of Ehrlich ascites tumour growth

is an accumulation of cells with a 4C DNA content that parallels the increase in tumour mass (Figure 3). Transplantation of large numbers of tumour cells from the plateau phase of growth results in a rapid rearrangement of the DNA distribution towards a dominance of cells with a 2C DNA content. The fact that vinblastine-treatment blocked more than 20% of the cells in mitosis within 6 h after transplantation shows that extensive cell division is initiated as a result of transplantation (Andersson and Kjellstrand, 1974). This suggests that a large number of the 4C cells, which had accumulated during plateau phase growth, are stimulated to enter mitosis directly from a G2 state.

In order to establish whether polyploid G1 cells are present among the 4C cells of a plateau phase tumour, cell populations were pulse-labelled with tritiated thymidine, ([^{3}H]TdR), immediately following transplantation and blocked in mitosis by repeated injections of vinblastine. In the presence of vinblastine, unlabelled diploid 4C cells accumulate in metaphase, whereas unlabelled tetraploid 4C cells proceed into the tetraploid cell cycle where they are easily quantifiable by means of microphotometry. The number of polyploid cells (including binucleated cells) in the plateau phase tumour amounts to approximately 20% of the total number of cells (12% are polyploid 4C cells and 6–8% are cells exhibiting a DNA content greater than 4C) (Andersson, 1977). It may be assumed that these polyploid cells continue to grow during plateau phase growth and that they eventually may generate a separate class, with a modal cell size larger than that of diploid 4C cells. This fraction of cells may constitute the third size class which appears towards the end of tumour growth (Andersson and Agrell, 1972).

The median durations of the cell cycle phases were determined from the per cent labelled mitoses curves (Quastler and Sherman, 1959). With increasing tumour mass there is a considerable increase in the mean duration of the cell cycle, mainly due to a lengthening of the S and G2 phases (Table 1). The growth fraction, i.e. the percentage of cycling cells, shows a significant but moderate decrease with increasing tumour mass. The majority of the out-of-cycle-cells (cells remaining unlabelled following repeated injections of [^{3}H]TdR for a period of 24 h) exhibit a 2C DNA content during the early phase of tumour growth. With increasing tumour mass, however, the 4C fraction expands and eventually dominates. Cell death, as determined by measuring the

Table 1 Growth parameters of the Ehrlich ascites tumour grown *in vivo*

	7-day tumour	10-day tumour	12-day tumour
Duration of the cell cycle phases (h)			
G1	10.8	14.0	n.d.[1]
S	26.8	52.0	n.d.
G2	5.7	10.0	n.d.
Cell proliferation rate (%h^{-1})	1.32	0.85	0.59
Cell death rate (% h^{-1})	0.39	0.39	0.39
Labelling index (%)	47	40	35
Growth fraction	0.77	0.60	n.d.
Cell loss factor[2]	0.30	0.46	0.66

[1]n.d. = not determined
[2]Cell loss factor = the ratio between the rate of cell loss and the rate of cell proliferation

loss of [^{125}I] from mice bearing [^{125}I]-iodo-2-2′-deoxyuridine-labelled ([^{125}I]UdR) tumour cells, occurs at a low and almost constant rate throughout tumour growth. Thus a lengthening of the cell cycle (mainly of the S and G2 phases) is the main determinant in growth rate deceleration of this tumour (Andersson and Heby, 1977). This decrease in the rate of the traverse of the cell cycle seems to be associated with an increased tendency to produce polyploid cells, including binucleated cells (Figure 4), as shown by Andersson and Agrell (1972).

III. POLYAMINE SYNTHESIS IN RELATION TO TUMOUR GROWTH KINETICS

As a general rule polyamines accumulate in cells and tissues that are in a state of rapid growth. A relationship between growth rate and polyamine metabolism was first observed in the Ehrlich ascites tumour (Andersson and Heby, 1972). The putrescine concentration was found to decrease in parallel with the decreasing growth rate; thus indicating that putrescine might play an

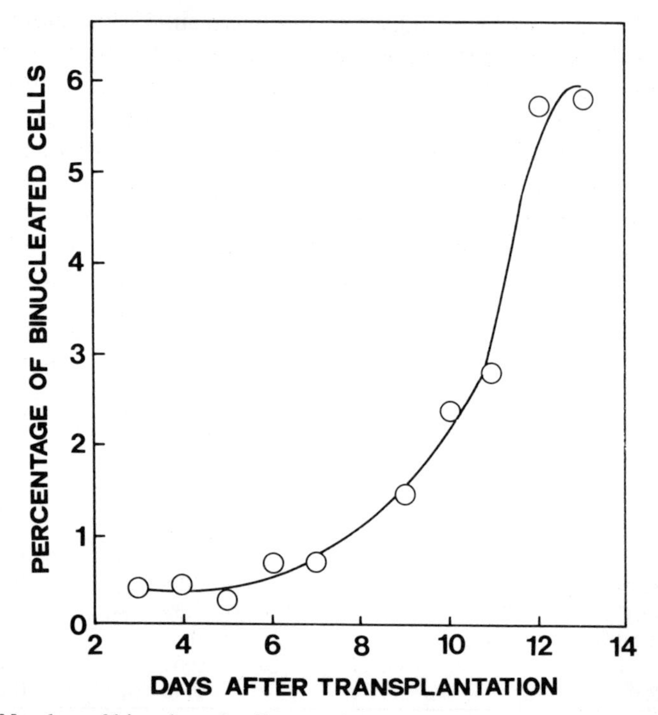

Figure 4 Number of binucleated cells at various times of Ehrlich ascites tumour growth

essential role in cell proliferation. It has recently been shown that polyamine synthesis is rapidly induced when Ehrlich ascites tumour cells are stimulated to proliferate by transplantation into new hosts (Harris and coworkers, 1975; Andersson, Österberg and Heby, 1978). Following the initial increase in the activities of ODC and SAMD (S-adenosylmethionine decarboxylase) there is a continuous decrease with time. Both enzymes show high positive correlations with the rate of cell proliferation throughout tumour growth (Figure 5). A similar relationship has been observed during *in vitro* growth of a rat brain tumour cell line (Heby and coworkers, 1975). Surprisingly, however, the decline in polyamine biosynthetic activity in the Ehrlich ascites tumour is not paralleled by a decrease in the cellular spermidine and spermine content. Instead, the cellular content of these polyamines increases with decreasing growth rate (Figure 2).

By studying the growth kinetics of the Ehrlich ascites tumour, it was possible to find a plausible explanation for the increase in cellular spermidine and spermine content during plateau phase growth. Thus, we propose that it is due to the observed accumulation of cells with a 4C DNA content (G2 phase cells, polyploid cells and binucleated cells), cells that are likely to have an elevated spermidine and spermine content as a result of their larger size. This

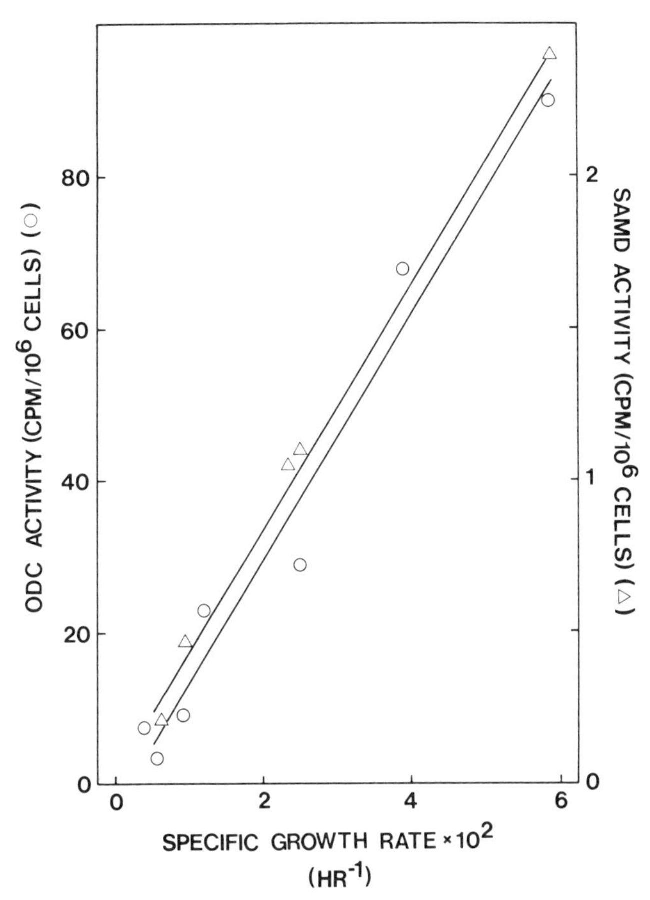

Figure 5 Specific growth rate of the Ehrlich ascites tumour linearly fitted to the activities of ornithine decarboxylase (ODC) and S-adenosylmethionone decarboxylase (SAMD)

idea is supported by the fact that the cellular polyamine content in AKR leukaemic cells increased in parallel with increasing cell size (Heby and coworkers, 1973). The fact that there was a high positive correlation between the cellular spermidine and spermine content and the cellular nucleic acid content during Ehrlich ascites tumour growth, suggests that there might be a functional coupling between the synthesis of polyamines and nucleic acids, and that polyamines may play a role in DNA replication and cell division (Chapters 1–3, 22).

According to studies on synchronously growing Chinese hamster ovary cell populations, polyamine synthesis may be resolved into two periods of increased activity; the first in late G1/early S and the second in late S/G2 (Heby and Andersson, Chapter 2). These presynthetic and premitotic activities are of particular interest in view of the fact that polyamines may play a role in DNA replication (Fillingame, Jorstad and Morris, 1975; Krokan and Eriksen, 1977) and in nuclear division (Sunkara, Seman and Rao, 1978). To evaluate the importance of the polyamines for progression through the Ehrlich ascites tumour cell cycle, we have used polyamine synthesis inhibitors as a means of partly depleting the cells of their polyamine content. Two types of inhibitors have been used: (1) DL-α-methylornithine (α-MO), which inhibits the synthesis of putrescine by acting as a competitive inhibitor of ODC, and (2) methyl-GAG, which inhibits the synthesis of spermidine and spermine, probably by acting as a competitive inhibitor for the substrate (S-adenosyl-L-methionine) and as an uncompetitive inhibitor for the activator (putrescine) in the SAMD catalysed reaction. Repeated injections of α-MO or methyl-GAG into tumour-bearing mice resulted in an accumulation of cells in the S and G2 phases (Heby, Andersson and Gray, 1978). This finding suggests that a normal polyamine complement is essential for progression through these phases of the cell cycle.

In the case of methyl-GAG the effect initially appeared to be limited mainly to a lengthening of the S phase (Heby, Andersson and Gray, 1978). This finding is consistent with results obtained with concanavalin A-stimulated lymphocytes following methyl-GAG treatment (Fillingame, Jorstad and Morris, 1975). It might be concluded from these experiments that methyl-GAG does not alter the rate of entry of cells into the S phase. Rather, methyl-GAG might inhibit the rate of DNA replication (Fillingame, Jorstad and Morris, 1975; Krokan and Eriksen, 1977), thus causing a lengthening of the S phase and a subsequent decrease in the number of cells entering mitosis. Presently, it appears that the decreased rate of DNA synthesis caused by polyamine depletion might be due to a reduction of the number of replicons (Fillingame, Jorstad and Morris, 1975; Krokan and Eriksen, 1977) or a reduced rate of DNA replication fork movement (Geiger and Morris, 1978; Chapter 1).

During unperturbed Ehrlich ascites tumour growth there is a progressive lengthening of the S phase and an accumulation of cells in the G2 phase. This kinetic behaviour is paralleled by a steady decrease in the ODC and SAMD activities. Therefore, it is interesting to note that by inhibiting the ODC and SAMD activities in rapidly growing Ehrlich ascites tumour cells, one generates a state of growth and a cell cycle distribution which is similar to the one observed during plateau phase growth.

IV. EFFECTS OF EXTRACELLULAR POLYAMINES ON TUMOUR GROWTH KINETICS

In the Ehrlich ascites tumour model the concentrations of putrescine and spermidine in cell-free ascites fluid and serum increase significantly with increasing tumour mass (Heby and Andersson, 1978). Since there is a continuous increase in the cell loss factor (Table 1) it seems highly probable that the polyamines are released into the extracellular fluids as a result of cell lysis (Heby and Andersson, 1978).

We have investigated further the role of cell death in the accumulation of extracellular polyamines. Thus, cell death was induced by heterologous transplantation of the mouse tumour into gerbils and was monitored by measuring the release of [^{125}I] from [^{125}I]UdR-labelled cells (Andersson and coworkers, 1978). The polyamine content of cell-free ascites fluid and serum was correlated with the activity of lactate dehydrogenase (LDH) in these fluids, because, according to a recent report on a heterologous tumour system, extracellular LDH is a reliable marker of tumour cell death (Pesce and coworkers, 1977). Furthermore, since LDH of the Ehrlich ascites tumour cells almost exclusively consists of isozyme LDH 5 and since LDH isozymes of the mouse are electrophoretically different from those of the gerbil, it is possible to identify the tumour-derived isozyme (LDH 5) in serum of the gerbil during the entire course of tumour regression. This study revealed a high positive correlation between tumour-specific serum LDH activity and polyamine content during tumour regression, and thus lends further support for the contention that an increased extracellular polyamine level is mainly due to their release from dead or dying tumour cells.

In the Ehrlich ascites tumour the continuously increasing concentration of extracellular polyamines might be involved in the depression of the growth rate which accompanies the increase in tumour mass. This proposal is based on the fact that administration of putrescine to actively proliferating tumour cells causes an effective block of their ODC activity and a considerable deceleration of the growth rate (Andersson and Heby, 1977; Linden, Andersson and Heby, 1978). The mechanism of action of extracellular putrescine has attracted much attention, because it has revealed a new means of enzyme regulation. Thus, recent reports indicate that the activity of ODC might be controlled by the end-product of its reaction, i.e. putrescine, not only by simple product inhibition (Heller, Fong and Canellakis, 1976; McCann, Tardif and Mamont, 1977; Chapter 7). Instead an increase in the extracellular concentration of putrescine might induce the synthesis of an 'antizyme' inhibitor of ODC (Chapter 9). In accordance with these observations we have found that the injection of putrescine into Ehrlich ascites tumour-bearing mice (25 μmoles h^{-1}) results in a specific and complete eradication of the ODC activity in the tumour cells within 4 h (Andersson and Heby, 1977). Utilizing Sephadex (G-75 Superfine) chromatography it was possible to demonstrate a significant accumulation of an ODC-antizyme during putrescine treatment (Linden, Andersson and Heby, 1978). Even though the dose used in this experiment generates an extracellular putrescine concentration which might exceed that normally occurring in body fluids, it might be assumed that the extracellular concentration of polyamines during plateau phase growth (15 nmoles ml^{-1}) is sufficient to induce antizyme formation and inhibition of the ODC activity. In fact, putrescine at a concentration of 10 nmoles ml^{-1} eliminated within 5 h almost all intracellular

ODC activity in rat hepatoma cells as a result of ODC-antizyme synthesis (McCann, Tardif and Mamont, 1977).

Since inhibition of ODC by α-MO blocks Ehrlich ascites tumour cell proliferation, we considered it important to determine whether the observed inhibition of the ODC activity by putrescine administration would have the same effect on the traverse of the cell cycle as did the synthetic ODC inhibitor. Interestingly, this was found to be the case; cells accumulated in the S and G2 phases of the cell cycle. Putrescine administration did not affect the total amount of DNA synthesized in the whole cell population (i.e. the total amount of [^{3}H]TdR incorporated into DNA was unaffected), but there were more cells synthesizing DNA (i.e. the labelling index increased) at a lower rate (i.e. the amount of [^{3}H]TdR incorporated per cell decreased) as shown by Linden, Andersson and Heby (1978). This implies that extracellular putrescine causes a lengthening of the S phase (without affecting the initiation of DNA synthesis) and consequently a decrease in the rate of cell proliferation.

The change in growth pattern of the Ehrlich ascites tumour generated by inhibiting the ODC activity by the addition of putrescine or α-MO is similar to that generated by inhibiting the SAMD activity by methyl-GAG treatment, namely a prolongation of the S phase with no major effect on the initiation of DNA synthesis. In view of the fact that there is a continuous decrease in the ODC and SAMD activities and a progressive lengthening of the S phase during unperturbed Ehrlich ascites tumour growth, it is tempting to suggest that active inhibition of polyamine synthesis might be involved in the deceleration of the growth rate of the tumour.

The observation that the rate of DNA synthesis in cultured bone marrow myelocytes is significantly depressed after the addition of cell-free ascites fluid from plateau phase Ehrlich ascites tumours (Lala, 1972), suggests that the change in cell cycle time during unperturbed tumour growth is determined mainly by changes in the local environment of the tumour cells. An exponential increase in population doubling time, like that occurring during Ehrlich ascites tumour growth (Figure 1), is most probably due to an actively increasing depression of the growth rate operating from the earliest stages of tumour growth. We propose that the continuous release of polyamines from dead or dying tumour cells, which generates a progressive increase in the polyamine concentration in the local environment of the cells, actively depresses the intracellular polyamine synthesis necessary for DNA replication and cell proliferation.

REFERENCES

Andersson, G. (1977). *J. Cell. Physiol.*, **90**, 329–336.

Andersson, G., and Agrell, I. (1972). *Virchows Arch. Abt. B. Cell Path.*, **11**, 1–10.

Andersson, G., Bengtsson, G., Albinsson, A., Rosén, S., and Heby, O. (1978). *Cancer Res.*, **38**, 3938–3943.

Andersson, G., and Heby, O. (1972). *J. Nat. Cancer Inst.*, **48**, 165–172.
Andersson, G., and Heby, O. (1977). *Cancer Res.*, **37**, 4361–4366.
Andersson, G., and Kjellstrand, P. (1974). *Virchows Arch. B. Cell Path.*, **16**, 311–318.
Andersson, G., Österberg, S., and Heby, O. (1978). *Int. J. Biochem.*, **9**, 263–267.
Bachrach, U., Don, S., and Wiener, H. (1974). *Cancer Res.*, **34**, 1577–1580.
Don, S., Wiener, H., and Bachrach, U. (1975). *Cancer Res.*, **35**, 194–198.
Fillingame, R. H., Jorstad, C. M., and Morris, D. R. (1975). *Proc. Nat. Acad. Sci. USA*, **72**, 4042–4045.
Gazdar, A. F., Stull, H. B., Kilton, L. J., and Bachrach, U. (1976). *Nature*, **262**, 696–698.
Geiger, L. E., and Morris, D. R. (1978). *Nature, Lond.*, **272**, 730–732.
Harris, J. W., Wong, Y. P., Kehe, C. R., and Teng, S. S. (1975). *Cancer Res.*, **35**, 3181–3186.
Hauschka, T. S., Grinnell, S. T., Révész, L., Klein, G. (1957). *J. Nat. Cancer Inst.*, **19**, 13–31.
Heby, O., and Andersson, G. (1978). *Acta Path. Microbiol. Scand. Sect. A*, **86**, 17–20.
Heby, O., Andersson, G., and Gray, J. W. (1978). *Exp. Cell Res.*, **111**, 461–464.
Heby O., Marton, L. J., Wilson, C. B., and Martinez, H. M. (1975). *J. Cell. Physiol.*, **86**, 511–522.
Heby, O., Sarna, G. P., Marton, L. J., Omine, M., Perry, S., and Russell, D. H. (1973). *Cancer Res.*, **33**, 2959–2964.
Heller, J. S., Fong, W. F., and Canellakis, E. S. (1976). *Proc. Nat. Acad. Sci. USA*, **73**, 1858–1862.
Krokan, H., and Eriksen, A. (1977). *Eur. J. Biochem.*, **72**, 501–508.
Lala, P. K. (1972). *Eur. J. Cancer*, **8**, 197–204.
Linden, M., Andersson, G., and Heby, O. (1978). Unpublished data.
McCann, P. P., Tardif, C., and Mamont, P. S. (1977). *Biochem. Biophys. Res. Commun.*, **75**, 948–954.
Pesce, A. J., Bubel, H. C., DiPersio, L. and Michael, J. G. (1977). *Cancer Res.* **37**, 1998–2003.
Quastler, H., and Sherman, F. G. (1959). *Exp. Cell Res.* **17**, 420–438.
Sunkara, P. S., Seman, G., and Rao, P. N. (1978). *Proc. Am. Assoc. Cancer Res.*, **19**, 62.

Chapter 5

The Possible Roles of Polyamines in Prereplicative Development and DNA Synthesis: A Critical Assessment of the Evidence

A. L. BOYNTON, J. F. WHITFIELD, P. R. WALKER

I. INTRODUCTION

Following activation by exogenous agents or other circumstances, the complex set of parallel and sequential reactions leading to chromosome replication, mitosis and cell division is promoted and coordinated by a variety of agents which briefly appear or accumulate at the appropriate points in the growth-division cycle (Figure 1). The polyamines, putrescine, spermidine and spermine are among the several small molecules which are believed to regulate this complex process mainly because their levels rise at specific points in the cell cycle (Chapters 1 and 2). In the present brief review, we will attempt to identify the points of action of these amines in the replicative (G1) and DNA synthetic (S) phases of the growth-division cycle.

II. DEFINITIONS

At the outset, it is necessary to present our view of the cell cycle and to define the terms we will use to describe it (see also Whitfield, 1979).

We suggest that all nucleated eukaryotic cells contain a set of coordinately regulated structural genes, the products of which give rise to the complex series of processes leading to the initiation of DNA synthesis, mitosis and cell division. Those cells in which these structural genes and/or their coordinators are inactivated (e.g. hepatocytes, lymphocytes and salivary gland acinar cells *in vivo*) are committed fully to non-proliferative functions and will be considered to be *proliferatively inactivated*. The process which causes such inactivated cells to initiate proliferative development will be termed *proliferative activation* (Figure 1). By contrast, cells with activated proliferogenic genes (e.g. all non-neoplastic, pre-neoplastic or neoplastic cells *in vitro*) which for some reason such as serum deprivation cannot actively proliferate (or actively 'cycle') will be considered to be in a resting or 'G0 state' (restriction point 2 of Figure 1). The process which causes such proliferatively quiescent 'G0' cells to resume 'cycling' is different from proliferative activation and will be termed proliferative stimulation (Figure 1).

It must be noted that this view of the 'G0 state' is different from the conventional one which defines a 'G0 cell' as any non-proliferating cell (be it an hepatocyte in the intact adult liver *in vivo* or a serum-deprived neoplastic SV-3T3 fetal mouse cell *in vitro*) which responds to an appropriate stimulus by re-entering the cell cycle usually at restriction point 2 in the scheme of Figure 1 (Lajtha, Oliver and Gurney, 1962; Rajewsky, 1974). According to some authors the cell cycle might also include a post-mitotic 'G0 pause' in which the cell decides whether to initiate another cycle or to inactivate its proliferogenic genes and then take the non-proliferative route to differentiation and specialized function (reviewed by Rajewsky, 1974). We feel it is important for the investigation of proliferative control mechanisms to distinguish as sharply as possible between a reversible non-proliferative state in which the proliferogenic genes are inactivated (restriction point 1 of Figure 1), and a reversible non-proliferative ('G0') state in which the proliferogenic genes have not been inactivated (i.e. restriction point 2 of Figure 1). As will be seen below, this concept has been useful for locating the point of action of putrescine in pre-replicative development.

III. POLYAMINES AND *IN VIVO* SYSTEMS

The most widely studied proliferative activation in the whole animal is the activation of the hepatocyte by partial resection of the adult rat liver. Removal of two-thirds of the liver activates each remaining hepatocyte's set of proliferogenic genes, the products of which induce a complex set of biochemical reactions culminating in DNA synthesis and cell division (Figure 2). One group of prereplicative reactions unleashed by partial resection is the synthesis of ornithine decarboxylase, putrescine and the polyamines, spermidine and spermine. The rate limiting enzyme in this sequence is

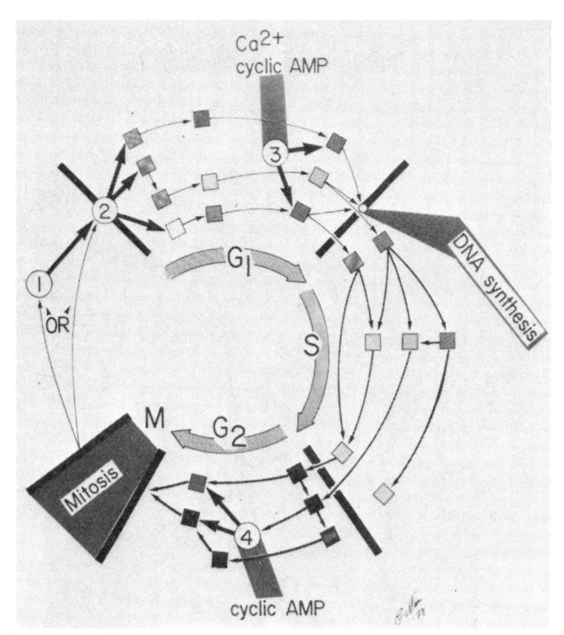

Restriction Points

1. Proliferatively inactivated differentiated cells.
   e.g. hepatocytes, lymphocytes, parotid gland acinar cells.
   Proliferatively activated by partial hepatectomy, concanavalin A, isoproterenol.
2. G_0 STATE (proliferatively quiescent) due to cell density, serum deprivation etc. Stimulated by serum factors, FGF etc.
3. G_1/S transition block induced by calcium deprivation. Stimulated by calcium.
4. G_2/M transition block induced by high cyclic AMP concentrations. Stimulated by a drop in the cyclic AMP level.

Figure 1 An impressionist's view of many (but still mostly unknown) components of the proliferogenic 'reaction package', and identification of at least four possible regulatory points within this ever-changing process. This complex sequence begins when a proliferogen, such as a concanavalin A for small lymphocytes or isoproterenol for salivary gland acinar cells, activates the regulatory set of proliferogenic genes at the first restriction point 1. The activation process derepresses the batteries of proliferogenic genes, thus generating new species of messenger RNA which start the prereplicative (G1) processes at restriction point 2. The progress through prereplicative development and the remainder of the cell cycle is then guided by brief surges of intracycle regulators such as cyclic AMP (restriction point 3 and 4). These various processes converge at appropriate points in the cell cycle to initiate DNA synthesis, redistribute the microtubules into a spindle apparatus, and finally initiate mitosis and cytokinesis. Each daughter cell might repress its proliferogenic genes and resume its specialized functions (restriction point 1), proceed to restriction point 2 and enter a reversible proliferatively quiescent (G0) state, or initiate another growth-division cycle.

If any sequence of the prereplicative network should fail or be blocked by specific inhibitors, the cell might functionally (and structurally) return to restriction point 2 and enter a G0 state without inactivating its proliferogenic genes. Similarly, a failure of one of the non-DNA-synthetic events in the late prereplicative S or G2 phases of the cycle may cause the cell to stop its proliferative development at the third or fourth restriction point

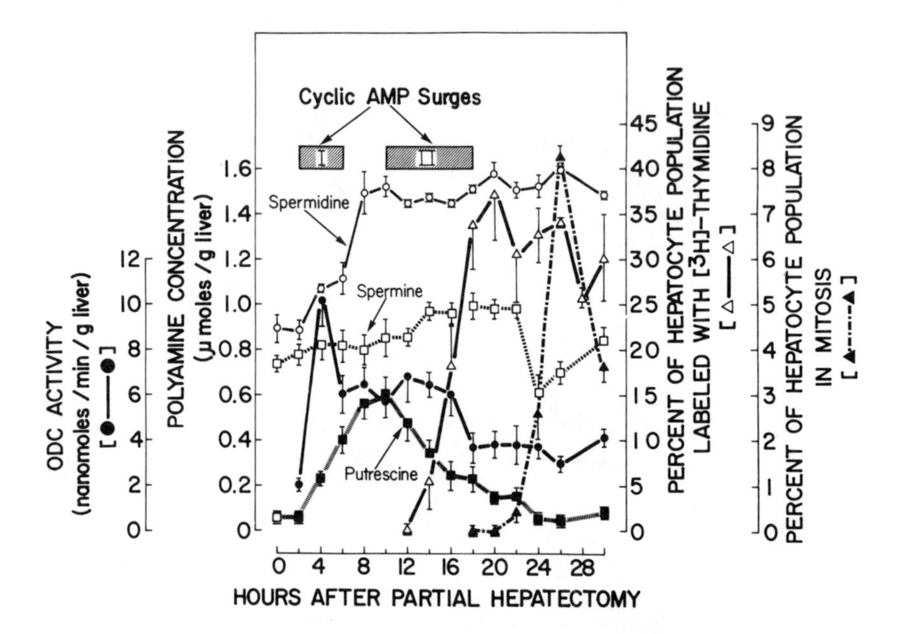

Figure 2 The prereplicative changes in ODC activity and cyclic AMP and polyamine syntheses in rat hepatocytes activated by partial hepatectomy. Samples of liver remnants were taken every two hours following partial hepatectomy and homogenates prepared for determination of ODC, putrescine, spermidine and spermine activities as described by Walker, Sikorska and Whitfield (1978). Cyclic AMP determinations were assayed according to MacManus and coworkers (1972), while DNA and mitotic activities were determined according to Walker, Sikorska and Whitfield, (1978). The shaded areas (I and II) represent the mean durations of the first and second prereplicative cyclic AMP surges. Each point is the mean ± s.e. mean of 9 to 12 separate animals. Taken in part from Walker, Sikorska and Whitfield, (1978). (Reproduced by permission of The Wistar Institute Press, Philadelphia, Penn.)

ornithine decarboxylase (ODC) which catalyzes the synthesis of putrescine (1,4-diaminobutane) from ornithine. Under our conditions, the activity of this pivotal enzyme begins to increase during the first h after surgery, peaks at 4 h and then falls to a lower value but which is still 10-fold greater than the pre-activation value (Figure 2: Walker, Sikorska and Whitfield, 1978). Others have observed a second surge of ODC activity shortly before the initiation of hepatocyte DNA synthesis which might be proliferatively irrelevant, since it results either from the use of much younger rats (e.g. 100–120 g; Gaza, Short and Lieberman, 1973; Hölttä and Jänne, 1972) or from the feeding schedule (Yanagi and Potter, 1977; Barbirolli and coworkers, 1975). As expected, the putrescine level rises immediately after the first ODC surge and peaks between 8 and 10 h after surgery. Under our conditions, the spermidine level (which is determined by the combined activities of the putrescine-stimulated S-adenosylmethionine decarboxylase and spermidine synthase) begins to rise immediately after the putrescine surge, in contrast to other reports of no detectable rise until 24 h after activation when the hepatocytes have already begun to replicate DNA (Dykstra and Herbst, 1965; Jänne, 1967; Pösö and Jänne, 1976). On the other hand, it seems to be generally agreed that the spermine content remains unchanged until 24 to 26 h after surgery, and then halves as the hepatocytes flow into mitosis (Figure 2; Walker, Sikorska and Whitfield, 1978; Dykstra and Herbst, 1965; Jänne, 1967; Russell, Medina and Snyder, 1970; Heby and Lewan, 1971; Pösö and Jänne, 1976).

The only other extensively studied *in vivo* proliferative activation system is the isoproterenol (IPR)-treated acinar cells of the mouse parotid gland (Inoue and coworkers, 1975; reviewed by Whitfield, 1979). Intraperitoneal injection of this proliferogen causes a massive secretion, and subsequent resynthesis, of α-amylase and starts prereplicative development (from restriction point 1 of Figure 1) which culminates in the initiation of DNA synthesis 20 to 24 h later (Figure 3: Inoue and coworkers; Baserga and Heffler, 1967; Durham, Baserga and Butcher, 1974). During the first hour after injection of IPR, there is a brief but proliferatively unnecessary (Durham, Baserga and Butcher, 1974; Whitfield, 1979) burst of cyclic AMP synthesis, followed by an increase in chromatin template activity and the synthesis of messenger RNA (i.e. poly(A)-associated) (Novi and Baserga, 1972a; Novi and Baserga, 1972b). During the next 7 h ODC activity, the putrescine and spermidine contents and ribosomal (18S and 28S) RNA syntheses reach their peak values (Figure 3). As in the case of proliferatively activated hepatocytes the spermine level does not increase. On the contrary, the spermine level began to decline slowly but steadily 5 hours after activation (Figure 3).

A possible link between increased ODC activity and the stimulation of cyclic AMP formation has been proposed by Beck, Bellantone and Callenakis (1972), and by Byus and Russell (1974). The single early surge of ODC activity, observed under our conditions in proliferatively activated rat liver (by partial resection), follows closely the first (but proliferatively unnecessary) of two surges of cyclic AMP accumulation (Figure 2: MacManus and coworkers, 1972; MacManus and coworkers, 1973). Similarly, the early surge of ODC activity in the isoproterenol-treated parotid gland is preceded by a 20-min burst of cyclic AMP synthesis (Figure 3: Durham, Baserga and Butcher, 1974). Other cyclic AMP elevators which may or may not be proliferogenic, such as adrenocorticotropic hormone, glucagon, the methylxanthines and the dibutyryl analogue of cyclic AMP itself, also increase ODC activity in the adrenal cortex and liver (Beck, Ballentone and Callenakis, 1972; Byus and Russell, 1974; Gaza, Short and Lieberman, 1973).

One argument against cyclic AMP being the stimulator of ODC synthesis (see Chapter 8) is the fact there are invariably two prereplicative bursts of

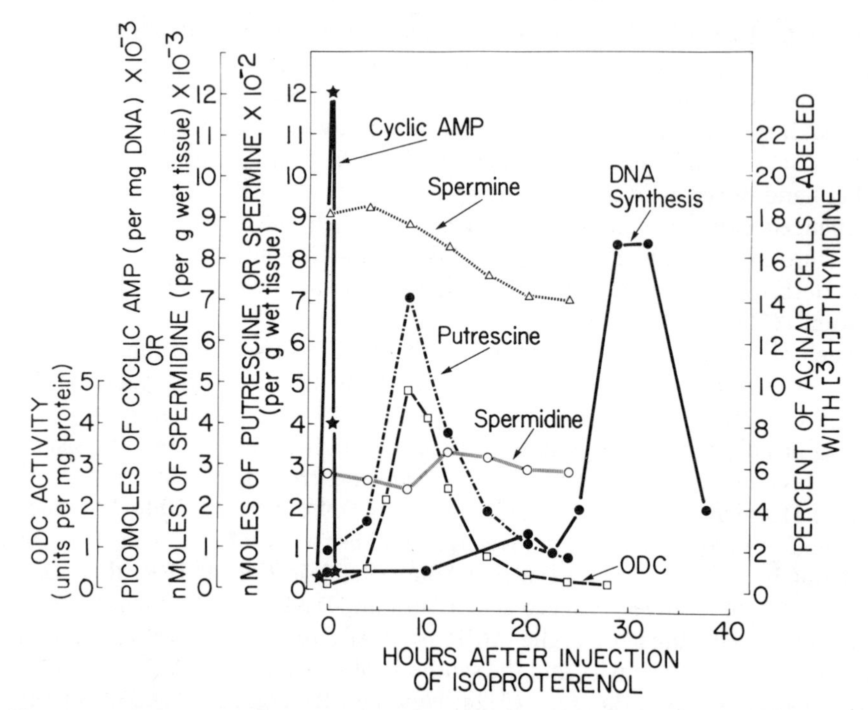

Figure 3 An example of the effects of a single injection of isoproterenol on the syntheses of ODC, putrescine, spermidine, spermine cyclic AMP and DNA in mouse parotid gland acinar cells. ODC activity and polyamine contents are taken from Inoue and coworkers, (1975). The percentage of cells in S phase is from Baserga and Heffler, (1967) and cyclic AMP levels are from Durham, Baserga and Butcher, (1974). It should be noted that it is not known whether there is a second late prereplicative cyclic AMP surge such as occurs in the partially resected liver, since no one has yet measured the cyclic AMP levels later than one hour after injection of isoproterenol. (These results are reproduced by permission of the Japanese Biochemical Society and R. Baserga respectively.)

cyclic AMP in proliferatively activated rat hepatocytes (MacManus and coworkers, 1972). These occur only under certain circumstances, such as the use of young rats or certain feeding schedules where there are two bursts of ODC activity (Barbirolli and coworkers, 1975; Gaza, Short and Lieberman, 1973; Hölttä and Jänne, 1972; Yanagi and Potter, 1977). Furthermore, Thrower and Ord (1974) claim that a 30 to 40% reduction, by exposure to the adrenergic antagonists pindolol and propanolol of the first cyclic AMP surge in activated hepatocytes, does not affect subsequent increase in ODC activity. Until we have some quantititave information on the sensitivity of the ODC activation-synthetic system to cyclic AMP, however, this is very weak contrary evidence since the residual cyclic AMP surge, or even an intracellular redistribution of pre-existing cyclic AMP, might be sufficient to trigger the full ODC response. Finally, Mufson and coworkers (1977) found that a single application of TPA (PMA) to mouse skin did not raise the tissues' cyclic AMP level during the first 30 min, but it did cause a large (200 to 400-fold) rise in the ODC activity 4 h later. The contradictory force of this evidence is also very weak, since the epidermal cyclic AMP content actually does not start to rise until 30 min after exposure to TPA (PMA), and then peaks 30 to 90 min afterwards (Grimm and Marks, 1974).

Since an early cyclic AMP surge in the activated hepatocyte or the salivary gland acinar cell is not needed for normal prereplicative development (Durham, Baserga and Butcher, 1974; MacManus and coworkers, 1973; Whitfield, 1979), and if cyclic AMP is indeed the stimulator of ODC formation, an increased ODC activity might not be needed for prereplicative development. Undoubtedly, an increase in ODC activity, and presumably the resulting polyamine syntheses, cannot by themselves lead to the initiation of DNA synthesis, since injection of non-proliferogenic cyclic AMP elevators such as glucagon, methylxanthines and dibutyryl cyclic AMP, increases ODC activity but without a later initiation of DNA synthesis (Beck, Bellantone and Canellakis, 1972; Byus and Russell, 1974; Gaza Short and Liberman, 1973).

It is possible that the increase in ODC activity and the consequent stimulation of polyamine synthesis are among the several different prereplicative processes which are unleashed in a coordinated manner only by proliferogens and *together*, but not separately, lead to the initiation of DNA synthesis (Figure 1: Whitfield, 1979). One way to assess the importance of the polyamine syntheses for prereplicative development is to determine whether their inhibition prevents an initiation of DNA synthesis. Inhibition of ODC activity in mouse parotid gland cells by administration of HAVA (DL-α-hydrazino-δ-aminovaleric acid) 4 h after injection of the proliferogenic isoproterenol does not affect spermidine or spermine synthesis. It does reduce the stimulation of putrescine synthesis by 50% and completely prevents the initiation of DNA synthesis. This effect must be due to the putrescine shortage since it can be reversed by injection of putrescine (Inoue and coworkers, 1975). As prereplicative development progresses, however, the sensitivity to the inhibitor declines and it no longer affects the subsequent initiation of DNA synthesis when injected at 14 h (Inoue and coworkers, 1975). Similarly, the inhibition of spermidine synthesis in the isoproterenol-activated parotid gland by injection (at 4 h after isoproterenol injection) of methyl-GAG (methylglyoxal-*bis* (guanylhydrazone) which inhibits S-adenosyl methionine decarboxylase activity (Williams-Ashman and Schenone, 1972) as well as polyamine function and transmembrane transport (Clark and Fuller, 1975; French and coworkers, 1960) stops the expected prereplicative increase in the spermidine level between 8 and 12 h after isoproterenol treatment. It also greatly impairs the initiation of DNA synthesis without affecting the levels of pre-existing spermine (Inoue and coworkers, 1975). Inhibition of ODC activity by multiple prereplicative injection of DAP (1,3-diaminopropane) also seems to inhibit the incorporation of tritiated thymidine into the DNA of liver tissue activated by partial hepatectomy (Pösö and Jänne,, 1976). These data must be treated cautiously because of their extreme variability. On the other hand, and in contrast to the response of the parotid gland acinar cells, injection of sub-lethal doses of methyl-GAG do not seem to affect hepatocyte DNA synthesis (Höltta and coworkers, 1973). (Unfortunately, these observations must be treated even more cautiously because the authors present no data to support their claim).

It is probably safe to conclude from all of these observations that the polyamines are important participants in the proliferative activation process (from restriction point 1 to restriction point 2) and subsequent proliferative development. Their importance seems to be greater for isoproterenol-activated parotid gland acinar cells than it is for hepatocytes activated by partial resection. This difference is probably due ultimately to the brevity of the acinar cell-activating stimulus from a single injection of isoproterenol, compared with the persistence of the hepatocyte-activating stimulus of partial resection, which presumably lasts until the replacement of the resected liver tissue. Thus, because of the transience of the isoproterenol stimulus a brief inhibition of an evanescent prereplicative process can permanently stop the acinar cell's DNA-synthetic response whilst the same inhibition would, at the most, only delay the initiation of hepatocyte DNA synthesis because of the persistent influence of the liver tissue deficit.

ODC and its product putrescine may be specifically involved in an RNA-synthetic component of the proliferative activation process, since the large ODC and putrescine surges following IPR treatment of parotid gland acinar cells coincide almost exactly with a surge of ribosomal (18S and 28S) RNA synthesis (Novi and Baserga, 1972a; Inoue and coworkers 1975). Neither the polyamine nor the RNA syntheses are due to the massive α-amylase secretion and resynthesis which is also induced by IPR, because pilocarpine causes the same enzyme secretion and resynthesis without increasing chromatic template activity, ODC synthesis, ribosomal RNA synthesis or initiating DNA synthesis (Durham, Baserga and Butcher, 1974; Inoue and coworkers, 1974; Novi and Baserga, 1972a; Novi and Baserga, 1972b). Support for an involvement of putrescine in proliferogenic RNA syntheses is provided by the observation of Pierce and Fausto (1978) that putrescine can stimulate RNA chain elongation when rat liver chromatin is exposed to *Escherichia coli* RNA polymerase.

IV. POLYAMINES AND *IN VITRO* SYSTEMS

Although it is rather fragmentary, the evidence obtained with the *in vivo* systems (see also Chapter 3) suggests an involvement of the polyamines in pre-replicative (G1) development. After proliferative activation (from restriction point 1 of Figure 1) or, more commonly, after proliferative stimulation (from restriction point 2 of Figure 1) of cultivated cells *in vitro*, there is usually a rise in the ODC activity followed by an increase in the cellular putrescine content (Clark and Fuller, 1975; Fillingame and Morris, 1973; Fillingame, Jorstad and Morris, 1975; Fuller, Gerner and Russell, 1977; Heby and coworkers, 1975; Russell and Stambrook, 1975). On the other hand, there may be no measurable increases in the spermidine and spermine levels before the initiation of DNA synthesis (Boynton, Whitfield and Isaacs, 1976; Clark and Duffy, 1976; Heby and coworkers, 1975).

The possiblity of cyclic AMP being the principal stimulator of ODC activity, which was raised during the discussion of the *in vivo* systems, is also suggested by some observations on *in vitro* systems. There is invariably a late prereplicative cyclic AMP surge which triggers the initiation of DNA synthesis probably by starting the syntheses of the four deoxyribonucleotide precursors of DNA (Whitfield, 1979). The importance of this surge is illustrated by the fact that its inhibition in cells such as spleen lymphocytes or rat hepatocytes prevents the initiation of DNA synthesis (MacManus and coworkers, 1973; Wang and Foker, 1978). In Chinese hamster CHO cells, this late prereplicative cyclic AMP surge is followed rapidly by an increase in ODC activity (Russell and Stambrook, 1975). Moreover, a continuous exposure of hamster BHK/21 cells to dibutyryl cyclic AMP or serum plus dibutyryl cyclic AMP (treatments which presumably raise the intracellular cyclic AMP content) increases ODC activity, but it does not stimulate the initiation of DNA synthesis (Hogan, Shields and Curtis, 1974). Rather than proving an irrelevance of polyamines to DNA synthesis, this lack of correlation between cyclic AMP-stimulated ODC activity and DNA synthesis is probably due to inhibition of prereplicative development by a premature and abnormally prolonged cyclic AMP surge (Whitfield, 1979; Whitfield and coworkers, 1976). Nevertheless, much more information must be obtained before we can decide whether cyclic AMP has a role in proliferatively relevant polyamine synthesis.

As was the case for the *in vivo* systems, a stimulation of polyamine synthesis by itself does not cause the initiation of DNA synthesis. Similarly, exposure of proliferatively quiescent (at restriction point 2 of Figure 1) confluent monolayers of fetal hamster cells *in vitro* to the non-proliferogenic (and non-cyclic AMP-elevating (Boynton, unpublished observations) PMA increases ODC activity and polyamine synthesis without initiating DNA synthesis. Thus, we must again conclude that the polyamine syntheses cannot by themselves initiate DNA synthesis because they are only part of a totally needed proliferogenic 'reaction package'.

A definite involvement of polyamines at some stage of prereplicative development has already been suggested in the preceding section. Suppression of ODC activity and putrescine synthesis soon after proliferative activation (from restriction point 1 of Figure 1) of parotid gland acinar cells and maybe hepatocytes *in vivo*, prevents the subsequent initiation of DNA synthesis (Inoue and coworkers, 1975; Pösö and Jänne, 1976). An early prereplicative (i.e. just beyond restriction point 2 of Figure 1) exposure of synchronized (by 'Mitotic shake-off') populations of already proliferatively activated Chinese hamster CHO cells to the ODC inhibitor DAP, lowers the cellular putrescine and spermidine levels without affecting the proportion of cells which enter the S (DNA-synthetic) phase of the first cycle (Sunkara, Rao and Nishioka, 1977). Similarly, exposure of serum-starved HTC neoplastic liver cells to other ODC inhibitors such as α-methylornithine (α-MO) or HAVA soon after proliferative stimulation by fresh serum (from blockage at restriction point 2 of Figure 1) prevents an increase in the cellular putrescine content but without affecting the completion of the first growth-division cycle. It does stop the progression of the cells, through the prereplicative phase of the next cycle, probably because of the eventual depletion of the spermidine and spermine stores (Mamont and coworkers, 1976; Harik, Hollenberg and Snyder, 1974). Finally, the suppression of putrescine synthesis by exposure to α-MO 6 hours after proliferative activation and/or stimulation (from restriction points 1, 2 or beyond (Figure 1: Mastro and Mueller, 1974; Whitfield and coworkers, 1974) by concanavalin A also does not affect flow of bovine suprapharyngeal lymphocytes into the S phase of the first cycle (Morris, Jorstad and Seyfried, 1977). Thus, it seems that putrescine is itself involved only in the proliferative activation process (i.e., the transition from restriction point 1 to 2 of Figure 1), but is indirectly involved in subsequent post-activation development through its involvement in spermidine and spermine production.

On the other hand, spermidine and spermine are probably involved in a post-activation phase of early prereplicative development. Hence, proliferatively quiescent BALB/3T3 fetal mouse reticuloendothelial cells, or WI-38 fetal human lung cells, in confluent monolayers start making DNA 10 to 12 h after proliferative stimulation (from restriction point 2 of Figure 1) elicited by a medium-serum change. They do not initiate DNA synthesis when exposed to methyl-GAG (at a concentration of 10 μM or more) at the beginning of pre-replicative development (Figures 4 and 5: Boynton, Whitfield and Isaacs, 1976). The polyamine requiring process must only occur early in the prereplicative phase because exposure of the stimulated cells to methyl-GAG near the end of their prereplicative development did not prevent the initiation of DNA synthesis (Figure 4: Boynton, Whitfield and Isaacs, 1976). Similarly, an early (0 or 15 h) exposure of phytohemagglutinin-activated (from restriction point 1 of Figure 1) guinea-pig lymphnode lymphocytes to methyl-GAG reduces their spermidine content and prevents the initiation of DNA synthesis (as measured by incorporation of [^{3}H]-thymidine into the acid-insoluble fraction), but a later exposure (i.e. 48 h) to the inhibitor has no effect on either the spermidine content or DNA synthesis (Otani and coworkers, 1974). Morris, Jorstad and Seyfried (1977) have confirmed the findings of Otani and coworkers using

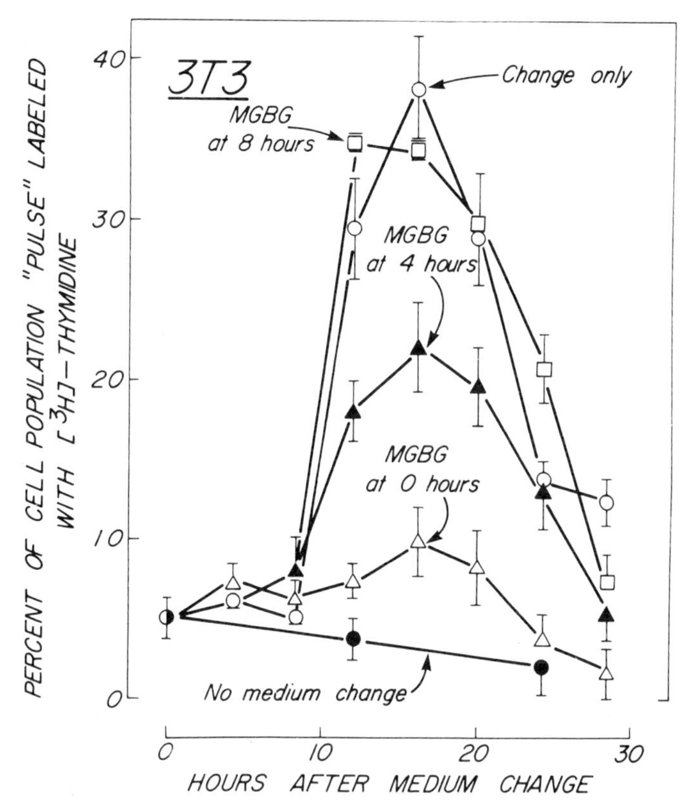

Figure 4 An example of the effects of methyl-GAG (MGBG) on the initiation of DNA synthesis by BALB/3T3 mouse cells. Cells were plated and allowed to become confluent and proliferatively quiescent. The medium was replaced to stimulate a new growth-division cycle (from restriction point 2 of Figure 1) and methyl-GAG (MGBG), 100 μM, was added at various times thereafter. The points are the means ± s.e. mean of the values from four cultures. From Boynton, Whitfield and Isaacs, (1976). (Reproduced by permission of the Wistar Institute Press, Philadelphia, Penn.)

concanavalin A-activated and/or stimulated populations of bovine suprapharyngeal lymphocytes (see Chapters 1 and 22). By contrast, exposure of serum-starved fetal rat fibroblast-like cells in sparse cultures to methyl-GAG at the time of proliferative stimulation (from restriction point 2 of Figure 1) by fresh serum does not affect the first growth-division cycle. It does block prereplicative development of the cells in the next cycle, probably because only then do the polyamine contents fall below a critical level (Rupniak and Paul, 1978a). The validity of attributing the suppression of the initiation of DNA synthesis by methyl-GAG to its ability to inhibit specifically spermidine and spermine function and synthesis is confirmed by the fact that this suppression (be it in the first BALB/3T3 and WI-38 cells or the second rat fetal fibroblast-like cells, post-stimulation cycle) is relieved by a brief (e.g. between the second and third hours of prereplicative development) or continuous exposure to spermidine or spermine, but not to putrescine (Figure 5; Boynton, Whitfield and Isaacs, 1976; Rupniak and Paul, 1978a; Rupniak and Paul, 1978b). At first sight, these observations seem to be contradicted by the fact that putrescine relieves the block induced by α-MO in the prereplicative phase of the second post-stimulation cycle of HTC hepatoma cells (Mamont

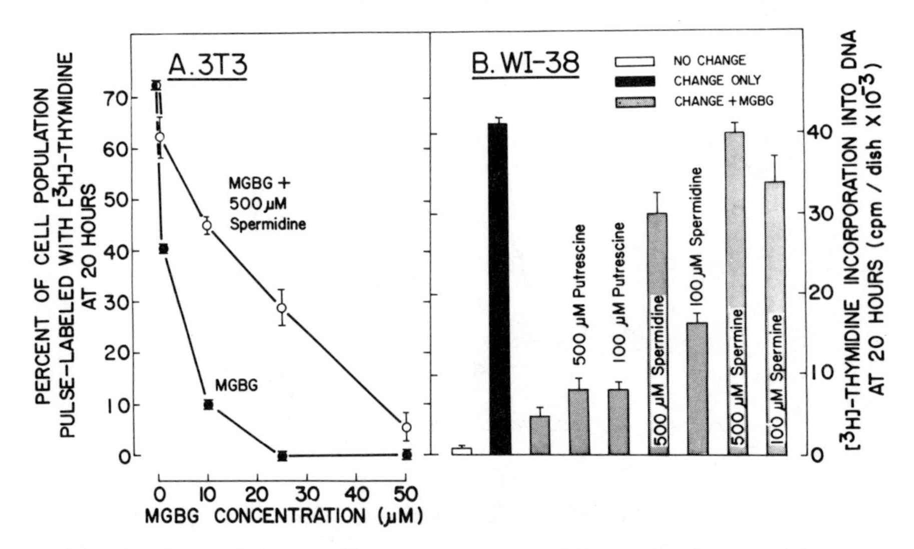

Figure 5 An example of the ability of spermidine or spermine, but not putrescine, to overcome the inhibition of initiation of DNA synthesis by methyl-GAG (MGBG). Cultures of BALB/3T3 fetal mouse cells or WI-38 fetal human cells were grown to confluence and became proliferatively quiescent (i.e. blocked at restriction point 2 of Figure 1). A new growth-division cycle was initiated by a fresh medium-serum change and the culture exposed to (A) various concentrations of MGBG or (B) 100 μM methyl-GAG (MGBG). At either 2 hours (A), or 3 hours (B), the methyl-GAG-containing medium was removed (and stored at 37°C) and replaced by 2.0 ml serum-free medium containing the appropriate polyamine. One hour later, this polyamine-containing medium was discarded and replaced by the original culture medium containing methyl-GAG. The percentage of cells in the S phase (A) or the incorporation of [^{3}H]-thymidine into DNA (B) was determined between 19 and 20 h after stimulation. The points are the means ± s.e. mean of the values from four cultures. From Boynton, Whitfield and Isaacs, 1976. (Reproduced by permission of The Wistar Institute Press, Philadelphia, Penn.)

and coworkers, 1976). However, α-MO, unlike methyl-GAG, inhibits putrescine synthesis and thereby blocks the entire polyamine synthetic system. Therefore, the spermidine and spermine stores of the α-MO-treated cell are progressively depleted and since all other enzymes are active they can be replenished equally effectively by provision of exogenous spermidine, spermine or putrescine.

Observations of Rupniak and Paul (1978b) suggest that neoplastic transformation by the SV-40 virus causes pre-neoplastic BALB/3T3 fetal mouse cells to lose the methyl-GAG sensitive step of prereplicative development. This cannot be a general consequence of neoplastic transformation because methyl-GAG completely, but reversibly, blocks the extremely neoplastic K-BALB/3T3 cells (originating from an infection of BALB/3T3 cells with the Kirsten murine sarcoma virus) in an early phase of their cycle (Figure 6).

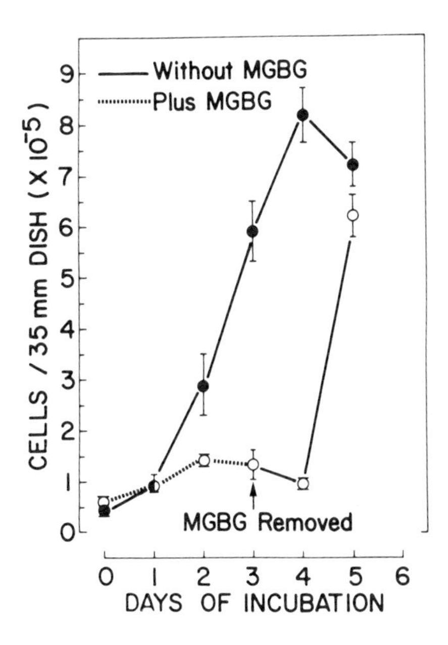

Figure 6 An example of the reversible inhibition by methyl-GAG (MGBG) of the proliferation of K-BALB/3T3 mouse cells. These cells originated from the infection of the pre-neoplastic BALB/3T3 fetal mouse cells with the Kirsten murine sarcoma virus. Cells were plated either in the presence or absence of 50 μM methyl-GAG (MGBG), and the MGBG containing medium was replaced with fresh medium 3 days later. The number of cells per dish was determined at 24 h intervals. The points are the means $\pm$ s.e. mean of the values from four cultures

Some investigators (Sunkara, Rao and Nishioka, 1977; Fillingame, Jorstad and Morris, 1975; Morris, Jorstad and Seyfried, 1977; Krokan and Eriksen, 1977) have suggested that spermidine and spermine are also directly involved in DNA synthesis. Thus, DAP-treated Chinese hamster CHO cells flow normally into the S phase of their cycle, but then the incorporation of [^{3}H]-

thymidine into the acid-insoluble fraction (presumably DNA) is reduced (Sunkara, Rao and Nishioka, 1977). This reduction of [^{3}H]-thymidine incorporation is certainly not due to an inhibition of DNA synthesis, since exposure of S-phase CHO cells to DAP greatly reduces their polyamine contents without impeding their progression through the S and G2 phases into mitosis (Sunkara, Rao and Nishioka, 1977). Indeed, our own observations (Boynton, unpublished observations) suggest that DAP reduces [^{3}H]-thymidine incorporation into DNA by inhibiting the uptake of the nucleotide by the cell. Lowering the spermidine and spermine contents of the neoplastic HeLa human cells (synchronized by blockage at the G1/S boundary by the DNA-synthetic inhibitor amethopterin) by prolonged (16-h) exposure to methyl-GAG also reduces the incorporation of [^{3}H]-thymidine into DNA by 60% while only slightly delaying their progression through the S and G2 phases into mitosis after removal of the amethopterin-induced block (Krokan and Eriksen, 1977). Obviously, such a profoundly reduced [^{3}H]-thymidine incorporation into DNA and a normal cell multiplication can co-exist only if the reduced thymidine incorporation is due to a proliferatively irrelevant inhibition of the cellular uptake of exogenous thymidine, or if there is an equally proliferatively irrelevant stimulation of endogenous thymidylate synthesis. On the other hand, Krokan and Eriksen (1977) did observe a lowering of DNA synthesis by polyamine-depleted nuclei isolated from methyl-GAG treated S-phase HeLa cells which seemed to be due to an inhibition of the initiation of DNA chain elongation, but not on-going chain elongation. This inhibition however, might only be an artifact of the isolation of methyl-GAG-treated nuclei rather than a valid indication of spermidine and spermine function, since it cannot be reversed by exposure of the nuclei to polyamines. Fillingame, Jorstad and Morris (1975) and Morris, Jorstad and Seyfried (1977) found that prereplicative exposure to methyl-GAG reduced the subsequent incorporation of [^{3}H]-thymidine into the acid-insoluble fraction of bovine suprapharyngeal lymphocytes. Because of their poorly synchronized, slowly responding cells, it is impossible to decide whether this inhibition is due to an inhibition of the subsequent prereplicative development of multiply cycling cells, the inhibition of DNA synthesis *per se*, an inhibition of thymidine uptake, or a stimulation by the spermidine analogue methyl-GAG of endogenous thymidylate synthesis. Finally, the clearest suggestion of a role for spermidine and spermine in DNA synthesis *seems* to be the observation of Boynton, Whitfield and Isaacs (1976) that methyl-GAG reduces the incorporation of [^{3}H]-thymidine into the acid-insoluble fraction and reduces the amount of autoradiographically demonstrable [^{3}H]-thymidine incorporation into the nuclei of S-phase BALB/3T3 and WI-38 cells. None the less, once again the inhibition of thymidine incorporation into DNA simply cannot be due to an inhibition of DNA synthesis because the methyl-GAG-treated cells progressed normally through the S phase of their cycle (Figure 4). Since the uptake of exogenous thymidine (into the acid-soluble fraction) is also normal under these circum-

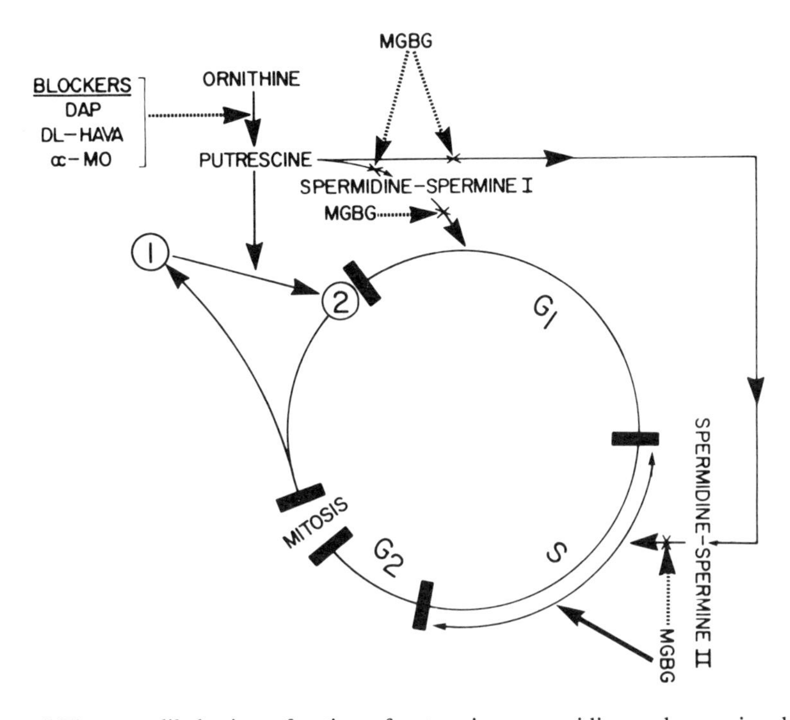

Figure 7 The most likely sites of action of putrescine, spermidine and spermine during the cell cycle. Putrescine appears to be the only polyamine involved in the proliferative activation process (i.e. the transition from restriction point 1 to point 2), because inhibitors of putrescine synthesis (such as DL-HAVA) only inhibit the initiation of DNA synthesis by proliferatively inactive cells such as salivary gland acinar cells. Since cultivated cells such as BALB/3T3, CHO, etc., are permanently proliferatively activated they never pass through the putrescine-dependent stage (from restriction point 1 to point 2). Therefore, these inhibitors cannot prevent such cells from initiating DNA synthesis via their ability to inhibit putrescine synthesis although they do block the next cycle because of an eventual depletion of the cellular spermidine and spermine levels. On the other hand, spermidine and spermine are definitely required at an early phase of prereplicative development (G1), because a specific inhibition of spermidine-spermine synthesis and function e.g. by methyl-GAG (MGBG) (*broken arrow*), causes an inhibition of DNA synthesis which can be reversed by spermidine or spermine but not by putrescine. In deference to the view of others, we have indicated by *broken arrows* that spermidine and spermine might be involved in the DNA synthetic S phase because the spermidine analogue methyl-GAG (MGBG) reduces the incorporation of [^{3}H]TdR into the acid-insoluble fraction of a variety of cell types. A *solid arrow* must also be included at the S phase because methyl-GAG (MGBG) does not affect the cell's progress through S phase and into mitosis. In our view it is far more likely that this spermidine analogue inhibits [^{3}H]-thymidine incorporation not by suppressing DNA synthesis, but by either inhibiting the transmembrane transport of thymidine or by actually increasing the endogenous pool of thymidine nucleotides by stimulating thymidylate synthesis. The abbreviations are: HAVA, DL-α-hydrazino-δ-aminovaleric acid; MGBG, methylglyoxal *bis*(guanylhydrazone). *Broken arrows* indicate inhibition and *solid arrows* indicate stimulation

stances, the reduced thymidine incorporation into DNA almost has to be due to a *stimulation* of endogenous thymidylate synthesis by methyl-GAG.

V. SUMMARY AND CONCLUSIONS

Proliferative development requires the *coordinated* functions of a battery of proliferogenic genes whose activation gives rise to a complex series of parallel and sequential processes which culminate in the initiation of DNA synthesis, mitosis and cytokinesis. Putrescine, produced by increased ODC activity, seems to be itself involved exclusively in the proliferative activation process (i.e. in the transition from restriction point 1 to point 2 [Figures 1 and 7]). Therefore, there is no specifically putrescine-dependent step in the proliferative development of cells such as those of the various established cell lines (e.g. BALB/3T3, CHO, HTC, WI-38) which have permanently activated proliferogenic genes (i.e., cells which are either actively 'cycling' or 'resting' at restriction point 2 of Figure 1).

Putrescine, spermidine and spermine are specifically involved in an early phase of prereplicative development (G1). Therefore, all active cells in cycle, be they *in vivo* or *in vitro* are vulnerable to inhibitors of putrescine synthesis (e.g. α-MO, HAVA) as well as to specific inhibitors (e.g. methyl-GAG) of the syntheses and functions of spermidine and spermine (Figure 7). Some workers believe that spermidine and spermine are also directly involved in DNA synthesis (Figure 7), because the spermidine analogue methyl-GAG reduces the incorporation of [^{3}H]-thymidine into the acid-insoluble fraction of various types of cells after they have entered the S (DNA-synthetic) phase of their cycle. In no case can this inhibition unequivocally be attributed to an inhibition of DNA synthesis since there is no other symptom of reduced DNA synthesis, such as either a prolongation of the S phase or a reduced flow of cells into mitosis. In fact, the inhibition could be due either to an inhibition of the cellular uptake of exogenous [^{3}H]-thymidine or to a stimulation by the spermidine-like methyl-GAG of endogenous thymidylate synthesis.

ACKNOWLEDGEMENTS

We thank R. Isaacs, M. Sikorska and R. Tremblay for technical assistance and D. Gillan for preparing the illustrations. Publication N.R.C.C. No. 17941.

REFERENCES

Barbirolli, B., Moruzzi, M. S., Tadolini, B., and Monti, M. G. (1975). *J. Nutr.*, **105**, 408–412.

Baserga, R., and Heffler, S. (1967). *Exp. Cell Res.*, **46**, 571–579.

Beck, U. T., Bellantone, R. A., and Canellakis, E. S. (1972). *Biochem, Biophys. Res. Commun.*, **48**, 1649–1655.

Boynton, A. L., Whitfield, J. F., and Isaacs, R. J. (1976). *J. Cell. Physiol.*, **89**, 481–488.
Byus, C. V., and Russell, D. H. (1974). *Science*, **241**, 650–652.
Clark, J. L., and Duffy, P. (1976). *Arch. Biochem. and Biophys.*, **172**, 551–557.
Clark, J. L., and Fuller, J. L. (1975). *Biochemistry*, **14**, 4403–4409.
Durham, J. P., Baserga, R., and Butcher, F. R. (1974). *Biochim. Biophys. Acta*, **372**, 196–217.
Dykstra, W. G., and Herbst, E. J. (1965). *Science*, **149**, 428–429.
Fillingame, R. H., Jorstad, C. M., and Morris, D. R. (1975). *Proc. Nat. Acad. Sci. USA*. **72**, 4042–4045.
Fillingame, R. H., and Morris, D. R. (1973). *Biochemistry*, **12**, 4479–4487.
French, F. A., Freelander, B. L., Hasking, A., and French, J. (1960). *Acta Unio Internat. Contra Cancrum*, **16**, 614–624.
Fuller, J. M., Gerner, E. W., and Russell, D. H. (1977). *J. Cell. Physiol.*, **93**, 81–88.
Gaza, D., Short, J., and Lieberman, I. (1973). *FEBS Lett.*, **32**, 251–253.
Grimm, W., and Marks, F. (1974). *Cancer Res.*, **34**, 3128–3134.
Harik, S. I., Hollenberg, M. D., and Snyder, S. H. (1974). *Nature, Lond.*, **249**, 250–251.
Heby, O., Marton, L. J., Zardi, L., Russell, D. H., and Baserga, R. (1975). *Exptl. Cell Res..*, **90**, 8–14.
Hogan, B., Shields, R., and Curtis, D. (1974). *Cell*, **2**, 229–233.
Hölttä, E., Hannonen, P., Pispa, V., and Jänne, J. (1973). *Biochem. J.*, **136**, 669–676.
Heby, O., and Lewan, L. (1971). *Virch. Arch. Abt. B. Zellpath.*, **8**, 58–66.
Hölttä, E., and Jänne, J. (1972). *FEBS Lett.*, **23**, 117–121.
Inoue, H., Kato, Y., Takigawa, M., Adachi, K., and Takeda, Y. (1975). *J. Biochem.* **77**, 879–893.
Inoue, H., Tanoika, H., Shiba, K., Asada, A., Kato, Y., and Takeda, Y. (1974). *J. Biochem.*, **75**, 679–687.
Jänne, J. (1967). *Acta. Physiol. Scand. Suppl*, **300**, 1–71.
Krokan, H., and Eriksen, A. (1977). *Eur. J. Biochem.*, **72**, 501–508.
Lajtha, L. G., Oliver, R., and Gurney, C. W. (1962). *Brit. J. Haemat.*, **8**, 442–460.
MacManus, J. P., Braceland, B. M., Youdale, T., and Whitfield, J. F. (1973). *J. Cell. Physiol.*, **82**, 157–164.
MacManus, J. P., Franks, D. J., Youdale, T., and Braceland, B. M. (1972). *Biochem. Biophys. Res. Commun.*, **49**, 1201–1207.
Mamont, P. S., Bohlen, P., McCann, P. P., Bey, P., Schuber, F., and Tardif, C. (1976). *Proc. Nat. Acad. Sci. USA.*, **73**, 1626–1630.
Mastro, A. M., and Mueller, G. C. (1974). *Exptl. Cell Res.*, **88**, 40–46.
Morris, D. R., Jorstad, C. M., and Seyfried, C. E. (1977). *Cancer Res.*, **37**, 3169–3172.
Mufson, R. A., Astrup, E. G., Simsiman, R. C., and Boutwell, R. K. (1977). *Proc. Nat. Acad. Sci. USA*, **74**, 657–661.
Novi, A. M. and Baserga, R. (1972a). *Lab. Invest.*, **26**, 540–547.
Novi, A. M., and Baserga, R. (1972b). *J. Cell. Biol.*, **55**, 554–562.
Otani, S., Yasuhiro, M., Matsui, I., and Morisawa, S. (1974). *Mol. Biol. Reports*, **1**, 431–436.
Pierce, D. A., and Fausto, N. (1978). *Biochemistry*, **17**, 102–109.
Pösö, H., and Jänne, J. (1976). *Biochem. J.*, **158**, 485–488.
Rajewsky, M. F. (1974). Proliferative properties of malignant cell systems. *Handbuch d. allg. Pathologie*, Bd VI/5-6, Springer Verlag, Berlin-Heidelberg-New York, pp. 289–325.
Rupniak, H. T., and Paul, D. (1978a). *J. Cell. Physiol.*, **94**, 161–170.

Rupniak, H. T., and Paul, D. (1978b). Inhibition of the cell cycle by polyamines in normal and transformed fibroblasts. In R. A. Campbell, D. R. Morris, D. Bartos, G. D. Daves and F. Bartos (Eds), *Advances in Polyamine Research*, Vol. 1, Raven Press, New York, pp. 117–126.

Russell, D. H., Medina, V. J., and Snyder, S. H. (1970). *J. Biol. Chem.*, **245**, 6732–6738.

Russell, D. H., and Stambrook, P. J. (1975). *Proc. Nat. Acad. Sci. USA.*, **72**, 1482–1486.

Sunkara, P. S., Rao, P. N., and Nishioka, K. (1977). *Biochem. Biophys. Res. Commun.*, **74**, 1125–1133.

Thrower, S., and Ord, M. (1974). *Biochem. J.*, **144**, 361–369.

Walker, P. R., Sikorska, M., and Whitfield, J. F. (1978). *J. Cell. Physiol.*, **94**, 87–91.

Wang, T. and Foker, J. (1978). Cyclic AMP during successive proliferation cycles. In W. J. George, L. G. Ignarro, P. Greengard and G. A. Robinson (Eds), *Advances in Cyclic Nucleotide Research*, Vol. 9, Raven Press, New York.

Whitfield, J. F. (1979). The roles of adrenergic agents, calcium ions and cyclic nucleotides in the control of cell proliferation. In L. Szekeres (Ed.), *Adrenergic Activators and Inhibitors, Handbook of Experimental Pharmacology*, Springer Verlag, Berlin, Heidelberg, New York (in press).

Whitfield, J. F., Boynton, A. L., MacManus, J. P., Rixon, R. H., Walker, P. R., and Armato, U. (1976). The positive regulation of cell proliferation by a calcium-cyclic AMP control couplet. In M. Abou-Sabe (Ed.), *Cyclic Nucleotides and the Regulation of Cell Growth*, Dowden, Hutchinson & Ross, Inc., Stroudsburg, Pennsylvania, pp. 97–130.

Whitfield, J. F., MacManus, J. P., Boynton, A. L., Gillan, D. J. and Isaacs, R. J. (1974). *J. Cell. Physiol.*, **84**, 445–458.

Williams-Ashman, H. G., and Schenone, A. (1972). *Biochem. Biophys. Res. Commun.*, **46**, 288–295.

Yanagi, S., and Potter, V. R. (1977). *Life Sciences*, **20**, 5109–1519.

Chapter 6

The Induction of Ornithine Decarboxylase in Normal and Neoplastic Cells

URIEL BACHRACH

I. INTRODUCTION

Interest in the regulation of polyamine biosynthesis in animal cells was triggered by the finding that growth processes are closely related to cellular polyamine synthesis (Cohen, 1971; Bachrach, 1973; Russell, 1973a,b; Heby and coworkers, 1975; Raina and Jänne, 1975; Tabor and Tabor, 1976). Numerous studies indicated that ornithine decarboxylase (ODC, L-ornithine carboxy-lyase EC 4.1.1.1.17), is the rate-limiting enzyme in the biosynthesis of polyamines (Chapter 2). The occurrence of ODC in animal tissues was reported independently by three laboratories in 1968 (Jänne and Raina, 1968; Pegg and Williams-Ashman, 1968; Russell and Snyder, 1968). Russell and Snyder (1968) also demonstrated the increase in ODC activity in rat liver after partial hepatectomy (Figure 1), during the development of chick embryo (Figure 2) and in fetal liver and different rat tumours (Figure 3).

Increase in ODC activity has been noted in various growing organs or tissues and during the different stages of differentiation and development. Some of these systems are listed in Table 1.

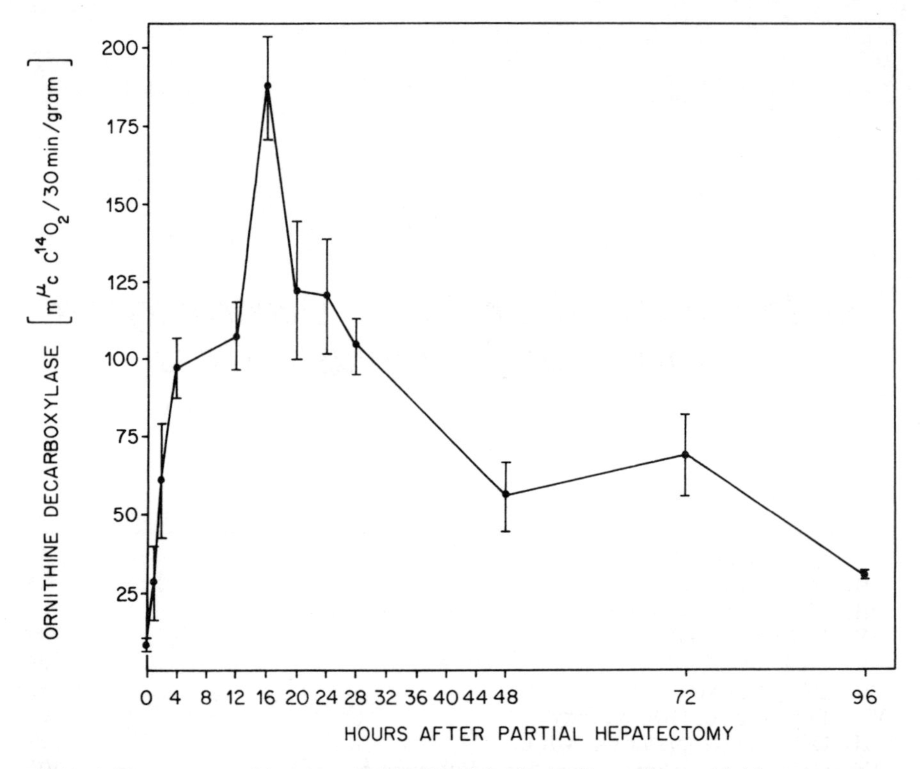

Figure 1 Time course of increase in ODC activity in rat liver after partial hepatectomy. (Taken from Russell and Snyder, 1968)

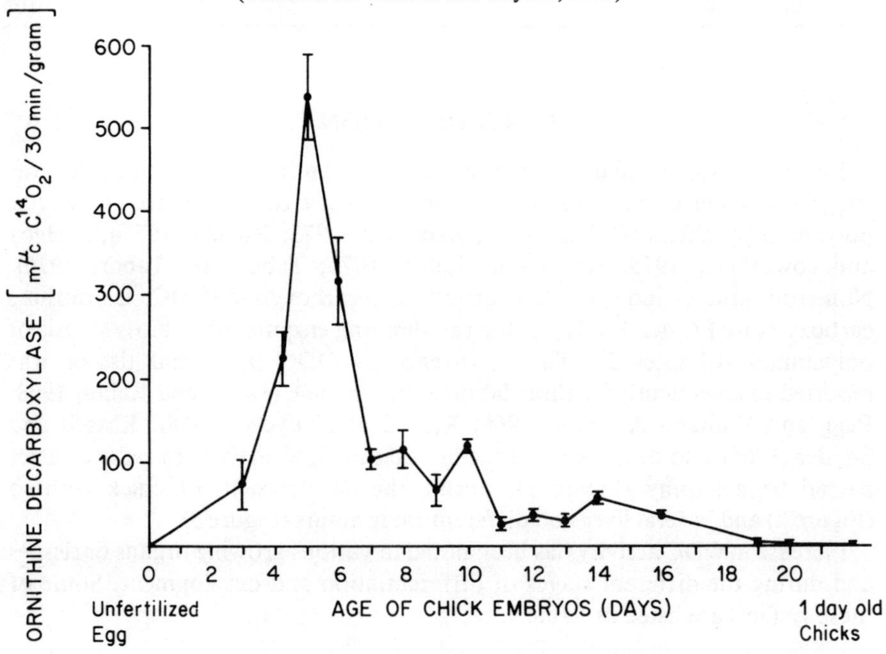

Figure 2 Changes in ODC activity during the development of the chick embryo. (Taken from Russell and Snyder, 1968).

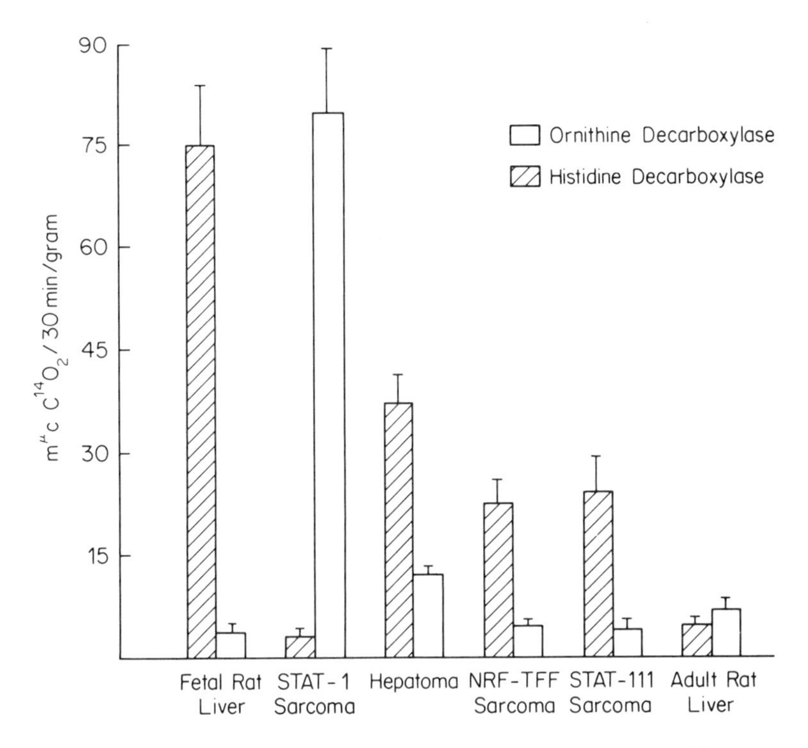

Figure 3 Ornithine decarboxylase activity in fetal and adult rat liver and in rat tumours. (Taken from Russell and Snyder, 1968)

In recent years, it has been suggested that mechanisms that regulate the activity of ODC are an integral part of growth-control programs. The enzyme also attracted considerable attention when Russell and Snyder (1969b) reported that ODC has an extremely short half-life. This result suggested that ODC also plays a role in regulating growth. Some of the properties of ODC were reviewed by Morris and Fillingame (1974).

II. INDUCTION OF ODC

A. Effect of Hormones

Virtually all hormones have been shown to increase ODC activity in appropriate target tissues (Panko and Kenney, 1971; Nicholson, Levine and Orth, 1976). Some of the most dramatic illustrations of the activation of ODC by hormones were studies with cultures of glioma and neuroblastoma cells (Bachrach, 1975). Thus, the addition of catecholamines and the phosphodies-

Table 1 Changes in ODC activity during development and differentiation

System	Reference
Rat brain	Anderson and Schanberg (1972)
intestine	Ball and Balis (1976)
brain, heart	Bartolome, Lau and Slotkin (1977)
liver	Ferioli, Ceruti and Camoli (1976)
placenta	Maudsley and Kobayashi (1977)
Rat uterus, pregnancy	Saunderson and Heald (1974)
ovary, pregnancy	Guha and Jänne (1976)
kidney hypertrophy	Brandt, Pierce and Fausto (1972)
regenerating liver	Jänne and Raina (1968)
	Raina and Jänne (1968)
	Russell and Snyder (1969a)
	Fausto (1969)
	Schrock, Oakman and Bucher (1970)
	Raina and coworkers (1970)
	Russell and McVicher (1971)
	Hölttä and Jänne (1972)
	Cavia and Webb (1972)
Rat heart constriction	Feldman and Russell (1972)
hypertrophy	Matsushita, Sogani and Raben (1972)
Human lymphocytes — stimulated	Kay and Lindsay (1973)
Guinea pig lymphocytes — stimulated	Mizoguch and coworkers (1975)
Chinese hamster cells, synchronized culture	Russell and Stambrook (1975)
Drosophila melanogaster larvae	Byus and Herbst (1976)
Chick embryo retina	DeMello, Bachrach and Nirenberg (1976)

terase inhibitor IBMX (3-isobutyl-1-methylxanthine) to glioma cells resulted in a 1,000-fold increase in ODC activity (Figure 4). A similar increase (Figure 5) was obtained by adding prostaglandin E_1 (PGE_1) and IBMX to cultures of neuroblastoma cells (Bachrach, 1975).

The induction of ODC by various hormones is summarized in Table 2.

B. Effect of Cyclic Nucleotides

In addition to hormones, cyclic nucleotides or dibutyryl cyclic AMP stimulated ODC activity in different organs and cells in tissue cultures (Table 3). It has been suggested that cyclic nucleotides mediate ODC induction (see review by Russell, Byus and Manen, 1976). This suggestion is most interesting, as ODC and the products of its action, the polyamines, may be the link between hormones, cyclic AMP — the 'second messenger' and RNA and protein biosynthesis.

Not all tissues responded, however, to administration of dibutyryl cyclic

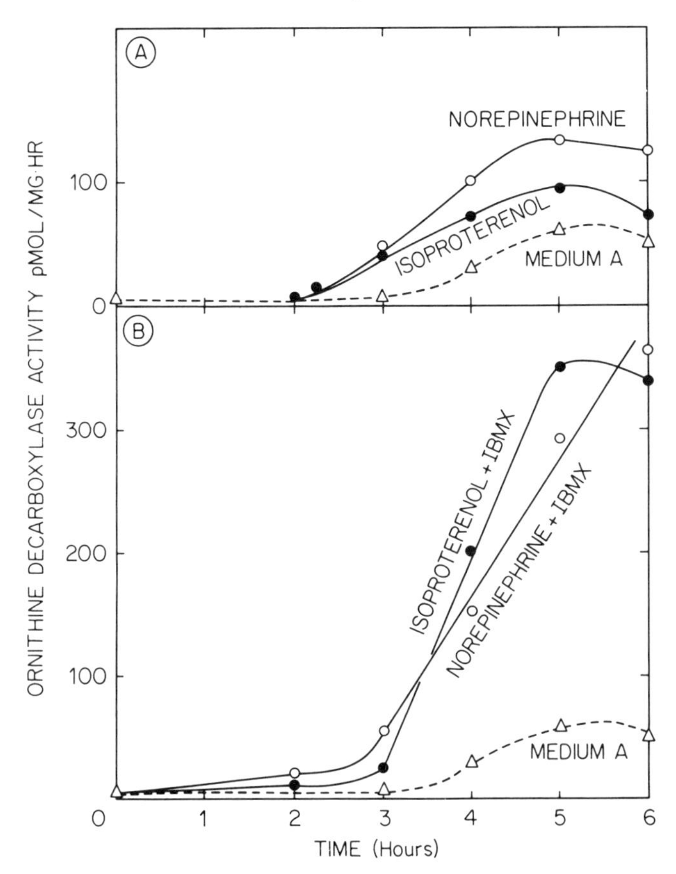

Figure 4 Effect of catecholamines and IBMX on ODC activity of C_6-BU-1 glioma cells. Isoproterenol (0.1 μM) or norepinephrine (1 μM) were added to the cells in the presence of absence of 0.5 ml IBMX in a synthetic medium (Medium A) (Taken from Bachrach, 1975)

AMP with increased ODC activity. Thus, treatment of rats with ACTH caused the increase in adrenal cyclic AMP without changing ODC activity (Levine, Leaming and Raskin, 1978). It has also been claimed that the administration of dibutyryl cyclic AMP to intact rats may simulate the release of pituitary hormones, including growth hormones, which may be responsible for the increase in ODC activity.

Despite this, there is ample circumstantial evidence that cyclic AMP activates cyclic AMP-dependent protein kinase, which in turn may trigger ODC induction by regulating gene expression (see review by Jungman and Russell, 1977).

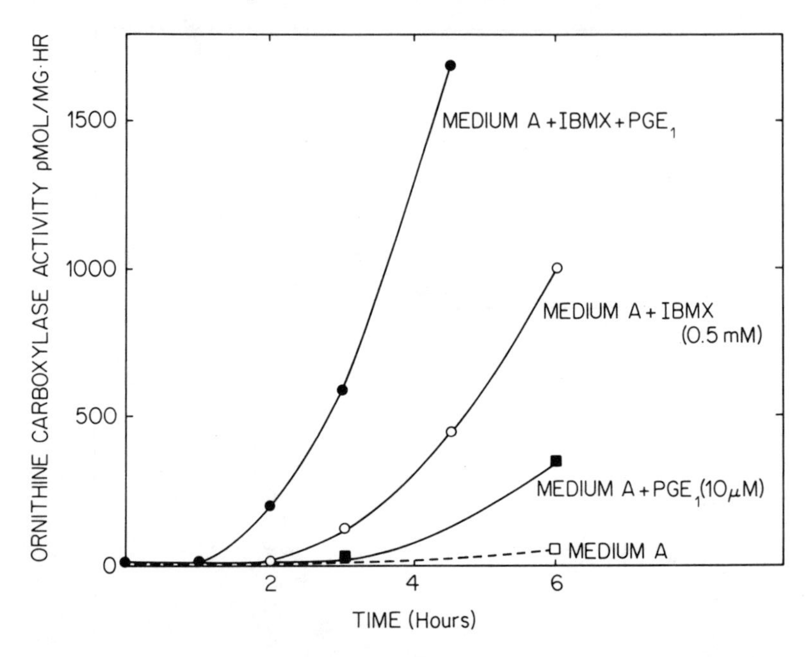

Figure 5 Effect of PGE_1 and IBMX on ODC activity of N115 neuroblastoma cells. Enzyme activity at zero time was 2 pmol mg^{-1} protein $hour^{-1}$. (Taken from Bachrach, 1975)

We have recently used a direct genetic approach to examine the involvement of cyclic AMP in the induction of ODC and added dibutyryl cyclic AMP to cultures of S49 lymphoma cells and to cultures of a protein kinase-negative mutant. We showed (Hochman, Plotek and Bachrach, 1978b) that dibutyryl cyclic AMP stimulated ODC activity in S49 lymphoma cells but not in the protein kinase-negative mutant. The following conclusions were drawn from those experiments:

(a) the induction of ODC is mediated by cAMP, at least in S49 lymphoma cells; (b) cyclic AMP is not the exclusive mediator for ODC induction, as this enzyme also occurs in protein kinase-negative cells which are not affected by cyclic AMP. Similar results were obtained independently by Insel and Fenno (1978), who reported that dibutyryl cyclic AMP caused an initial increase in the activities of ODC and S-adenosyl-L-methionine decarboxylase (SAM decarboxylase, S-adenosyl-L-methionine carboxy-lyase EC 4.1.4.50, which catalyses the synthesis of spermidine) in cultures of the wild type S49 lymphoma. This increase in enzyme activity was followed by a steep decrease in activity in S49 cells, but not in the protein kinase-negative mutant.

Table 2 Induction of ODC by various hormones and drugs

System	Hormone	Reference
Rat, liver	Growth hormone	Jänne, Raina and Siimes (1968)
Rat, liver	Growth hormone	Russell and Snyder (1969a)
Rat, liver	Growth hormone	Russell, Snyder and Medina (1970)
Rat, liver	Growth hormone	Richman and coworkers (1971)
Rat, liver	Growth hormone	Panko and Kenney (1971)
Rat, kidney, heart	Growth hormone	Sogani and coworkers (1972)
Rat, kidney, heart	Growth hormone	Brandt, Pierce and Fausto (1972)
Rat, kidney, heart	Growth hormone	Levine and coworkers (1973)
Rat, brain	Growth hormone	Roger, Schanberg and Fellows (1974)
Glioma cells	Catecholamines	Bachrach (1975)
Mouse, parotid glands	Catecholamines	Inoue and coworkers (1974)
Baby hamster kidney cells	Prostaglandin E_1	Hogan, Shields and Curtis (1974)
Neuroblastoma cells	Prostaglandin E_1	Bachrach (1975)
Mouse, mammary epithelium	Prostaglandin E_1	Oka and Perry (1976)
Rat, ovary	Prostaglandin E_2	Lamprecht and coworkers (1973)
Rat, uterus	Luteinizing hormone	Cohen, O'Malley and Stastny (1970)
Rat, ovary	Luteinizing hormone	Kobayashi, Kupelian and Maudsley (1971)
Rat, ovary	Luteinizing hormone	Kaye and coworkers (1973)
Rat, ovary	Luteinizing hormone	Lamprecht and coworkers (1973)
Rat, ovary	Luteinizing hormone	Maudsley and Kobayashi (1974)
Rat, ovary	Gonadotropic hormone	Maudsley and Kobayashi (1977)
Rat, testes	Gonadotropic hormone	Reddy and Villee (1975)
Mouse, kidney	Testosterone	Grahn and coworkers (1973)
Rat, liver	Hydrocortisone	Richman and coworkers (1971)
Rat, liver	Hydrocortisone	Panko and Kenny (1971)
Rat, kidney	Hydrocortisone	Brandt, Pierce and Fausto (1972)
Rat, brain	Hydrocortisone	Anderson and Schanberg (1975)
Rat, adrenal	Adrenocorticotropic hormone	Richman and coworkers (1973)
Rat, adrenal	Adrenocorticotropic hormone	Levine and coworkers (1973)
Rat, adrenal	Adrenocorticotropic hormone	Levine and coworkers (1975)
Rat, liver	Glucagon	Gaza, Short and Lieberman (1973)
Rat, liver	Glucagon	Mallette and Exton (1973)
Mouse mammary gland	Insulin	Aisbitt and Barry (1973)
Baby hamster kidney cells	Insulin	Hogan, Shields and Curtis (1974)
Rat, small intestine	Insulin	Maudsley, Leif and Kobayashi (1976)
Rat, liver	Insulin-like activity factor	Haselbacher and Humbel (1976)
Rat, thyroid	Thyrotropin	Matsuzaki and Suzuki (1974)
Rat, thyroid	Thyrotropin	Richman and coworkers (1975)
Rat, brain	Thyroxin	Anderson and Schanberg (1975)
Rat, kidney	Parathyroid hormone	Scalabrino and Ferioli (1976)
Rat, brain	Nerve growth factor	Roger, Schanberg and Fellows (1974)
Rat, ganglia	Nerve growth factor	Macdonell and coworkers (1977)
Chick embryo fibroblast	Epidermal growth factor	Stastny and Cohen (1972)
Mouse mammary gland	Epidermal growth factor	Aisbitt and Barry (1973)
Rat, liver	Growth factor	Sogani and coworkers (1972)
3T3 cells	Multiplication-stimulating activity factor MSA	Nissley, Passamani and Short (1977)
Rat, liver	Phenobarbital	Byus and coworkers (1976)
Silkmoth pupal tissue	Ecdysone	Wyatt and coworkers (1973)

Table 3 Effect of cyclic nucleotides on ODC activity

System	Nucleotide	Reference
Rat, liver	Dibutyryl cAMP	Beck, Bellantone and Canellakis (1972)
Rat, liver	Dibutyryl cAMP	Hölttä and Raina (1973)
Rat, liver	Dibutyryl cAMP	Eloranta and Raina (1975)
Rat, adrenal	Dibutyryl cAMP	Richman and coworkers (1973)
Rat, liver	Dibutyryl cAMP	Levine, Leaming and Raskin (1978)
Rat, testes	Dibutyryl cAMP	Reddy and Villee (1975)
Mouse, mammary glands	Dibutyryl cAMP	Aisbitt and Barry (1973)
Mouse, mammary epithelium	Dibutyryl cAMP	Oka and Perry (1976)
Coturnix quail	Dibutyryl cAMP	Preslock and Hampton (1973)
Baby hamster kidney cells	Dibutyryl cAMP	Hogan, Shields and Curtis (1974)
Neuroblastoma, glioma cells	Dibutyryl cAMP	Bachrach (1975)
Rat hepatoma cells	Dibutyryl cAMP	Canellakis and Theoharides (1976)
Rat, adrenal medulla	Aminophylline	Byus and Russell (1974)
Rat, cortex liver & kidney	IBMX	Byus and Russell (1975)
Guinea-pig lymphocytes	Dibutyryl cGMP	Mizoguchi and coworkers (1975)

C. Effect of Glutamine and Asparagine

Hogan and Murden (1974) found that glutamine and asparagine were active in stimulating ODC activity of cultured hepatoma cells. These findings were also confirmed by Prouty (1976), who found that glutamine increased ODC activity in HeLa cells by approximately 100-fold, and that the half-life of the enzyme changed from 12 to 55 min after incubation for 4 h. Putrescine prevented the induction of ODC by glutamine (Prouty, 1976).

The effect of glutamine and asparagine on ODC induction was also studied by Chen and Canellakis (1977), who showed that these two amino acids increased the ODC activity of cultured neuroblastoma cells suspended in salt-glucose or complete medium. Actinomycin D did not inhibit the induction by asparagine but a substantial increase in half-life was observed when cycloheximide was added to the culture along with asparagine.

III. CHANGES IN ODC ACTIVITY DURING GROWTH AND DEVELOPMENT

Significant changes in ODC activity have been noticed during the development of the brain of fetal and neonatal rats (Anderson and Schanberg, 1972). A similar increase in ODC activity has also been described during the development of the retina of chick embryos (DeMallo, Bachrach and Nirenberg, 1976) and the uterus of pregnant rats (Saunderson and Heald, 1974). Guha and Jänne (1976) reported that the activity of ODC fluctuated in rat ovaries during the normal oestrous cycle. The activity of the enzyme was associated with the growth of the ovarian tissue, was low at the beginning of the pregnancy and rose when the placenta was formed in the rat. ODC activity was also high in the developing rat placenta (Maudsley and Kobayashi, 1974).

The ODC activity of rat intestine, liver, and brain was also found to vary dramatically as animals develop and age (Fenioli, Ceruti and Camoli, 1976; Ball and Balis, 1976), as did the ODC during the development of rat epididymis (Majumder, MacIndoe and Turkington, 1974).

During fetal and neonatal brain development, the highest ODC levels paralleled the period of most rapid cellular growth and replication. Disturbances in ODC activity during fetal and postnatal life can represent an early index of disturbed development. Thus, maternal morphine administration has been shown to cause a delay in the normal maturational decreases of brain ODC activity of the offspring (Butler and Schanberg, 1975).

The effect of another opiate, methadone, on the developmental pattern of ODC and on organ weights, has also been examined by Slotkin, Lau and Bartolome (1976). Exposure to methadone in the postnatal period either directly or via the mother (rat) resulted in delays in maturation and decreases in brain ODC. This was accompanied by deficits in the weight of brain and heart. In addition to opiates, ethanol was also shown to cause alterations in the developmental pattern of ODC in the brain and heart. Thadani and coworkers (1977) have recently shown that chronic ethanol ingestion by pregnant rats affected the weight and ODC pattern of the heart and brain of the offspring. The effect of maternal behaviour on the ODC activity of the brain and heart of developing rat pups has also been investigated (Butler, Susskind and Schanberg, 1978). These interesting studies showed that rat pups removed from their mother and placed in a warm incubator for several hours showed a significant decrease in ODC activity in the brain and heart. It has been suggested that active maternal behaviour is necessary to maintain normal synthesis of polyamines in the brain and heart of the pup during development. If these results hold also for other systems, then behavioural and psychological factors may control ODC activity, in addition to its regulation by various drugs. It is obvious that such an approach bears far-reaching consequences. It would imply that these environmental factors affect the mental and physical development of the newborn.

IV. PROPERTIES OF ODC

A. Half-life of ODC

Russell and Snyder (1969b) determined the half-life of ODC and used compounds which inhibit the synthesis of proteins (such as cycloheximide or puromycin), or RNA (actinomycin D). They demonstrated that ODC had an exceptionally short half-life which ranged from 15 to 45 min — the shortest half-life of any enzyme in eukaryotes. Although it has been generally accepted that ODC has a very short half-life, the use of cycloheximide has been criticized, as this compound may also inhibit protein degradation (Canellakis and coworkers, 1978).

The half-life of ODC varies during the growth cycle and is affected by various factors. Thus, when stationary HTC cells were diluted into fresh medium, the half-life of ODC increases from 11 to 60 min (Hogan and Murden, 1974). In this case, too, the shorter half-life has been associated with low ODC activity and with slow growth rates. A similar increase (from 11 to 60 min) in the half-life was also described by Prouty (1976), who studied the induction of ODC by glutamine in HeLa cells. In 3T3 cells, on the other hand, the half-life of ODC dropped during the induction from 240 to 40 min, indicating an opposite correlation (Clark, 1974).

The activity of ornithine decarboxylase also rose during the transformation of chick embryo fibroblasts by Rous sarcoma virus. In this case, too, the elevation of ODC activity was associated with the increase of the half-life (Bachrach, 1976b). It should be noted that the increase in SAM decarboxylase activity during the transformation of chick embryo fibroblasts by Rous sarcoma virus was not associated with the increase in half-life of SAM decarboxylase.

In all these experiments, the half-life of ODC has been calculated by using inhibitors which block the synthesis of new enzyme. This approach has definite drawbacks. To overcome them, attempts have been made to determine the half-life of ODC by alternative procedures. Hölttä (1975) succeeded in preparing antiserum against ODC, purified from rat liver. Immunoglobulins from the immune sera were covalently coupled to agarose by cyanogen bromide activation. It has been shown that the increase in ODC activity, after growth hormone administration, or after partial hepatectomy, corresponded with a similar increase in the immunoreactive enzyme protein. In addition, the rapid decay in ODC activity in regenerating rat liver after cycloheximide injection, was accompanied by a decrease in the immunoreactive protein. These results confirmed previous findings that ODC has a very short half-life and that cycloheximide might be used to determine its stability.

Obenrader and Prouty (1977b), who used a 7,000-fold purified ODC preparation from thioacetamide-treated rat liver, found that the half-life of ODC was 19 to 24 min, while the half-life of disappearance of antigen was 28 to 33 min. The discrepancy has been explained by differences in the activity and the antigenic properties of the enzyme.

An alternative method for the determination of the half-life of ODC was applied by Ben Hur, Prager and Riklis (1978), who found that protein synthesis was arrested by heating Chinese hamster cells at 42°C. When the activity of ODC was determined in the heated cells, a half-life of 19 min was obtained. Prouty (1976) studied the decay of ODC activity in HeLa cells and found that the inactivation process required energy.

Little is known about the mechanism of ODC inactivation. Immunological studies (Hölttä, 1975) suggested that inactivation involves protein breakdown. On the other hand, TPCK, which inhibits general protein breakdown in hepatoma cells, had no effect on ODC half-life (McIlhinney and Hogan, 1974; Prouty, 1976). It thus appears that the mechanism of inactivation of ODC is different from that responsible for the degradation of most proteins.

B. Heat Stability

Obenrader and Prouty (1977a) studied the stability of ODC obtained from the liver of thioacetamide-treated rats. They found that at 45°C the enzyme was stable for at least 30 min. At 52°C, however, the activity decayed with a half-life of approximately 10 min. A biphasic inactivation curve was obtained at 48°C, indicating the existence of a multiform enzyme.

C. Molecular Weight

The purification of ODC from various sources permitted the calculation of its molecular weight. Thus Ono and coworkers (1972) purified ODC about 5,400-fold from the soluble fraction of liver from thioacetamide-treated rats. The purified enzyme showed a single protein band on polyacrylamide gel electrophoresis. Ono and coworkers (1972) estimated the molecular weight of the enzyme to be approximately 100,000.

The molecular weight of purified ODC from cultured 3T3 mouse fibroblast was also determined by Boucek and Lembach (1977), who used sodium dodecyl sulphate gel electrophoresis. A molecular weight of approximately 55,000 was calculated for the subunit of the purified enzyme, while gel filtration on Bio-Gel gave a molecular weight of 110,000 daltons. It has been suggested that the active enzyme has a molecular weight of approximately 55,000, but forms an inactive dimer in the absence of dithiothreitol. A similar inactivation by dimerization has also been reported for ODC from rat prostate (Jänne and Williams-Ashman, 1971). The existence of two forms of ODC has also been reported by Mitchell and coworkers (1978), who isolated monomer and dimer forms from plasmodial homogermates of *Physarum polycephalum*. These forms had molecular weights of 80,000 and 160,000 respectively, and differed in their activities. In this case, too, the activity of ODC could be modulated by controlling the transition of the enzyme between the two forms.

Theoharides and Cannelakis (1976) also analysed $[C^{14}]$-labelled ODC, purified from regenerating rat liver, and estimated its molecular weight as 90,000 using sodium dodecyl sulphate polyacrylamide gel electrophoresis. Obenrader and Prouty (1977b) determined the molecular weight of a 7,000-fold purified ODC, derived from rat liver, by first labelling the enzyme *in vitro* by reductive methylation using formaldehyde and sodium-$[^3H]$-borohydride followed by polyacrylamide gel electrophoresis. They came to the conclusion that the subunit of the enzyme had a molecular weight of approximately 50,000 daltons, and a dimeric form of 100,000 daltons. In addition, a peak molecular weight of approximately 75,000 daltons was also observed. It has been speculated that this is a complex of the ODC subunit with a molecule of antizyme (see Chapter 9), which apparently has a molecular weight of 25,000.

Heat inactivation curves also suggested that more than one protein was responsible for the enzymatic activity of rat liver ODC (Obenrader and Prouty, 1977a). The same authors also used affinity chromatography (activated thiol-Sepharose 4B) and polyacrylamide gel electrophoresis to demonstrate that ODC exists in multiple forms.

D. Co-factors

According to Friedman, Halpern and Canellakis (1972) and Ono and coworkers (1972), the purified ODC from rat liver required thiol groups for maximal activity and was inhibited by inhibitors of pyridoxal enzymes, such as L-canaline or isonicotinic acid hydrazide. Polyamines at millimolar concentrations had little effect on purified rat ODC. The inactivation of ODC by removing thiol groups has also been reported for the rat prostate enzyme (Jänne and Williams-Ashman, 1971), and for that purified from 3T3 mouse fibroblasts (Boucek and Lembach, 1977). The importance of thiol groups in the activity of ODC was also demonstrated by Obenrader and Prouty (1977a), who succeeded in purifying hepatic ODC by affinity or activated thiol Sepharose. Friedman, Halpern and Canellakis (1972), who used partially purified ODC, found that at low substrate concentrations there was a 55% increase in the activity by dithiothreitol (up to 25 mM).

Pyridoxal phosphate is also a co-factor for ODC. According to Clark and Fuller (1976), there are two forms of ODC with respect to pyridoxal phosphate affinity in exponentially growing 3T3 mouse fibroblasts, one with a Km of 10 μM and the other with a Km of 0.4 μM.

E. pH Optimum

The occurrence of a double pH optimum for ODC was first noted by Raina and Jänne (1968), who showed that increasing the mercaptoethanol concentration from 1 to 10 mM shifted the pH optimum of crude enzyme from 7.4 to 8.1. On the other hand, Friedman, Halpern and Canellakis (1972) found that the purified enzyme exhibited a single optimum at pH 7.7 and according to Ono and coworkers (1972) the optimal pH was 7.0. It is quite possible that the crude enzyme consisted of ODC in multiple forms, while the purified enzyme was a more homogenous preparation.

F. Isoelectric Point

ODC purified from thioacetamide-treated rat liver (purified 5,400-fold) had an isoelectric point of pH 4.1 (Ono and coworkers, 1972).

G. Km Value

According to Ono and coworkers (1972), a 5,400-fold purified liver enzyme had a Km value for L-ornithine of $2 \cdot 10^{-4}$ M at pH 7.0.

V. INHIBITORS OF ODC

Williams-Ashman, Corti and Tadolini (1976) have recently reviewed the structure and mode of action of compounds that inhibit ODC in eukaryotes. The following compounds received special attention (see also Chapter 11).

A. α-Methyl Ornithine

This compound, which was employed by Abdel-Monem, Newton and Weeks (1974), competed with L-ornithine with a Ki of 20×10^{6} M. Mamont and coworkers (1976) and Chapter 10 studied the effect of α-methyl ornithine on cultured rat hepatoma cells and found that it inhibited putrescine and spermidine synthesis and prevented cell proliferation and DNA replication. This inhibitor, similar to other ornithine analogues, increased the half-life of ODC, apparently by interacting with the active site (McCann and coworkers, 1977).

B. α-Hydrazino Ornithine

This inhibitor was used by Harik and Snyder (1973), who found it to be relatively selective; other pyridoxal phosphate-dependent enzymes were not as sensitive as rat ventral prostate ODC (Ki 2×10^{-6} M). This ornithine analogue also inhibited putrescine accumulation in regenerating rat liver and in rat hepatoma cells in tissue culture (Harik, Hollenberg and Snyder, 1974a). The inhibition by α-hydrazino ornithine, however, was greatly reduced at high pyridoxal phosphate concentrations, so that the drug apparently reacted not only with L-ornithine but also with pyridoxal phosphate. It was also shown (Harik, Hollenberg and Snyder, 1974b) that when added to rat HTC hepatoma cells at the time of dilution with medium, α-hydrazino ornithine evoked a dose-related increase in ODC activity and a doubling in half-life. This observation raises the question as to whether α-hydrazino ornithine (and α-methyl ornithine) might be used as anti-tumour drugs as the inhibition of ODC activity is only temporary, and removal of the drug would cause an apparent increase in decarboxylase activity.

Inoue and coworkers (1975) and Sawayama, Kinugasa and Nishimura (1976) synthesized an isomer of α-hydrizino ornithine (HAVA), DL-α-hydrazino-δ aminovaleric acid, and reported its effect on putrescine synthesis in mouse parotid glands. Treatment with HAVA *in vivo* greatly diminished the enhancement of putrescine levels in mouse parotid glands evoked by administration of isoproterenol. This drug did not affect the isoproterenol-dependent increase in SAM decarboxylase.

C. N-(5′ Phosphopyridoxyl)-ornithine

Heller and coworkers (1975) synthesized several N-(5′ phosphopyridoxyl)-amino acids and found that rat liver ODC was strongly inhibited by N-(5′ phosphopyridoxyl)-ornithine. The phosphate group was essential for ODC inhibition, which was non-competitive with respect to both ornithine and pyridoxal phosphate.

As pyridoxal phosphate is an essential co-factor for ODC, the activity of the latter may be regulated nutritionally. Indeed, Sturman and Kremzner (1974) found that putrescine levels were significantly reduced in the liver and brain of vitamin B_6-deficient rats.

D. 8-Azaguanine, 5-Azacytidine

These compounds, which delay DNA synthesis, also delayed the induction of ODC in regenerating liver (Cavia and Webb, 1972). There are reports that the concentrations of polyamines and the activity of ODC in mouse L1210 leukaemic cells were depressed by administration of anti-neoplastic drugs, such as methotrexate, cytosine arabinoside (Russell, 1972) and 5-azacytidine (Heby and Russell, 1973).

E. Polyamines

At 1 mM concentrations they caused a 40% inhibition of ODC of rat intestine (Ball and Balis, 1976), and potassium chloride inhibited the activity of purified ODC by 50% at 270 mM (Oberlander and Prouty, 1977b).

F. 1,4-Diamino-butyne

It has been speculated by Williams-Ashman, Corti and Tadolini (1976) that unsaturated putrescine analogues should inhibit ODC activity. This has been confirmed experimentally by Relyea and Rando (1976), who found that 1,4-diamino-2-butene inhibited ODC from rat livers. The inhibition was competitive with a Ki of 2.10^{-6} M. Another putrescine analogue, 1,4-diamino-2-butyne, was synthesized by Wright, Rennert and Seiler (1977), and was found to inhibit ODC from fibroblasts.

VI. ODC ANTIZYME

One of the most outstanding advances in the study of ODC regulation has been the recent demonstration of ODC antizyme (Canellakis and coworkers, 1978; Chapter 9). In elegant studies, Canellakis and his associates showed the presence of a protein of a molecular weight of approximately 26,500 daltons in various cells exposed to polyamines (Fong, Heller and Canellakis, 1976). This ODC antizyme reacted specifically with ODC, neutralized its activity, and had a relatively short half-life. The rat and glioma antizyme neutralized preparations of ODC from a variety of sources, and might serve as an 'antibody' for the assay of ODC under different physiological conditions (Heller, Fong and Canellakis, 1976). The induction of antizyme by 10 mM putrescine is illustrated in Figure 6 (taken from Heller, Fong and Canellakis, 1976) and the separation of ODC from ODC antizyme by gel filtration is shown in Figure 7. The enzyme-antizyme complex may be dissociated by 10% ammonium sulphate or KCl, and the two components can then be recovered (Figure 8).

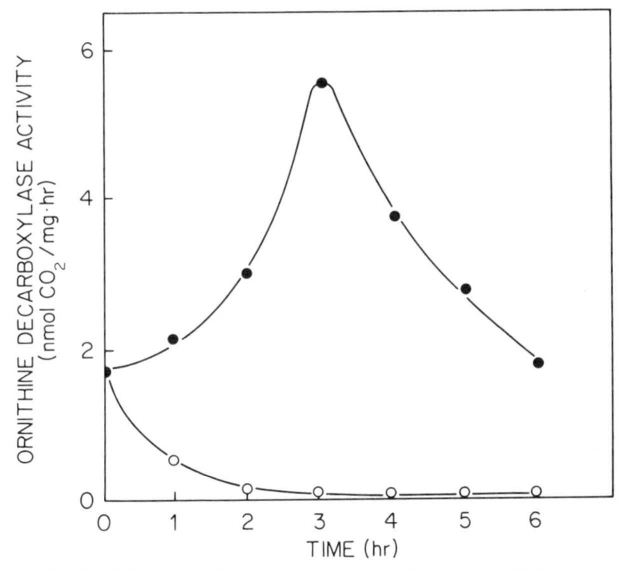

Figure 6 Effect of 10 mM putrescine on the stimulation of ornithine decarboxylase in L1210 cells. The L1210 cells (12×10^6 cells ml^{-1}) were diluted with fresh medium plus serum to a concentration of 4×10^5 cells ml^{-1} in the presence (○——○) or absence (●——●) of 10 mM putrescine (Taken from Heller, Fong and Canellakis, 1976)

The evidence to date indicates that the synthesis of ODC antizyme is inhibited by cycloheximide, but only partially by actinomycin D. It has recently been postulated that ODC antizyme is normally attached to subcellular components (mainly the nucleus and ribosomes) and that polyamines at millimolar

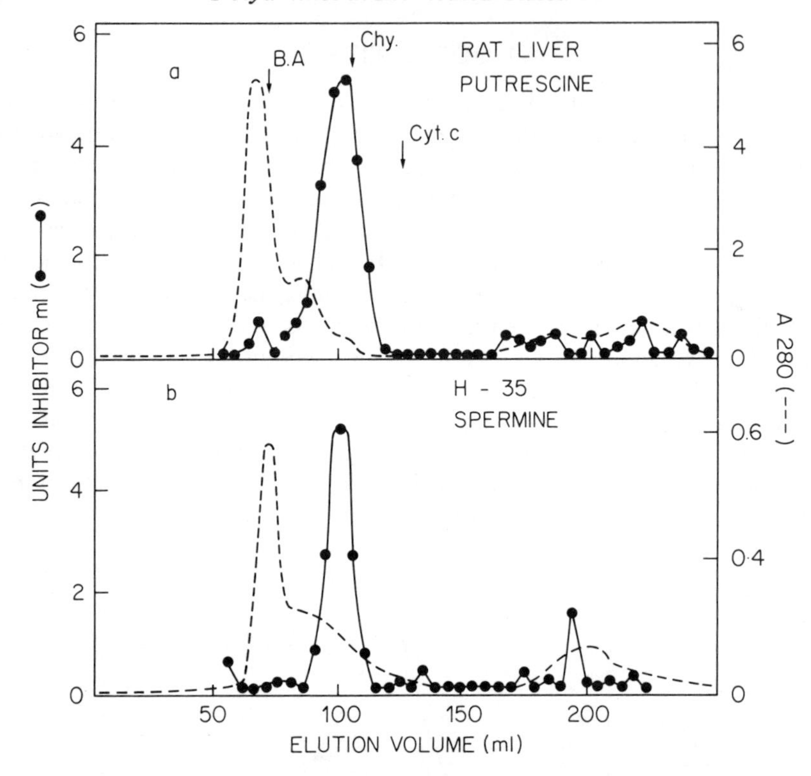

Figure 7 Sephadex G-75 chromatography of ODC inhibitor (a). Antizyme was induced by putrescine in rat liver and separated by gel filtration using 50 mM Tris buffer, pH 7.8, containing 50 mM KCl, 0.1 mM EDTA, and 5 mM mercaptoethanol. (b) Antizyme was induced by 5 mM spermine in H-35 cells and fractioned as in (a). Molecular weight standards used were: B.A.; bovine serum albumin (67,500); Chy; chymotrypsinogen A (25,000) and Cyt. c. cytochrome C (12,400). (Taken from Heller, Fong and Canellakis, 1976)

concentrations cause its release (Heller and coworkers, 1977). Polyamines may thereby regulate ODC activity and control additional biosynthesis. It has also been suggested that ODC antizyme binds to its template, acting as a cytoplasmic repressor (Heller and coworkers, 1977).

It is most likely that the inhibitory effect of 1,3-diaminopropane on ODC induction (Kallio and coworkers, 1977) is due to antizyme formation.

A. Properties

The molecular weight of ODC antizyme has been determined mainly by gel filtration using G75 or G100 Sephadex columns. The inhibitor is heat-labile and is sensitive to proteases, including chymotrypsin or trypsin (Heller, Fong and Canellakis, 1976).

It appears that ODC antizyme has an extremely short half-life, ranging from 11 to 24 min. This value was obtained by treating rats with cycloheximide, after first injecting polyamines, and assaying ODC antizyme levels in the liver. Similar results were obtained when cultures of L1210, H-35 or neuroblastoma cells were treated with polyamines and cycloheximide (Fong, Heller and Canellakis, 1976; Heller, Fong and Canellakis, 1976; Canellakis and coworkers, 1978).

Very recently, Canellakis and coworkers (1978) purified ODC antizyme from rat liver 120-fold, using column chromatography on DEAE-Sephadex and various forms of Sephadex. The final product was heterogenous as determined by acrylamide gel electrofocusing. It is of special interest that a similar heterogeneity was also obtained with polyacrylamide gel electrophoresis, in analogy to the heterogeneity of ODC after polyacrylamide gel electrophoresis (Oberlander and Prouty, 1977a). These findings led to the assumption (Canellakis and coworkers, 1978) that ODC antizyme exists in multiple forms, and that each form neutralizes the corresponding form of ODC. This suggestion is supported by the finding that the 120-fold purified ODC antizyme did not neutralize all the ODC extracted from cells while *in vivo*, putrescine induced the formation of ODC antizyme, which completely neutralized cellular ODC (Canellakis and coworkers, 1978).

Antizyme is not the only agent that controls cellular ODC activity. A compound similar to ODC antizyme, has previously been described by Icekson and Kaye (1976), who isolated an ornithine decarboxylase inactivating factor

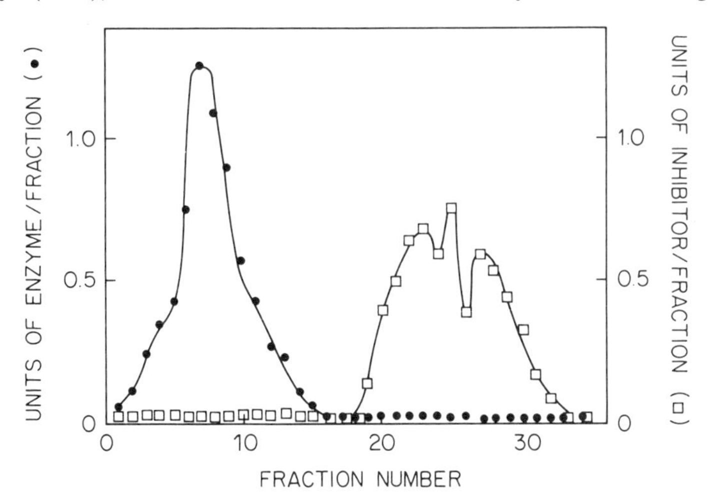

Figure 8 Chromatography of ammonium sulphate treated ODC-'ODC antizyme' complex. Approximately 9 units of ODC was mixed with an equivalent amount of antizyme in 0.5 ml buffer and 0.1 ml of saturated ammonium sulphate. The mixture was applied to a Sephadex G-75 superfine column. ODC activity (● —— ●); antizyme activity (□ —— □). (Taken from Heller, Fong and Canellakis, 1976)

(ODIF) from rat ventral prostate. This factor was active against ODC, but had no effect on other cellular enzymes tested. Pyridoxal phosphate protected ODC against inactivation by ODIF and this factor was separated from ODC by precipitation at pH 4.6. It has been suggested that ODIF did not contain proteolytic activity and that the inactivation of ODC by this factor was not simply due to proteolytic digestion. It remains to be seen whether ODIF is identical with antizyme or whether it represents a new group of regulatory agents.

VII. REGULATION OF ODC ACTIVITY

As ODC plays a central role in regulating cellular growth, its synthesis and activity should be controlled rigorously (see also Chapters 6–8). One important factor in this control is the decay of enzyme activity. This process, which is relatively fast, is poorly understood as yet, and apparently varies during the growth cycle and as the result of neoplastic transformation (see above). An additional important regulatory factor is ODC antizyme, which is activated (or synthesized) when cellular polyamine levels reach a definite threshold level. The function of this inhibitor is to neutralize cellular ODC rather than to inhibit its synthesis. The synthesis of ODC, on the other hand, is also subject to regulation by polyamines.

It is most likely that the major (but not exclusive) process of ODC induction involves the following steps:

(a) binding of hormone to membrane receptors;
(b) activation of adenylate cyclase and formation of cyclic AMP;
(c) activation of cyclic AMP-dependent protein kinase;
(d) phosphorylation (of an inhibitory protein?).

It is conceivable that ODC induction can be inhibited by interfering with each of these steps. Canellakis and coworkers (1978) have recently suggested that when the cellular level of polyamines increases, some of it interacts with specific sites on the membrane. It has been postulated that tumour cells, which are rich in polyamines, have lost some of these receptor sites. If this suggestion is correct, then the binding of polyamines to the membrane might have a dual effect: (a) induce antizyme formation; (b) alter the charge of specific sites on the membranes and possibly interfere with the binding of the hormone or of a stimulatory protein (Morley, 1972).

The importance of membrane integrity in controlling ODC activity has also been stressed by Friedman and coworkers (1977), who reported that the hypotonicity of the incubation medium stimulated thyroid ODC activity. It has been proposed that the activation and the effect of cations (Chen, Heller and Canellakis, 1976b) is related to changes in cell membrane conformation.

Colchicine and vinblastine, in micromolar concentrations, were found to inhibit the activity of ODC of mouse leukaemia L1210 cells (Chen, Heller and

Canellakis, 1976a). The inhibition has also been attributed to membrane effects and it has been suggested that these agents cause the disruption of the microtubule system and thereby abolish the information mediated by the cell surface. Colchicine also inhibited the induction of ODC when injected 11-12 h before treatment with tumour promoters (O'Brien, Simsiman and Boutwell, 1976).

Adenylate cyclase is also inhibited by colchicine, but the effective concentrations were 100 times higher than those which inhibited ODC activity. It is therefore unlikely that the inhibition of ODC by colchicine is primarily due to the inhibition of adenylate cyclase and other related membrane enzymes.

Step (b), namely activation of adenylate cyclase can definitely be regulated by polyamines. Polyamines were shown to inhibit the formation of cyclic AMP in *Physarum polycephalum* (Atmar and coworkers, 1976), myocardial cell cultures (Clô and coworkers, 1977) and human cultured fibroblasts (Wright, Buehler and Rennert, 1976).

Polyamines also inhibit cyclic AMP-dependent protein kinase (step (c), above). Thus the protein kinase from mouse epidermis (Murray, Froscio and Rogers, 1976), silkworm (Takai and coworkers, 1976), cultured glioma cells (Bachrach, Katz and Hochman, 1978) and rat liver (Hochman, Katz and Bachrach, 1978a) were all inhibited by polyamines at millimolar concentrations. It should be noted that spermine was found to interact with the catalytic and not the regulatory subunit of the enzyme (Hochman, Katz and Bachrach, 1978a), thus exhibiting a selective binding which is rather unusual for an organic cation like spermine.

A schematic representation of the various possible regulatory steps for putrescine synthesis is given in Figure 9.

VIII. ODC AND NEOPLASTIC GROWTH

The activity of ODC has been shown to be markedly elevated in tumour cells (see Bachrach, 1976a; Chapter 15). In addition, a close correlation between the level of ODC and the growth rate has been demonstrated in a series of Morris rat hepatomas (Williams-Ashman, Coppoc and Weber, 1972).

To further study the differences in ODC activity in normal and neoplastic cells, the following experimental approaches have been employed: (a) transformation by oncogenic viruses; (b) tumour promotion in carcinogen-initiated mouse skin; (tumour induction in animals maintained on a carcinogenic diet.

A. Transformation by Oncogenic Viruses

A direct evidence for the correlation of ODC activity and neoplastic growth has been obtained by comparing enzyme activity of normal and virus-transformed cells. Lembach (1974) used normal mouse 3T3 fibroblasts and the SV40-transformed cell-line SV101 and found that under certain experimental conditions, the transformed cells maintained ODC levels approximately 200-fold higher than those found in the untransformed cells. More recently, ODC has been purified from both 3T3 mouse fibroblasts and SV40-transformed cultures. It has been demonstrated (Boucek and Lembach, 1977) that the enzyme from transformed fibroblasts exhibited a significant higher specific activity than that purified from rat liver by Ono and coworkers (1972). This difference may be due either to the transformation process, or alternatively, to an intrinsic difference in ODC from cultured cells as compared with tissues. It would be most interesting to compare the specific activity of ODC of normal and transformed cultured cells or normal and neoplastic tissues.

Heby, Goldstein and Marton (1978) studied the rate of synthesis and the concentrations of polyamines in primary cultures of mouse kidney cells infected with a polyoma virus. They showed that the infection with this oncogenic virus induced a biphasic production of polyamines. The major production occurred during the virus-induced synthesis of host cell DNA.

A different system was used by Bachrach, Don and Wiener (1974), who studied polyamine and ODC levels in normal and Rous sarcoma virus (RSV)-transformed chick embryo fibroblasts. It has been shown that elevations in polyamine levels and ODC activities were closely related to the transformation

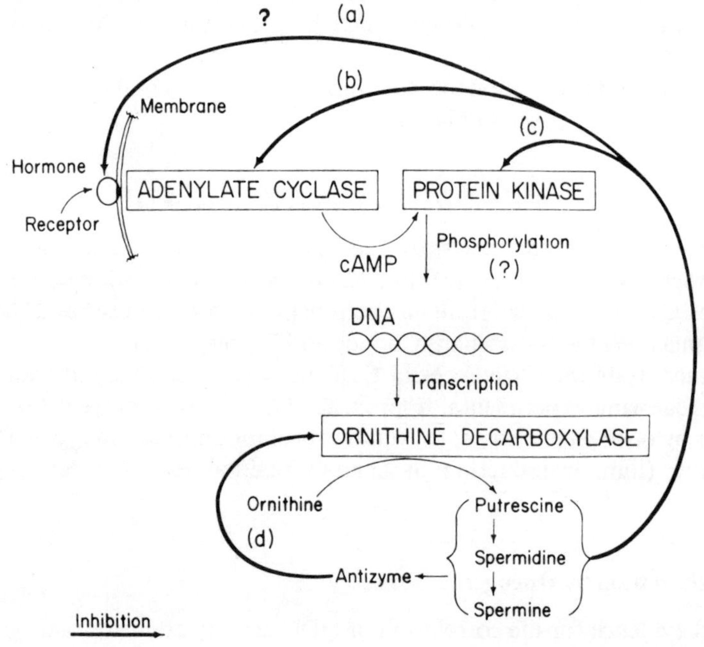

Figure 9 A schematic representation of the regulation of putrescine biosynthesis

process. Thus, infection with unrelated viruses or with a temperature-sensitive mutant of RSV under restrictive conditions did not induce an increase in ODC activity (Don, Wiener and Bachrach, 1975; Don and Bachrach, 1975). The increase in ODC activity during the transformative process appears to be a very early event and is expressed approximately 20 min after shifting the temperature of RSV-infected culture from restrictive to the permissive temperature (Bachrach, 1978). It thus appears that the increase in ODC activity is an early marker for the expression of malignancy.

ODC activity was also elevated during the infection of BALB/3T3 cells with murine sarcoma virus (MSV). The rise in ODC was due to the transforming virus component of the MSV complex, as infection with a non-transforming strain of murine leukaemia virus (MuLV) had no effect on ODC activity (Gazdan and coworkers, 1976).

B. Tumour Promotion

Tumours can be efficiently produced in mouse skin by treatment with a subcarcinogenic dose of a compound such as dimethylbenzanthracene, followed by application of a promoting agent (O'Brien, Simsiman and Boutwell, 1975; Chapter 15). These findings led to the conclusion that the induction of ODC is an early, possibly obligatory event in mouse skin carcinogenisis (O'Brien, 1976). Canellakis and coworkers (1978) came to a similar conclusion. Attempts to demonstrate the involvement of cyclic AMP, cyclic GMP and the β-agonist isoproterenol in the induction of ODC by the tumour promoters, gave negative results (Mufson and coworkers, 1977) and the role of cyclic nucleotides in the induction of ODC during skin tumour promotion has been disputed (Verma and Boutwell, 1977). On the other hand, recent studies by Verma, Rice and Boutwell (1977) suggested that prostaglandins may play a role in the induction of ODC. This suggestion is based on the finding that compounds which inhibit prostaglandin synthesis (such as indomethacin and acetylsalicyclic acid) prevented the induction of epidermal ODC *in vivo*.

The induction of epidermal ODC in mice after treatment with a tumour promoter was significantly decreased in mice maintained on a Vitamin B_6-deficient diet (Murray and Froscio, 1977). These results stressed again the importance of pyridoxal phosphate as a co-factor of ODC.

C. Tumour Induction in Animals Maintained on Carcinogenic Diet

Recent studies by Scalabrino and coworkers (1978) demonstrated that ODC activity rose significantly in the livers of rats fed 4-dimethylaminoazobenzene. This increase in ODC activity seemed to be specific and the levels of tyrosine aminotransferase were not affected by the carcinogen.

IX. ORNITHINE DECARBOXYLASE AND RNA POLYMERASE

It has been suggested by Manen and Russell (1977) that ODC does not act exclusively as a catalyst in putrescine synthesis but also serves as an initiation factor for RNA polymerase I. It has been known for a long time that the control of RNA polymerase I (E.C. 2.7.7.6) activity, is not through increased synthesis of the enzyme, but rather through a modification of the enzyme structure, apparently by the presence of labile protein. Manen and Russell (1977) suggested that ODC has the properties of such a labile protein and thus might regulate the activity of RNA polymerase I. This was shown indirectly (Manen and Russell, 1975) by injecting methyl-isobutyl xanthine (a phosphodiesterase inhibitor) into rats and comparing the rise and fall of ODC and RNA polymerase I activities. These studies demonstrated that the activities of both enzymes followed a similar time course and that they declined with similar rates after cycloheximide administration. In addition, the activity of RNA polymerase I (isolated from rat liver nuclear preparations) increased by the addition of a 40-fold purified preparation of ODC (Manen and Russell, 1975). In a recent report, Manen and Russell (1977) described the purification of ODC more than a thousand-fold by coupling RNA polymerase I to Sepharose columns. It has been implied from these experiments that RNA polymerase I and ODC interact physically and it has been postulated that ODC increases the initiation of rate of RNA polymerase I, being the labile protein which regulates the activity of RNA polymerase I.

X. SUMMARY AND CONCLUSIONS

ODC is the rate-limiting enzyme in the biosynthesis of polyamines. The activity of this enzyme is high in rapidly growing organs and in tumour cells and its activity also rises after administering hormones to various animals. In tissue cultures, serum, cyclic AMP, glutamine or asparagine also cause the activation of this enzyme.

ODC has the shortest half-life of any known enzyme in eukaryotes. The half-life of the enzyme, however, varies during the cell cycle, being maximal when the growth rate increases. A similar increase in the half-life of the enzyme is also noticed during cell transformations by oncogenic viruses or after applying tumour promoting agents to mouse skin.

ODC has a molecular weight of approximately 55,000 and is inactivated by ornithine decarboxylase antizyme, which has a molecular weight of 25,000 daltons. Various ornithine analogues, such as α-methyl ornithine inhibit the enzyme, apparently by interacting with its active site.

There is indirect evidence that ODC also regulates the activity of RNA polymerase I, serving as an initiation factor.

ODC plays a central role in differentiation and in regulating growth. Its synthesis is controlled by various mechanisms, including inhibition of adenylate cyclase and protein kinase by polyamines, which are the end-products of the reaction catalysed by ODC. Fluctuations in the half-life of enzyme and the induction (or activation) of ODC antizyme also cause changes in ODC activity.

REFERENCES

Abdel-Monem, M. M., Newton, N. E. and Weeks, C. E. (1974). *J. Med. Chem.*, **17**, 447–457.

Aisbitt, R. P. G. and Barry, J. M. (1973). *Biochim. Biophys. Acta*, **320**, 610–616.

Anderson, T. R. and Schanberg, S. M. (1972). *J. Neurochem.*, **19**, 1471–1481.

Anderson, T. R. and Schanberg, S. M. (1975). *Biochem. Pharmacol.*, **24**, 495–501.

Atmar, V. J., Westland, J. A., Gracia, G. and Kuehn, G. D. (1976). *Biochem. Biophys. Res. Commun.*, **68**, 561–568.

Bachrach, U. (1973). In *Function of Naturally Occurring Polyamines*, Academic Press, New York and London.

Bachrach, U. (1975). *Proc. Nat. Acad. Sci. USA*, **72**, 3087–3091.

Bachrach, U. (1976a). *Ital. J. Biochem.*, **25**, 77–93.

Bachrach, U. (1976b). *Biochem. Biophys. Res. Commun.*, **72**, 1008–1013.

Bachrach, U. (1978). Polyamine synthesis in normal and neoplastic cells. In R. A. Campbell, D. R. Morris, D. Bartos, G. D. Daves and F. Bartos (Eds), *Advances in Polyamine Research*, Vol. 1, Raven Press, New York, pp. 83–91.

Bachrach, U., Don, S. and Wiener, H. (1974). *Cancer Res.* **34**, 1577–1580.

Bachrach, U., Katz, A. and Hochman, J. (1978). *Life Sci.*, **22**, 817–822.

Ball, W. J. Jr., and Balis, M. E. (1976). *Cancer Res.*, **36**, 3312–3316.

Bartolome, J., Lau, C. and Slotkin, T. A. (1977). *J. Pharmacol.*, **202**, 510–518.

Beck, W. T., Bellantone, R. A. and Canellakis, E. S. (1972). *Biochem. Biophys. Res. Commun.*, **48**, 1649–1655.

Ben-Hur, E., Prager, A. and Riklis, E. (1978). *Proc. Israel Biochem. Soc.*, p.181.

Boucek, R. J. Jr., and Lembach, K. (1977). *Arch. Biochem. Biophys.*, **184**, 408–415.

Brandt, J. T., Pierce, D. A. and Fausto, N. (1972). *Biochim. Biophys. Acta*, **279**, 184–193.

Butler, S. R. and Schanberg, S. M. (1975). *Biochem. Pharmacol.*, **24**, 1915–1918.

Butler, S. R., Suskind, M. R. and Schanberg, S. M. (1978). *Science*, **199**, 445–447.

Byus, C. V., Costa, M., Sipes, I. G., Brodie, B. B. and Russell, D. H. (1976). *Proc. Nat. Acad. Sci. USA*, **73**, 1241–1245.

Byus, C. V. and Herbst, E. J. (1976). *Biochem. J.*, **154**, 31–33.

Byus, C. V. and Russell, D. H. (1974). *Life Sci.*, **15**, 1991–1997.

Byus, C. V. and Russell, D. H. (1975). *Science*, **187**, 650–652.

Canellakis, E. S., Heller, J. S., Kyriakidis, D. and Chen, K. Y. (1978). 'Intracellular levels of ornithine decarboxylase, its half-life, and a hypothesis relating polyamine-sensitive membrane receptors to growth' In *Advances In Polyamine Research, Vol. I* (Eds R. A. Campbell, D. R. Morris, D. Bartos, G. D. Daves and F. Bartos), pp. 17–30, Raven Press, New York.

Canellakis, Z. N. and Theoharides, T. C. (1976). *J. Biol. Chem.*, **251**, 4436–4411.

Cavia, E. and Webb, T. E. (1972). *Biochim. Biophys. Acta*, **262**, 546-554.

Chen, K. Y. and Canellakis, E. S. (1977). *Proc. Nat. Acad. Sci. USA*, **74**, 3791–3795.

Chen, K., Heller, J. and Canellakis, E. S. (1976a). *Biochem. Biophys. Res. Commun.*, **68**, 401–408.

Chen, K., Heller, J. S. and Canellakis, E. S. (1976b). *Biochem. Biophys. Res. Commun.*, **70**, 212–220.

Clark, J. L. (1974). *Biochemistry*, **13**, 4668–4674.

Clark, J. L. and Fuller, J. L. (1976). *Eur. J. Biochem.*, **67**, 303–314.

Clô, C., Orlandini, G., Guarnieri, C. and Caldarera, C. M. (1977). *Bull. Mol. Biol. Med.*, **2**, 48–57.

Cohen, S. S. (1971). In *Introduction to the Polyamines*, Prentice Hall, Englewood Cliffs, N. J.

Cohen, S., O'Malley, B. W. and Stastny, M. (1970). *Science*, **170**, 336–338.

DeMello, F. G., Bachrach, U. and Nirenberg, M. (1976). *J. Neurochem.* **27**, 847–851.

Don, S., and Bachrach, U. (1975). *Cancer Res.*, **35**, 3618–3622.

Don, S., Wiener, H. and Bachrach, U. (1975). *Cancer Res.*, **35**, 194–198.

Eloranta, T. and Raina, A. (1975). *FEBS Lett.*, **55**, 22–24.

Fausto, N. (1969). *Biochim. Biophys. Acta*, **190**, 193–201.

Feldman, M. J. and Russell, D. H. (1972). *Am. J. Physiol.*, **222**, 1199–1203.

Fenioli, M. E., Ceruti, G. and Camoli, R. (1976). *Exp. Geront.*, **11**, 153–156.

Fong, W. F., Heller, J. S. and Canellakis, E. S. (1976). *Biochim. Biophys. Acta*, **428**, 456–465.

Friedman, S. J., Halpern, K. V. and Canellakis, E. S. (1972). *Biochim. Biophys. Acta*, **261**, 181–187.

Freidman, Y., Parks, S., Levasseur, S., and Burke, G. (1977). *Biochem. Biophys. Res. Commun.*, **77**, 57–64.

Gaza, D. J., Short, J. and Lieberman, I. (1973). *FEBS Lett.*, **32**, 251–253.

Gazdar, A. F., Stull, H. B., Kilton, L. J. and Bachrach, U. (1976). *Nature, Lond.*, **262**, 696–698.

Grahn, B., Henningsson, S. G., Kahlson, G. and Rosengren, E. (1973). *Brit. J. Pharmacol.*, **48**, 113–120.

Guha, S. K. and Jänne, J. (1976). *Acta Endocrinol.*, **81**, 793–800.

Harik, S. I., Hollenberg, M. D. and Snyder, S. H. (1974a). *Nature, Lond.*, **249**, 250–251.

Harik, S. I., Hollenberg, M. D. and Snyder, S. H. (1974b). *Mol. Pharmacol.*, **10**, 41–47.

Harik, S. I. and Snyder, S. H. (1973). *Biochim. Biophys. Acta*, **327**, 501–509.

Haselbacher, G. K. and Humbel, R. E. (1976). *J. Cell. Physiol.*, **88**, 239–246.

Heby, O., Goldstein, D. A. and Marton, L. J. (1978). 'Regulation of cellular polyamine synthesis during lytic infection with polyoma virus' Superinduction with actinomycin D. In R. A. Campbell, D. R. Morris, D. Bartos, G. D. Dave and F. Bartos (Eds) *Advances in Polyamine Research Vol. I.* Raven Press, New York, pp. 133–152.

Heby, O., Marton, L. J., Wilson, C. B. and Martinez, H. M. (1975). *J. Cell. Physiol.*, **86**, 511–521.

Heby, O. and Russell, D. H. (1973). *Cancer Res.*, **33**, 159–165.

Heller, J. S., Canellakis, E. S., Bussolotti, D. L. and Coward, J. K. (1975). *Biochim. Biophys. Acta*, **403**, 197–207.

Heller, J. S., Fong, W. F. and Canellakis, E. S. (1976). *Proc. Nat. Acad. Sci.*, **73**, 1858–1862.

Heller, J. S., Kyriakidis, D., Fong, W. F. and Canellakis, E. S. (1977). *Eur. J. Biochem.*, **81**, 545–550.

Hochman, J., Katz, A. and Bachrach, U. (1978a). *Life Sci.*, **22**, 1481–1482.

Hochman, J., Plotek, Y. and Bachrach, U. (1978b). *Proc. Israel Biochem. Soc.*, p.102.

Hogan, B. L. M. and Murden, S. (1974). *J. Cell. Physiol.*, **83**, 345–352.

Hogan, B., Shields, R. and Curtis, D. (1974). *Cell*, **2**, 229-233.
Hölttä, E. (1975). *Biochim. Biophys. Acta*, **399**, 420-427.
Hölttä, E. and Jänne, J. (1972). *FEBS Lett.*, **23**, 117-121.
Hölttä, E. and Raina, A. (1973). *Acta Endocrinol.*, **73**, 794-800.
Icekson, I. and Kaye, A. M. (1976). *FEBS Lett.*, **61**, 54-58.
Inoue, H., Kato, Y., Takigawa, M., Adachi, K. and Takeda, Y. (1975). *J. Biochem.*, **77**, 879-893.
Inoue, H., Tanioka, H., Shiba, K., Asada, A., Kato, Y. and Takeda, Y. (1974). *J. Biochem.*, **75**, 679-687.
Insel, P. A. and Fenno, J. (1978). *Proc. Nat. Acad. Sci.*, **75**, 862-865.
Jänne, J. and Raina, A. (1968). *Acta Chem. Scand.*, **22**, 1349-1351.
Jänne, J., Raina, A. and Siimes, M. (1968). *Biochim. Biophys. Acta*, **166**, 419-426.
Jänne, J. and Williams-Ashman, H. G. (1971). *J. Biol. Chem.*, **246**, 1725-1732.
Jungman, R. A. and Russell, D. H. (1977). *Life Sci.*, **20**, 1787-1798.
Kallio, A., Löfman, M., Pösö, H. and Jänne, J. (1977). *FEBS Lett.*, **79**, 195-199.
Kay, J. E. and Lindsay, V. J. (1973). *Biochem. J.*, **132**, 791-796.
Kaye, A. M., Icekson, I., Lamprecht, S. A., Gruss, R., Tsafriri, A. and Lindner, H. R. (1973). *Biochem.*, **12**, 3072-3076.
Kobayashi, Y., Kupelian, J. and Maudsley, D. V. (1971). *Science*, **172**, 379-380.
Lamprech, S. A., Zor, U., Tsafriri, A. and Lindner, H. R. (1973). *J. Endocr.*, **57**, 217-233.
Lembach, K. L. (1974). *Biochim. Biophys. Acta*, **354**, 88-100.
Levine, J. H., Leaming, A. B. and Raskin, P. (1978). On the mechanism of activation of hepatic ornithine decarboxylase by cyclic AMP. In R. A. Campbell, D. R. Morris, D. Bartos, G. D. Daves and F. Bartos (Eds), *Advances in Polyamine Research*, Vol. 1, Raven Press, New York, pp. 51-58.
Levine, J. H., Nicholson, W. E., Liddle, G. W. and Orth, D. N. (1973). *Endocrinology*, **92**, 1089-1095.
Levine, J. H., Nicholson, W. E., Peytremann, A. and Orth, D. N. (1975). *Endocrinology*, **97**, 136-144.
MacDonnell, P. C., Nagaiah, K., Lakshmanan, J. and Guroff, S. (1977). *Proc. Nat. Acad. Sci., USA*, **74**, 4681-4684.
Majumder, G. C., MacIndoe, J. H. and Turkington, R. W. (1974). *Life Sci.*, **15**, 45-55.
Mallette, L. E., and Exton, J. H. (1973). *Endocrinology*, **93**, 640-644.
Mamont, P. S., Böhlen, P., McCann, P. P., Bey, P., Schuber, F. and Tardif, C. (1976). *Proc. Nat. Acad. Sci., USA*, **73**, 1626-1630.
Manen, C. A. and Russell, D. H. (1975). *Life Sci.*, **17**, 1769-1776.
Manen, C. A. and Russell, D. H. (1977). *Science*, **195**, 505-506.
Matsushita, S., Sogani, R. K., and Raben, M. S. (1972). *Circ. Res.*, **31**, 699-709.
Matsuzaki, S. and Suzuki, M. (1974). *Endocr. Japonica*, **21**, 529-537.
Maudsley, D. V. and Kobayashi, Y. (1974). *Biochem. Pharmacol.*, **23**, 2697-2703.
Maudsley, D. V. and Kobayashi, Y. (1977). *Biochem. Pharmacol.*, **26**, 121-124.
Maudsley, D. V., Leif, J. and Kobayashi, Y. (1976). *Am. J. Physiol.*, **231**, 1557-1561.
McCann, P. P., Tardif, C., Duchesne, M. C. and Mamont, P. S. (1977). *Biochem. Biophys. Res. Commun.*, **76**, 893-899.
McIlhinney, A., and Hogan, B. L. M. (1974). *Biochim. Biophys. Acta*, **372**, 358-365.
Mitchell, J. L. A., Anderson, S. N., Carter, D. D., Sedory, M. J., Scott, J. F. and Varland, D. A. (1978). Role of modified forms of ornithine decarboxylase in the control of polyamine synthesis. In R. A. Campbell, D. R. Morris, D. Bartos, G. D.

Daves and F. Bartos (Eds), *Advances in Polyamine Research*, Vol. I, Raven Press, New York, pp. 39-50.
Mizoguchi, Y., Otani, S., Matsui, I. and Morisawa, S. (1975). *Biochem. Biophys. Res. Commun.*, **66**, 328-335.
Morley, C. G. D. (1972). *Biochem. Biophys. Res. Commun.*, **49**, 1530-1535.
Morris, D. R. and Fillingame, R. H. (1974). *Ann. Rev. Biochem.*, **43**, 303-325.
Mufson, R. A., Astrup, E. G., Simsiman, R. C. and Boutwell, R. K. (1977). *Proc. Nat. Acad. Sci. USA*, **74**, 657-661.
Murray, A. W. and Froscio, M. (1977). *Biochem. Biophys. Res. Commun.*, **77**, 693-699.
Murray, A. W., Froscio, M. and Rogers, A. (1976). *Biochem. Biophys. Res. Commun.*, **71**, 1175-1681.
Nicholson, W. E., Levine, J. H. and Orth, D. N. (1976). *Endocrinology*, **98**, 123-128.
Nissley, P. S., Passamani, J. and Short, P. (1977). *J. Cell. Physiol.*, **89**, 392-402.
Obenrader, M. F. and Prouty, W. F. (1977a). *J. Biol. Chem.*, **252**, 2860-2865.
Obenrader, M. F. and Prouty, W. F. (1977b). *J. Biol. Chem.*, **252**, 2866-2872.
O'Brien, T. G. (1976). *Cancer Res.*, **36**, 2644-2653.
O'Brien, T. G., Simsiman, R. C. and Boutwell, R. K. (1975). *Cancer Res.*, **35**, 1662-1670.
O'Brien, T. G., Simsiman, R. C. and Boutwell, R. K. (1976). *Cancer Res.*, **36**, 3766-3770.
Oka, T. and Perry, J. W. (1976). *J. Biol. Chem.*, **251**, 1738-1744.
Ono, M., Inoue, H., Suzuki, F. and Takeda, Y. (1972). *Biochim. Biophys. Acta*, **284**, 285-297.
Panko, W. B. and Kenney, F. T. (1971). *Biochem. Biophys. Res. Commun.*, **43**, 346-350.
Pegg, A. E. and Williams-Ashman, H. G. (1968). *Biochem. J.*, **108**, 533-539.
Preslock, J. P. and Hampton, J. K. Jr. (1973). *Am. J. Physiol.*, **225**, 903-907.
Prouty, W. E. (1976). *J. Cell. Physiol.*, **89**, 65-76.
Raina, A. and Jänne, J. (1968). *Acta Chem. Scand.*, **22**, 2375-2378.
Raina, A. and Jänne, J. (1975). *Med. Biol.*, **53**, 121-147.
Raina, A. Jänne, J., Hannonen, P., and Hölttä, E. (1970). *Ann. N.Y. Acad. Sci.*, **171**, 697-708.
Reddy, P. R. K. and Villee, C. A. (1975). *Biochem. Biophys. Res. Commun.*, **65**, 1350-1354.
Relyea, N. and Rando, R. R. (1976). *Biochem. Biophys. Res. Commun.*, **67**, 392-402.
Richman, R., Dobbins, C., Voina, S., Underwood, L., Mahaffee, D., Gitelman, H. J., Van Wyk, J. and Ney, R. L. (1973). *J. Clin. Invest.*, **52**, 2007-2015.
Richman, R., Park, S., Akbar, M., Yu, S. and Burke, G. (1975). *Endocrinology*, **96**, 1403-1412.
Richman, R. A., Underwood, L. E., Van Wyk, J. J. and Voina, S. J. (1971). *Proc. Soc. Exp. Biol. Med.*, **138**, 880-884.
Roger, L. J., Schanberg, S. M. and Fellows, R. E. (1974). *Endocrinology*, **95**, 904-911.
Russell, D. H. (1972). *Cancer Res.*, **32**, 2459-2469.
Russell, D. H. (Ed) (1973a) In *Polyamines in Normal and Neoplastic Growth*, Raven Press, New York.
Russell, D. H. (1973b). *Life Sci.*, **13**, 1635-1647.
Russell, D. H., Byus, C. V. and Manen, C. A. (1976). *Life Sci.*, **19**, 1297-1306.
Russell, D. H. and McVicker, T. A. (1971). *Biochim. Biophys. Acta*, **244**, 85-93.
Russell, D. H. and Snyder, D. H. (1968). *Proc. Nat. Acad. Sci. USA*, **60**, 1420-1427.
Russell, D. H. and Snyder, S. H. (1969a). *Endocrinology*, **84**, 223-228.

Russell, D. H. and Snyder, D. H. (1969b). *Mol. Pharmacol.*, **5**, 253–262.
Russell, D. H., Snyder, S. H. and Medina, V. J. (1970). *Endocrinology*, **86**, 1414–1419.
Russell, D. H. and Stambrook, P. J. (1975). *Proc. Nat. Acad. Sci., USA*, **72**, 1482–1486.
Saunderson, R. and Heald, P. J. (1974). *J. Reprod. Fert.*, **39**, 141–143.
Sawayama, T., Kinugasa, H. and Nishimura, H. (1976). *Chem. Pharm. Bull.*, **24**, 326–329.
Scalabrino, G. and Ferioli, M. E. (1976). *Endocrinology*, **99**, 1085–1090.
Scalabrino, G., Pösö, H., Hölttä, E., Hannonen, P., Kallio, A., and Jänne, J. (1978). *Int. J. Cancer*, **21**, 239–245.
Schrock, T. R., Oakman, N. J. and Bucher, N. L. R. (1970). *Biochim. Biophys. Acta*, **204**, 564–577.
Slotkin, T. A., Lau, C. and Bartolome, M. (1976). *J. Pharmacol.*, **199**, 141–148.
Sogani, R. K., Matsushita, S., Mueller, J. F. and Raben, M. S. (1972). *Biochim. Biophys. Acta*, **279**, 377–386.
Stastny, M. and Cohen, S. (1972). *Biochim. Biophys. Acta*, **204**, 578–589.
Sturman, J. A. and Kremzner, L. T. (1974). *Life Sci.*, **14**, 977–983.
Tabor, C. W. and Tabor, H. (1976). *Ann. Rev. Biochem.*, **45**, 285–306.
Takai, Y., Nakaya, S., Inoue, M., Kishimoto, A., Nishiyama, K., Yamamura, H. and Nishizuka, Y. (1976). *J. Biol. Chem.*, **251**, 1481–1487.
Thadani, P. V., Lau, C., Slotkin, T. A. and Schanberg, S. M. (1977). *Biochem. Pharmacol.*, **26**, 523–527.
Theoharides, T. C. and Canellakis, Z. N. (1976). *J. Biol. Chem.*, **251**, 1781–1784.
Williams-Ashman, H. G., Coppoc, G. L. and Weber, G. (1972). *Cancer Res.*, **32**, 1924–1932.
Williams-Ashman, H. G., Corti, A. and Tadolini, B. (1976). *Ital. J. Biochem.*, **25**, 5–32.
Wright, R., Buehler, B. A. and Rennert, O. M. (1976). *Ped. Res.*, **10**, 373.
Wright, R. K., Rennert, O. M. and Seiler, N. (1977). *Fed. Proc.*, **36**, 798.
Wyatt, G. R., Rothaus, K., Lawler, D. and Herbst, E. J. (1973). *Biochim. Biophys. Acta*, **304**, 482–494.
Verma, A. K. and Boutwell, R. K. (1977). *Cancer Res.*, **37**, 2196–2201.
Verma, A. K., Rice, H. M. and Boutwell, R. K. (1977). *Biochem. Biophys. Res. Commun.*, **79**, 1160–1166.

Chapter 7

Regulation of Ornithine Decarboxylase in Eukaryotes

PETER P. MCCANN

I. INTRODUCTION

As has been pointed out many times and aptly stated by Jänne, Pösö and Raina (1978) in their recent, comprehensive review, ornithine decarboxylase is a unique mammalian enzyme because of several unusual properties, including its striking inducibility and very short half-life (see also Chapter 6). Other extensive reviews have been written about ornithine decarboxylase (ODC) and its regulation (Morris and Fillingame, 1974), and its relationship to all of polyamine biosynthesis as a whole (Tabor and Tabor, 1976). These, along with other reviews (Cohen, 1971; Bachrach, 1973; Russell, 1973; Raina and Jänne, 1975), have given a comprehensive picture of the 'state of the art'. Indeed, the investigation of mammalian ODC is the most active area of polyamine research (Tabor and Tabor, 1976).

This brief review is not intended to be a comprehensive synopsis of the subject of ornithine decarboxylase, but rather an overview of the various mechanisms for the regulation of mammalian ODC. These are in some cases quite complex. It may be that attempting to describe an overall unified mechanism for ODC regulation is like trying to explain the specific role of polyamines in a cell. In fact, the reason why there are so many mechanisms of ODC regulation is undoubtedly related to the ubiquity of polyamines themselves.

II. INDUCTION AND BIPHASIC EXPRESSION OF ODC

Very rapid induction of ODC was almost simultaneously described by Russell and Snyder (1968) and Jänne and Raina (1968) in the regenerating rat liver where dramatic increases of enzyme activity occurred very shortly after partial hepatectomy. Later many different stimuli were shown to cause early and large increases of ODC in numerous eukaryotic systems, often concurrently with the onset of cell proliferation. They have been extensively discussed by Morris and Fillingame (1974) and Tabor and Tabor (1976). The stimuli range from simple stress of animals by cold, etc., to various complex additions of hormones and growth factors to cells in culture. The work of Hogan (1971), Hogan and Murden (1974a) and Hogan, McIlhinney and Murden (1974b), for example, demonstrated the dependence of the induction of ODC in proliferating rat hepatoma tumour (HTC) cells on amino acids, serum, insulin and cell density. Also experiments in mouse fibroblast (3T3) cells and chick embryo fibroblasts have shown that ODC is increased by factors such as Multiplication Stimulating Activity (MSA) (Nissley, Passamani and Short, 1976) and Non-Suppressible Insulin-Like Activity (NSILA) (Haselbacher and Humbel, 1976) which stimulate cell multiplication and DNA synthesis. Another perhaps related phenomenon is the increase in ODC during viral induced cell transformation. For example, when non-tumorigenic mouse fibroblast (BALB/3T3) cells are infected by murine sarcoma virus, an immediate marked elevation of ODC is seen (Gazdar and coworkers, 1976).

There is evidence that this early induction of ODC may be regulated or mediated by cyclic-AMP (Chapter 6). Canellakis and Theoharides (1976) have shown a correlation between the rise of cAMP and the induction of ODC in HTC cells, while simultaneous fluctuations of cAMP and ODC were seen in Chinese hamster cells (Russell and Stambrook, 1975). *In vivo*, simultaneous rises in cAMP and ODC were noted in rat adrenals after treatment with aminophylline (Byus and Russell, 1975) and a stimulation of ODC was seen in rat liver after treatment with dibutyryl cAMP (Beck, Bellantone and Canellakis, 1972). Indeed, it has been proposed (Russell and coworkers 1976; Byus and coworkers, 1976) that the early cell growth phenomena, i.e. the pleotypic response, and early expression of ODC may be related in that ODC is induced and dependent upon activation of cAMP-dependent protein kinase. Recently, however, studies with perfused normal rat liver treated with dibutyryl cAMP have shown that increases in cAMP levels do not stimulate ODC activity and, therefore, generalizations about the relationship between the two phenomena may be premature (Levine, Leaming and Raskin, 1978). In any case, the early stimulation of ODC can be elicited by numerous factors. This argues for a general mechanism of induction which perhaps could be linked to a cAMP function activated in a growth stimulation situation.

Interestingly enough it was noted in several different systems that the initial rise in ODC was precisely regulated and was expressed as biphasic or double peaks of ODC activity seen during one cell doubling time. This was first described in regenerating rat liver (Gaza, Short and Lieberman, 1973; Hölttä and Jänne, 1972) and in kidney after nephrectomy (Brandt, Pierce and Fausto, 1972). The phenomenon has also been widely noted in cells in culture: in hamster lung fibroblast (Don C) cells (Friedman, Bellantone and Canellakis, 1972), in mouse mammary cells (Oka and Perry, 1976), in mouse fibroblast (3T3) cells (Clark, 1974), in Chinese hamster ovary (CHO) cells (Heby and coworkers, 1976), in mouse kidney cells infected with polyoma virus (Goldstein, Heby and Marton, 1976), in *Physarum polycephalum* (Mitchell and Rusch, 1973; Sedory and Mitchell, 1977), in Ehrlich ascites cells (Harris and coworkers, 1975) and in HTC cells (McCann and coworkers, 1975). It was found in HTC cells (McCann and coworkers, 1975), Don C cells (Friedman, Bellantone and Canellakis, 1972), Ehrlich ascites cells (Harris and coworkers, 1975), CHO cells (Heby and coworkers, 1976) and in polyoma infected kidney cells (Goldstein, Heby and Marton, 1976), that the S phase of DNA synthesis fell just between the two ODC peaks in late G_1 and in G_2 respectively. This could indicate specific requirements for higher levels of putrescine just before DNA synthesis and later during mitosis.

III. CONTROL OF ENZYMATIC ACTIVITY

The question that must be asked is how fluctuations and specific biphasic changes in ODC activity are regulated. Rises in ODC activity require protein synthesis; this is evidenced by the sensitivity of ODC to protein synthesis inhibitors such as cycloheximide and trichodermin, e.g. in regenerating rat liver (Russell and Snyder, 1969; Gaza, Short and Lieberman, 1973; Jänne and Hölttä, 1974) and in several cell culture systems (Hogan and Murden, 1974a; Clark, 1974; Prouty, 1976; Goldstein, Heby and Marton, 1976; McCann, Tardif and Mamont, 1977a). Significantly, an actual decay in immunoreactive ODC protein was seen after cycloheximide treatment in regenerating rat liver (Hölttä, 1975; Kallio and coworkers, 1977; Obenrader and Prouty, 1977b) and in HTC cells (Canellakis and Theoharides, 1976). There is no simple answer, however, to the question at what level — transcriptional, post-transcriptional or post-translational — are the biphasic changes of ODC during cell division regulated?

A. The Role of RNA-synthesis

The data of Gaza, Short and Lieberman (1973) from regenerating rat liver, using actinomycin D, which inhibits DNA-dependent synthesis of RNA, showed that RNA formation was essential for the first of the biphasic peaks of ODC activity. The second increase was dependent upon RNA made during the period of the first rise in activity. The results of Clark (1974) in 3T3 cells using cordycepin, which inhibits the appearance of nascent RNA in the cytoplasm, suggested the same thing. Somewhat contrary results were obtained by Oka and Perry (1976) in mouse mammary cells where experiments with actinomycin D suggested that the first increase in ODC activity was independent of new RNA synthesis or controlled at a post-transcriptional level, whereas the second ODC peak was regulated at both a transcriptional and a post-transcriptional level. Canellakis and Theoharides (1976) also saw differential effects of actinomycin D on ODC in HTC cells depending on the time and method of induction of the enzyme. Other experiments (McCann and coworkers, 1977b) showed that when actinomycin D was added to HTC cells at the peak of initial ODC activity (3 h), the second increase was not observed (Figure 1). When added during the decay of the first peak of ODC activity (5 h), actinomycin D progressively blocked ODC activity, but some activity remained 8 h later. If actinomycin D was added 7 h after dilution, the second

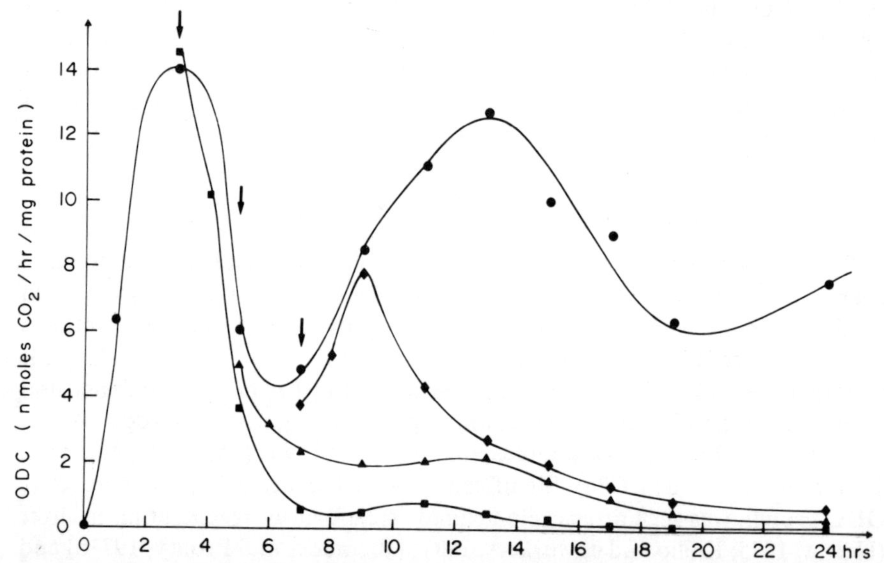

Figure 1 Effects of actinomycin D on enzyme activity in growing HTC cells after induction of cell proliferation by dilution into fresh medium with 10% calf-serum (time zero). Aliquots of cells were assayed for ODC at the indicated times. Enzyme activities are averages from at least three experiments on different cultures. Actinomycin D (1 μg ml^{-1}) was dissolved in phosphate-buffered saline and brought to pH 7.2 before addition to spinner cultures at times indicated by the arrows. Control (●); addition of actinomycin D 3 h after dilution (■); addition of actinomycin D 5 h after dilution (▲); addition of actinomycin D 7 h after dilution (♦)

increase of ODC was abbreviated and largely suppressed. Actinomycin D added at 13 h had no effect. Although stimulation of ODC activity was not entirely dependent on RNA synthesis, synthesis was required to fully achieve the first peak of activity. Similarly, most of the second peak of activity was also dependent on continued *de novo* RNA synthesis contrary to the observations in regenerating rat liver (Gaza, Short and Lieberman, 1973) and in 3T3 cells (Clark, 1974). The decrease of ODC activity to zero when actinomycin D was added at 3 h (Figure 1) suggests that RNA synthesis was normally sufficient at that time to achieve the full expression of ODC activity for the first phase. When actinomycin D was added at 5 or 7 h, RNA synthesis was already turned on again since residual ODC was present for some time afterwards. Thus, it seems there could be a biphasic synthesis of RNA necessary for expression of ODC in HTC cells, indicating that RNA synthetic events play some role in the regulation of ODC during the cell cycle and corroborating findings in other cell systems (Canellakis and Theoharides, 1976; Oka and Perry, 1976). A good deal of evidence, however, does show non-effectiveness or partial effectiveness of RNA inhibitors on the expression of ODC. This was noted in HTC cells (McCann and coworkers, 1977b) as well as in other cells (Kay and Lindsay, 1973; Goldstein, Heby and Marton, 1976) and perhaps argues for a greater importance of a post-transcriptional control of ODC after RNA synthesis.

B. The Role of Polyamines

Although at this point it is difficult to propose a specific mechanism of how RNA synthesis for ODC would be regulated, there is considerable evidence which indicates that one of the key regulatory roles for a post-transcriptional control of ODC is played by the polyamines themselves. Results in human oral carcinoma (KB) cells (Pett and Ginsberg, 1968) and in lymphocytes (Kay and Lindsay, 1973), and also *in vivo* in regenerating rat liver (Schrock, Oakman and Bucher, 1970; Jänne and Hölttä, 1974; Pösö and Jänne, 1976a and b) all demonstrated the rapid post-transcriptional effect of exogenous putrescine, spermidine or spermine in decreasing ODC activity. Further experimental evidence in three different cell culture systems (3T3 cells, Clark and Fuller, 1975; H-35 rat hepatoma cells, Fong, Heller and Canellakis, 1976; HTC cells, McCann, Tardif and Mamont, 1977a) showed that even though putrescine is a weak competitive inhibitor (Morris and Fillingame, 1974) (HTC cell ODC Ki = 2×10^{-3} M; McCann, Tardif and Mamont, 1977a) the low concentrations of putrescine used to block ODC *in situ* in these cells were not sufficient to cause a direct *in vitro* inhibition of ODC activity. Addition of putrescine together with cycloheximide did not increase the rate of ODC inactivation compared to cycloheximide alone, demonstrating that putrescine did not cause an increase in the rate of ODC degradation (Clark and Fuller, 1975; McCann, Tardif and Mamont, 1977a).

A mechanism for the post-transcriptional control of ODC by polyamines was then proposed by Clark and Fuller (1975) who had indirect evidence to show that polyamines specifically blocked new enzyme synthesis in 3T3 cells. The results of Clark and Fuller (1975) were supported by Canellakis and Theoharides (1976), whose work indicated that treatment of HTC cells with low concentrations of putrescine and other polyamines caused disappearance of ODC immunoreactive protein. A similar finding was made by Kallio and coworkers (1977) in regenerating rat liver after injection of putrescine or 1,3 diaminopropane.

C. The Antizyme

Seemingly in opposition to the evidence that polyamines controlled ODC synthesis, was the discovery by J. S. Heller, W. F. Fong and E. S. Canellakis (Heller, Fong and Canellakis, 1976; Fong, Heller and Canellakis, 1976; Chapter 9) of the antizyme which was induced by the presence of putrescine or other polyamines, and which specifically interacted with ODC to form an inactive enzyme-antizyme complex. The ODC antizyme was detected in H-35 rat hepatoma cells, L 1210 mouse leukaemia cells, N-18 mouse neuroblastoma cells and rat liver (Heller, Fong and Canellakis, 1976) and subsequently was confirmed to be present in HTC cells (McCann, Tardif and Mamont, 1977a), rat liver (Jefferson and Pegg, 1977; Pegg, Canover and Wrona, 1978; Pösö, Guha and Jänne, 1978) and rat thyroid (Friedman and coworkers, 1977). The antizyme, however, could not be detected in several other rat tissues, including kidney, prostate and heart, after various polyamine and diamine treatments, even though ODC was totally blocked (Pegg, Canover and Wrona, 1978). Moreover, the decrease of immunoreactive protein seen after polyamine treatment in HTC cells (Canellakis and Theoharides, 1976) and regenerating rat liver (Kallio and coworkers, 1977) still supported the existence of a polyamine ODC regulatory mechanism other than the induction of the antizyme.

Results in HTC cells demonstrated that varying concentrations of putrescine and 1,3 diaminopropane proportionately decreased cellular ODC activity and subsequently induced free antizyme (Figure 2). Even cell extracts which had been treated with 10^{-5} M putrescine, however, contained ODC and antizyme as bound complex, which could be separated by column chromatography with assay buffer containing 250 mM NaCl (McCann, Tardif and Mamont, 1977a). These particular results, therefore, did not rule out that the antizyme was the only polyamine regulatory mechanism for ODC.

Recently, however, HMO_A cells, a clone of HTC cells with greatly stabilized ODC (McCann and coworkers, 1977b), have been used as a probe to investigate more precisely the polyamine regulation of ODC (McCann and coworkers, 1978). These results have shown that there are indeed two distinct polyamine-controlled mechanisms. As shown in Figure 3, the slow partial effect of 10^{-5} M putrescine on HMO_A cell ODC is clearly similar to the effect seen with the general protein synthesis inhibitor cycloheximide and suggests that both are working by a similar mechanism, that is by blocking new ODC synthesis.

When 10^{-2} M putrescine was used to block HMO_A cell ODC, the effect was so rapid as compared to either 10^{-5} M putrescine or cycloheximide, that clearly another mechanism was being brought into play. This was induction of the antizyme. It had previously been shown (Fong, Heller and Canellakis, 1976; Heller, Fong and Canellakis, 1976; Pösö, Guha and Jänne, 1978) that the induction of antizyme was dependent on new protein synthesis and, indeed, when cycloheximide and 10^{-2} M putrescine were added together (Figure 3B) the immediate blocking effect of 10^{-2} M putrescine was greatly suppressed. The fact that cycloheximide did not completely block the 10^{-2} M putrescine effect can be explained by the fact that some preformed membrane-bound

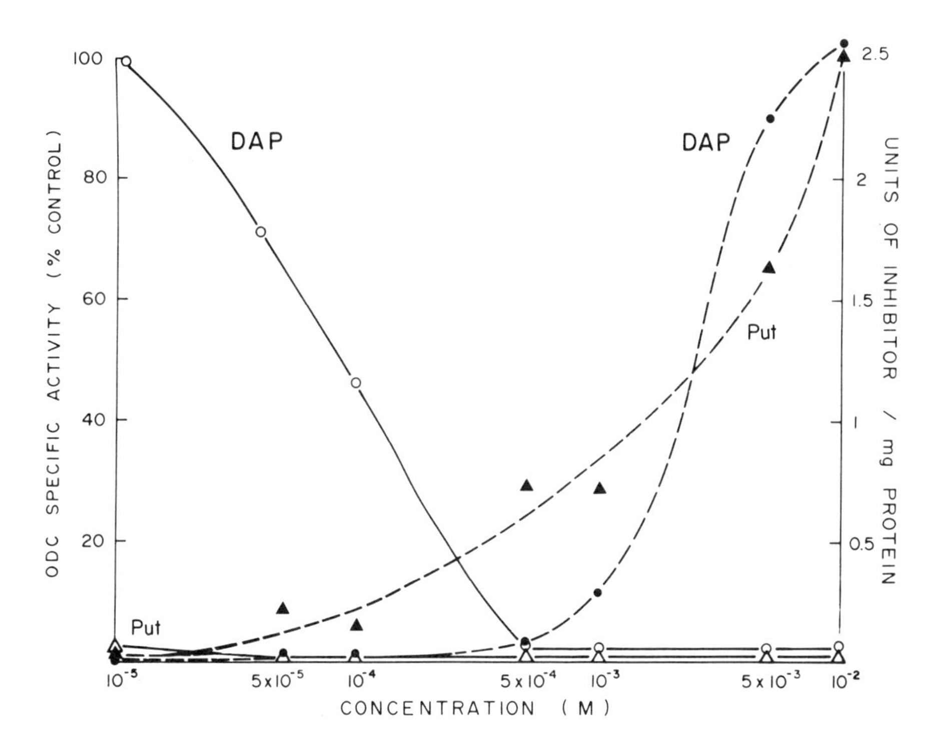

Figure 2 Dose response of varying concentrations of putrescine and 1,3 diaminopropane on cellular ODC and on the induction of a soluble ODC inhibitor, the antizyme. HTC cells were incubated for eight hours after induction of ODC by fresh medium. They were then incubated for an additional five hours in the presence or absence of increasing concentrations of polyamines. Aliquots of cells were taken and extracts were prepared and assayed for ODC. Activities are expressed relative to the specific activity of control cells (5.7 nmoles CO_2 hour^{-1} mg^{-1} protein) grown at the same time. Effect of putrescine △ and 1,3 diaminopropane ○ on ODC activity, and the effect of putrescine ▲ and 1,3 diaminopropane ● on the appearance of ODC antizyme

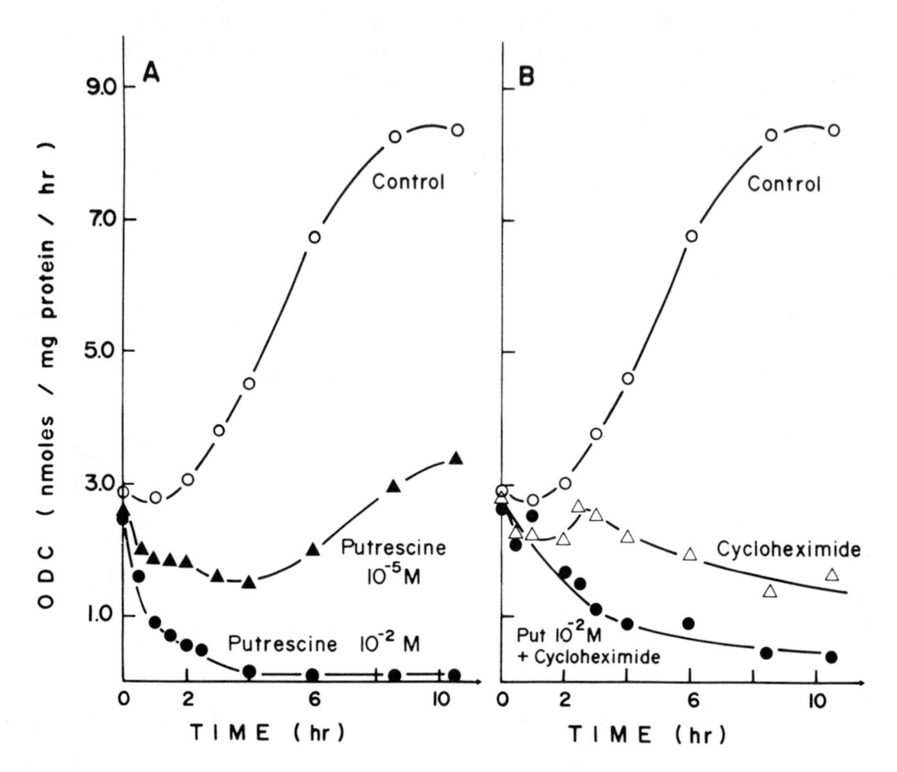

Figure 3 Effect of cycloheximide, putrescine and cycloheximide plus putrescine on ornithine decarboxylase activity in HMO_A cells. The cells were diluted into fresh medium in spinner culture (1.5×10^5 cells ml^{-1}) and incubated for 12 h at 37°C. The cells were then divided into five individual cultures and putrescine and/or cycloheximide were added as below. After three hours, aliquots of cells (2.0×10^6) were taken at the indicated times and washed by centrifugation twice with phosphate-buffered saline at 4°C, sonicated, and then assayed for ODC. Activities are means from at least three different experiments using different cultures. (A): control (○); + 10^{-5} M putrescine (▲); + 10^{-1} M putrescine (●). (B): control (○); + 50 μg/ml^{-1} cycloheximide (△); + 50 μg/ml^{-1} cycloheximide and 10^{-2} M putrescine (●)

antizyme was already present and was released by the high concentration of putrescine *in situ* (Heller and coworkers, 1977; McCann and coworkers, 1979). Thus, it can be argued that the low concentration of putrescine (10^{-5} M) only blocks ODC synthesis and is not sufficient either to induce new antizyme or to release the preformed membrane-bound antizyme from HMO_A cells. In contrast, 10^{-2} M putrescine induces and releases antizyme, which explains why ODC activity rapidly dropped to zero from the partially stable level seen after cycloheximide or 10^{-5} M putrescine treatment (Figure 3).

Further evidence that a low concentration of putrescine blocked ODC synthesis was that in 10^{-5} M putrescine-treated HMO_A cells no antizyme could be detected, unlike in HTC cells. High concentrations of 10^{-2} M putrescine, however, did induce antizyme and formation of the ODC-antizyme complex in the HMO_A clone cells. In striking contrast, similarly treated HTC cells contained only free antizyme and no ODC-antizyme complex (McCann and coworkers, 1979). This certainly suggests that in HTC cells even under conditions of high exogenous putrescine (10^{-2} M), blocking enzyme synthesis was the primary means by which ODC was decreased.

By using the HMO_A cell line, it has been possible to demonstrate that two different mechanisms exist for the polyamine regulation of ODC at a post-transcriptional level. Another recent hypothesis (Canellakis and coworkers, 1978; Heller and coworkers, 1978) has also suggested that intracellular and exogenous polyamines may regulate ODC by different mechanisms. This hypothesis is based on the idea that changes were seen in induced ODC activities in several cell lines after treatments with very low concentrations of exogenous polyamines, even though these concentrations (10^{-5} M) and less) were considerably less than the measured intracellular levels of putrescine, spermidine and spermine. Thus the exogenous amines could perhaps regulate ODC by interaction with special external cell membrane sites (Canellakis and coworkers, 1978; Heller and coworkers, 1978). There was already some immunological data which suggested the existence of membrane putrescine sites in normal rat kidney (NRK) cells (Quash and coworkers, 1978), and there was also other evidence for the presence of membrane-associated sites which could affect ODC activity in L 1210 cells (Chen, Heller and Canellakis, 1976a). Again the effect of some cations such as NA^+, K^+, or MG^{2+} on the induction of ODC in L 1210 cells, albeit only at fairly high concentrations, may be related to an interaction at an ODC membrane site (Chen, Heller and Canellakis, 1976b).

Besides the possible interaction of polyamines at external cell membrane sites, it was proposed (Canellakis and coworkers, 1978; Heller and coworkers, 1978) that the other regulatory mechanism would be the intracellular induction of the ODC antizyme which would respond only to fairly high *in situ* levels of polyamines. The results of McCann, Tardif and Mamont (1977a) and McCann and coworkers (1979) would fit this hypothesis as the evidence shows that polyamines induce antizyme and suggests that they can also block ODC synthesis. There is no reason to suppose that a block of synthesis could not be effected via an interaction of exogenous amines at a specific membrane site.

D. Changes in Ornithine Decarboxylase Turnover

Other models for post-translational control of ODC, apart from complexation with the antizyme, have been suggested by Morris and Fillingame (1974) and by Clark (1974) who pointed out that the turnover of ODC in all eukaryotic systems was rapid and at times changed under several specific conditions. Clark (1974) reported, for example, that ODC half-life was longer in quiescent 3T3 cells than in growing cells. A change in ODC turnover was also seen in HTC cells by Hogan and Murden (1974a) who noted the reverse, that is a decrease in the rate of apparent enzyme degradation when the cells were stimulated to grow. Similar results to Hogan and Murden (1974a) were obtained by Prouty (1976) in HeLa cells. It should be noted that in neither HTC nor HeLa cells, could longer half-lives account for the total observed rises in ODC activity. In an interesting study, Chen and Canellakis (1977) found that enormous changes in ODC half-life can be obtained by manipulating asparagine and glutamine concentrations in the medium of N 18 mouse neuroblastoma cells. They suggested, therefore, that some caution must be exercised in interpreting various measures of ODC half-lives. They also suggested that asparagine may be important *in situ* in regulating ODC activity by stabilizing the enzyme. Munro and coworkers (1975) working with HeLa cells, noted similarly that changes in external osmolality (NaCl) brought about a substantial change in ODC half-life. Again, it was recently found in rat liver treated with thioacetamide which causes necrosis, that the substantial rises seen in ODC activity were due to significant slowing of the rate of ODC turnover (Pösö, Guha and Jänne, 1978). Bachrach (1976) also observed that the elevated rises of ODC seen after cell transformation by rous sarcoma virus were due in part to a decreased rate of ODC degradation.

A relationship between change in ODC turnover and regulation of total ODC activity was also seen in experiments involving specific ODC inhibitors. Mechanistically this concept is interesting because the inhibitors (ornithine analogues) compete with the substrate ornithine to enter the active site of the enzyme. By an unknown mechanism this slows ODC degradation by the degradative enzyme(s) and thus infers an interrelationship between the active site and proteolytic degradation which might be important for the normal regulation of the enzyme in the cell. This phenomenon was first described by Harik, Hollenberg and Snyder (1974), using the competitive inhibitor α-hydrazino-ornithine in HTC cells and rat liver. While putrescine accumulation was effectively blocked, a dramatic rise of ODC activity was measured in cell extracts after the inhibitor was washed out. A similar phenomenon was seen in mouse parotid glands and in mouse sarcoma (Inoue and coworkers, 1975; Kato and coworkers, 1976). Enzyme turnover was considerably slowed in all of these cases. α-Methyl ornithine, another more specific inhibitor of ODC, blocks putrescine accumulation and also proliferation of HTC cells (Mamont and coworkers, 1976). ODC activity measured in washed, inhibitor-free sonicates of HTC cells treated with 5 mM α-methyl ornithine increased to reach a new steady state that was three times above control (Figure 4). When cycloheximide was added to cells cultured in the presence of 5 mM α-methyl

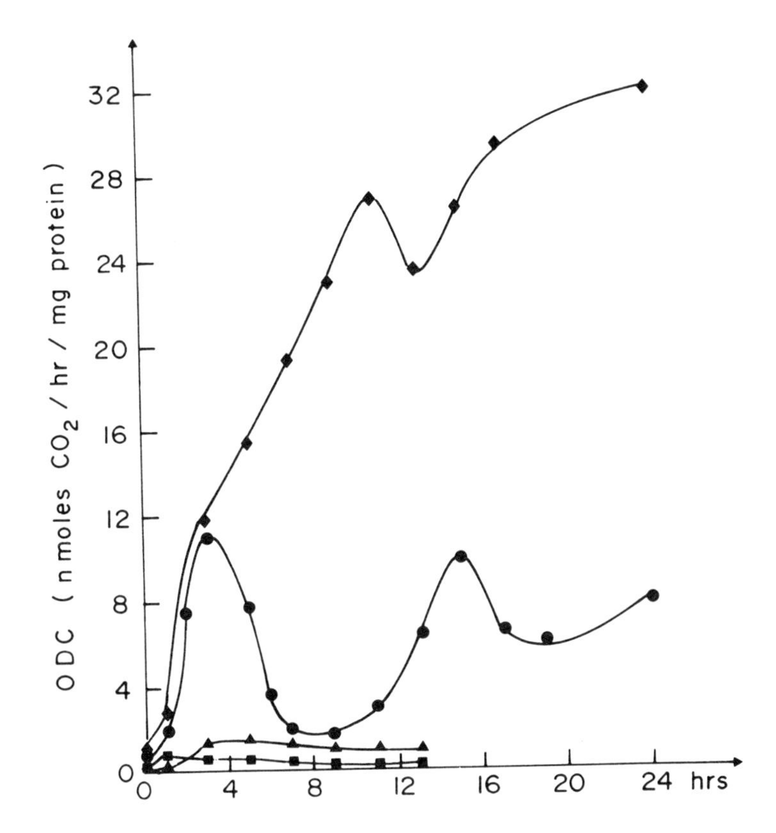

Figure 4 Enzyme activity in growing HTC cells after initiation of cell proliferation by dilution into fresh medium at time zero. Aliquots of cells were taken and assayed for ODC at the indicated times. Activities are averaged from at least three experiments using different cultures. Substances being investigated were dissolved in phosphate-buffered saline and brought to pH 7.2 before addition to spinner cultures. Control (●); addition of 5 mM α-methyl ornithine at time zero (◆); dilution into medium without serum (0.1% bovine serum albumin) (■); dilution into medium without serum but with 5 mM α-methyl ornithine (▼)

ornithine, the half-life of the cellular ODC was 57 min. In the absence of α-methyl ornithine the half-life was 19 min, i.e. three times less (McCann and coworkers, 1977b). Other evidence suggests, however, that part of the total increase of activity seen in Figure 4 was due to an increase in net synthesis of ODC resulting from a drastic depletion of putrescine (McCann and coworkers, 1977b). This again argues for the mechanism, discussed previously, that polyamines can control amounts of ODC in a cell, particularly at the level of synthesis.

E. Active and Non-active Forms of ODC

Another post-translational mechanism for regulation of ODC activity has been described by Mitchell and coworkers (Mitchell and Carter, 1977; Sedory and Mitchell, 1977; Mitchell and coworkers, 1978) in *Physarum polycephalum*. They found two physically and kinetically distinct forms of ODC (described in Chapter 6), and presented evidence that fluctuations of activity during the cell cycle of *Physarum* are the result of transition of the enzyme between the active and the relatively less active form. They also suggest that actual ODC enzyme synthesis is limited to a short period in early to mid-S phase notwithstanding the biphasic pattern of enzyme activity normally seen in this organism (Sedory and Mitchell, 1977).

Obenrader and Prouty (1977a) have detected two forms of ODC in rat liver, but it is unclear if the two forms represent a post-translational regulatory mechanism for ODC activity during the cell cycle. Clark and coworkers (Clark and Fuller, 1976; Fuller, Greenspan and Clark, 1978) detected multiple forms of ODC in 3T3 cells but argued that changes from one form to another could account for only a small part of the total increase of ODC seen in cells stimulated to grow. They also demonstrated by the Dixon analysis that the actual concentration of enzyme molecules increases during this period (Fuller, Greenspan and Clark, 1978).

IV. SUMMARY AND CONCLUSIONS

Obviously a number of factors can influence changes in ODC activity in different types of cells. For regulation of ODC:

(1) RNA synthesis,
(2) translational control by polyamines,
(3) antizyme induction or release by polyamines,
(4) increased or decreased enzyme turnover,
(5) transition between an active and a less active form

are important in one system or another, and, in many cases, simultaneously or consecutively. The overall importance of each mechanism can, and does, vary from cell to cell. One general conclusion is that existence of a tremendous variety of complex ODC regulatory mechanisms reinforces even more the critical importance of this enzyme in its rate-limiting role of putrescine and consequently new polyamine biosynthesis.

In the context of this discussion of the regulation of one enzyme, it must be remembered that the above ODC regulatory mechanisms are interrelated with and dependent upon numerous other factors. For example, one must consider the levels of arginase which control the amount of ornithine synthetized. Also the levels of ornithine-δ-aminotransferase, which convert ornithine to citrulline, may have a direct or indirect effect on ODC. The size of intracellular ornithine pools may be important in the regulation of ODC, although at present there is no real experimental evidence for this. Regarding this point, it must also be remembered that changes in ODC enzymatic activity are not necessarily always reflected by changes in putrescine levels (McCann and coworkers, 1975). As has been pointed out by Maudsley, Leif and King, (1978), with a restricted supply of ornithine a 'spare' enzyme capacity may exist which quantitatively may have no relation to the amount of putrescine formed.

REFERENCES

Bachrach, U. (1973). In *Function of Naturally Occurring Polyamines*, Academic Press, New York.

Bachrach, U. (1976). *Biochem. Biophys. Res. Comm.*, **72**, 1008–1013.

Beck, W. T., Bellantone, R. A., and Canellakis, E. S. (1972). *Biochem. Biophys. Res. Comm.*, **48**, 1649–1655.

Brandt, J. T., Pierce, D. A., and Fausto, N. (1972). *Biochim. Biophys. Acta*, **279**, 184–193.

Byus, C. V., Costa, M., Sipes, I. G., Brodie, B. B., and Russell, D. H. (1976). *Proc. Natl. Acad. Sci. USA*, **73**, 1241-1245.

Byus, C. V., and Russell, D. H. (1975). *Science*, **187**, 650–652.

Canellakis, Z. N., and Theoharides, T. C. (1976). *J. Biol. Chem.*, **251**, 4436–4441.

Canellakis, E. S., Heller, J. S., Kyriakidis, D., and Chen, K. Y. (1978). Intracellular levels of ornithine decarboxylase, its half-life, and a hypothesis relating polyamine-sensitive membrane receptors to growth. In D. R. Morris, D. Bartos, G. D. Daves R. A. Campbell and F. Bartos (Eds) *Advances in Polyamine Research* Vol. 1, Raven Press, New York, pp. 17–30.

Chen, K. Y., and Canellakis, E. S. (1977). *Proc. Natl. Acad. Sci. USA*, **74**, 3791-3795.

Chen, K., Heller, J. S., and Canellakis, E. S. (1976a). *Biochem. Biophys. Res. Commun.*, **68**, 401-408.

Chen, K., Heller, J. S., and Canellakis, E. S. (1976b). *Biochem. Biophys. Res. Commun.*, **70**, 212–220.

Clark, J. L. (1974). *Biochemistry*, **13**, 4668–4674.

Clark, J. L., and Fuller, J. L. (1975). Biochemistry, **14**, 4403–4409.

Clark, J. L., and Fuller, J. L. (1976). *Eur.. J. Biochem.*, **67**, 303–314.

Cohen, S. S. (1971). In *Introduction to the Polyamines*, Prentice-Hall, Englewood Cliffs, New Jersey.

Fong, W. F., Heller, J. S., and Canellakis, E. S. (1976). *Biochim. Biophys. Acta*, **428**, 456–465.

Friedman, S. J., Bellantone, R. A., and Canellakis, E. S. (1972). *Biochim. Biophys. Acta*, **261**, 188–193.

Friedman, Y., Park, S., Levasseur, S., and Burke, G. (1977). *Biochim. Biophys. Acta*, **500**, 291-303.

Fuller, J. L., Greenspan, S., and Clark, J. L. (1978). Regulation of ornithine decarboxylase activity in 3T3 cells stimulated to grow. In R. A. Campbell, D. R. Morris, D. Bartos, G. D. Dowes and F. Bartos (Eds), *Advances in Polyamine Research*, Vol. 1, Raven Press, New York, pp. 31–38.

Gaza, D. J., Short, J. and Lieberman, I. (1973). *Biochem. Biophys. Res. Commun.*, **54**, 1482–1488.

Gazdar, A. F., Stull, H. B., Kitton, L. J., and Bachrach, U. (1976). *Nature, Lond.*, **262**, 696–698.

Goldstein, D. A., Heby, O., and Marton, L. J. (1976). *Proc. Natl. Acad. Sci. USA*, **73**, 4022–4026.

Harik, S. I., Hollenberg, M. D., and Snyder, S. H. (1974). *Mol. Pharmacol.*, **10**, 41–47.

Harris, J. W., Wong, Y. P., Kehe, C. R., and Teng, S. S. (1975). *Cancer Res.*, **35**, 3181-3186.

Haselbacher, G. K., and Humbel, R. E. (1976). *J. Cell. Physiol.*, **88**, 239–246.

Heby, O., Gray, J. W., Lindl, P. A., Marton, L. J., and Wilson, C. B. (1976). *Biochem. Biophys. Res. Commun.*, **71**, 99–105.

Heller, J. S., Chen, K. Y., Kyriakidis, D. A., Fong, W. F., and Canellakis, E. S. (1978). *J. Cell. Physiol.* In press.

Heller, J. S., Fong, W. F., and Canellakis, E. S. (1976). *Proc. Natl. Acad. Sci. USA*, **73**, 1858–1862.

Heller, J. S., Kyriakidis, D., Fong, W. F., and Canellakis, E. S. (1977). *Eur. J. Biochem.*, **81**, 545–550.

Hogan, B. L. M. (1971). *Biochem. Biophys. Res. Commun.*, **45**, 301–307.

Hogan, B. L. M., and Murden, S. (1974a). *J. Cell. Physiol.*, **83**, 345–352.

Hogan, B. L. M., McIlhinney, A., and Murden, S. (1974b). *J. Cell. Physiol.*, **83**, 353–358.

Hölttä, E. (1975). *Biochim. Biophys. Acta*, **399**, 420–427.

Hölttä, E., and Jänne, J. (1972). *FEBS Lett.*, **23**, 117–121.

Inoue, H., Kato, Y., Tukigawa, M., Adachi, K., and Takeda, Y. (1975). *J. Biochem.*, **77**, 879–893.

Jänne, J., and Hölttä, E., (1974). *Biochem. Boiphys. Res. Commun.*, **61**, 449–456.

Jänne, J., and Raina, A. (1968). *Acta Chem. Scand.*, **22**, 1349–1351.

Jänne, J., Pösö, H., and Raina, A. (1978). *Biochim. Biophys. Acta (Cancer Rev.)*, **473**, 241–293.

Jefferson, L. S., and Pegg, A. E. (1977). *Biochim. Biophys. Acta*, **484**, 177–187.

Kallio, A., Löfman, M., Pösö, H., and Jänne, J. (1977). *FEBS Lett.*, **79**, 195–199.

Kato, Y., Inoue, H., Gohda, E., Tamada, F., and Takeda, Y. (1976). *Gann*, **67**, 569–576.

Kay, J. E., and Lindsay, V. J. (1973). *Biochem. J.*, **132**, 791–796.

Levine, J. H., Leaming, A. B., and Raskin, P. (1978). On the mechanism of activation of hepatic ornithine decarboxylase by cyclic AMP. In R. A. Campbell, D. A. Morris, D. Bartos, G. D. Daves and F. Bartos (Eds), *Advances in Polyamine Research*, Vol. 1, Raven Press, New York, pp. 51–58.

Mamont, P. S., Böhlen, P., McCann, P. P., Bey, P., Schuber, F., and Tardif, C. (1976). *Proc. Natl. Acad. Sci. USA*, **73**, 1626–1630.

Maudsley, D. V., Leif, J., and King, J. J. (1978). Polyamine biosynthesis in serum-stimulated HeLa cells pulse-labelled with ^{3}H-ornithine. In R. A. Campbell, D. R. Morris, D. Bartos, G. D. Daves and F. Bartos (Eds), *Advances in Polyamine Research*, Vol. 1, Raven Press, New York, pp. 93–100.

McCann, P. P., Tardif, C., Mamont, P. S., and Schuber, F. (1975). *Biochem. Biophys. Res. Commun.*, **64**, 336–341.

McCann, P. P., Tardif, C., and Mamont, P. S. (1977a). *Biochem. Biophys. Res. Commun.*,**75**, 948–954.

McCann, P. P., Tardif, C., Duchesne, M. C., and Mamont, P. S. (1977b) *Biochem. Biophys. Res. Commun.*, **76**, 893–899.

McCann, P. P., Tardif, C., and Hornsperger, J. M. (1978). Two distinct mechanisms for ornithine decarboxylase regulation by polyamines in rat hepatoma cells. *Submitted.*

Mitchell, J. L. A., and Rusch, H. P. (1973). *Biochim. Biophys. Acta*, **297**, 503–516.
Mitchell, J. L. A., and Carter, D. D. (1977). *Biochim. Biophys. Acta*, **483**, 425–434.
Mitchell, J. L. A., Anderson, S. N., Carter, D. D., Sedory, M. J., Scott, J. F., and Varland, D. A. (1978). Role of modified forms of ornithine decarboxylase in the control of polyamine biosynthesis. In R. A. Campbell, D. R. Morris, D. Bartos, G. D. Daves and F. Bartos (Eds), *Advances in Polyamine Research*, Vol. 1, Raven Press, New York, pp. 39–50.
Morris, D. R., and Fillingame, R. H. (1974). *Ann. Rev. Biochem.*, **43**, 303–325.
Munro, G. F., Miller, R. A., Bell, C. A., and Verderber, E. L. (1975). *Biochim. Biophys. Acta*, **411**, 263–281.
Nissley, S. P., Passamani, J., and Short P. (1976). *J. Cell. Physiol.*, **89**, 393–402.
Obenrader, M. F., and Prouty, W. F. (1977a). *J. Biol. Chem.*, **252**, 2860–2865.
Obenrader, M. F., and Prouty, W. F. (1977b). *J. Biol. Chem.*, **252**, 2866–2872.
Oka, T., and Perry, J. W. (1976). *J. Biol. Chem.*, **251**, 1738–1744.
Pegg, A. E., Conover, C., and Wrona, A. (1978). *Biochem. J.*, **170**, 651–660.
Pett, D. M., and Ginsberg, H. S. (1968). *Fed. Proc.*, **27**, 615.
Pösö, H., Guha, S. K., and Jänne, J. (1978). *Biochim. Biophys. Acta*. In press.
Pösö, H., and Jänne, J. (1976a). *Biochem. Biophys. Res. Commun.*, **69**, 885–892.
Pösö, H., and Jänne, J. (19976b). *Biochem. J.*, **158**, 485–488.
Prouty, W. F. (1976). *J. Cell. Physiol.*, **89**, 65–76.
Quash, G., Bonnefoy-Roch, A. M., Gazzolo, L., and Niveleau, A (1978). 'Cell membrane polyamines: a qualitative and quantitative study using immunolatex and tritiated antipolyamine antibodies'. In R. A. Campbell, D. R. Morris, D. Bartos, G. D. Daves and F. Bartos (Eds), *Advances in Polyamine Research*, Vol. 1, Raven Press, New York, pp. 85–92.
Raina, A., and Jänne, J. (1975). *Med. Biol.*, **53**, 121–147.
Russell, D. H. (Ed) (1973). In *Polyamines in Normal and Neoplastic Growth*, Raven Press, New York.
Russell, D. H., and Snyder, S. H. (1968). *Proc. Natl. Acad. Sci. USA*, **60**, 1420–1427.
Russell, D. H., and Stambrook, P. J. (1975). *Proc. Natl. Acad. Sci. USA*, **72**, 1482–1486.
Russell, D. H., and Snyder, S. H. (1969). *Mol. Pharmacol.*, **5**, 253–262.
Russell, D. H., Byus, C. V., and Manen, C. A. (1976). *Life Sci.*, **19**, 1297–1306.
Schrock, T. R., Oakman, N. J., and Bucher, N. L. R. (1970). *Biochim. Biophys. Acta*, **204**, 564–577.
Sedory, M. J., and Mitchell, J. L. A. (1977). *Exptl. Cell Res.*, **107**, 105–110.
Tabor, C. W., and Tabor, H. (1976). *Ann. Rev. Biochem.*, **45**, 285–306.

Chapter 8

Induction of Ornithine Decarboxylase and Cyclic AMP Levels during the Cell Cycle of Synchronized BHK Cells

HIROSHIGE HIBASAMI, MINORU TANAKA, JUN NAGAI AND TADAO IKEDA

I. INTRODUCTION

Polyamine biosynthesis is greatly increased in a variety of biological systems that have been stimulated to undergo rapid growth (Hibasami and coworkers, 1976: Chapters 1–4). During the past decade, considerable information has been obtained regarding the stimulation of ornithine decarboxylase (ODC) (E.C. 4.1.1.17), the rate-limiting enzyme in polyamine biosynthesis. Some studies have correlated increased tissue cyclic AMP levels following hormonal stimulation with subsequent rises in ODC activity (Richman and coworkers, 1973).

We have studied ODC induction in synchronous BHK cells during the cell cycle traverse in relation to cellular cyclic AMP levels and to serum-addition after a short period of serum-starvation.

II. CELL CYCLE AND ODC ACTIVITY

In our experiments BHK (21/C 13) cells were grown as monolayer cultures in Eagle's minimal essential medium (MEM) containing 10% (v/v) calf serum (CS). From the exponentially growing cultures synchronous cell populations were prepared by the thymidine block method (Hibasami and coworkers, 1977).

Immediately after release of the thymidine block (zero time) the cells were in early S phase; 2 h later 90% of the cells entered the S phase. At 8 h 20% of the cells were in S phase. From 10 through 16 h the majority of the cells were progressed through G_2 + M phase. At 22 h 15% of the cells entered the second round S phase and at 26 h, 46% (Figure 1A). Duration of the phases was determined by counting the mitotic cells and the cells labelled with tritiated thymidine. The duration of the phases sometimes fluctuated somewhat with the batch of calf serum used.

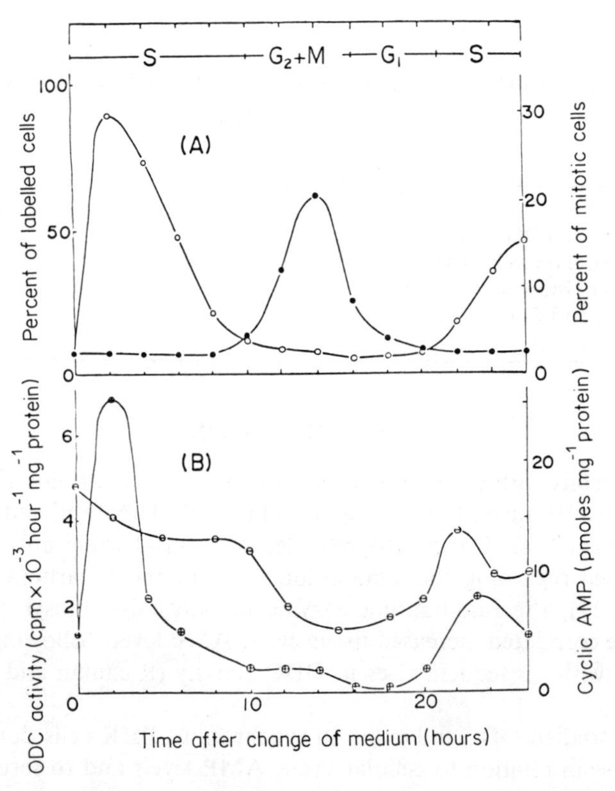

Figure 1(A) Changes in per cent of labelled cells (○) and mitotic cells (●), and (B) ODC activities (⊕) and cAMP levels (⊖) during the cell cycle of BHK cells. The cells were planted at a density of 10^5 cells ml^{-1} in MEM containing 10% v/v calf serum. Thymidine was added to two-day-old cultures in a final concentration of 2.5 mM. After 15 h, the monolayer cells were washed three times with pre-warmed MEM to remove thymidine, incubated again in normal medium at zero time, and analyzed at 2 or 3 h intervals. Each point represents the mean of 3 determinations

A marked increase in ODC activity was noted in the early S phase. The activity decreased during G_2+M and G_1 phase, and then increased again during late $G_1\sim$ the second round early S phase, showing a second peak (Figure 1B). ODC activity was determined by the method of Kay and Lindsay (1973).

Synchronously growing Don C cells (Friedman, Bellantone and Canellakis, 1972) and Chinese hamster ovary fibroblasts (Heby and coworkers, 1976) display two peaks of ODC activity, one at late G_1 and the other, which was not observed in the BHK cells, at G_2+M phase. In these studies synchronous populations of the cells were obtained by selective detachment and collection of mitotic cells, and the cell cycle was started at M phase, while in our studies BHK cells were synchronized by the thymidine block method and the cell cycle was begun at early S phase. The changes in ODC activity during the cell cycle traverse may depend largely on the method for synchronization and/or the cell-line used for the experiments.

III. CELL CYCLE AND CYCLIC AMP LEVELS

Cyclic AMP has been implicated as a possible mediator for ODC induction. Kuo and Greengard (1969) have postulated that cyclic AMP exerts its influence on cellular metabolism through the activation of cyclic AMP-dependent protein kinase. Byus and Russell (1974) were the first to suggest a role for a cyclic AMP-dependent protein kinase in the regulation of ODC activity.

In the synchronous BHK cells, the concentration of cyclic AMP gradually decreased during $S\sim G_2+M$ phase and then increased to a maximum during late $G_1\sim$ the second round early S phase. This increase preceded the rise in ODC activity (Figure 1B), a fact which suggests an ODC induction mediated by cyclic AMP. A marked increase in ODC activity shortly after release of the thymidine block could not be interpreted readily by such a mechanism, however.

In HeLa cells synchronized by a double thymidine block, maximal levels of cyclic AMP are also observed at the $G_1\sim S$ border; the level of cyclic AMP increased during the transition from M phase to the G_1 phase and rose continually as the cells traversed the G_1 phase (Zeilig and coworkers, 1974). Russell and Stambrook (1975) revealed an elevation in cyclic AMP level during the G_1 phase which is followed by an increase in ODC activity in Chinese hamster V_{79} cells synchronized by mitotic selection; cyclic AMP increased within 3 h (G_1 phase) and reached a level 2-fold higher than that in the early G_1 phase within 6 h (early S phase), and ODC activity increased within 4 h, attaining a maximum at 8 h.

On the basis of the increase in ODC activity of rat liver after administration of dibutyryl cyclic AMP, it was speculated that increases in intracellular cyclic AMP may mediate all hormonal stimulation of ODC activity. This had been advanced by several authors (Hölttä and Raina, 1971; Beck, Bellantone and Canellakis, 1972). Hepatic ODC induction in normal rats and an increase of adrenal ODC activity in hypophysectomized rats, after administration of dibutyryl cyclic AMP, were also reported by Richman and coworkers (1973) and by Beck, Bellantone and Canellakis (1972), respectively. Not all tissues have responded, however, to administration of dibutyryl cyclic AMP with increases in ODC activity (Preslock and Hampton, 1973; Richman, 1973; Mizoguchi and coworkers, 1975; Eloranta and Raina, 1975).

Dibutyryl cyclic AMP produced an increase in ODC activity when added to rat hepatoma (Theoharides and Canellakis, 1975), to rat adrenal carcinoma in tissue culture (Richman and coworkers, 1973) and to confluent mouse L cells (Yamasaki and Ichihara, 1976), potentiated the action of calf serum in inducing ODC activity of cultured BHK cells (Hogan, Shields and Curtis, 1974), and also increased ODC activity in confluent C6-BU-1 glioma cells when added together with theophylline (Bachrach, 1975). The activity of ODC also increased when the glioma and N 115 neuroblastoma cells were treated with several compounds that increase cellular cyclic AMP levels (Bachrach, 1975). Administration of phosphodiesterase inhibitors, methylxanthine derivatives, results in an increase of cyclic AMP level in several tissues of the rat, which is followed by a rapid increase in ODC activity (Byus and Russell, 1974). Chen and Canellakis (1977) observed in neuroblastoma cell cultures that dibutyryl cyclic AMP and some compounds that can increase intracellular cyclic AMP level have no effect on the induction of ODC activity unless suboptimal concentrations of L-asparagine are present in the medium.

In the synchronized BHK cells, activation of cyclic AMP-dependent protein kinase, as determined by the method of Corbin and Reimann (1974), changed almost in parallel with cyclic AMP concentration, determined directly (Hibasami and coworkers, 1977), as shown in Figure 2. Accordingly, a rise in this activation, beginning at late G_1, also preceded the elevation of ODC activity in the second round early S phase.

Costa and coworkers (1976) reported that administration of a single dose of Aroclo-1254, a polychlorinated biphenyl mixture, results in a biphasic response of the activation of cyclic AMP-dependent protein kinase and ODC induction in rat liver. Byus and coworkers (1976) described the parenteral administration of a single dose of 3-methylcholanthrene to rats causes an increase of cyclic AMP concentration and activation of cyclic AMP-dependent protein kinase followed by an increased activity of ODC in the liver.

Hence we suggest that the ODC activity during late $G_1 \sim$ the second round early S phase might be induced mainly by cyclic AMP-dependent mechanism through the activation of cyclic AMP-dependent protein kinase, but that of the initial early S phase might be caused in a different way.

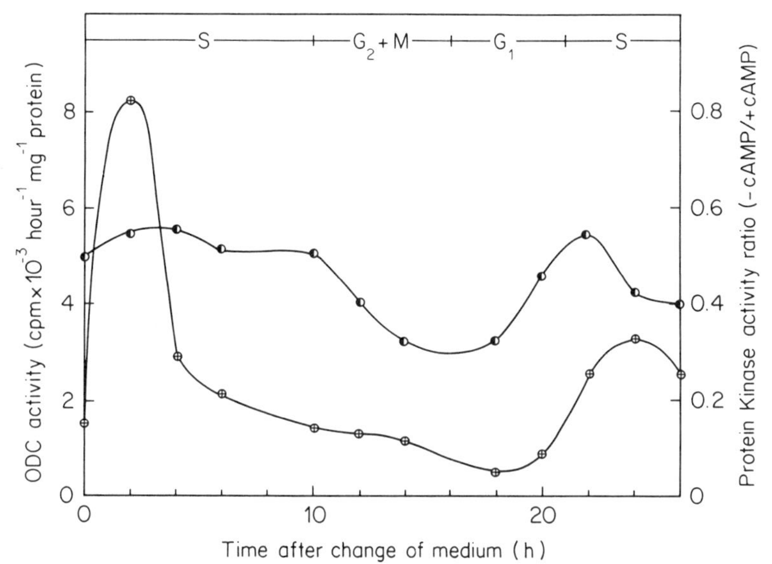

Figure 2 Changes in activation of cyclic AMP-dependent protein kinase (○) and ODC activity (⊕) during the cell cycle of BHK cells. Experimental conditions are the same as described in Figure 1. Determinations were made at zero time and then 2 or 4 h intervals

IV. TEMPORARY INDUCTION OF ODC

In contrast to the ODC induction mediated by cyclic AMP and cyclic AMP-dependent protein kinase, temporary inductions of ODC activity due to some stimuli through changes in circumstances have been reported. This enzyme in rat liver increases by intraperitoneal injection of celite and mannitol, stimuli which do not induce growth (Schrock, Oakman and Bucher, 1970). A sudden decrease in NaCl concentration of the growth medium also produces rapid increase in ODC activity and putrescine concentration in HeLa cells. In this case, there are no obvious effects on RNA or protein content, nor on growth rates (Munro and coworkers, 1975). A similar increase in ODC activity occurs after dilution of cells into fresh medium (Hogan, 1971; McCann and coworkers, 1975) or after addition of fresh serum (Hogan, Schields and Curtis, 1974; Howard and coworkers, 1974; McCann and coworkers, 1975; Hibasami and coworkers, 1976) and amino acids (Kay, Lindsay and Cook, 1972; Hogan and Murden, 1974; Chen and Canellakis, 1977; Yamasaki and Ichihara, 1977) to the cultures.

We have confirmed thereafter that a temporary increase of ODC activity always occurs at any phase of synchronized BHK cells (Figure 3). The induction depended on the duration of serum starvation and on the amount of serum added (Tables 1 and 2). In addition, ODC activity of synchronous BHK cells 3 h after release of the thymidine block was 5-fold that at zero time, but

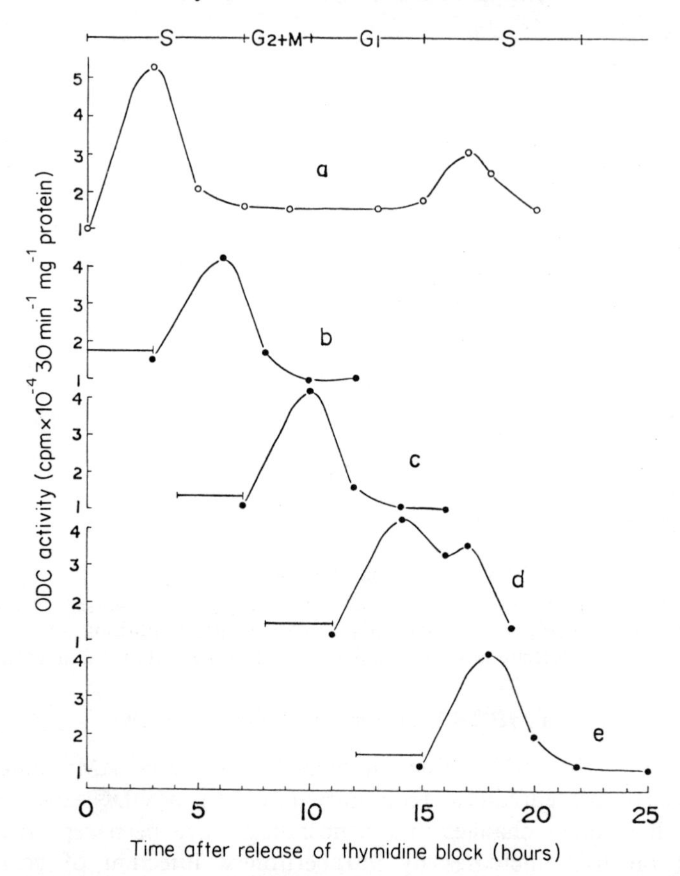

Figure 3 ODC induction by the addition of CS after a short period of CS-starvation at different phases of the cell cycle. After the thymidine treatment for 15 h the medium was replaced by a medium containing 10% CS at zero time and the culture incubated for (a) 20, (b) 0, (c) 4, (d) 8 and (e) 12 h. Then the medium was replaced with that containing 0.5% CS for 3 h (indicated by bars), except for (a)

Table 1 Effects of the period of CS-starvation treatment. At 7 h after release of thymidine block the medium was replaced with that containing 0.5% CS for indicated periods. After the period, CS was added to 10% and ODC activities were determined 3 h after CS addition

CS-starvation (h)	ODC activity (cpm × 10^{-4}/30 min/mg protein)
0	1.42
1	1.94
3	2.55
5	2.48

Table 2 Effects of CS concentrations on ODC induction. Experimental conditions were as in Figure 3 except that various concentrations of CS were added after the CS-starvation treatment between 7 and 10 h after release of thymidine block. ODC activities were measured 3 h after CS addition

Added serum (%)	ODC activity (cpm × 10^{-4}/30 min/mg protein)
0	1.03
2	1.22
5	2.45
10	2.74

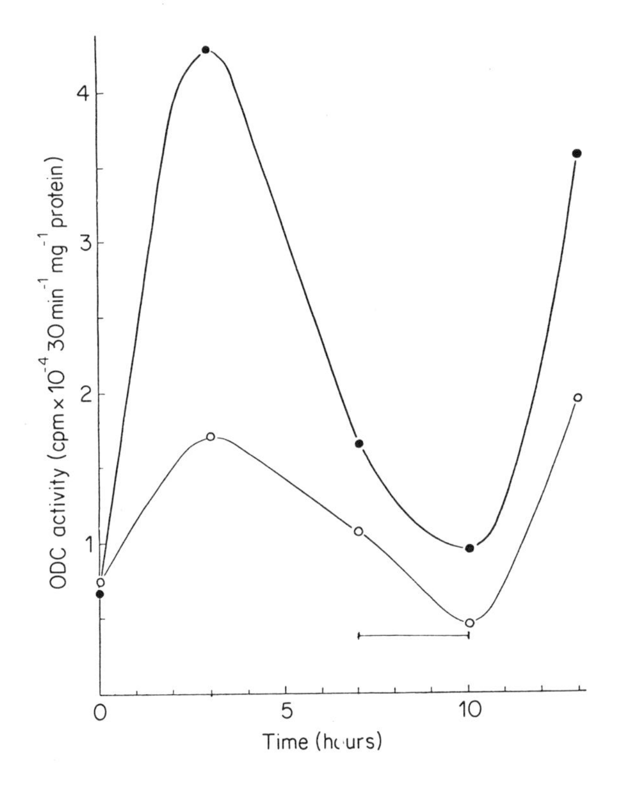

Figure 4 ODC induction in synchronous and asynchronous cultures. ●, synchronous cultures; ○, asynchronous cultures grown in Eagle's MEM containing 10% CS without thymidine after CS-starvation for 3 h. Both cultures were treated with CS-starved medium for 3 h as in Figure 3 (indicated by a bar). Abscissa represents hours after release of the thymidine block for synchronous cultures and those after the preliminary CS-starvation for asynchronous cultures

that of asynchronous cells which had been serum-starved for 3 h before zero time was only doubled (Figure 4). This difference cannot be fully interpreted at present, since it could not be reduced to the difference in synchrony of the cultures and in the phase of major cell populations. We have to assume, therefore, that the treatment with an excess of thymidine might facilitate the ODC induction (Hibasami and coworkers, 1978). It is tempting to speculate that the initial increase of ODC activity in early S phase might be controlled by many factors including cyclic AMP, although the effect of the treatment with thymidine followed by exchange of the medium with fresh one may be predominant.

V. SUMMARY AND CONCLUSIONS

Synchronized BHK (21/C 13) cells display a marked increase in ODC activity shortly after release of the thymidine block (early S phase), and a less pronounced one in the second round early S phase. A maximal cyclic AMP level, consistent with a maximal activation of cyclic AMP-dependent protein kinase, precedes the second peak of ODC activity, a fact which suggests an ODC induction mediated by cyclic AMP through cyclic AMP-dependent protein kinase. The initial rise in ODC activity might depend largely on pretreatment of the cells, and an exchange of the medium.

REFERENCES

Bachrach, U. (1975). *Proc. Natl. Acad. Sci. USA*, **72**, 3087–3091.

Beck, W. T., Bellantone, R. A. and Canellakis, E. S. (1972). *Biochem. Biophys. Res. Commun.*, **48**, 1649–1655.

Byus, C. V. and Russell, D. H. (1974). *Life Sci.*, **15**, 1991–1997.

Byus, C. V. and Russell, D. H. (1975). *Science*, **187**, 650–652.

Byus, C. V., Costa, M., Sipes, I. G., Brodie, B. B. and Russell, D. H. (1976). *Proc. Natl. Acad. Sci. USA*, **73**, 1241–1245.

Chen, K. Y. and Canellakis, E. S. (1977). *Proc. Natl. Acad. Sci. USA*, **74**, 3791–3795.

Corbin, J. D. and Reimann, E. M. (1974). Assay of cyclic AMP-dependent protein kinases. In J. G. Hardman and B. W. O'Malley (Eds), *Methods in Enzymology*, Vol. 38, Academic Press, New York, pp. 287–290.

Costa, M., Costa, E. R., Manen, C. A., Sipes, I. G. and Russell, D. H. (1976). *Mol. Pharmacol.*, **12**, 871–878.

Eloranta, T. and Raina, A. (1975). *FEBS Lett.*, **55**, 22–24.

Friedman, S. J., Bellantone, R. A., and Canellakis, E. S. (1972). *Biochim. Biophys. Acta*, **261**, 188–193.

Heby, O., Gray, J. W., Lindl, P. A., Marton, L. J., and Wilson, C. B. (1976). *Biochem. Biophys. Res. Commun.*, **71**, 99–105.

Hibasami, H., Tanaka, M., Nagai, J., and Ikeda, T. (1976). *Mie Med. J.*, **25**, 223–230.

Hibasami, H., Tanaka, M., Nagai, J., and Ikeda, T. (1977). *Aust. J. Exp. Biol. Med. Sci.*, **55**, 379–383.

Hibasami, H., Tanaka, S., Nagai, J., and Ikeda, T. (1978). *Aust. J. Exp. Biol. Med. Sci.*, **56**, 279–285.

Hogan, B. L. M. (1971). *Biochem. Biophys. Res. Commun.*, **45**, 301-307.
Hogan, B. L. M., and Murden, S. (1974). *J. Cell Physiol.*, **83**, 345-352.
Hogan, B., Shields, R., and Curtis, D. (1974). *Cell*, **2**, 229-233.
Hölttä, E., and Raina, A. (1971). *Scand. J. Clin. Lab. Invest.*, Suppl., **116**, 54.
Howard, D. K., Hay, J., Melvin, W. T. and Durhum, J. P. (1974). *Exp. Cell Res.*, **86**, 31-42.
Kay, J. E., and Lindsay, V. J. (1973). *Exp. Cell Res.*, **77**, 428.
Kay, J. E., Lindsay, V. J., and Cooke, A. (1972). *FEBS Lett.*, **21**, 123-126.
Kuo, J. F., and Greengard, P. (1969). *Proc. Natl. Acad. Sci. USA*, **64**, 1349-1353.
Lowry, O. H., Rosebrough, N. J., Farr, A. C., and Randall, R. J. (1951). *J. Biol. Chem.*, **193**, 265-275.
McCann, P. P., Tardif, C., Mamont, P. S., Schuber, F. (1975). *Biochem. Biophys. Res. Commun.*, **64**, 336-341.
Mizoguchi, Y., Otani, S., Matsui, I., and Morisawa, S. (1975). *Biochem. Biophys. Res. Commun.*, **66**, 328-335.
Munro, G. F., Miller, R. A., and Bell, C. A. (1975). *Biochim. Biophys. Acta*, **411** (2), 263-281.
Preslock, J. P., and Hampton, J. K. Jr. (1973). *Am. J. Physiol.*, **225**, 903-907.
Richman, R., Dobbins, C., Voina, S., Underwood, L., Mahaffee, D., Gitelman, H. J., Van Wyk, J., and Ney, R. L. (1973). *J. Clin. Invest.*, **52**, 2007-2015.
Russell, D. H., and Stambrook, P. J. (1975). *Proc. Natl. Acad. Sci. USA*, **72**, 1482-1486.
Schrock, T. R., Oakman, N. J. and Bucher, N. L. R. (1970) *Biochim. Biophys. Acta*, **204**, 564-577.
Theoharides, T. C., and Canellakis, Z. N. (1975). *Nature, Lond.*, **255**, 733-734.
Yamasaki, Y. and Ichihara, A. (1976), *J. Biochem.*, **80**, 557-562.
Yamasaki, Y. and Ichihara, A. (1977). *J. Biochem.*, **81**, 461-465.
Zeilig, C. E., Johnson, R. A., Sutherland, E. W., and Friedman, D. L. (1974). *Fed. Proc.*, **33**, 1391.

Chapter 9

Minimal Requirements for the Induction of the Antizyme to Ornithine Decarboxylase

JOHN S. HELLER AND E. S. CANELLAKIS

I. INTRODUCTION

Some of the strengths and limitations inherent in animal experiments are indicated in a publication from this laboratory by Beck, Bellantone and Canellakis (1973), which showed that puromycin administration to rats induced liver ornithine decarboxylase activity (ODC; L-ornithine carboxy-lyase, E.C. 4.1.1.17). The authors suggested that perhaps their results could be best interpreted by assuming the existence of a protein inhibitor to ODC whose synthesis was differentially sensitive to inhibition by puromycin. The nature of the original observation, however, was such that the authors could also foresee other plausible interpretations; i.e. that there occurs an accumulation of ODC mRNA resulting in enhanced ODC synthesis after the drug wears off, that puromycin may be stimulating the release of adrenal cortical hormones or that puromycin may increase intracellular cAMP levels with subsequent increases in ODC activity, or that the inhibition of protein synthesis resulted in increased amino acid levels which could act as inducers of ODC. Later, Levine, Nicholson and Orth (1975), upon repeating these experiments, suggested that the puromycin effect could be mediated through the hypophysis. Subsequent work by Fong, Heller and Canellakis (1976) and Heller, Fong and Canellakis (1976) showed that a non-competitive inhibitor of ODC existed in liver as well as in all other eukaryote cells studied; they isolated and partially purified the ODC antizyme. Although this finding was in keeping with the foremost suggestion of Beck, Bellantone and Canellakis (1973) mentioned above, it is still not clear today that the ODC antizyme is in fact related to the mechanism by which puromycin induced ODC activity.

The interpretive latitude that necessarily exists in animal experiments is inherent in the large number of variables to be considered; the nutritional state of the animal, the diurnal rhythms, the possible metabolic modification of the inducing stimulus by the animal's tissues, the effect of stress or the ability of the inducing stimulus to release hormones. The presence of many such natural modifiers of ODC activity, and their probable elicitation during the stress imposed upon the animals by the experimental conditions, differentially affects the primary stimulus under study. Consequently, the superimposition of such ill-defined variables precludes the definition of a cause-effect relationship between the inducer and the inducible macromolecule. Therefore, although the information that can be derived from animal experiments can be very valuable, we find it difficult to rely on such studies to elucidate the sequence of molecular reactions that lead to the increase of ODC activity.

The use of cells in culture eliminates many of these variables as well as the influences from other tissues, nevertheless, the same central difficulty confronts the investigator. The difficulty is that of distinguishing a primary causal agent from one that initiates a sequence of reactions which indirectly causes an increase in ODC activity.

Among the assumptions that are inherent in experiments performed with cells in tissue culture is that the inducing agents affect the cell directly, and that they are not exerting an indirect effect as a result of an undefined interaction between the inducing agent, medium components or serum. Another assumption is that metabolic inhibitors have only one predominant site of action, that this is specifically related to the reaction under study and furthermore that their action is not affected by the composition of the medium. The experiments reported by Chen and Canellakis (1977) suggest that these various assumptions may not be completely valid; they furthermore emphasize that the induction of ODC activity and the effect of inhibitors on this induction is greatly dependent upon the composition of the medium.

In these studies, Chen and Canellakis (1977) minimized the components of the medium and developed a salts/glucose medium which permitted the induction of ODC; in such a minimal medium the effects of a single metabolite at a time could be studied. The use of this salts/glucose medium emphasized the essential nature of asparagine (or of glutamine) to the induction of ODC activity. An important conclusion of this work was that neither the addition of cAMP nor of any of the compounds tested that were known to raise endogenous levels of cAMP, i.e. cholera toxin, N^6, O^2-dibutyryl adenosine 3′:5′-cyclic monophosphate (Bt_2cAMP), prostaglandin E_1 plus isobutylmethylxanthine, would induce ODC activity in neuroblastoma cells in the salts/glucose medium unless asparagine (or glutamine) were present. These results indicated that the induction of ODC activity in growth medium by cAMP could be mediated through these amino acid amides. This experimental finding, which we have since reproduced in a number of cells in culture, could not have been elucidated easily by animal experiments.

In the present investigation we have taken advantage of the experimental simplicity afforded by the salts/glucose minimal medium and have extended its use to the determination of the minimal requirements for the induction of ODC antizyme. This is done as a preliminary to undertaking the evaluation of the molecular mode of induction of this protein which may be involved in the regulation of ODC activity *in vivo* (see Chapter 7).

II. RESULTS

A. Definition of the Experimental Approach

In these experiments, neuroblastoma cells are grown as previously described in growth media (Bachrach, 1975; Heller, Fong and Canellakis, 1976). When these cells have become confluent, they are transferred either to salts/glucose medium or to fresh growth medium and the effects of a variety of superimposed variables upon them are studied.

ODC antizyme isolated from neuroblastoma cells under these conditions was characterized by determining some of the general properties that we have previously provided. Its molecular weight, as determined by sephadex chromatography was 26,000 ± 1,500. This compares closely with the previously reported molecular weight of 26,500 daltons for ODC antizyme isolated from a variety of cells (Heller, Fong and Canellakis, 1976). Furthermore, if heated to 55°C for various times, or if treated with trypsin or chymotrypsin, the induced protein loses its ability to inhibit ODC activity; the activity is not affected by RNase (Fong, Heller and Canellakis, 1976; Heller, Fong and Canellakis, 1976).

B. The Effect of Putrescine on the Induction of ODC Antizyme

Figure 1 shows that ODC antizyme can be detected when neuroblastoma cells, in the presence of salts/glucose medium are exposed to as little as 10^{-7} M putrescine; the activity of free ODC antizyme increases, as the putrescine levels are raised further.

In contrast, if the neuroblastoma cells are transferred to fresh growth medium, ODC antizyme is not detected, even with 10^{-4} M putrescine. In this medium, ODC antizyme can be first detected at 10^{-3} M putrescine. At 10^{-2} M putrescine, a sharp increase in the induction of ODC antizyme is seen in both media.

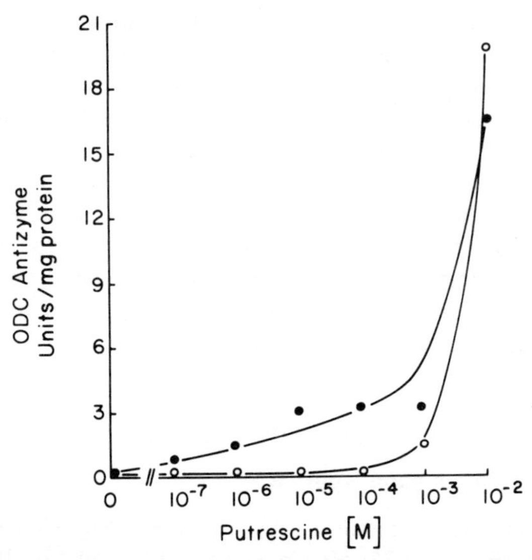

Figure 1 Effect of increasing concentrations of putrescine on ODC antizyme induction. Confluent N18 neuroblastoma cells (Amano, Richelson and Nirenberg, 1972) were rinsed 3 times with saline and transferred either to salts/glucose medium (earle's balanced salt solution — Gibco) or to Dulbecco's growth medium (Gibco) fortified with 10% fetal calf serum. Neutralized putrescine in saline was added to give the indicated concentrations. Extracts of the cells were prepared as previously described by Chen and Canellakis (1977) and ODC and ODC antizyme activities assayed as described by Heller and coworkers (1977). One unit of ODC antizyme activity is defined as the amount which will inhibit one unit of ODC activity, i.e. 1 nmole CO_2 h^{-1}. Salts/glucose (●), Dulbecco's growth medium (○)

A probable reason for this difference is that transfer of the cells to a fresh growth medium in the absence of putrescine induces ODC activity (Bachrach, 1975; Chen and Canellakis, 1977); in a separate experiment, we have in fact determined that 5 units of ODC activity are induced under such conditions. Consequently, the induction of ODC antizyme that occurs by the addition of low concentrations of putrescine, is probably juxtaposed by the concurrent induction of ODC; this would result in a neutralization of the activity of the ODC antizyme. Only when the amount of ODC antizyme induced surpasses that of ODC would it be possible to detect the presence of free ODC antizyme.

These results therefore indicate the much greater sensitivity of induction of ODC antizyme in the salts/glucose medium; however, they also indicate that when a high putrescine concentration (10^{-2} M) is used as an inducer, the level of ODC antizyme attained by the cells in the growth medium is higher than that attained by the cells in the salts/glucose medium. This may be related to the better nutritional state of the cells, but we believe that the underlying cause may be much more complex.

In the salts/glucose medium, the induction of ODC antizyme activity in neuroblastoma cells which have no basal ODC activity, indicates that the elicitation of ODC antizyme activity is not dependent upon the concurrent existence of ODC activity or upon the prior induction of ODC activity. This could not have been presumed from earlier experiments (Fong, Heller and Canellakis, 1976; Heller, Fong and Canellakis, 1976). In these, the experimental conditions were such that ODC and ODC antizyme were being induced simultaneously and free ODC antizyme could only be assayed for after the ODC was completely neutralized by ODC antizyme.

C. The Time Course of Appearance of ODC Antizyme

The time course of appearance of ODC antizyme is shown in Figure 2. The activity of ODC antizyme reached a peak in 5 to 7 h after transferring the neuroblastoma cells to a salts/glucose medium plus 10 mM putrescine and was maintained for up to 10 h. The time necessary to reach maximal activity was somewhat variable; we have indications that this time may be related to the time the cells reached confluency; however, the maximal ODC antizyme activity was always relatively constant (17-18 units mg^{-1} protein). On the other hand, neuroblastoma cells transferred to complete medium plus 10 mM

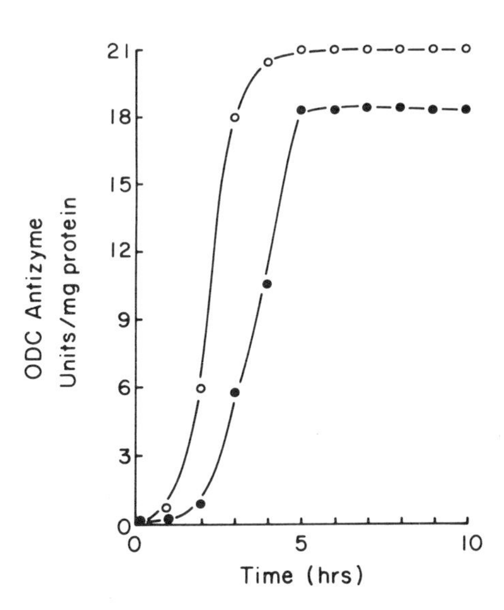

Figure 2 Time course of induction of ODC antizyme. Confluent neuroblastoma cells were treated as described in Figure 1 except that 10 mM putrescine was added to both sets of cultures at time zero. All other experimental details are as described in Figure 1. Salts/glucose (●). Dulbecco's growth medium (○)

putrescine, generally reached peak values of ODC antizyme activity earlier and attained a peak activity that was approximately 20% higher (21-22 units mg^{-1} protein).

D. Effect of Inhibitors of RNA Synthesis and of Protein Synthesis on ODC Antizyme Induction

Concurrent RNA synthesis does not appear to be required for the induction of ODC antizyme activity. Actinomycin D added at time of transfer delays the increase in ODC antizyme activity; however, the peak activity is only inhibited approximately 20% (Figure 3). This lack of inhibition of the induction of ODC antizyme activity by actinomycin D has also been reported to occur in cell cultures maintained in growth media (Fong, Heller and Canellakis, 1976; McCann, Tardif and Mamont, 1977), as well as in animal studies (Friedman and coworkers, 1977).

In contrast to actinomycin D, cycloheximide inhibits the induction of ODC antizyme in the salts/glucose medium (Figure 3); the same inhibitory effect of cycloheximide on the induction of ODC antizyme activity is observed in cells maintained in a complete growth medium (Fong, Heller and Canellakis, 1976).

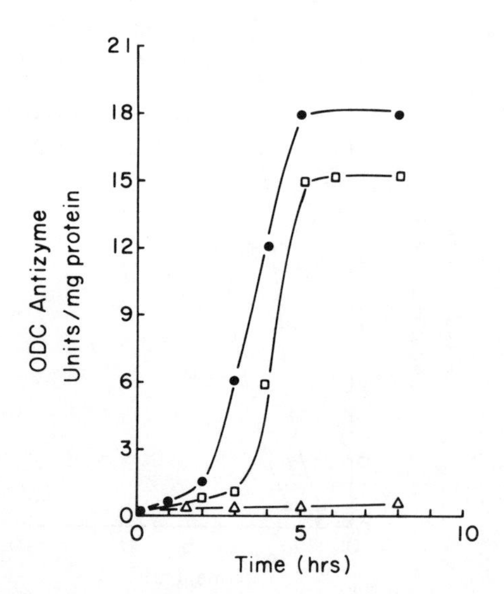

Figure 3 Effect of antinomycin D and cycloheximide on the induction of ODC antizyme in salts/glucose medium. Confluent neuroblastoma cells were rinsed as in Figure 1 and transferred to salts/glucose medium containing 10 mM putrescine (●); 10 mM putrescine plus actinomycin D (5μg/ml) (□) or 10 mM putrescine plus cycloheximide (50 μg/ml) (△). All additions were completed in 10 min. All other experimental details are as described in Figure 1

E. The $t_{1/2}$ of ODC Antizyme

The addition of cycloheximide at various times after the transfer of neuroblastoma cells to salts/glucose medium plus 10 mM putrescine inhibits any further induction of ODC antizyme activity as shown in Figure 4. This figure furthermore emphasizes that cycloheximide maintains the existing intracellular level of ODC antizyme activity. If cycloheximide is added at zero time, before ODC antizyme activity is induced, the activity of ODC antizyme remains at basal levels. If cycloheximide is added at various times along the ODC antizyme induction curve, the ODC antizyme remains at that existing level of activity.

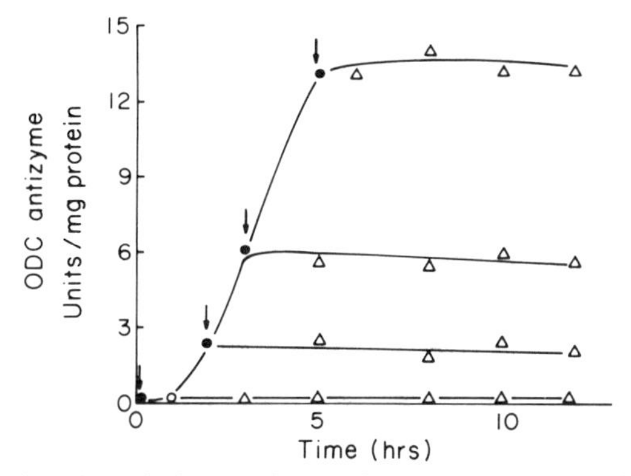

Figure 4 Effect of cycloheximide on induced ODC antizyme activity in salts/glucose medium. Confluent neuroblastoma cells were transferred to salts/glucose medium containing 10 mM putrescine (●) as described in Figure 1. Cycloheximide (50 μg/ml) (△) was added at the times indicated by the arrows and the cells harvested at the times indicated. All experimental details are as described in Figure 1

The ODC antizyme activity shows only a slow decrease with time and the $t_{1/2}$ for induced ODC antizyme is greater than 24 h; similar results were obtained at the various time points along the induction curve. In contrast, the addition of cycloheximide to neuroblastoma cells in fresh growth medium in which ODC antizyme activity had also been induced with 10 mM putrescine, showed a rapid decline of ODC antizyme activity with a $t_{1/2} = 66$ min (Figure 5).

The maintenance of ODC antizyme activity can be shown to be completely dependent upon the presence of putrescine (Figure 6). ODC antizyme activity was induced for 5 h in confluent neuroblastoma cells in salts/glucose medium plus 10 mM putrescine; upon replacement of the inducing medium with a putrescine-free salts/glucose medium, the further induction of ODC antizyme activity ceased and the ODC antizyme activity declined rapidly in the first 90 min (Figure 6). The $t_{1/2}$ calculated from the decline in activity after removal of putrescine was approximately 60 min.

These results therefore emphasize that the addition of cycloheximide to a salts/glucose medium containing putrescine did not lower the existing activity of ODC antizyme, nor did it alter the rate of decline of ODC antizyme activity subsequent to the removal of putrescine. Cycloheximide therefore appears to inhibit the further induction of ODC antizyme but does not precipitate a decline in ODC antizyme activity. These results have a counterpart in the induction of ODC by asparagine in the salts/glucose medium. Chen and Canellakis (1977) showed that asparagine is sufficient and necessary for the induction of ODC activity, and that the addition of cycloheximide to a salts/glucose medium containing asparagine did not lower the existing ODC activity, nor did it alter the rapid loss of ODC activity that occurred subsequent to the removal of asparagine.

III. DISCUSSION

We have previously shown that spermidine, spermine and a number of diamines, including putrescine, will induce ODC antizyme in L-1210 cells (Heller and coworkers, 1978). Our understanding of the mechanism of the induction of ODC antizyme activity is now being greatly facilitated by the use of the salts/glucose medium. It has been possible to show that ODC antizyme activity can be elicited by concentrations of putrescine as low as 10^{-7} M. This demonstration resolves many of the questions that may have existed pertaining to the biological relevance of the induction of ODC antizyme activity because of the high putrescine concentrations that were necessary to elicit this activity in a complete growth medium. It therefore becomes quite plausible that ODC antizyme can act as a modulator of ODC activity *in vivo*. This would be in agreement with our previous results which have shown that similarly low levels of putrescine (10^{-6} M to 10^{-5} M) added to the medium, inhibit intracellular ODC activity (Heller and coworkers, 1978). The notable characteristic of this phenomenon is that it can occur despite the presence of the normally higher intracellular levels of putrescine (1.5×10^{-4} M).

Of particular interest is the correspondence of the conditions for the induction of ODC activity to those of the induction of ODC antizyme activity. In salts/glucose medium both activities are induced by high concentrations of a single metabolite, asparagine and putrescine respectively. The maintenance of their activity is dependent upon the continued presence of that metabolite. In the presence of the respective inducer, cycloheximide in both cases, does not alter the existing biological activity. Removal of the inducer causes a rapid loss of the corresponding activities, resulting in short half-lives (12 1/2 min in the case of ODC and 60 min in the case of ODC antizyme). The induction of neither activity is inhibited by actinomycin D.

In growth medium asparagine and putrescine again act as the respective inducers of these activities; however, only the induction of ODC antizyme is

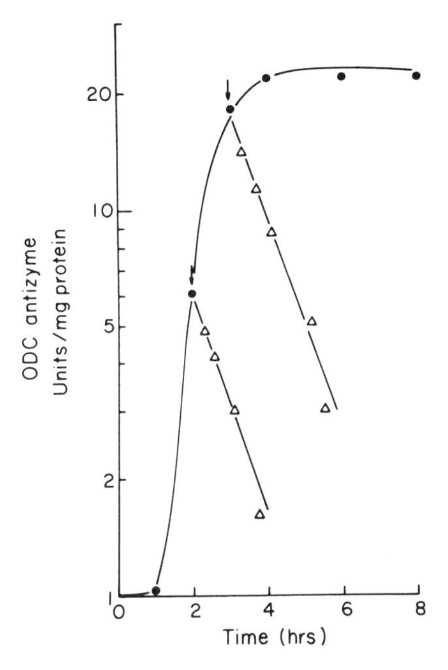

Figure 5 Effect of cycloheximide on induced ODC antizyme activity in growth medium. Confluent neuroblastoma cells were transferred to Dulbecco's growth medium fortified with 10% fetal calf serum containing 10 mM putrescine (●). Cycloheximide (50 μg/ml) (△) was added at the times indicated by the arrows and the cells harvested at the times indicated. All experimental details are as described in Figure 1

solely dependent upon the addition of its inducer. Growth medium contains glutamine as well as a number of alternate inducers of ODC. In such growth media, the addition of cycloheximide decreases the activity of both ODC and of ODC antizyme despite the presence of the respective inducers. It is apparent therefore, that the effect of cycloheximide on these two opposing biological activities is dependent upon the composition of the media and that cycloheximide has no effect in minimal media. A most interesting conclusion that may therefore be suggested, is that cycloheximide acts on alternate complex pathways of induction and does not interfere with the essential minimal mechanism required for the maintenance of the activities of ODC and of ODC antizyme. A comparable suggestion may be made for actinomycin D, because actinomycin D does not inhibit the induction of ODC antizyme; it also does not inhibit the induction of ODC if this is induced by asparagine. If ODC is induced by cAMP or prostaglandin E_1 plus isobutylmethylxanthine, however, then actinomycin D inhibits its induction (Chen and Canellakis, 1977).

At this time it is very difficult to comment definitively on the molecular

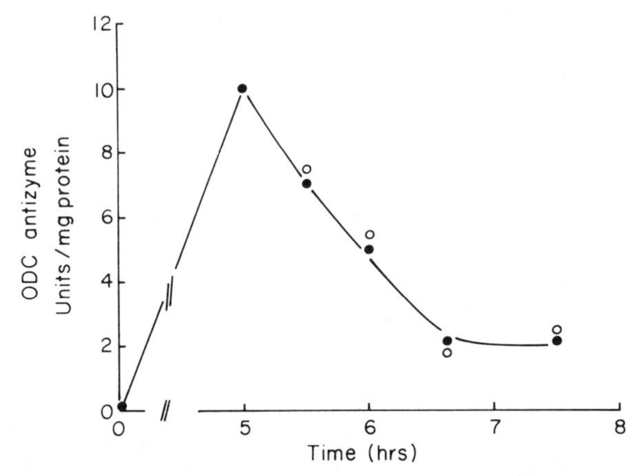

Figure 6 Effect of removing putrescine on ODC antizyme activity in salts/glucose medium. Confluent neuroblastoma cells were transferred to salts/glucose medium containing 10 mM putrescine as described in Figure 1. After 5 h the cells were rinsed twice with salts/glucose and transferred to salts/glucose medium. Cycloheximide (50 μg/ml) was added where indicated. Cells were harvested at the times indicated and all other experimental details are as described in Figure 1. Plus cycloheximide (○); minus cycloheximide (●)

mechanism of induction of these opposing activities. These results, as well as our unpublished data, imply that a most complex interrelationship exists and that this may involve many more factors than we have so far determined.

In our studies with *Escherichia coli*, we have been able to prove definitively that, in addition to the existence of an ODC antizyme, there exists a macromolecular activator of ODC (Kyriakidis, Heller and Canellakis, 1978). It also appears that much of the control of ODC in *E. coli* is exerted at the post-transitional level through the interaction of the macromolecular ODC activators and inhibitors.

This problem is presently being carefully unravelled in our laboratory from a number of points of view and with the use of a variety of cellular systems. The basic fascination of the problem is the multiplicity and the complexity of controls that appear to be related to the regulation of the activity of ornithine decarboxylase.

IV. SUMMARY AND CONCLUSION

The induction of the macromolecular inhibitor (antizyme) to ornithine decarboxylase (ODC; L-ornithine carboxylase, E.C. 4.1.1.17) in neuroblastoma cells maintained in salts/glucose medium or in growth medium has been studied. Putrescine is necessary and sufficient to induce ODC antizyme activity; the removal of putrescine results in the loss of this activity. In the salts/glucose medium, as little as 10^{-7} M putrescine is adequate to induce ODC antizyme activity in neuroblastoma cells; this induction is not inhibited by the presence of actinomycin D. Cycloheximide inhibits the further induction of ODC antizyme activity; it does not cause a decline in ODC activity. The addition of cycloheximide retains the existing level of ODC antizyme activity and does not alter the rate of loss activity that occurs upon removal of putrescine. In the growth medium neuroblastoma cells respond similarly, except that higher concentrations of putrescine are required for the induction of ODC antizyme activity and that cycloheximide causes a rapid fall in ODC antizyme activity even in the presence of putrescine. These characteristics of ODC antizyme induction have their counterparts to those seen in the induction of ODC by asparagine in salts/glucose and in growth medium.

ACKNOWLEDGEMENTS

The authors thank Ms B. E. Stanley and Ms L. Marshall for technical assistance and Ms M. Auslander for transcription of the manuscript. This work was supported in part by research grants from the American Cancer Society; E.S.C. is a U.S. Public Health Service Research Career Professor.

REFERENCES

Amano, T., Richelson, E. and Nirenberg, M. (1972). *Proc. Nat. Acad. Sci. USA*, **69**, 258-263.

Bachrach, U. (1975). *Proc. Nat. Acad. Sci. USA*, **72**, 3087-3091.

Beck, W. T., Bellantone, R. A. and Canellakis, E. S. (1973). *Nature, Lond.*, **241**, 275-277.

Chen, K. Y. and Canellakis, E. S. (1977). *Proc. Nat. Acad. Sci. USA*, **74**, 3791-3795.

Fong, W. F., Heller, J. S. and Canellakis, E. S. (1976). *Biochim. Biophys. Acta*, **428**, 456-265.

Friedman, Y., Park, S., Levasseur, S. and Burke, G. (1977). *Biochim. Biophys. Acta*, **500**, 291-303.

Heller, J. S., Fong, W. F. and Canellakis, E. S. (1976). *Proc. Nat. Acad. Sci. USA*, **73**, 1858-1862.

Heller, J. S., Kyriakidis, D., Fong, W. F. and Canellakis, E. S. (1977). *Eur. J. Biochem.*, **81**, 545-550.

Heller, J. S., Chen, K. Y., Kyriakidis, D. A., Fong, W. F. and Canellakis, E. S. (1978). *J. Cell Physiol.*, **96**, 225-234.

Kyriakidis, D. A., Heller, J. S. and Canellakis, E. S. (1978). *Proc. Nat. Acad. Sci. USA.*, **15**, 4699-4703.

Levine, J. H., Nicholson, W. E. and Orth, D. N. (1975). *Proc. Nat. Acad. Sci. USA*, **72**, 2279-2283.

McCann, P. P., Tardif, C. and Mamont, P. S. (1977). *Biochim. Biophys. Res. Commun.*, **75**, 948-954.

Chapter 10

Biochemical Consequences of Drug-induced Polyamine Deficiency in Mammalian Cells

PIERRE S. MAMONT, PHILIPPE BEY AND JAN KOCH-WESER

I. INTRODUCTION

In mammalian cells many biochemical processes including the membrane transport of glucose, amino acids and nucleotides and the overall rates of protein and RNA synthesis and of protein degradation are influenced by withdrawal or addition of growth-promoting agents. The coordinated response of these processes to environmental changes which affects growth of untransformed but not of transformed or malignant cells, has been designated by Tomkins 'the pleiotypic response' (Hershko and coworkers, 1971). Another early change associated with the transition of cells from quiescence to proliferation is an increase of activity of the two decarboxylases involved in polyamine biosynthesis, L-ornithine decarboxylase (E.C. 4.1.1.17) (ODC: Chapter 6) and S-adenosylmethionine decarboxylase (E.C. 4.1.1.50) (SAMD). This increase of activities is followed by concomitant intracellular accumulation of putrescine and spermidine and sometimes of spermine (Tabor and Tabor, 1976; Jänne, Pösö and Raina, 1978). Even though stimulation of the rate of polyamine biosynthesis always appears to accompany increased cell proliferation, however, it remained uncertain whether this increase is circumstantial or essential for the cells to enter into mitotic activity.

In mammalian cells, two experimental approaches have been employed to create polyamine deficiency: use of more or less specific inhibitors of the polyamine biosynthetic enzymes and modulation of ODC activity by means of putrescine analogues. We have concentrated primarily on the first approach and have synthetized and studied novel inhibitors of ODC.

II. INHIBITION OF POLYAMINE BIOSYNTHETIC ENZYMES

In mammalian cells polyamine biosynthesis involves the sequential action of two decarboxylases and two transferases. Inhibitors of polyamine biosynthesis that have been developed thus far (Table 1) in attempts to block the accumulation of polyamines *in vivo* are directed towards the decarboxylases. Only one inhibitor of the transferases has been reported (Coward, Motola and Moyer, 1977), and its activity is still controversial. Two factors explain the preferential development of decarboxylase inhibitors. First, decarboxylation of ornithine is believed to be the rate-limiting step of the polyamine biosynthetic pathway, though in special cases, ornithine synthesis may also be a limiting factor (Williams-Ashman, Pegg and Lockwood, 1968; Oka and Perry, 1974a; McCann and coworkers, 1975; Klein and Morris, 1978). Secondly the complexity of the catalytic mechanisms of the transferases is still incompletely understood, whereas the mechanism of action of pyridoxal phosphate (PLP)-dependent and to a lesser extent, pyruvate-dependent α-amino acid decarboxylases is well established (Boecker and Snell, 1972). This knowledge combined with previous experience from the extensive work aimed at the inactivation of PLP-dependent decarboxylases such as 3,4-dihydroxyphenylalanine and histidine decarboxylases (Sourkes, 1966; Shepherd and Mackay, 1967) permitted rational approaches to the design of inhibitors of ODC. With the exception of DL-α-difluoromethylornithine (DFMeOrn) and α-acetylenic putrescine, the action of such inhibitors and of the SAMD inhibitors have been discussed in detail in two recent reviews (Williams-Ashman, Corti and Tadolini, 1976; Jänne and coworkers, 1978) and are summarized in Table 2 (see also Chapter 11).

A. Enzyme-Activated Irreversible Inhibitors of ODC

DFMeOrn and α-acetylenic putrescine are the first known synthetic irreversible inhibitors of ODC (Metcalf and coworkers, 1978). They belong to the novel class of enzyme-activated irreversible inhibitors known as kcat inhibitors (Rando, 1974) or suicide enzyme inactivators (Abeles and Maycock, 1976). In contrast to the classical irreversible inhibitors, which usually contain a reactive function, these compounds are substrate or product analogues of the target enzyme and are chemically inert. Their action requires an enzyme-catalysed activation which generates an electrophilic form of the substrate analogue inside the active site of the enzyme. A subsequent reaction of this

Table 1 Inhibitors of Ornithine Decarboxylase and S-Adenosylmethionine Decarboxylase

ODC				References
2-hydrazino-5-aminopentanoic acid	α-hydrazino ornithine	*E. coli*	Ki = 2.0[a]	(Harik and Snyder, 1973)
	(α-HO)	Rat liver	Ki = 0.5	(Harik and Snyder, 1973)
DL-2-hydrazino-5-aminopentanoic acid	(DL-HAVA)	Mouse parotid gland	Ki = 0.52	(Inoue and coworkers, 1975)
DL-2-hydrazino-2-methyl-5-aminopentanoic acid	α-hydrazino-α-methyl ornithine	Rat prostate	Ki = 3.0	(Abdel-Monem and coworkers, 1975b)
N-(5′-phosphopyridoxyl) ornithine		Rat liver	Ki = 66	(Heller and coworkers, 1975)
DL-2,5-diamino-2-*trans*-3-pentenoic acid	dehydro ornithine	Rat liver	Ki = 4.4	(Relyea and Rando. 1975)
DL-2,5-diamino-2-methyl-pentanoic acid	α-methylornithine (MeOrn)	HTC cells	Ki = 40	(Mamont and coworkers, 1976; Bey and coworkers, 1978)
		Rat liver	Ki = 40—80	(Bey and coworkers, 1978; Abdel-Monem and and coworkers, 1975a)
		Rat prostate	Ki = 20—40	(Abdel-Monem and coworkers, 1974; Abdel-Monem and coworkers, 1975a)
		Mouse spleen	Ki = 26	(Abdel-Monem and coworkers, 1975a)
DL-2,5-diamino-2-difluoromethyl pentanoic acid	α-difluoromethylornithine (DFMeOrn)	Rat liver	Ki = 39; $t_{1/2}$[b] = 3.1 min	(Metcalf and coworkers, 1978)
1,4-diamino-2-butanone		*Aspergillus nidulans*	Ki = 0.91	(Stevens, McKinnon and Winther, 1977)
1,4-diamino-*trans*-2-butene	dehydro-putrescine	Rat liver	Ki = 2	(Relyea and Rando, 1975)
DL-5-hexyne-1,4-diamine	α-acetylenic putrescine	Rat liver	Ki = 2.3; t $t_{1/2}$ = 9.7 min	(Metcalf and coworkers, 1978)
SAMD				
1,1′-[(methylethane diylidene)- dinitrilo] diguanidine	methylglyoxal *bis*(guanyl hydrazone) (Methyl-GAG)	Rat prostate	K = 1	(Williams-Ashman and Schenone, 1972; Corti and coworkers, 1974)
		Rat liver	K = 1	(Hölttä and coworkers, 1973)
1,1′-[methylethane diylidene)-dinitrilo]-*bis*-(3-aminoguanidine)	(MBAG)	Rat liver	Ki = 54; $t_{1/2}$ = 6 min	(Pegg and Conover, 1976; Pegg, 1978)

a: μM
b: Half-life at infinite concentration of inhibitor.

Table 2 Effects of Inhibitors of Polyamine Biosynthesis on Intracellular Polyamines

	Putrescine	*Spermidine*	*Spermine*		*References*
α-hydrazino ornithine	↓	ND	ND	HTC cells, Regenerating rat liver	(Harik, Hollenberg and Snyder, 1974b)
DL-HAVA	↓	0	0	Mouse parotid gland	(Inoue and coworkers, 1975)
	↓	0	0	Mouse sarcoma-180	(Kato and coworkers, 1976)
α-hydrazino-α-methyl ornithine	↓	ND	ND	Bovine lymphocytes	(Abdel-Monem, Newton and Weeks, 1975b)
↓-methylornithine	↓	↓	0	HTC, L1210 cells	(Mamont and coworkers, 1976; Mamont and coworkers, 1978; Newton and Abdel-Monem, 1977)
	↓	↓	0	Bovine lymphocytes	(Abdel-Monem and coworkers, 1975a; Morris, Jorstad and Seyfried, 1977)
	↓	0	0	Polychaete eggs	(Emanuelson and Heby, 1978)
	↓	↓	0	Rat prostate	(Danzin and coworkers, unpublished)
	↓	0	0	Rat testis	(Danzin and coworkers, unpublished)
α-difluoromethylornithine	↓	↓	0↓	HTC, L1210 cells	(Mamont and coworkers, 1978a)
	↓	↓	↓	Rat prostate	(Danzin and coworkers, 1979d)
	↓	↓	0	Rat thymus, rat testis	(Danzin and coworkers, 1979c)
	↓	↓	0	Leukaemic mouse spleens	(Prakash and coworkers, 1978)
1,4-diamino-2-butanone	↑	↓	↓	*Aspergillus nidulans*	(Stevens, McKinnon and Winther, 1977)
α-acetylenic putrescine	↓	↓	0	Rat prostate	(Danzin and coworkers, 1979b)
methylglyoxal *bis*(guanyl hydrazone)	↑	↓	↓	Bovine and guinea pig lymphocytes	(Fillingame and Morris, 1973b; Morris, Jorstad and Seyfried, 1977) Otani and coworkers, 1974)
				L1210 cells	(Newton and Abdel-Monem, 1977; Dave, Pathak and Porter, 1978)
	↑	↓	↓	Hela S_3 cells	(Krokan and Eriksen, 1977)
				Rat brain tumour cells	(Heby and coworkers, 1978)
				Rat embryo fibroblasts	(Rupniak and Paul, 1978)
	↑	0	0	Rat liver, rat thymus	(Hölttä and coworkers, 1973; Pegg, 1973)
	↑	0	0	Rat kidney	(Pegg, 1973)
	↑	0	0	Rat liver, rat thymus	(Hölttä and coworkers, 1973; Pegg, 1973)
	↑	0	0	Rat kidney	(Pegg, 1973)
	↑	↓	↓	Leukaemic mouse spleen	(Heby and Russell, 1974)
	ND	↓	ND	Mouse mammary epithelium	(Oka and Perry, 1974b)
MBAG	↑	0	0	Rat liver	(Pegg, 1978)
	↑	↓	↓	Rat kidney	(Pegg, 1978)

O: No change
ND: Not determined

activated intermediate with a nucleophilic residue of the active site leads by covalent linkage to irreversible inactivation of the enzyme. The specificity of these molecules is therefore determined both by their binding affinity and by their effectiveness as substrates. The rational design of such inhibitors requires not only that the mechanism of action of the enzyme is established, but also that the steric, electronic and conformational changes of the natural substrate which can be tolerated by the enzyme's active site are known. By systematic substitution we have delineated some of the structural features required for binding of L-ornithine to the active site of mammalian ODC (Bey and coworkers, 1978).

Preincubation of liver ODC from thioacetamide-treated rats (Ono and coworkers, 1972) with DFMeOrn or α-acetylenic putrescine results in a time-dependent loss of enzyme activity which follows pseudo-first order kinetics for at least two half-lives (Metcalf and coworkers, 1978). By plotting the half-life of enzyme activity as a function of the reciprocal of the inhibitor concentration (Kitz and Wilson, 1962) an apparent dissociation constant Ki and an inactivation rate constant kcat have been determined (DFMeOrn : Ki $= 39 \mu M$, kcat $= 3.7 \times 10^{-3}$ sec^{-1}; α-acetylenic putrescine: Ki $= 2.3$ μM, kcat $= 1.2 \times 10^{-3}$ sec^{-1}). Prolonged dialysis (24 h) of the inactivated enzyme against incubation buffer fails to restore enzymatic activity, suggesting that the inhibition is irreversible. The protective effects of L-ornithine, of the competitive inhibitor MeOrn and of putrescine demonstrate that the inhibition is active site-directed. Moreover, the activity of DFMeOrn and α-acetylenic putrescine resides only with the (—) optical isomers. As anticipated from the postulated inhibition mechanism depicted in Figure 1, DFMeOrn does not irreversibly inhibit glutamic acid and aromatic amino acid decarboxylases *in vitro*. It also appears to be devoid of activity against L-ornithine aminotransferase *in vivo* (Jung and Seiler, 1978).

Although α-acetylenic putrescine appears *in vitro* to be more potent than DFMeOrn, the latter was preferred for studies of the consequences of *in situ* inhibition of ODC. This choice was dictated mainly by the fact that α-acetylenic putrescine is metabolized *in vivo* into γ-acetylenic-γ-aminobutyric acid (Danzin and coworkers, 1979a), a potent irreversible inhibitor of γ-aminobutyrate α-oxoglutarate aminotransferase (Jung and coworkers, 1977) and to a lesser extent of glutamic acid decarboxylase (Jung and coworkers, 1978). Moreover, the structural analogy of this compound with putrescine could entail effects other than inhibition of ODC.

B. Effects of DFMeOrn on HTC Cell ODC Activity and Polyamine Synthesis

The cell model system which we chose to test the effect of ODC inhibitors is an established line of rat hepatoma (HTC) cells which has been derived from the Morris hepatoma 7288C (Thompson, Tomkins and Curran, 1966). These

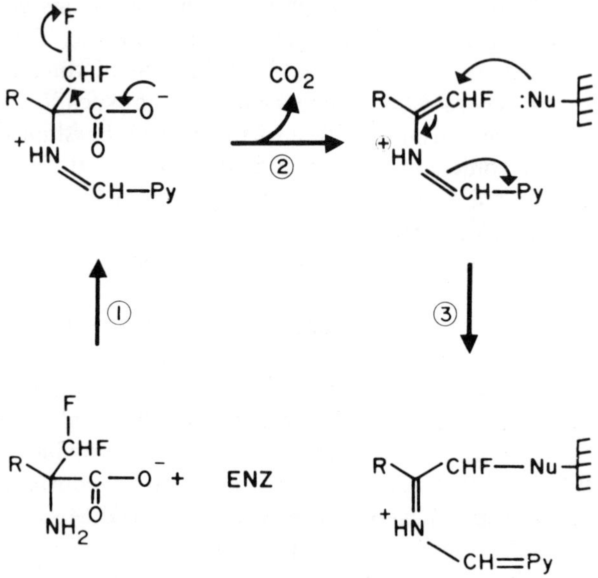

Figure 1 Hypothetical mechanism of the irreversible inactivation of ODC by DFMeOrn. After formation of a Schiff base between the inhibitor and the pyridoxal-phosphate co-factor in the enzyme's active site (step 1), loss of CO_2 and elimination of a fluorine atom, possibly via a concerted mechanism (step 2), leave a highly reactive electrophilic intermediate which can alkylate a nucleophilic residue (Nu) from the active site to irreversibly inactivate the enzyme (step 3). R = $H_2N{-}(CH_2)_3{-}$; Py = Pyridoxal-phosphate ring system

cells when grown in suspension culture show a logarithmic growth for 3 days with a population doubling time of about 24 h.

Induction of cell proliferation by dilution of a high density cell culture with fresh medium is accompanied by a biphasic increase of ODC activity, the peaks occurring just before and after the parasynchronous wave of DNA synthesis (McCann and coworkers, 1975). Addition of DFMeOrn to the culture medium, immediately after dilution, blocks increase of ODC activity (Mamont and coworkers, 1978a). When the compound is added 3 h after dilution, ODC is inactivated in a dose-dependent manner (Figure 2).

Depletion of cellular putrescine content is correlated with the inhibition of ODC activity (Mamont and coworkers, 1978a). As illustrated in Figure 3, 5 mM DFMeOrn depleted intracellular putrescine within 6 h and spermidine within one generation period. Spermine accumulation was blocked on day 2. Prolonged drug-treatment of HTC cells and L1210 mice leukaemia cells, however, never led to spermine depletion (Mamont and coworkers, 1978a). This finding suggests that the blockade of ODC activity by DFMeOrn was not complete.

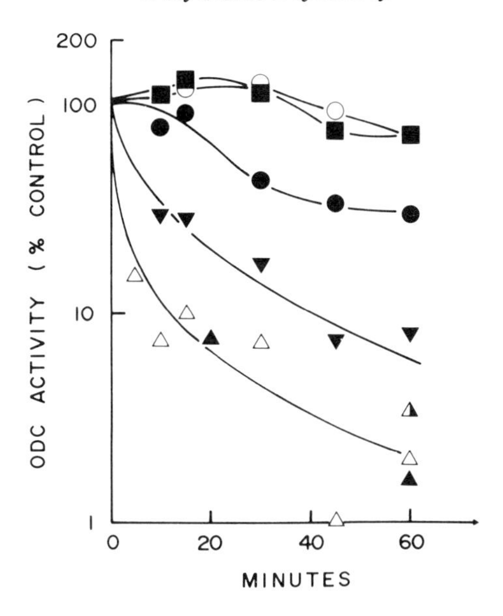

Figure 2 Dose-dependent *in situ* inactivation of HTC cell ODC by DFMeOrn. HTC cell proliferation was induced by dilution of high density cell cultures with fresh medium supplemented with 10% calf serum (Mamont and coworkers, 1976). Three hours after dilution, aliquots were further incubated in the presence or absence of DFMeOrn for an additional 1 h in an Aquatherm water bath shaker (New Brunswick Scientific). At the indicated time, after extensive washing of the cells with cold phosphate-buffered saline, cell extracts were prepared and ODC activities measured (Mamont and coworkers, 1978a). Results are expressed as % of the control (3.8 nmoles CO_2 $hour^{-1}$ mg^{-1} protein). (○) 10^{-6} M, (■) 10^{-5} M, (●) 10^{-4} M (▼) 10^{-3} M and (△,▲,◭) 5×10^{-3} M DFMeOrn (3 experiments)

HTC cells, incubated 3 days in the presence or absence of DFMeOrn, were labelled with trace amounts of tritiated ornithine (1 μM). As shown in Table 3, incorporation of radioactivity into putrescine and spermidine was reduced by 98–99% in DFMeOrn-treated cells. In contrast, the ornithine and spermine fractions were unchanged, though control cells incorporated 5 times more radioactive ornithine than DFMeOrn-treated cells. This result indicates that the reduced amount of putrescine synthetized in the presence of DFMeOrn allows accumulation of spermine. Furthermore, it suggests that the compound does not inhibit the synthetic reactions for spermidine and spermine.

C. Effects of Combined Treatment with 1,3-diaminopropane and DFMeOrn on HTC Cell Polyamine Synthesis

Putrescine added to the culture medium of growing cells decreases ODC activity (Pett and Ginsberg, 1968; Kay and Lindsay, 1973; Clark and Fuller,

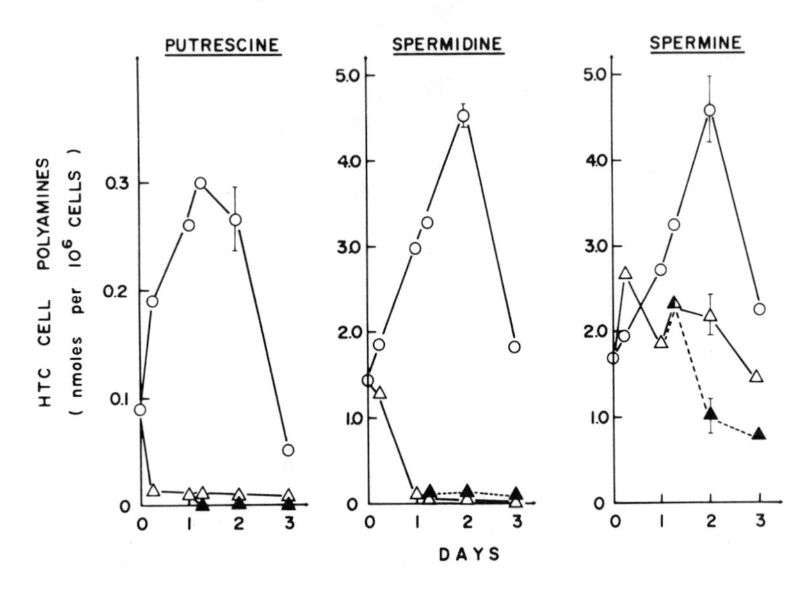

Figure 3 Effects of DFMeOrn plus 1,3-diaminopropane (DAP) on HTC cell polyamines. HTC cells (1×10^5 cells ml^{-1}) in Swim's 77 medium supplemented with 10% dialysed horse serum were incubated from time 0 in the absence (○) or in the presence of 5 mM DFMeOrn (△). At day 1, 1 mM DAP (▲) was added to DFMeOrn-treated cell cultures. Polyamines were measured at the indicated time (Mamont and coworkers, 1978a). Results are expressed as nmoles/10^6 cells and are the means of three determinations ± s.e. mean

1975; McCann, Tardif and Mamont, 1977a). 1,3-Diaminopropane (DAP) and cadaverine have similar effects as putrescine (Pösö and Jänne, 1976; Kallio, Pösö and Jänne, 1977b; McCann, Tardif and Mamont, 1977a). The actions of these amines may be post-transcriptional (Kallio and coworkers, 1977a; Clark and Fuller, 1975) and (or) result from the induction of a macromolecular inhibitor of ODC named 'antizyme' (Chapters 7 and 9). This type of inhibition, called 'gratuitous' repression of polyamine synthesis, has been suggested as a means of producing polyamine deficiency in organs and cell systems (Pösö and Jänne, 1976).

High amounts (13 mM) of DAP added to Chinese hamster ovary cell cultures deplete putrescine and variably reduce spermidine and spermine levels (Sunkara, Rao and Nishioko, 1977). Repeated injections of DAP or cadaverine to hepatectomized rats (Pösö and Jänne, 1976; Kallio, Pösö and Jänne, 1977b) or to mice bearing the Ehrlich ascites carcinoma cells (Kallio and coworkers, 1977c) block putrescine and spermidine accumulation. Consequently, we used DAP and cadaverine in combination with DFMeOrn in order to produce a more stringent polyamine deficiency in HTC cells.

Table 3 Polyamine biosynthesis in the presence of DFMeOrn

	Incorporation				
	Total	Ornithine Fraction	Putrescine	Spermidine	Spermine
Control	507,340	62,450	142,530	275,030	19,030
+ DFMeOrn	91,070	62,510	1,730	3,540	19,760

HTC cell spinner cultures were incubated for 3 days in the presence or absence of 5 mM DFMeOrn. On day 3, cells were gently centrifuged (200 × g) and resuspended in fresh medium in the absence or presence of the drug at a cell density of 1.5×10^5 cells ml^{-1} At that time, cells were pulse-labelled with 5 μCi ml^{-1} DL—(5—^{3}H)ornithine dihydrochloride (5 Ci $mmol^{-1}$, Amersham) for 4 h. Perchloric acid cell extracts were then analysed by Dowex Column chromatography (Inoue and Mizutani, 1973) followed by subsequent fractionation of the polyamine fraction with the use of a Durrum D-500 amino acid analyser (Marton and Lee, 1975)

Addition of 1 mM DAP to HTC cell cultures elicits the induction of free antizyme and blocks ODC activity (McCann, Tardif and Mamont, 1977a). As illustrated in Figure 3, DAP added to HTC cells incubated for one generation with DFMeOrn depleted the trace amounts (6 pmoles/10^6 cells) of residual putrescine further to undetectable levels, doubled the reduced amount (40 pmoles/10^6 cells) of spermidine and decreased spermine levels by 50%. 1 mM cadaverine used in the same experimental condition decreased spermine levels to the same extent. Cellular concentrations of DAP and cadaverine were 10-15 nmoles per 10^6 cells which is about 40 to 100 times higher than the usual HTC cell putrescine content. Only trace amounts (0.1%) of N-3-aminopropyl-1,3-diaminopropane were formed in these cells after addition of DAP. In contrast, cadaverine elicited the formation of two new amines with slightly lower and higher retention times than spermine on ion-exchange chromatography. Structural identification of these new amines will most likely show them to be N-3-aminopropyl cadaverine and N,N′-*bis* (3-aminopropyl)-cadaverine. This would account for the capability of cadaverine to substitute putrescine in the aminopropyl transferase reactions as recently suggested by indirect *in vitro* evidence (Hibasami and Pegg, 1978). The further reduction of spermine content by adding DAP to DFMeOrn may result from a more complete blockade of ODC. An inhibitory effect of DAP on spermine synthase, however, cannot be excluded. In agreement with this possibility is the recent demonstration of an *in vitro* inhibitory action of DAP on rat prostate spermine synthase (Hibasami and Pegg, 1978). Although in that study competitive inhibition by DAP could not be demonstrated, inhibition by DAP may nevertheless be important when there is spermidine deficiency as in HTC cells treated with DFMeOrn. Our results confirm that DAP and cadaverine can be used to manipulate polyamine metabolism. High amounts of these amines, however, immediately accumulate intracellularly and reverse the antiproliferative effects of DFMeOrn (Figures 4B and 5).

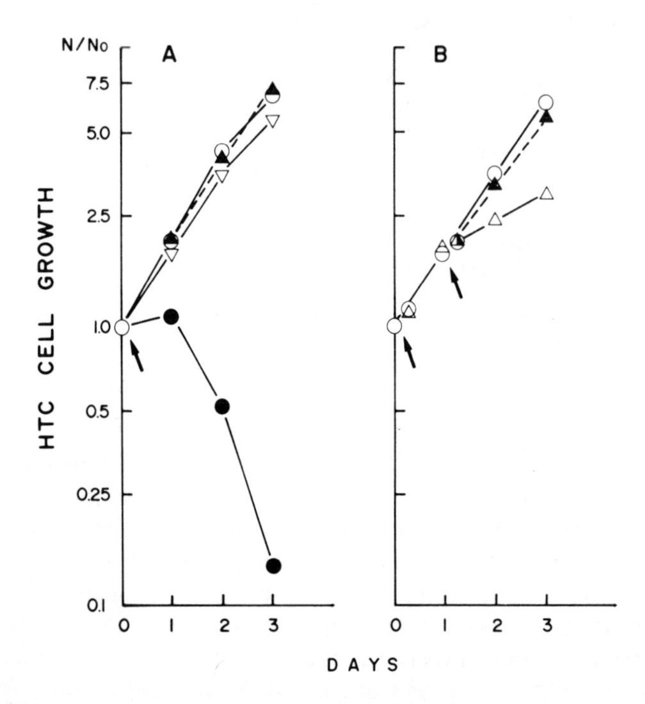

Figure 4 Effects of DAP (A) and DFMeOrn plus DAP (B) on HTC cell growth. HTC cells in Swim's 77 medium supplemented with 10% dialysed horse serum were incubated A) from time 0 in the absence (○) or in the presence of 1 mM (▲), 5 mM (▽), 10 mM (●) DAP; B) from time 0 (first arrow) in the absence (○) or in the presence of 5 mM DFMeOrn (△). At day 1 (second arrow), 1 mM DAP (▲) was added to DFMeOrn-treated cell cultures. Results are expressed as the ratio N/No where N = the number of viable cells ml^{-1} at day 1,2,3 and No = number of cells at day 0

III. BIOCHEMICAL CONSEQUENCES OF POLYAMINE DEFICIENCY

A. Cell Replication

Inhibition of HTC cell polyamine accumulation by MeOrn, a competitive inhibitor of ODC (Abdel-Monem, Newton and Weeks, 1974; Abdel-Monem and coworkers, 1975b) affects neither DNA synthesis nor the mitotic activity during the first round of cell division. Once spermidine depletion is achieved, i.e. within one generation period in the case of HTC cells, a striking decrease of DNA synthesis and cell multiplication rates occurs (Mamont and coworkers, 1976). The delay in the decrease of the spermidine level may explain previous failures to demonstrate an effect of α-hydrazino ornithine on HTC cell RNA and DNA synthesis (Harik, Hollenberg and Snyder, 1974b) and of MeOrn on leukaemia cell DNA content (Newton and Abdel-Monem, 1977). Other experiments (Mamont and coworkers, unpublished) have shown that partial inhibition (30%) or incorporation of tritiated uridine into total RNA occurs after a 48 h period incubation with MeOrn. At the same time, incorporation of tritiated leucine into total proteins, as well as induction of tyrosine aminotransferase (TAT) by dexamethasone, were reduced by 30–50%.

A functional relationship between spermidine levels and growth rates in HTC and L1210 leukaemia cells is suggested by the concomitant action of DFMeOrn on these parameters (Mamont and coworkers, 1978a). It is further confirmed by comparing the effects of both MeOrn and DFMeOrn on an HTC cell variant which overproduces putrescine and spermidine relative to the parental counterpart (Mamont and coworkers, 1978b). The over-accumulation of putrescine and spermidine leads to an initial resistance of these cells to the anti-proliferative drugs. Significantly, when spermidine depletion is finally established (within 2.5 - 3 generations), cell growth slows.

That the anti-proliferative effects of ornithine analogues are related to their ODC-inhibitory activities is further demonstrated by the fact that the (—) enantiomer of MeOrn which exhibits a very low *in vitro* ODC-inhibitory activity (1/70th of that of the (+) antipode) (Bey and coworkers, 1978) has no effect on HTC cell proliferation. Furthermore, addition of L-ornithine to the culture medium of inhibited cells results in a complete or partial resumption of cell proliferation in the case of MeOrn and DFMeOrn respectively.

Putrescine, spermidine or spermine added to HTC cells previously treated for 3 days with DFMeOrn, immediately restore nearly maximum rates of cell proliferation. As illustrated in Figure 5, the reversal effect is dose-dependent, maximum effect being achieved with 10 μM putrescine or spermidine and 1 μM spermine. Higher concentrations of these amines lead to a progressive diminution of their reversal action. 10 μM putrescine replenishes the initial intracellular levels of putrescine and spermidine (Mamont and coworkers, 1978a). Added spermine doubled the intracellular spermine content in 24 h with no accumulation of spermidine or putrescine (Mamont and coworkers, 1978c). However, it cannot be considered established that the recently discovered polyamine oxidase (Hölltä, 1977) or possible other pathways that convert spermine into spermidine and putrescine (Siimes, 1967; Siimes and Jänne, 1967; Seiler, 1973) are non-functional in HTC cells.

The structural specificity of the reversal effect of the polyamines was also investigated. 1 μM DAP or cadaverine could not overcome the inhibition of cell proliferation (Mamont and coworkers, 1976), but at 0.1–1.0 mM they reversed inhibited growth (Figures 4B and 5). Higher concentrations were not tested since 10 mM DAP alone blocked HTC cell proliferation, without the lag time of one generation period usually observed with ODC inhibitors (Figure 4B), and became subsequently cytotoxic (Figure 4A). This effect does not appear to depend on the presence of diamine oxidase activity, since the same results were obtained using calf or horse sera.

B. Regulation of Polyamine Biosynthetic Enzymes

α-Hydrazino ornithine (Harik, Hollenberg and Snyder, 1974a; Inoue and coworkers, 1975), MeOrn (McCann and coworkers, 1977b; Mamont and coworkers, 1978b) and methyl-GAG (Pegg, Corti and Williams-Ash, 1973; Fillingame and Morris, 1973a; Pegg and Jefferson, 1974) increase the activity of their respective target enzymes, ODC and SAMD, when organs or cells are treated with these inhibitors. In all cases, the apparent degradative rates of the enzymes are slowed. A possible explanation of this increased half-life is the ability of these inhibitors to bind to the enzyme and thereby to render the enzymic protein less susceptible to proteolytic digestion (Schimke, 1973). It is also possible that the inhibitors affect the expression of ODC and SAMD indirectly via their polyamine-depleting properties. Increasing evidence appears to indicate that putrescine exerts multiple negative controls on ODC (see Chapter 7 and Chapter 9). Similarly, spermidine may regulate SAMD (Mamont and coworkers, 1978b). Comparative studies of the effect of MeOrn and DFMeOrn on SAMD activity in HTC and its polyamine-overproducing variant show an inverse relationship between intracellular spermidine levels and SAMD (Mamont and coworkers, 1978b). This relationship is further substantiated by the use of the (+) and the (—) optical isomers of MeOrn: (+)-α-MeOrn depletes spermidine levels and increases HTC cell SAMD activity, whereas the (—) enantiomer is inactive. Furthermore, 10 μM spermidine added to DFMeOrn-pretreated cells reduced elevated SAMD activity to control levels. Other polyamines also lowered elevated SAMD with the following order of effectiveness: spermidine $\geqslant$ spermine > putrescine $\gg$ cadaverine. 1 mM DAP exerted no effect for at least 36 h. Given by itself, 1 mM DAP like other ODC inhibitors increased SAMD activity above control presumably because of its polyamine-depletion property (Mamont and coworkers, in preparation).

Whatever the molecular mechanism by which polyamines affect ODC and SAMD activities, these actions make it difficult to create and sustain a stringent polyamine deficiency by ODC and SAMD inhibitors. This is true of cell cultures but especially *in vivo*.

IV. PERSPECTIVES

Our studies suggest that polyamines are required for maximum rates of cell proliferation. This finding, based on experiments using rat hepatoma cells as a model system appears to apply to rapidly growing cells in general. Antiproliferative effects of ODC inhibitors have been confirmed in many cell lines including rat Zajdela hepatoma and C_6 astrocytoma, mouse L1210 leukaemia and EMT6 sarcoma, human MA160 prostatic adenoma (Mamont and coworkers, 1978a and unpublished results) WI38 (Dufy and Kremzner, 1977) and human fibroblasts (Hölttä, Pohjanpelto and Jänne, 1979).

In HTC cells the most immediate and predominant consequence of putrescine and spermidine depletion is a decrease in DNA synthesis. This is followed by effects on general cell metabolism as manifested by reduced rates of leucine and uridine incorporation into total protein and RNA respectively and (or) by reduced membrane transport of these precursors. Complete arrest of cell multiplication has not been observed, however, even after prolonged putrescine and spermidine depletion (Mamont and coworkers, 1978a). The question remains whether this residual growth results from the maintenance of initial spermine levels or whether it is spermine-independent.

The effect of inhibition of spermidine and spermine synthesis by methyl-GAG, suggests also a specific role of these amines in the S phase of lymphocytes (Otani and coworkers, 1974; Fillingame, Jorstad and Morris, 1975; Knutson and Morris, 1978), Balb/3T3 and WI38 fibroblasts (Boynton, Whitfield and Isaacs, 1978). No effect on overall RNA and protein synthesis has been detected in these cell systems. In HeLa cells decrease of DNA synthesis parallels the decrease of protein synthesis, although these effects do not appear to be causally connected (Kokran and Eriksen, 1977). In the presence of methyl-GAG, rat embryo fibroblasts (Rupniak and Paul, 1978) and rat brain tumour cells accumulate in the G_1 and Ehrlich ascites cells in the G_2 phase of the cell cycle (Heby, Andersson and Gray, 1978a: Heby and coworkers, 1978b), a situation which has been compared to that resulting from nutrient starvation.

Taken together the available information may suggest that polyamine deficiency induces or potentiates a negative pleiotypic response even in malignant cells where the transformed phenotype appears to be under constant pleiotypic activation (Hershko and coworkers, 1971). In fact, it was suggested (Clô and coworkers, 1976) that polyamines are part of the constant pleiotypic-activated mechanism of the malignant cells.

At present this possibility cannot be fully substantiated for several reasons. None of the inhibitors of polyamine biosynthesis leads to complete polyamine depletion (Table 2). The best compound presently available for the depletion of putrescine and spermidine is DFMeOrn, but at the concentrations used marked spermine deficiency has not been achieved. The SAMD inhibitor methyl-GAG limits spermidine and spermine synthesis but leads to a striking accumulation of putrescine which can be reduced but not totally abolished by MeOrn (Morris, Jorstad and Seyfried, 1977). The same is true for 1,1-[(methyl-ethane diylidene)-dinitrilo]-*bis*-(3-aminoguanidine) (MBAG) a recently discovered irreversible inhibitor of SAMD (Pegg, 1978). Increasing evidence indicates that not all effects of methyl-GAG result solely from SAMD inhibition.

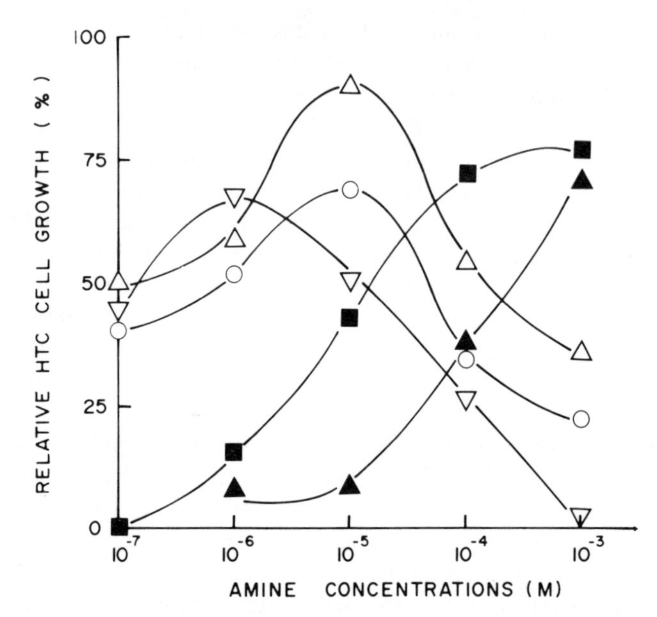

Figure 5 Reversal of inhibition of HTC cell division by polyamines. HTC cell spinner cultures were incubated for 3 days in Swim's 77 supplemented with 10% dialysed horse serum and in the presence or absence of 5 mM DFMeOrn. At that time, cells were gently centrifuged (200 × **g**) and resuspended in fresh medium in the presence or absence of the drug at a cell density of 1.5 × 10^5 cells ml^{-1}. Increasing concentrations of amines were added to DFMeOrn-treated cells and cell number were determined 2 days later. Growth is expressed as % relative to the control (100%) and DF-MeOrn treated cells arbitrarily set at 0%. (0) putrescine; (△) spermidine; (▽) spermine; (■) cadaverine and (▲) DAP

Methyl-GAG is structurally similar to spermidine and spermine and can compete with these polyamines for their cell membrane transport (Clark and Fuller, 1975) for binding to DNA (Sartorelli and coworkers, 1965) for catabolism (Hölltä and coworkers, 1973; Hölltä, 1977). It can also interfere with fundamental metabolic processes by mechanisms which may not be polyamine-related (Otani and coworkers, 1977; Pathak, Porter and Dave, 1977; Dave, Pathak and Porter, 1978).

All present experimental approaches used to create polyamine deficiency in mammalian cells have certain shortcomings. In the normal polyamine environment control mechanisms operating intracellularly and, as suggested by Canellakis and coworkers, 1978, at the cell membrane probably maintain a critical intracellular polyamine concentration, essential for optimal rates of the metabolic processes. As an immediate consequence of polyamine deficiency, ODC and SAMD activities increase and oppose the action of the inhibitors. In this regard, irreversible inhibitors fortunately have great advantages over competitive inhibitors.

The combination of catalytic inhibitors of ODC and SAMDC with putrescine analogues that serve as ODC 'repressors' is not promising, because these analogues can fulfil the requirement for natural endogenous polyamines. Thus, dose-dependent inactivation of ODC parallels the dose-dependent reactivation of inhibited growth. The strong inhibitory activity of DAP in Chinese hamster ovary cells (Sunkara, Rao and Nishioko, 1977) and in HTC cells (Figure 4A) cannot be attributed solely to the blockade of ODC activity. It may depend in part on other effects which occur at high concentrations. Even for the usual polyamines, the reversal of the anti-proliferative effects of ODC inhibitors declines beyond 10 μM (Figure 5).

Our work demonstrates that both DAP and cadaverine can substitute for putrescine and spermidine in HTC cells. Formation of DAP by mammalian tissues has never been demonstrated. Cadaverine appears to be a normal mammalian tissue constituent (Caldarera, Barbiroli and Moruzzi, 1965; Dolezavola, Stepito-Klanco and Fairweather, 1974; Schmidt-Glenewinckel, Nomura and Giacobini, 1977; Henningsson, Persson and Rosengren, 1976). The fact that cadaverine can substitute for putrescine and may possibly be metabolized into its spermidine and spermine homologues makes it worthwhile to search for this amine in eukaryotic systems, particularly in conditions of putrescine and spermidine deficiency which might elicit its appearance. More importantly, comparison of the structural features of polyamines required for cell replication processes with those required for inactivation of ODC and SAMD may differentiate these functions. This possibility is enhanced by the finding that DAP can substitute for putrescine and spermidine to restore maximum rates of cell proliferation without controlling SAMD, presumably because it cannot form the corresponding spermidine analogue.

V. SUMMARY AND CONCLUSIONS

In conclusion, the ODC inhibitors have been useful tools to show that polyamines play important roles in cell replication processes. These compounds permit the selection of mammalian cells with altered but stable polyamine metabolism which will permit comparative studies on the genetic regulation of the polyamine biosynthetic enzymes and more generally on polyamine cell metabolism. They may be of great value in assessing sequential events occurring in embryonic development (Manen and Russell, 1973; Emanuelson and Heby, 1978; Alexandre, 1978; Brachet and coworkers, 1978) in cell differentiation (Reuben and coworkers, 1976) and during virus infection (Williamson, 1976) or viral cell transformation (Bachrach, 1978). *In vivo* studies with ODC inhibitors have just started. At the present time it is premature to confidently attribute to polyamine deficiency the increase of survival time of L1210 leukaemia mice (Prakash and coworkers, 1978), the blockade of Ehrlich ascites cells in S and G_2M (Heby, Andersson and Gray, 1978a) or the slowed weight gain of mouse sarcoma tumour (Kato and coworkers, 1976) and of the prostate of castrated rats treated with testosterone (Danzin and coworkers, 1979d). These results are of sufficient interest to be pursued further and to encourage the design of further potent, specific and irreversible inhibitors of the mammalian polyamine biosynthetic enzymes.

ACKNOWLEDGEMENTS

The work presented was accomplished with the collaborative efforts of Drs. B. W. Metcalf (Chemistry), C. Danzin and M. J. Jung (Enzymology), J. Grove and P. Böhlen (Analytical Biochemistry), P. J. Schechter and N. J. Prakash (Experimental Therapeutics). The authors wish to thank Dr. N. Seiler for his helpful comments. We dedicate this review to the memories of G. M. Tomkins and Philippe S. Mamont.

REFERENCES

Abeles, R. H., and Maycock, A. L. (1976). *Account of Chem. Res.*, **9**, 313–319

Abdel-Monem, M. M., Newton, N. E., and Weeks, C. E. (1974). *J. Med. Chem.*, **17**, 447–451.

Abdel-Monem, M. M., Newton, N. E., Ho, B.C., and Weeks, C. E. (1975a). *J. Med. Chem.*, **18**, 600–604.

Abdel-Monem, M. M., Newton, N. E., and Weeks, C. E. (1975b). *J. Med. Chem.*, **18**, 945–948.

Alexandre, H. (1978). *C. R. Acad. Sci.*, **286**, 1215–1217.

Bachrach, U. (1978). Polyamine synthesis in normal and neoplastic cells. In R. A. Campbell, D. R. Morris, D. Bartos, G. D. Daves and F. Bartos (Eds), *Advances in Polyamine Research*, Vol. 1, Raven Press, New York, pp. 83–91.

Bey, P., Danzin, C., Van Dorsselaer, V., Mamont, P. S., Jung, M. J., and Tardif, C. (1978). *J. Med. Chem.*, **21**, 50–55.

Boecker, E. A., and Snell, E. E. (1972). Aminoacid decarboxylases. In P. D. Boyer (Ed) *The Enzymes*, Vol. 7, Academic Press, pp. 217–253.

Boynton, A. L., Whitfield, J. F., and Isaacs, R. (1976). *J. Cell Physiol.*, **89**, 481–488.

Brachet, J., Mamont, P., Boloukhère, M., Baltus, E., and Hanocq-Quertier, J. (1978). *C. R. Acad. Sci.*, **287**, 1289–1292.

Caldarera, C. M., Barbiroli, B., and Moruzzi, G. (1965). *Biochem. J.*, **97**, 84–88.

Canellakis, E. S., Heller, J. S., Kyriakidis, D., and Chen, K. Y. (1978). Intracellular levels of ornithine decarboxylase, its half-life, and a hypothesis relating polyamine-sensitive membrane receptors to growth. In R. A. Campbell, D. R. Morris, D. Bartos, G. D. Daves and F. Bartos (Eds), *Advances in Polyamine Research*, Vol. 1, Raven Press, New York, pp. 17–30.

Clark, J. L., and Fuller, J. L. (1975). *Biochemistry*, **14**, 4403–4409.

Clô, C., Orlandini, G. C., Casti, A., Guarnieri, C. (1976). *Ital. J. Biochem.*, **25**, 94–114.

Corti, A., Dave, C., Williams-Ashman, H. G., Mihich, E., and Schenone, A. (1974). *Biochem. J.*, **139**, 351–357.

Coward, J. K., Motola, N. C. and Moyer, J. D. (1977). *J. Med. Chem.*, **20**, 500–505.

Danzin, C., Jung, M. J., Seiler, N., and Metcalf, B. W. (1979a). *Biochem. Pharmacol.*, **28**, 633–639.

Danzin, C., Jung, M. J., Metcalf, B. W., Grove, J., and Casara, P. (1979b). *Biochem. Pharmacol.* **28**, 627-631.
Danzin, C., Jung, M. J., Grove, J., and Bey, P. (1979c). *Life Sciences*, **24**, 519-524.
Danzin, C., Jung, M., Claverie, N., Grove, J., Sjoerdsma, A., and Koch-Weser, J. (1979d). *Biochem. J.*, **180**, 507-513.
Dave, C., Pathak, S. N., and Porter, C. W. (1978). Studies in the mechanism of cytotoxicity of Methylglyoxal *Bis*-(guanylhydrazone) in cultured leukemia 1210 cells. In R. A. Campbell, D. R. Morris, D. Bartos, G. D. Daves and F. Bartos (Eds), *Advances in Polyamine Research*, Vol. 1, Raven Press, New York, pp. 153-171.
Dolezalova, H., Stepita-Klauco, M., and Fairweather, R. (1974). *Brain Res.*, **77**, 166-168.
Dufy, P. E., and Kremzner, L. T. (1977). *Exp. Cell Res.*, **108**, 435-439.
Emanuelsson, H., and Heby, O. (1978). *Proc. Nat. Acad. Sci. USA*, **75**, 1039-1042.
Fillingame, R. H., and Morris, D. R. (1973a). *Biochem. Biophys. Res. Commun.*, **52**, 1020-1025.
Fillingame, R. H., and Morris, D. R. (1973b). *Biochemistry*, **12**, 4479-4487.
Fillingame, R. H., Jorstad, C. M., and Morris, D. R. (1975). *Proc. Nat. Acad. Sci. USA*, **72**, 4042-4045.
Harik, S. I., and Snyder, S. H. (1973). *Biochim. Biophys. Acta*, **327**, 501-509.
Harik, S. I., Hollenberg, M. D., and Snyder, S. H. (1974a). *Mol. Pharmacol.*, **10**, 41-47.
Harik, S. I., Hollenberg, M. D., and Snyder, S. H. (1974b). *Nature, Lond.*, **249**, 250-251.
Heby, O., and Russell, D. H. (1974). *Cancer Res.*, **34**, 886-892.
Heby, O., Andersson, G., and Gray, J. W. (1978a). *Exp. Cell Res.*, **111**, 461-464.
Heby, O., Andersson, G., Gray, J. W., Marton, L. J. (1978b). Accumulation of rat brain tumor cells in G_1 and Ehrlich ascites tumor cells in G_2 by partial deprivation of cellular polyamine content. In R. A. Campbell, D. R. Morris, D. Bartos, G. D. Daves and F. Bartos (Eds), *Advances in Polyamine Research*, Vol. 1, Raven Press, New York, pp. 127-131.
Heller, J. S., Canellakis, E. S., Bussolotti, D. L., and Coward, J. K. (1975). *Biochim. Biophys. Acta*, **403**, 197-207.
Hershko, A., Mamont, P. S., Shields, R., and Tomkins, G. M. (1971). *Nature (New Biol.)*, **232**, 206-211.
Henningsson, S., Persson, L., and Rosengren, E. (1976). *Acta Physiol.*, **98**, 445-449.
Hibasami, H., and Pegg, A. E. (1978). *Biochem. Biophys. Res. Commun.*, **81**, 1398-1405.
Hölttä, E., Hannonen, P., Pispa, J., and Jänne, J. (1973). *Biochem. J.*, **136**, 669-676.
Hölttä, E. (1977). *Biochemistry*, **16**, 91-99.
Hölttä, E., Pohjanpelto, P., and Jänne, J. (1979) *Febs. Lett.*, **97**, 9-13.
Inoue, H., and Mizutani, A. (1973). *Anal. Biochem.*, **56**, 408-416.
Inoue, H., Kato, Y., Takigawa, M., Adachi, K., and Takeda, Y. (1975). *J. Biochem.*, **77**, 879-893.
Jänne, J., Pösö, H., and Raina, A. (1978). *Biochim. Biophys. Acta*, **473**, 241-293.
Jung, M. J., Lippert, B., Metcalf, B. W., Schechter, P. J., Böhlen, P., and Sjoerdsma, A. (1977). *J. Neurochem.*, **28**, 713-723.
Jung, M. J., Metcalf, B. W., Lippert, B., and Casara, P. (1978). *Biochemistry*, **17**, 2628-2632.
Jung, M. J., and Seiler, N. (1978). *J. Biol. Chem.* **253**, 7431-7439.
Kallio, A., Löfman, M., Pösö, H., and Jänne, J. (1977a). *Febs., Lett.*, **79**, 195-199.
Kallio, A., Pösö, H., and Jänne, J. (1977b). *Biochim. Biophys. Acta*, **479**, 345-353.
Kallio, A., Pösö, H., Guha, S. K., and Jänne, J. (1977c). *Biochem. J.*, **166**, 89-94.

Kato, Y., Inoue, H., Gohda, E., Tamada, F., and Takeda, Y. (1976). *Gann.*, **67**, 569–576.
Kay, J. E., and Lindsay, V. J. (1973). *Biochem. J.*, **132**, 791–796.
Kitz, R., and Wilson, I. B. (1962). *J. Biol. Chem.*, **237**, 3245–3249.
Klein, D., and Morris, D. R. (1978). *Biochem. Biophys. Res. Commun.*, **81**, 199–204.
Knutson, J. C., and Morris, D. R. (1978). 'DNA synthesis in nuclei isolated from cells inhibited in polyamine accumulation'. In R. A. Campbell, D. R. Morris, D. Bartos, G. D. Daves and F. Bartos (Eds), *Advances in Polyamine Research*, Vol. 1, Raven Press, New York, pp. 181–187.
Krokan, H., and Eriksen, A. (1977). *Eur. J. Biochem.* **72**, 501–507.
McCann, P. P., Tardif, C., Mamont, P. S., and Schuber, F. (1975). *Biochem. Biophys. Res. Commun.*, **64**, 336–341.
McCann, P. P., Tardif, C., and Mamont, P. S. (1977a). *Biochem. Biophys. Res. Commun.*, **75**, 948–954.
McCann, P. P., Tardif, C., Duchesne, M. C., and Mamont, P. S. (1977b). *Biochem. Biophys. Res. Commun.*, **76**, 893–899.
Mamont, P. S., Böhlen, P., McCann, P. P., Bey, P., Schuber, F., and Tardif, C. (1976). *Proc. Nat. Acad. Sci. USA*, **73**, 1626–1630.
Mamont, P. S., Duchesne, M. C., Grove, J., and Bey, P. (1978a). *Biochem. Biophys. Res. Commun.*, **81**, 58–66.
Mamont, P. S., Duchesne, M. C., Grove, J., and Tardif, C. (1978b). *Exp. Cell Res.* **115**, 387–393.
Mamont, P. S., Duchesne, M.-C., Joder-Ohlenbusch, A.-M., and Grove, J. (1978c). In N. Seiler, M. J. Jung and J. Koch-Weser (Eds), *Substrate-induced Irreversible Inhibition of Enzymes*, Elsevier/North-Holland Biomedical Press, pp. 43–54.
Manen, C. A., and Russell, D. H. (1973). Polyamines in marine vertebrates. In D. H. Russell (Ed), *Polyamines in Normal and Neoplastic Growth*, Raven Press, New York, pp. 277–288.
Marton, L. J., Lee, P. L. Y. (1975). *Clin. Chem.*, **21**, 1721–1724.
Metcalf, B. W., Bey, P., Danzin, C., Jung, M. J., Casara, P., and Vevert, J. P. (1978). *J. Amer. Chem. Soc.*, **100**, 2551–2553.
Morris, D. R., Jorstad, C. J., and Seyfried, C. E. (1977). *Cancer Res.*, **37**, 3169–3172.
Newton, N. E., and Abdel-Monem, M. M. (1977). *J. Med. Chem.*, **20**, 249–253.
Ono, M., Inoue, H., Suzuki, F., and Takeda, Y. (1972). *Biochim. Biophys. Acta*, **284**, 285–297.
Oka, T., and Perry, J. W. (1974a). *Nature, Lond.*, **250**, 660–661.
Oka, T., and Perry, J. W. (1974b). *J. Biol. Chem.*, **249**, 7646–7652.
Otani, S., Mizoguchi, Y., Matsui, I., and Morisawa, S. (1974). *Mol. Biol. Reports*, **1**, 431–436.
Otani, S., Matsui, I., and Morisawa, S. (1977). *Biochim. Biophys. Acta*, **478**, 417–427.
Pathak, S. N., Porter, C. W., and Dave, C. (1977). *Cancer Res.*, **37**, 2246–2250.
Pegg, A. E. (1973). *Biochem. J.*, **132**, 537–540.
Pegg, A. E., Corti, A., and Williams-Ashman, H. G. (1973). *Biochem. Biophys. Res. Commun.*, **52**, 696–701.
Pegg, A. E., and Jefferson, L. S. (1974). *Febs. Lett.*, **40**, 321–324.
Pegg, A. E., and Conover, C. (1976). *Biochem. Biophys. Res. Commun.*, **69**, 766–774.
Pegg, A. E. (1978). *J. Biol. Chem.*, **253**, 539–542.
Pett, D. M., and Ginsberg, H. S. (1968). *Fed. Proc.*, **27**, 615.
Pösö, H., and Jänne, J. (1976). *Biochem. J.*, **158**, 485–488.
Prakash, N. J., Schechter, P. J., Grove, J., and Koch-Weser, J. (1978). *Cancer Res.*, **38**, 3059–3062.
Rando, R. R. (1974). *Science*, **185**, 320–324.

Relyea, N., and Rando, R. R. (1975). *Biochem. Biophys. Res. Commun.*, **67**, 392–402.
Reuben, R. C., Wife, R. L., Breslow, R., Rifkind, R. A., and Marks, P. A. (1976). *Proc. Nat. Acad. Sci. USA*, **73**, 862–866.
Rupniak, H. R., and Paul, D. (1978). *J. Cell Physiol.*, **94**, 161–170.
Sartorelli, A. C., Ianotti, A. T., Booth, B. A., Schneider, F. H., Bertino, J. R., and Johns, D. G. (1965). *Biochim. Biophys. Acta*, **103**, 174–176.
Schimke, R. T. (1973). *Advances in Enzymology*, **37**, 135–187.
Schmidt-Glenewinckel, T., Nomura, Y., and Giacobini, E. (1977). *Neurochem. Res.* **2**, 619–637.
Seiler, N. (1973). Polyamine metabolism in the brain. In D. H. Russell (Ed), *Polyamines in Normal and Neoplastic Growth*, Raven Press, New York, pp. 137–156.
Shepherd, D. M., and MacKay, D. (1967). The histidine decarboxylases. In G. P. Ellis and G. B. West (Eds), *Progress in Medicinal Chemistry*, Vol. 5, Butterworths, London, pp. 199–250.
Siimes, M. (1967). *Acta Physiol. Scand. Suppl.*, **298**, 1–66.
Siimes, M., and Jänne, J. (1967). *Acta Chem. Scand.*, **21**, 815–817.
Sourkes, T. L. (1966). *Pharmacol. Rev.*, **18**, 53–60.
Stevens, L., McKinnon, I. M., and Winther, M. (1977). *Febs. Lett.*, **75**, 180–182.
Sunkara, P. S., Rao, P. H., and Nishioka, K. (1977). *Biochem. Biophys. Res. Commun.*, **74**, 1125–1133.
Tabor, H., and Tabor, C. W. (1972). *Advances in Enzymology*, **36**, 203–268.
Tabor, C. W., and Tabor, H. (1976). *Ann. Rev. Biochem.*, **45**, 285–306.
Thompson, E. B., Tomkins, G. M., and Curran, J. F. (1966). *Proc. Nat. Acad. Sci. USA*, **56**, 296–303.
Whitney, P. A., and Morris, D. R. (1978). *J. Bacteriol.*, **134**, 214–220.
Williams-Ashman, H. G., Pegg, A. E., and Lockwood, D. H. (1968). Mechanisms and regulation of polyamine and putrescine biosynthesis in male genital glands and other tissues of mammals. In G. Weber (Ed), *Advances in Enzyme Regulation*, Vol. 7, Pergamon Press, London, pp. 291–323.
Williams-Ashman, H. G., and Schenone, A. (1972). *Biochem. Biophys. Res. Commun.*, **46**, 288–295.
Williams-Ashman, H. G., Corti, A., and Tadolini, B. (1976). *Ital. J. Biochem.*, **25**, 5–32.
Williamson, J. D. (1976). *Biochem. Biophys. Res. Commun.*, **73**, 120–126.
Winther, M., and Stevens, L. (1978). *Febs Lett.*, **85**, 229–232.

Chapter 11

Inhibitors of the Biosynthesis of Putrescine, Spermidine and Spermine

LEWIS STEVENS AND EVELYN STEVENS

I. INTRODUCTION

This review is concerned with inhibitors of oligoamine* biosynthesis and does not consider the subsequent effects on other cell processes which have been reviewed (Chapter 10). During the past few years there has been much interest in developing inhibitors of oligoamine synthesis. The reasons for this are that oligoamines are synthesized rapidly at the onset of cell proliferation and may therefore be intimately involved in this process (Chapters 1–9). Oligoamine function may become more clearly defined when the synthesis of oligoamines is inhibited experimentally so that the cells become depleted of these compounds. Inhibition of the synthesis of oligoamines results in the arrest of eukaryotic and prokaryotic cell proliferation which may have important medical applications.

The typical eukaryote biosynthetic pathway for spermidine and spermine is shown in Figure 1. There are several sites at which it is possible to inhibit the pathway, and different types of inhibitor have been used to inhibit oligoamine

*Oligoamine is the collective name used in this review to include diamines, triamines and tetra-amines. It is in keeping with the way in which the prefix oligo- is used in other biochemical contexts, e.g. oligonucleotide, oligopeptide, oligosaccharide, and is therefore preferable to the term polyamine.

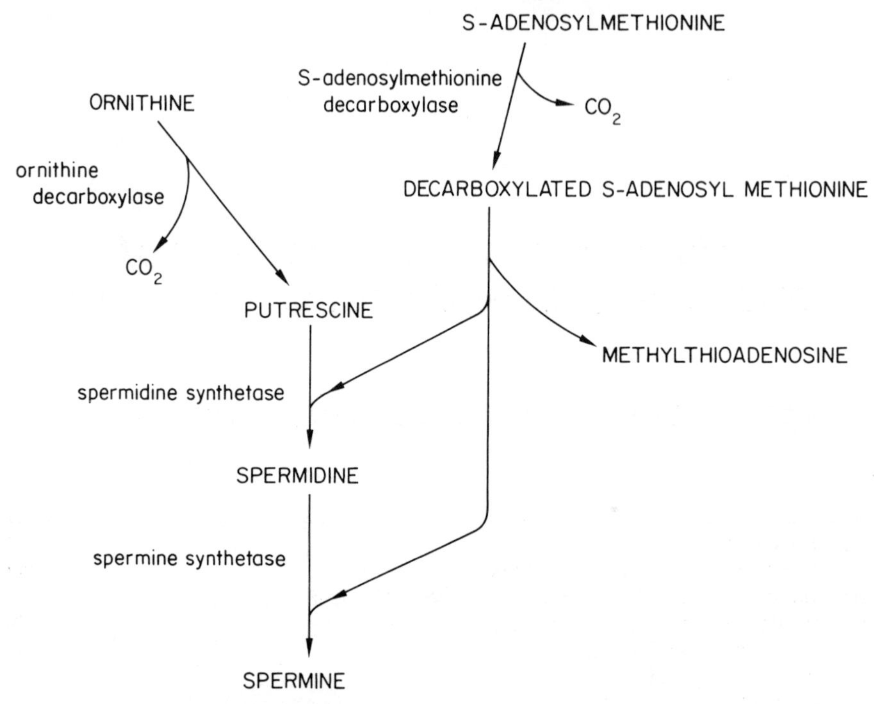

Figure 1 Biosynthetic pathway for oligoamines in eukaryotes

biosynthesis. The main differences between the eukaryotes and prokaryote pathways are the following: (1) prokaryotes have an additional route to putrescine from arginine; (2) prokaryote S-adenosylmethionine decarboxylases (SAMD) require Mg^{2+} for activity but are not activated by putrescine; (3) prokaryotes seem generally incapable of spermine biosynthesis.

There are therefore a number of features of the biosynthesis of oligoamines in eukaryotes which it is important to consider in designing inhibitors: (1) the rate-limiting steps in the pathway are generally considered to be ornithine decarboxylase (ODC) and SAM decarboxylase (SAMD); (2) both of these enzymes have short half-lives in contrast to spermidine and spermine synthetase; (3) the co-factor for ODC is pyridoxal phosphate whereas that for SAMD is covalently bound pyruvate (Pegg, 1977; Demetrion and coworkers, 1978); (4) the synthesis of ODC is suppressed by increasing intracellular concentrations of putrescine (Kay and Lindsay, 1973; Clark and Fuller, 1975; Stevens, McKinnon and Winther, 1976); (5) putrescine, and to a lesser extent other aliphatic diamines, are activators of SAMD; (6) although both ODC and SAMD turn over rapidly, the rate of turnover of spermidine and spermine is slow and this leads to a slow depletion of intracellular spermidine and spermine following enzyme inhibition.

Almost all of the research until the present time has been concerned with seeking inhibitors of ODC and SAMD. The inhibitors used fall broadly into the following categories: (1) structural analogues of ornithine and putrescine which may act as competitive inhibitors of ODC (see also Chapter 10); (2) compounds competing with the co-factors of these enzymes; (3) diamines acting as repressors of the synthesis of ODC; (4) inhibitors of SAMD.

Most of the inhibitors used so far appear to bind reversibly to the enzymes they inhibit but some, e.g. 1,1′-(methylethanediyledenedinitrilo)-*bis*(3-aminoguanidine) and DL-α-difluoromethylornithine, bind irreversibly and are thus likely to be more effective.

II. INHIBITORS OF ORNITHINE DECARBOXYLASE

A. Structural Analogues of Ornithine and Putrescine which Inhibit Ornithine Decarboxylase Activity

Most ornithine decarboxylases have Km values for ornithine between 0.01 mM and 0.5 mM and in many cases putrescine acts as a weak competitive inhibitor. Several structural analogues of both ornithine and putrescine have been tested as potential inhibitors of the enzyme (Table 1). ODC from rat prostate and regenerating liver are commonly used to test inhibitors as detailed below. The majority of the structural analogues are reversible competitive inhibitors of ornithine. Another feature of many inhibitors e.g. α-methylornithine, α-hydrazinoornithine and 1,4-diaminobutanone, is that *in vivo* they cause an increase in ODC provided the activity is then measured in dialysed tissue extracts. This is the result of stabilization of the enzyme (Chapter 10). ODC has a short half-life but this can be prolonged by binding inhibitors to it. The inhibitors vary in extent to which their effect may be reversed *in vitro* by increasing the pyridoxal phosphate concentration. This is discussed further with individual inhibitors.

1. Alkyl and Aryl Substituted Ornithines

A range of alkyl and aryl substituted ornithines have been tested as potential inhibitors of ODC. Skinner and Johansson (1972) tested a number of α-N and ε-N substituted ornithines but none caused inhibition. In contrast, the α-C substituted ornithines tested by Abdel-Monem and coworkers (1975a) showed some inhibitory activity. The larger the alkyl substituent on the α-C the less inhibitory activity up to C_4 but the hexyl and octyl derivatives showed more inhibitory activity. The benzyl derivative showed very low inhibitory activity. Of these derivatives the most potent, α-methylornithine, has been used extensively to inhibit putrescine synthesis (Chapter 10). High concentrations of pyridoxal phosphate do not reverse the inhibitory effect. α-Methylornithine

Table 1 Structural analogues of ornithine and

Compound	Structure	Source of ornithine decarboxylase	K_m ornithine (mM)	K_i inhibitor (mM)	K / K
α-methyl(±)-ornithine	$H_2N(CH_2)_3$—C(CH_3)(NH_2)—COOH	Rat prostate	0.13	0.04	3.
		Mouse spleen	0.09	0.026	3.
		Regenerating liver	0.1	0.2	0.
		Hepatoma (HTC)		0.04	
		Lactobacillus			
		Rat liver	0.087	0.04	2.
		Hepatoma (HTC)	0.09	0.04	2.
α-methyl(+)-ornithine		Rat liver	0.087	0.019	4.
		Hepatoma (HTC)	0.09	0.02	4.
		Bull prostate	0.072	0.02	3.
α-methyl(—)-ornithine		Rat liver	0.087	1.3	0.
		Hepatoma (HTC)	0.09	1.05	0.
		Bull prostate	0.072	1.0	0.
α-ethyl(±)ornithine	$CH_3NH(CH_2)_3$—C(CH_3)(NH_2)—COOH	Rat prostate	0.13	5.3	0.
α-n-propyl(±)ornithine		Rat prostate	0.13	7.8	0
α-n-butyl(±)ornithine		Rat prostate	0.13	11.0	0
α-n-hexyl(±)ornithine		Rat prostate	0.13	4.7	0
α-n-octyl(±)ornithine		Rat prostate	0.13	2.3	0.
α-benzyl(±)ornithine		Rat prostate	0.13	5.3	0
δ-N-methyl-α-methyl- -ornithine		Rat liver	0.087	0.5	0
		Bull prostate	0.072	0.74	0
δ-N-aminopropylornithine	$H_2N(CH_2)_3NH(CH_2)_3CH(NH_2)COOH$	Bull prostate	0.072	4.4	0
δ-N-aminopropyl-α- -methylornithine	$H_2N(CH_2)_3NH(CH_2)_3C(CH_3)(NH_2)COOH$	Bull prostate	0.072	5.5	0
tetrazolylornithine	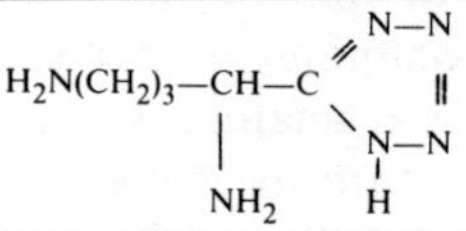 $H_2N(CH_2)_3$—CH(NH_2)—C(=N—N=N—NH)	Rat liver	0.087	0.12	0
		Hepatoma (HTC)	0.09	0.14	0
		Bull prostate	0.072	0.07	
trans 1,4-diaminocyclohexane- -1-carboxylic acid	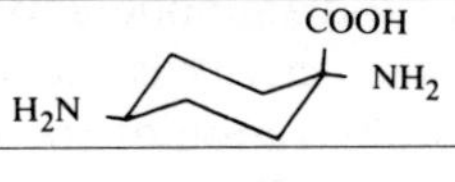 H_2N—(cyclohexane)—C(COOH)NH_2	Rat liver	0.087	0.07	
		Hepatoma (HTC)	0.09	0.07	
α-hydrazinoornithine	$H_2N(CH_2)_3$—CH($NHNH_2$)—COOH	*E.coli*	14.0	5×10^{-4}	2
		Rat ventral prostate	0.08	0.002	4
DL-α-hydrazino-δ-aminovaleric acid		Rat prostate		5.2×10^{-4}	
		Rat liver		5.2×10^{-4}	
(∓) 5-amino-2-hydrazino -methylpentanoic acid	$H_2N(CH_2)_3$—C($NHNH_2$)(CH_3)—COOH	Rat prostate	0.06	0.003	
DL-α-difluoromethyl ornithine	$H_2N(CH_2)_3$—C(NH_2)(CHF_2)—COOH	Hepatoma HTC			
trans-3dehydro-D,L-ornithine	$H_2NCH_2\,CH{=}CH\,CH(NH_2)COOH$	Rat liver	0.2	4.4×10^{-3}	
		Chick embryo muscle	0.15	2.8×10^{-3}	
trans-1,4-diamino-2-butene	$H_2N\,CH_2CH{=}CHCH_2NH_2$	Rat liver	0.2	2×10^{-3}	
1,4-diaminobutanone	$H_2N\,CH_2COCH_2CH_2NH_2$	Rat liver	0.2	0.046	
		Rat prostate	0.25	0.072	
		E. coli	4.4		
		Dicytostelium discoideum	0.67		
		Aspergillus nidulans	0.06	0.91×10^{-3}	
N-(5^1-phosphopyridoxal) -ornithine	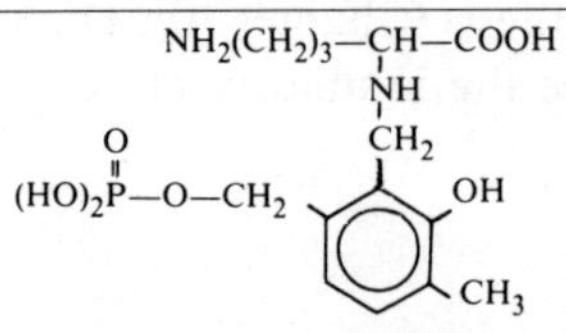 $NH_2(CH_2)_3$—CH(COOH)—NH—CH_2—(ring: $(HO)_2P(O)$—O—CH_2, OH, CH_3)	Rat liver	0.07		

utrescine which inhibits ornithine decarboxylase

Type of inhibition	Reversal of inhibition by pyridoxal phosphate	Synthesis (Reference)	Reference
ompetitive	—	Abdel Monem, Newton and	Abdel Monem, Newton and
ompetitive	—	Weeks (1974)	Weeks (1974)
ompetitive	—	Bey and Vevert (1977)	Abdel-Monem and
ompetitive	—	Ellington and Honigberg (1974)	coworkers (1975a)
ompetitive	—	Hoppe (1975)	Mamont and coworkers (1976)
ompetitive	—		O'Leary and Herreid (1978)
ompetitive	—		Bey and coworkers (1978)
ompetitive			Bey and coworkers (1978)
mpetitive			
ompetitive			
mpetitive			Bey and coworkers (1978)
mpetitive			
mpetitive			
mpetitive			
mpetitive			
mpetitive		Abdel Monem and coworkers	Abdel Monem and coworkers
mpetitive		(1975a)	(1975a)
mpetitive			
mpetitive			
mpetitive			
mpetitive			Bey and coworkers (1978)
mpetitive			Bey and coworkers (1978)
mpetitive			Bey and coworkers (1978)
mpetitive			Bey and coworkers (1978)
mpetitive			
mpetitive			
npetitive			Bey and coworkers (1978)
npetitive			
mpetitive			
mpetitive	+		Harik and Snyder (1973)
npetitive	+	Sawayama and coworkers (1976)	Inoue and coworkers (1975)
npetitive	+	Sawayama and coworkers (1976)	
mpetitive	+	Abdel Monem, Newton and Weeks (1975b)	Abdel Monem, Newton and Weeks (1975b)
eversible -competitive			Mamon and coworkers (1978)
npetitive	Non-competitive	Rando (1974)	Relyea and Rando (1975)
npetitive	Non-competitive		
npetitive	+	Karlen and Lindeke (1969)	Relyea and Rando (1975)
npetitive	—	Macholán (1965)	Stevens and coworkers (1978)
npetitive			
Mixed			
Mixed			
npetitive			
Non-petitive	Non-competitive	Heller and coworkers (1975)	Heller and coworkers (1975)

stabilizes ODC against degradation but McCann and coworkers (1977a) suggest that it may also have an indirect effect on the synthesis of ODC.

2. 5-Amino-2-hydrazinovaleric Acid (α-Hydrazinoornithine)

Harik and Snyder (1973) first demonstrated the compound as an ODC inhibitor and refer to the compound as α-hydrazinoornithine, but do not give a systematic name or structure. It is assumed that the compound to which they refer is the α-hydrazino derivative of 5-aminovaleric acid and not of ornithine (Inoue and coworkers, 1975). The compound used by Harik and Snyder (1973) differs from DL-α-hydrazino-δ-valeric acid used by Inoue and coworkers (1975) in that the former is optically active. In this section we use the names adopted by the authors in their papers.

The mechanism of inhibition appears more complex than that of α-methylornithine. It behaves competitively with respect to ornithine in both *E.coli* and rat prostate but the inhibition is greatly reduced or abolished in the presence of high concentrations of pyridoxal phosphate which is a non-competitive inhibitor with respect to α-hydrazinoornithine. Hydrazine, itself, is an inhibitor of ODC and is competitive with respect to pyridoxal phosphate, probably due to its interaction with the carbonyl group on pyridoxal phosphate. α-Hydrazinoornithine is a more effective inhibitor of ODC than hydrazine. It is also fairly specific for ODC, being much less inhibitory towards other pyridoxal phosphate dependent decarboxylases such as L-DOPA decarboxylase or L-glutamate decarboxylase. It also inhibits in the presence of a twenty-fold excess of pyridoxal phosphate. Thus though part of its inhibitory properties may be due to its ability to react with pyridoxal phosphate, its high selective potency might be due to its resemblance to the substrate.

Inoue and coworkers (1975) have demonstrated that DL-α-hydrazino-δ-aminovaleric acid is also a competitive inhibitor of ODC and that the inhibition is reduced by addition of pyridoxal phosphate.

When α-hydrazinoornithine is added to the culture medium of hepatoma cells, like α-methylornithine, it increases the ODC activity measured in dialysed cell extracts and increases the half-life of the enzyme from 10 to 28 min (Harik, Hollenberg and Snyder, 1974).

3. (±)-5-Amino-2-hydrazino-2-methylpentanoic Acid

Since both α-methylornithine and α-hydrazino-δ-aminovaleric acid are potent competitive inhibitors of ODC, Abdel-Monem, Newton and Weeks (1975b) tested the effects of combining both α-hydrazino- and α-methyl-functions within one analogue. (±)-5-Amino-2-hydrazino-2-methylpentanoic acid proved to be a competitive inhibitor of rat prostate ODC having a K_i value intermediate between those of α-methylornithine and α-hydrazino-δ-aminovaleric acid. When it was tested *in vivo*, however, it proved less effective than α-methylornithine in blocking an increase in putrescine levels in transformed lymphocytes. Pyridoxal phosphate reduces its inhibition *in vitro*.

4. Other Ornithine Analogues

The unsaturated analogue, *trans*-3-dehydroornithine is a strong competitive inhibitor of ODC (Relyea and Rando, 1975). More recently Mamont and coworkers (1978) have reported DL-α-difluoromethylornithine as being an irreversible inhibitor of ODC in rat hepatoma cells. No enzyme activity could be detected in dialysed cell extracts after the inhibitor treatment and 0.01mM α-difluoromethylornithine almost completely prevented putrescine and spermidine accumulation.

DL-α-hydrazino-δ-benzylaminovalerate is also reported to cause some inhibition of liver but DL-*erythro*- and DL-*threo*-β hydroxyornithines had no effect (Inoue and coworkers, 1975).

5. Analogues of Putrescine

The analogues of putrescine which have been found to inhibit ODC are the α,β-unsaturated analogues and 1,4-diaminobutanone. 1,4-Diamino-*trans*-butene is a very potent competitive inhibitor of the enzyme, 1,4-diamino-2-butyne is not as potent. The inhibition produced by both analogues is reversed by pyridoxal phosphate. It is thought that the complexes formed between the amines and pyridoxal phosphate may be the actual inhibitors (Relyea and Rando, 1975).

Many analogues of putrescine were tested as potential inhibitors of ODC from *Aspergillus nidulans*, but only 1,4-diaminobutanone inhibited strongly when added at 1.3 mM concentration (Stevens, McKinnon and Winther, 1977). 1,4-Diaminobutanone is a reversible competitive inhibitor of ornithine decarboxylase and was most effective against ODC from *A.nidulans* and least effective against the *E.coli* enzyme (see Table 1 and Stevens and coworkers, 1978).

In *A.nidulans* the ODC inhibition is completely reversed by dialysis of the crude extract. The inhibition by 0.1 mM 1,4-diaminobutanone is not appreciably reversed by pyridoxal phosphate concentrations up to 1 mM (i.e. 30 times the optimum concentration used in the assay). Although 1,4-diaminobutanone is a potent competitive inhibitor of ODC, a disadvantage is its instability in solution. In an attempt to produce a more stable inhibitor we made the oxime and semicarbazone derivatives of 1,4-diaminobutanone, but this resulted in the loss of inhibitory properties. We found that 2-hydroxyputrescine is also a competitive inhibitor of the *A.nidulans* enzyme, but it is much weaker having a K_i value comparable with that of putrescine (K_i = 0.06 mM).

1,4-Diaminobutanone induces an increased amount of ODC in germinating conidia of *A.nidulans* by stabilizing the enzyme against breakdown and thereby increasing its half-life (Stevens and McKinnon, 1977).

6. N-(5′-Phosphopyridoxyl)-ornithine

All ornithine decarboxylases so far investigated use pyridoxal phosphate as a coenzyme and probably involve as an intermediate the Schiff's base between ornithine and pyridoxal phosphate. Heller and coworkers (1975) therefore investigated the use of a stable adduct between ornithine and pyridoxal phosphate as a potential inhibitor of ODC. They synthesized N-(5′-phosphopyridoxyl)-ornithine, N-(5′-pyridoxyl)-ornithine and N-(5′-phosphopyridoxyl)-lysine and tested them as potential inhibitors of ODC, lysine decarboxylase, SAMD and tyrosine aminotransferase. No inhibitory activity was demonstrated for the dephosphorylated derivative. N-(5′-phosphopyridoxyl)-ornithine inhibited both ODC and lysine decarboxylase, however, the former was more strongly inhibited than the latter, whereas the reverse held for N-(5′-phosphopyridoxyl)-lysine. The adduct, N-(5′-phosphopyridoxyl)-ornithine thus sho a degree of specificity for ODC but not absolute specificity. The kinetics of inhibition of ODC by N-(5′-phosphopyridoxyl)-ornithine are complex. The inhibitor is non-competitive with respect to both ornithine and pyridoxal phosphate. Heller and coworkers (1975) suggest that N-(5′-phosphopyridoxyl)-ornithine binds to the site on the enzyme normally occupied by both ornithine and pyridoxal phosphate.

B. Inhibitors Competing for Pyridoxal Phosphate

Less specific inhibitors of ODC are those reagents which combine with pyridoxal phosphate preventing it from functioning as a co-factor. They would be expected to inhibit a variety of pyridoxal phosphate requiring enzymes and any specificity they show may depend on how strongly pyridoxal phosphate binds to the apoenzymes. Examples are 4-bromo-4-hydroxybenzyloxyamine (Leinweber, 1968) and the ornithine analogue canaline (Rahiala and coworkers, 1971). Although canaline ($H_2N.O.CH_2CH_2CH(NH_2)COOH$) appears to be a structural analogue of ornithine, it inhibits a number of transaminases and decarboxylases by complex formation between canaline and pyridoxal phosphate. There is spectroscopic evidence for complex formation between canaline and pyridoxal phosphate (Rahiala and coworkers, 1971). The inhibition of ODC by canaline can be readily reversed by pyridoxal phosphate whereas the inhibition of ornithine transaminase can not. Rahiala and coworkers (1971) suggest that this is due to the ready dissociation of the holoenzyme in the case of ODC as compared with that of transaminase.

C. Diamines Acting as Repressors of Ornithine Decarboxylase

Putrescine plays a key role in the control of spermidine and spermine biosynthesis. It acts at at least four sites in the regulation of spermine and spermidine biosynthesis: (1) it is a weak competitive inhibitor of ODC — this is probably not of physiological significance; (2) it represses the synthesis of ODC; (3) it activates SAMD; (4) it is the substrate for spermidine synthetase (Figure 2). An analogue of putrescine which could act at site (2) but not at site (4) might be useful in reducing intracellular oligoamine concentrations. If it were able to act at site (4) as a substrate for spermidine synthetase it might allow the production of spermidine analogues which might be able to substitute for spermidine *in vivo*. If, however, it affected only site (2) to a large degree, then it would prevent the production of tri- and tetra-amines. The diamine in question might then fulfil some of the roles of putrescine.

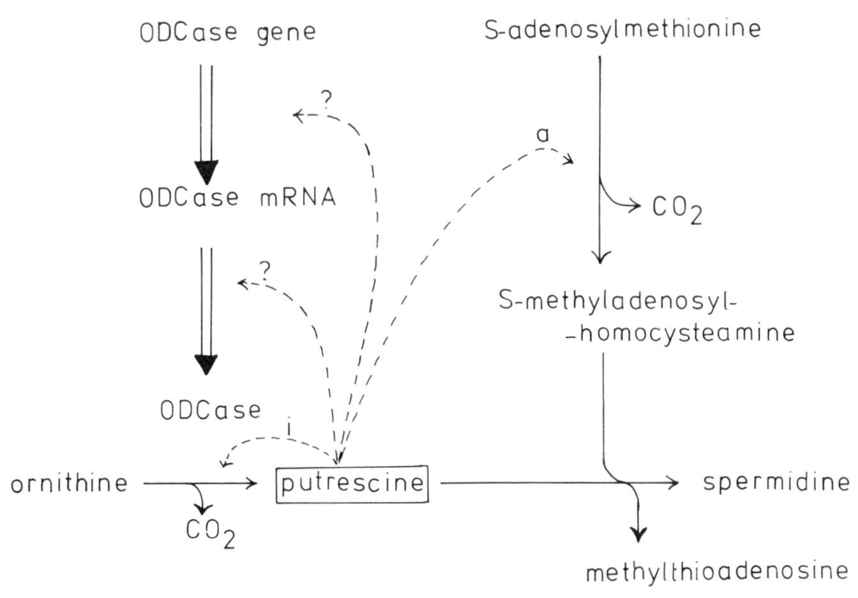

Figure 2 Role of putrescine in the control of spermidine biosynthesis
i = inhibitor, a = activator

This type of inhibitor was discovered by Pösö and Jänne (1976) and termed a 'gratuitous' repressor; it is the counterpart of the gratuitous inducer in inducible enzyme systems, e.g. isophenylthiogalactoside for β-galactosidase.

They found that when partially hepatectomized rats were given intraperitoneal injections of 1,3-diaminopropane, this suppressed the increase in ODC which normally occurs during the early stages of liver regeneration. 1,3-Diaminopropane did not affect the activity of ODC when measured *in vitro*. Kallio and coworkers (1977b) have shown that 1,3-diaminopropane given *in vivo* causes a reduction in the amount of immunoreactive protein produced and suggest that the amine affects the synthesis of ODC. But there is also evidence for the presence of an inhibitor produced in response to diamine injections in regenerating liver (Kallio and coworkers, 1977b), hepatoma cells (McCann, Tardif and Mamont, 1977b; Chapter 7) and in rat liver after growth hormone or thioacetamide treatments (Pegg, Conova and Wrona, 1978). Generally the inhibitor is only manifest when all ODC activity has been suppressed. There does not appear to be a parallel production of inhibitor and loss of ODC and Pegg, Conova and Wrona (1978) could not detect any inhibitor in kidney extracts when ODC activity had been completely suppressed. Thus some evidence suggests that 1,3-diaminopropane inhibits the synthesis of ODC but conversely other evidence suggests that it causes inhibitor production.

Kallio and coworkers (1977a) suggested that synthesis of ODC in regenerating liver is regulated in at least two different ways. At the early stages of liver regeneration the activation of the ODC gene occurs, whereas at later stages the enzyme level is probably controlled post-transcriptionally. They found that during the early stages of regeneration the effects of 1,3-diaminopropane are similar to those of α-amanitin, whereas at the later stages 1,3-diaminopropane or cycloheximide cause a similar rate of loss of ODC giving half-lives of 12 and 14 min respectively.

Guha and Jänne (1977) have also tested injecting diamines into rats and measuring ovarian ODC activity after gonatrophin treatment. 1,3-Diaminopropane and putrescine both partially prevent the rise in activity and, surprisingly, cadavarine and diaminohexane suppress it completely. None of these amines affected ODC activity *in vitro*. Guha and Jänne suggested that the differences in effectiveness of the diamines may be due to differential rates of uptake and also that the diamines do not take over the roles of the naturally-occurring amines. Injections of 1,2-diaminoethane, 1,3-diaminopropane or putrescine into rats also prevent an increase in ODC on ventral prostate and seminal vesicle following hormone replacement therapy (Piik and coworkers, 1977). 1,3-Diaminopropane has also been shown to affect the intracellular putrescine concentration and presumably therefore to inhibit ODC production in Chinese hampster ovary cells (Sunkara, Rao and Nishioka, 1977).

III. INHIBITORS OF S-ADENOSYLMETHIONINE DECARBOXYLASE

S-Adenosylmethionine decarboxylases appear to fall into three categories: (1) those which require the presence of putrescine or some other diamine for maximum activity; (2) those which require Mg^{2+} and are insensitive to putrescine; and (3) those not activated by either putrescine or Mg^{2+}. Although no extensive systematic study has been made of SAMD it seems likely that the enzymes from mammalian sources will fit into the first category. The bacterial enzymes are likely to be predominantly of the second category. In this section we shall be concerned mainly with the mammalian type.

A. (1,1′-[Methylethanediylidene)dinitrilo]diguanidine

This compound has been most widely used to inhibit SAMD. It is usually known by its semi trivial name methylglyoxal *bis*(guanylhydrazone) and abbreviated to either methyl-GAG or MGBG. This is unfortunate since, as can be seen from the structure, it implies a relationship to the guanine which it does not possess.

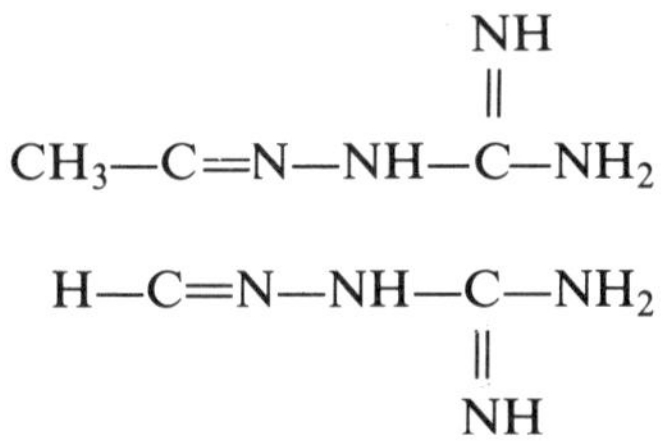

```
                        NH
                        ||
CH3—C=N—NH—C—NH2

     H—C=N—NH—C—NH2
                        ||
                        NH
```

At the suggestion of Dr. H. B. F. Dixon, (University of Cambridge), we use the semi trivial name methylglyoxal *bis* (amidinohydrazone) abbreviated to MGBA as more accurately conveying the structure.

Williams-Ashman and Schenone (1972) discovered that it is a potent inhibitor of SAMD. The mechanism of inhibition of SAMD has been investigated by Hölttä and coworkers (1973) and Corti and coworkers (1974).

MGBA is a much more potent inhibitor of the putrescine-dependent SAMD from rat prostate and from yeast than of the putrescine-insensitive SAMD of *E.coli* (Williams-Ashman and Schenone, 1972). The latter is only affected by concentrations of MGBA two orders of magnitude higher than that required for yeast and prostate SAM decarboxylases. The inhibition in all cases reported is reversible and is greatest when measured in the presence of optimum putrescine concentrations. When the kinetics of SAMD inhibition were studied Hölttä and coworkers (1973) and Corti and coworkers (1974) found that MGBA is competitive with respect to S-adenosylmethionine ($K_i < 1\ \mu M$ with saturating putrescine) and uncompetitive with respect to the activator putrescine.

Corti and coworkers (1974) also tested a number of analogues of MGBA as potential inhibitors of SAMD from rat liver. Dimethylglyoxal *bis*(amidinohydrazone), ethylglyoxal *bis*(amidinohydrazone) and di-N^1 methylglyoxal *bis*-

(amidinohydrazone) are potent inhibitors, whereas propanedialdehyde *bis*-(amidinohydrazone), pentanedialdehyde *bis*(amidinohydrazone) and di-N^{111}-methylglyoxal *bis*(amidinohydrazone), are relatively weak inhibitors. The primary amino groups at the ends of the molecules therefore appear important for inhibitory activity, whereas substitution on the N^{11} amino groups has much less effect.

The property of strong binding of MGBA to SAMD has been successfully used to make MGBA-sepharose columns for the purification of SAMD. The enzymes from rat liver (Pegg, 1974), baker's yeast (Pösö, Sinervirta and Jänne, 1975) and mouse mammary gland (Oka, Kano and Perry, 1978) have been purified using this method.

When MGBA is administered *in vivo* to animals it causes enhancement of SAMD when the latter is measured in dialysed extracts of kidney, prostate and testis (Pegg, Corti and Williams-Ashman, 1973). This is a phenomenon similar to that observed with inhibitors of ODC in that MGBA protects SAMD from degradation. In kidney it can increase the half-life of the enzyme from 2 to 20 h. Pegg and Jefferson (1974) have shown that this is due to stabilization of the enzyme by MGBA and not due to the build up of putrescine which occurs when SAMD is inhibited.

MGBA is the most widely used inhibitor of spermidine and spermine synthesis (Pegg, 1973; Kay and Pegg, 1973; Fillingame and Morris, 1973; Heby, Sauter and Russell, 1973; Caldarera and coworkers, 1975 and Sturman, 1976).

B. 1,1′-(Methylethanediylidenedinitrilo)-*bis*(3-aminoguanidine)

Although MGBA is a potent inhibitor of SAMD its effects when given *in vivo* are complicated by the fact that it stabilizes the enzymes against breakdown and causes an increase in the amounts of SAMD present in a tissue. Pegg and Comover (1976) synthesized a related compound 1,1′-(methyl-ethanediylidenedinitrilo)-*bis*(3-aminoguanidine) (abbreviated to MBAG) which differs from MGBA by two amino groups and by binding reversibly to inactivate SAMD. Significant inhibition of liver SAMD is achieved with only 0.2 μM MBAG. MBAG effectively inhibits liver and yeast but not *E.coli* SAMD.

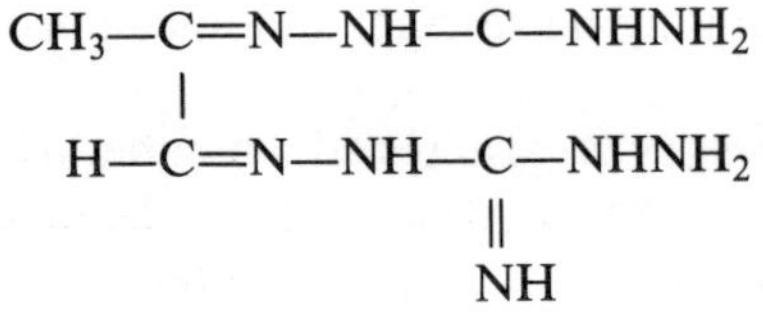

```
                        NH
                        ||
CH3—C=N—NH—C—NHNH2
        |
   H—C=N—NH—C—NHNH2
                        ||
                        NH
```

All three enzymes have a pyruvate bound co-factor (Pegg, 1977; Cohn, Tabor and Tabor, 1977 and Wickner, Tabor and Tabor, 1970) but the *E. coli* enzyme differs in not being sensitive to putrescine activation. When either liver or yeast enzymes were incubated in the presence of MBAG and a large excess of pyruvate, the latter afforded very little protection against inhibition. Inactivation was greater in the presence of putrescine. A more detailed study of the inhibition kinetics revealed that during the early stages of incubation of SAMD with MBAG, competitive reversible inhibition is observed and this is followed by irreversible inhibition (Pegg, 1978). Pegg (1978) suggests that this is consistent with the idea that MBAG first binds reversibly and then once positioned on the active site combines with an essential carbonyl group.

IV. INHIBITORS OF SPERMIDINE SYNTHETASE

Very little work has been carried out on potential inhibitors of spermidine synthetase. Spermidine synthetase has a much longer half-life than ODC or SAMD, and it is usually present in excess as compared with the decarboxylases. It would therefore be expected that higher concentrations of inhibitor would be required but that any resulting inhibition might be more prolonged. The two substrates for spermidine synthetase are decarboxylated S-adenosylmethionine and putrescine, and thus analogues of either might inhibit activity. The only analogue of decarboxylated S-adenosylmethionine so far tested is the 7-deaza analogue and Coward, Motola and Moyer (1977) have shown that this acts as a substrate of spermidine synthetase, though it is less efficient than the normal substrate. One end-product of spermidine synthetase is methylthioadenosine which is cleaved by a phosphorylase. If the 7-deaza analogue of decarboxylated S-adenosylmethionine is used as a substrate for spermidine synthetase, 5′-methylthiotubericidin would be expected as the product. This compound is not cleaved by ventral prostate phosphorylase. Coward, Motola and Moyer (1977) suggest that this might be a method of end-product regulation of spermidine and spermine biosynthesis.

We have tested the putrescine analogue, 1,4-diaminobutanone, as an inhibitor of spermidine synthetase. This was carried out using supernatants from germinating conidia of *A. nidulans* and the results are shown in Figure 3. Although 1,4-diaminobutanone can almost completely inhibit spermidine synthetase, it requires concentrations 10 to 100 times greater than that required to inhibit ODC to the same extent.

V. COMBINATIONS OF INHIBITORS

When an inhibitor of oligoamine synthesis is administered *in vivo*, the aim is often to reduce the overall concentration of oligoamines and then to study other cell processes in the oligoamine depleted cells. Spermine and spermidine

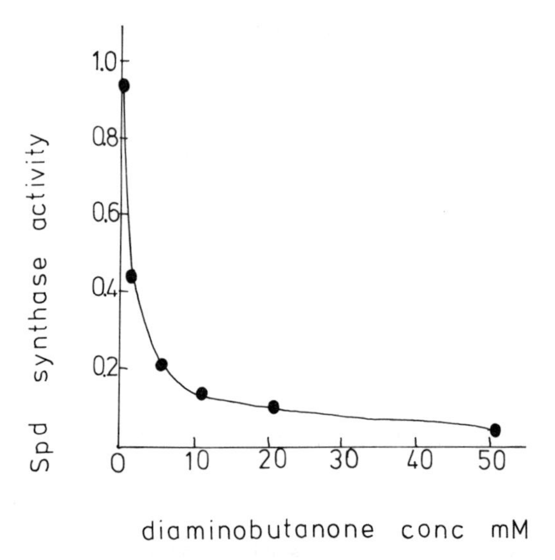

Figure 3 The effect of 1,4-diaminobutanone on spermidine synthetase activity in *Aspergillus nidulans.* Spermidine synthetase was assayed in the following incubation mixture:— 100 mM potassium phosphate buffer pH 7.4, 5 mM dithiothreitol, 0.4 mM ^{14}C-putrescine (specific activity 2.8 mCi/mmol), 0.05 mM decarboxylated S-adenosyl-methionine, 20-50 μl enzyme extract in a total volume of 100 μl

turn over relatively slowly and therefore it requires a relatively long period in order to facilitate this reduction. This often means that repeated doses of inhibitor have to be administered to be effective, especially if the inhibitor is readily eliminated by the cells. Even if an inhibitor were given which could completely inhibit either ODC or SAMD it would not prevent some oligo-amine accumulation from occurring. Inhibition of ODC alone would still enable spermine to be formed from spermidine, whereas inhibition of SAMD alone would allow putrescine accumulation. Thus it may be advantageous to use a combination of inhibitors.

Morris, Jorstad and Seyfried (1977) found that in transforming lymphocytes 5 mM α-methylornithine alone prevented the increase in putrescine and considerably reduced the rise of spermidine but the spermine increase was unaffected. If 5 mM α-methylornithine was given together with 8 μM MGBA, however, spermine was maintained at the level present in normal lymphocytes though there was some increase in the putrescine level.

Wiegand and Pegg (1978) used the combination of injections of 1,3-diamino-propane and MBAG to attempt to prevent the increase in oligoamine levels that occur during liver regeneration. If 1,3-diaminopropane alone is used, it is necessary to administer large doses every 3 h in order to block accumulation of putrescine and spermidine over a 24 h period. If 1,3-diaminopropane is given together with MBAG, however, a single injection is sufficient to lower the intracellular spermidine concentrations and to have marked inhibitory effect on DNA synthesis.

Much progress has been made in developing inhibitors of oligoamine synthesis but a single ideal inhibitor or combination of inhibitors has yet to emerge.

Note added in proof

New inhibitors and combinations of inhibitors have been used to prevent accumulation of spermidine and spermine in cells. Piik, Pösö and Jänne (1978) have shown that 1,3-diamino-2-propanol causes a more prolonged inhibition of ODC *in vivo* in regenerating rat liver than does 1,3-diaminopropane. Höltta, Pohjanpelto and Jänne (1979) have compared the effectiveness of DL-α-difluoromethylornithine and methylglyoxal *bis*(amidinohydrazone) in reducing oligoamine levels and in inhibiting DNA synthesis. Although both effectively prevent oligoamine accumulation, MGBA has a much earlier inhibitory effect on DNA synthesis suggesting this latter inhibition may be unrelated to oligoamine metabolism. Kallio and coworkers (1979) have found that the inhibition of ODC in rat liver induced by 1,3-diaminopropane is the result of production of macromolecular inhibitors which bind ODC. 2,5-Diamino-2-(cyanomethyl) pentanoic acid, an analogue of ornithine, was found to be inactive as an inhibitor of ODC from rat prostate (Abdel-Monem and Mikhail, 1978).

Additional References

Abdel-Monem, M. M. and Mikhail, E. A. (1978) *J. Pharm. Sciences*, **67**, 1174–1175.
Höltta, E., Pohjanpelto, P. and Jänne, J. (1979) FEBS Lett., **97**, 9–14.
Kallio, A., Löfman, M., Pösö, H. and Jänne, J. (1979) *Biochem. J.*, **177**, 63–69.
Piik, K., Pösö, H. and Jänne, J. (1978) FEBS Lett., **89**, 307–312.

ACKNOWLEDGEMENTS

Much of our work cited in this review was carried out with the support of grants from the Science Research Council. We are grateful to Dr. J. Jänne for the gift of decarboxylated S-adenosylmethionine.

REFERENCES

Abdel-Monem, M. M., Newton, N. E. and Weeks, C. E. (1974). *J. Med. Chem.* **17**, 447–451.
Abdel-Monem, M. M., Newton, N. E., Ho, B. C. and Weeks, C. E. (1975a). *J. Med. Chem.* **18**, 600–604.

Abdel-Monem, M. M., Newton, N. E. and Weeks, C. E. (1975b). *J. Med. Chem.* **18**, 945–948.

Bey, P. and Vevert, J. P. (1977). *Tetrahedron Lett.* **17**, 1455-1458.

Bey, P., Danzin, C., Dorsselaer, V. V., Mamont, P., Jung, M. and Tardif, C. (1978). *J. Med. Chem.* **21**, 50–55.

Caldarera, C. M., Casti, A., Guarnieri, C. and Moruzzi, G. (1975). *Biochem. J.* **152**, 91–98.

Clark, J. L. and Fuller, J. L. (1975). *Biochemistry* **14**, 4403–4409.

Cohn, M. S., Tabor, C. W. and Tabor, H. (1977). *J. Biol. Chem.* **252**, 8212–8216.

Corti, A., Dave, C., Williams-Ashman, H. G., Mihich, E. and Schenone, A. (1974). *Biochem. J.* **139**, 351–357.

Coward, J. K., Motola, N. C. and Moyer, J. D. (1977). *J. Med. Chem.* **20**, 500–505.

Demetrion, A. A., Cohn, M. S., Tabor, C. W. and Tabor, H. (1978). *J. Biol. Chem.* **253**, 1684–1686.

Ellington, J. and Honigberg, I. L. (1974). *J. Org. Chem.* **39**, 104–106.

Fillingame, R. H. and Morris, D. R. (1973). *Biochemistry*, **12**, 4479–4487.

Guha, S. K. and Jänne, J. (1977). *Biochem. Biophys. Res. Commun.* **75**, 136–142.

Harik, S. I. and Snyder, S. H. (1973). *Biochim. Biophys. Acta* **327**, 501–509.

Harik, S. I., Hollenberg, M. D. and Snyder, S. H. (1974). *Mol. Pharmacol.* **10**, 41–47.

Heby, O., Sauter, S. and Russell, D. H. (1973). *Biochem. J.* **136**, 1121–1124.

Heller, J. S., Canellakis, E. S. Bussolotti, D. L. and Coward, J. K. (1975). *Biochem. Biophys. Acta*, **403**, 197–207.

Hölttä, E., Hannonen, P., Pispa, J. and Jänne, J. (1973). *Biochem. J.* **136**, 669–676.

Hoppe, D. (1975). *Angew. Chem. Int. Ed. Engl.* **14**, 424.

Inoue, H., Kato, Y., Takigawa, M., Adachi, K. and Takeda, Y. (1975). *J. Biochem. (Japan)*, **77**, 879-893.

Kallio, A., Pösö, H., Scalabrino, G. and Jänne, J. (1977a). *FEBS Lett.*, **73**, 229–234.

Kallio, A., Löfman, M., Pösö, H. and Jänne, J. (1977b). *FEBS Lett.* **79**, 195–199.

Karlen, B. and Lindeke, B. (1969). *Acta. Pharm. Suecica*, **6**, 613–616.

Kay, J. E. and Lindsay, V. J. (1973). *Biochem. J.* **132**, 791–796.

Kay, J. E. and Pegg, A. E. (1973). *FEBS Lett.* **29**, 301–304.

Leinweber, F. J. (1968). *Mol. Pharmacol.* **4**, 337–348.

Macholan, L. (1965). *Coll. Czech. Chem. Commun.* **30**, 2074–2079.

Macholan, L. (1969). *Arch. Biochem. Biophys.* **134**, 302–307.

Mamont, P. S., Böhlen, P., McCann, P. P., Bey, P., Schuber, F. and Tardif, C. (1976). *Proc. Nat. Acad. Sci. USA*. **73**, 1626–1630.

Mamont, P. S., Duchesne, M., Grove, J. and Bey, P. (1978). *Biochem. Biophys. Res. Commun.* **81**, 58–66.

McCann, P. P., Tardif, C., Duchesne, M. and Mamont, P. S. (1977a). *Biochem. Biophys. Res. Commun.* **76**, 893–899.

McCann, P., Tardif, C. and Mamont, P. (1977b). *Biochem. Biophys. Res. Commun.* **75**, 948–954.

Morris, D. R., Jorstad, C. M. and Seyfried, C. E. (1977). *Can. Res.* **37**, 3169–3172.

Oka, T., Kano, K. and Perry, J. W. (1978). *Advances in Polyamine Research*, Vol. 1. (Ed. D. H. Russell) 59–67.

O'Leary, M. H. and Herreid, R. M. (1978). *Biochemistry*, **17**, 1010–1014.

Pegg, A. E. (1973). *Biochem. J.* **132**, 537–540.

Pegg, A. E. (1974). *Biochem. J.* **141**, 581–583.

Pegg, A. E. (1977). *FEBS Lett.*, **84**, 33–36.

Pegg, A. E. (1978). *J. Biol. Chem.* **253**, 539–542.

Pegg, A. E. and Conover, C. (1976). *Biochem. Biophys. Res. Commun.* **69**, 766–774.

Pegg, A. E. and Jefferson, L. S. (1974). *FEBS Lett.* **40**, 321–324.

Pegg, A. E., Corti, A. and Williams-Ashman, H. G. (1973). *Biochem. Biophys. Res. Commun.* **52**, 696–701.
Pegg, A. E., Conover, C. and Wrona, A. (1978). *Biochem. J.* **170**, 651–660.
Piik, K., Rajamäki, P., Guha, S. K. and Jänne, J. (1977). *Biochem. J.* **168**, 379–385.
Pösö, H. and Jänne, J. (1976). *Biochem. Biophys. Res. Commun.* **69**, 885–892.
Pösö, H., Sinervirta, R. and Jänne, J. (1975). *Biochem. J.* **151**, 67–73.
Rahiala, E., Kekomäki, M., Jänne, J., Raina, A. and Räihä, N. C. R. (1971). *Biochim. Biophys. Acta*, **227**, 337–343.
Rando, R. R. (1974). *Biochemistry*, **13**, 3859–3863.
Relyea, N. and Rando, R. R. (1975). *Biochem. Biophys. Res. Commun.* **67**, 392–402.
Sawayama, T., Kinugasa, H. and Nishimura, H. (1976). *Chem. Pharm. Bull.* **24**, 326–329.
Skinner, W. A. and Johansson, J. G. (1972). *J. Med. Chem.* **15**, 427–428.
Stevens, L. and McKinnon, I. M. (1977). *Biochem. J.* **166**, 635–637.
Stevens, L., McKinnon, I. M. and Winther, M. (1976). *Biochem. J.* **158**, 235–241.
Stevens, L., McKinnon, I. M. and Winther, M. (1977). *FEBS Lett.* **75**, 180–182.
Stevens, L., McKinnon, I. M. Turner, R. and North, M. J. (1978). *Trans. Biochem. Soc.* **6**, 407–409.
Sturman, J. A. (1976). *Life Sciences*, **18**, 879–886.
Sunkara, P. S., Rao, P. N. and Nishioka, K. (1977). *Biochem. Biophys. Res. Commun.* **74**, 1125–1133.
Wickner, R. B., Tabor, C. W. and Tabor, H. (1970). *J. Biol. Chem.* **245**, 2132–2139.
Wiegand, L. and Pegg, A. E. (1978). *Biochim. Biophys. Acta.* **517**, 169–180.
Williams-Ashman, H. G. and Schenone, A. (1972). *Biochem. Biophys. Res. Commun.* **46**, 288–295.

Chapter 12

Inhibition of Tumour Promoter-induced Mouse Epidermal Ornithine Decarboxylase Activity and Prevention of Skin Carcinogenesis by Vitamin A Acid and its Analogues (Retinoids)

AJIT K. VERMA AND R. K. BOUTWELL

ABBREVIATIONS

TPA, 12-*O*-tetradecanoylphorbol-13-acetate; ODC, ornithine decarboxylase; SAMD, S-adenosyl-L-methionine decarboxylase; TMMP, trimethylmethoxyphenyl; TMHP, trimethylhydroxyphenyl; DMBA, 7,12-dimethylbenz[a]-anthracene.

I. INTRODUCTION

The process of mouse skin carcinogenesis can be divided into at least two defined stages, initiation and promotion (Berenblum and Shubik, 1947; Mottram, 1944; Boutwell, 1964, 1974). Initiation is accomplished by a single application of a subcarcinogenic dose of an aromatic polycyclic hydrocarbon such as 7,12-dimethylbenz[a]anthracene (DMBA). Perhaps the critical biochemical event occurring during the initiation stage of skin tumour formation is the interaction of metabolite(s) of the chemical carcinogen with cellular macromolecules, especially DNA (Miller and Miller, 1977; Sims and Grover, 1974). Following initiation, few, if any, tumours will appear throughout the life span of the animal unless the initiated skin is repeatedly treated with one of a number of promoting agents (promotion stage). 12-0-Tetradecanoylphorbol-13-acetate (TPA) is the most potent tumour promoter among a number of phorbol diesters isolated from croton oil (Hecker and Schmidt, 1974; Van Duuren, 1969). The structure of the parent alcohol, phorbol, and certain of its esters, together with their tumour promoting activity, is depicted in Figure 1. Application of tumour promoters alone do not lead to the development of skin tumours; it is only their application following initiation that elicits tumours. Thus, the two-stage model of skin carcinogenesis provides a useful system in which biochemical events unique to either initiation or promotion can be studied and related to cancer formation.

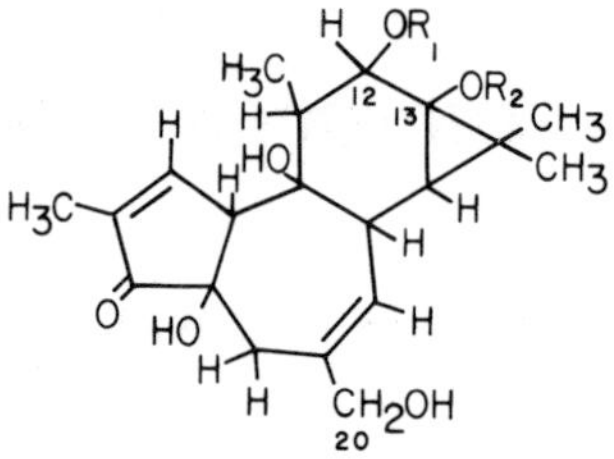

COMPOUND	R_1	R_2	ACTIVITY
Tetradecanoyl-phorbol-acetate	Tetradecanoate	Acetate	+ + + +
Phorbol-didecanoate	Decanoate	Decanoate	+ +
Phorbol-dibenzoate	Benzoate	Benzoate	+
Phorbol-diacetate	Acetate	Acetate	0
Phorbol	H	H	0

Figure 1 The structure of phorbol and certain of its esters, together with their promoting activity

In recent years, remarkable progress in the understanding of the biochemical mechanism of the promotion phase of skin carcinogenesis has been accomplished (see reviews, Boutwell, 1974, 1976, 1977). Application of TPA to mouse skin leads to enhanced incorporation of [^{32}P] into phospholipids (Rohrschneider and Boutwell, 1973) and to sequential activation of RNA, protein, and DNA synthesis (Baird, Sedgwick and Boutwell, 1971; Boutwell, 1974; Kreig, Kühlmann and Marks, 1974). Other biochemical changes observed following TPA treatment include enhanced phosphorylation of nuclear histones (Raineri, Simsiman and Boutwell, 1973), decreased histidase activity (Colburn, Lau and Head, 1975), and altered cyclic nucleotide metabolism in mouse epidermis (Belman and Troll, 1978; Grimm and Marks, 1974; Mufson, Simsiman and Boutwell, 1977; Murray, Verma and Froscio, 1976; Verma, Froscio and Murray, 1976a; Verma and coworkers, 1976b). Most of these biochemical changes are observed in many cell and tissue systems stimulated to proliferate and, thus, cannot be causally related to skin tumour promotion. Recently, induction of epidermal ornithine decarboxylase (EC 4.1.1.17; L-ornithine carboxy-lyase) (ODC) activity by tumour-promoting agents has been shown to be unique and essential for skin tumour promotion (Boutwell, 1977; O'Brien, Simsiman and Boutwell, 1975a, 1975b; O'Brien, 1976; Verma, Rice and Boutwell, 1977; Verma and coworkers, 1978).

The activities of the polyamine biosynthetic enzymes, especially ODC, and the levels of their biosynthetic products putrescine, spermidine, and spermine are elevated in various tissues stimulated to growth (Chapters 6–8). Several lines of evidence suggest that increased ODC activity and polyamine accumulation are important for the regulation of the synthesis of nucleic acid and protein (Tabor and Tabor, 1972, 1976; Chapters 1–8).

Numerous studies indicate that the polyamines and the enzymes that synthesize polyamines are implicated in neoplastic growth, such as in hepatomas and in the liver of mice inoculated with L 1210 leukaemic tumours or Ehrlich ascites cells (Chapter 4).

In view of the potential significance of ODC and polyamines in malignant transformation, their role in mouse skin carcinogenesis was investigated. Topical application of a tumour promoter to mouse skin leads to a dramatic induction of epidermal ODC activity between 4 and 6 h following promoter treatment. The evidence in support of the suggestion that this phenotypic change may be a marker for promotion has been reviewed (Boutwell, 1977; O'Brien, 1976), but for the purpose of this chapter, the relevant information is summarized below.

(1) Application of 17 nmoles of TPA in 0.2 ml of acetone to the shaved skin of the back of a mouse leads to about a 200-fold increase in epidermal ODC activity at 4.5 h following treatment. The magnitude of enzyme induction is dose dependent, and correlates with the ability of the dose to promote skin tumour formation (O'Brien, Simsiman and Boutwell, 1975a).

(2) The degree of induction of ODC activity correlates well with the tumour promoting ability of a number of structurally unrelated tumour promoters. Furthermore, the ability of a series of phorbol esters to induce ODC activity correlates with their ability to promote skin tumour formation (O'Brien, Simsiman and Boutwell, 1975b; O'Brien, 1976).

(3) Epidermal ODC activity is induced only following treatment of mouse skin with tumour promoters and not after treatment with non-promoting hyper-

plastic agents. In contrast, both tumour promoters and hyperplastic agents induce S-adenosyl-L-methionine decarboxylase (SAMD) activity, an enzyme involved in the biosynthesis of spermidine. The latter induction may be related to the hyperplasia commonly observed following their application to mouse skin (Clark-Lewis and Murray, 1978; O'Brien, Simsiman and Boutwell, 1975b: O'Brien, 1976).

(4) The level of ODC is elevated in skin papillomas and carcinomas, produced by the initiation-promotion procedure (O'Brien, 1976).

Further support to the proposal that TPA-induced ODC activity may be an essential component of the mechanism of skin tumour promotion was provided by our recent work on retinoids and skin tumour promotion (Verma and Boutwell, 1977; Verma, Rice and Boutwell, 1977; Verma and coworkers, 1978), which will be the focus of this chapter. Retinoic acid or certain of its analogues, applied 1 h before TPA treatment, inhibit both the induction of ODC activity and the formation of skin papillomas elicited by TPA. Data showing the effects of various retinoids (Verma, Rice and Boutwell, 1977; Verma and coworkers, 1978; Verma and coworkers, 1979) on ODC induction and formation of skin papillomas by TPA, will be presented and the results implying that prostaglandins may play a role in the induction of ODC activity by TPA will be summarized as well (Verma, Rice and Boutwell, 1977; Verma and coworkers, 1980).

II. EXPERIMENTS AND RESULTS

A. Effect of Retinoic Acid on TPA-induced ODC Activity and Skin Tumour Formation

As shown in Figure 2, application of 17 nmoles of TPA to mouse skin causes a pronounced increase in epidermal ODC activity. Maximum activity, about 200-fold above basal level, was observed between 4 and 8 h after TPA treatment. The enzyme activity returned to the original level at about 12 h. Treatment of mouse skin with 1.7 nmoles of retinoic acid 1 h prior to TPA treatment resulted in a dramatic reduction in the degree of induction of ODC activity by TPA. Retinoic acid treatment did not inhibit the induction of SAMD activity by TPA (Figure 2). The inhibition of the induction of ODC activity by retinoic acid was dose dependent (Figure 3). Enzyme induction was suppressed at a dose above 1.7 pmoles; 57% inhibition was observed after pretreatment with 0.17 nmoles of retinoic acid, whereas 3.4 nmoles completely inhibited the induction of ODC activity by TPA.

Since retinoic acid treatment blocks the induction of ODC without affecting SAMD induction by TPA, the effect of retinoic acid treatment on TPA-caused accumulation of their products, putrescine, spermidine, and spermine, was determined. The results are shown in Figure 4. Application of 17 nmoles

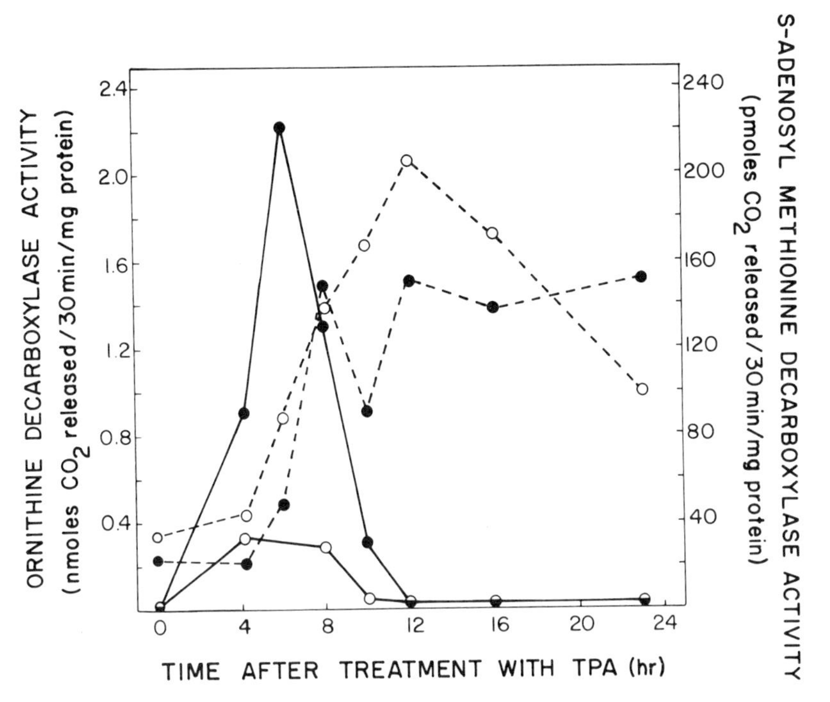

Figure 2 The effect of retinoic acid treatment on TPA-induced epidermal ODC and S-adenosyl-L-methionine decarboxylase activities. Groups of 4 mice were treated with 1.7 nmoles of retinoic acid or acetone 1 h before application of 17 nmoles of TPA. Mice were killed at the indicated times after TPA treatment, and ornithine and S-adenosyl-methionine decarboxylase activities from the same soluble epidermal extracts were determined. Each point represents the mean of triplicate determinations of enzyme activity. ODC activity: retinoic acid (○——○), acetone (●——●); S-adenosyl-L-methionine decarboxylase activity: retinoic acid (○– – –○), acetone (●– – –●)

of TPA resulted in an enhanced accumulation of putrescine; a maximum level of about 3- to 4-fold above that of the untreated control was found between 6 and 10 h following TPA treatment. Spermine level was not affected up to 24 h, whereas spermidine level was increased at 24 h following TPA treatment. Treatment with 1.7 nmoles of retinoic acid 1 h before treatment with 17 nmoles of TPA depressed the TPA-caused increased accumulation of putrescine, whereas the levels of spermidine and spermine were not altered. These results indicate the specificity of action of retinoic acid on ODC induction and accumulation of its product, putrescine, by TPA.

If induction of epidermal ODC activity is required for skin tumour formation, then inhibition of this activity by retinoic acid should depress skin tumour formation. Results illustrated in Figure 5 indicate that this was found to be the case. In the experiment described in Figure 5, Panels A and B, mice were

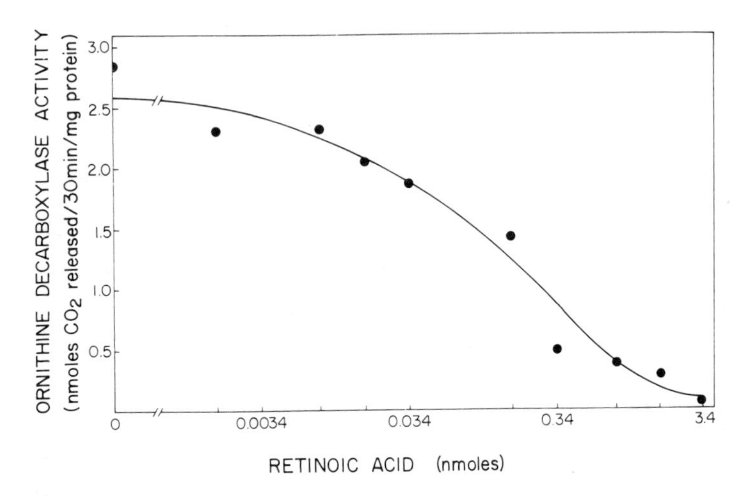

Figure 3 The effect of the dose of retinoic acid on the induction of epidermal ODC activity by TPA. Mice were treated topically with retinoic acid 1 h before application of 17 nmoles of TPA, and soluble epidermal extracts were prepared 4.5 h after TPA treatment. Each point represents the mean of triplicate determinations of enzyme activity from soluble epidermal extracts prepared from 4 mice

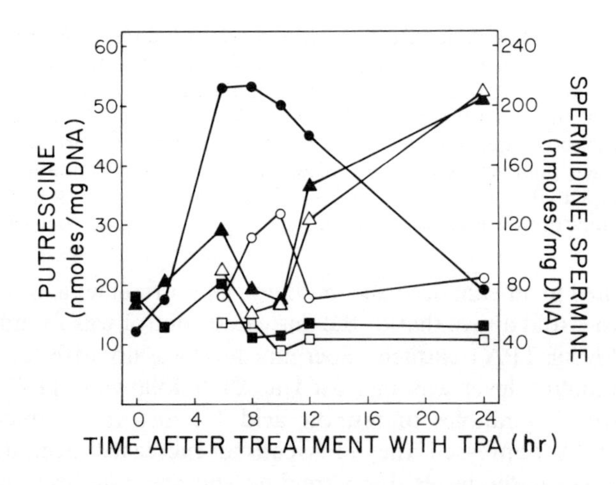

Figure 4 Time course of epidermal polyamine accumulation following TPA treatment to mouse skin. The mice were treated topically with either acetone or 1.7 nmoles retinoic acid in acetone 1 h before application of 17 nmoles TPA. Mice were killed at the indicated times after TPA treatment, and the levels of polyamines in the epidermis were determined. Each point represents the mean of determinations carried out on 2 groups of mice, with 3 mice in each group. Putrescine, acetone (●) and retinoic acid (○); spermidine, acetone (▲) and retinoic acid (△); spermine, acetone (■) and retinoic acid (□)

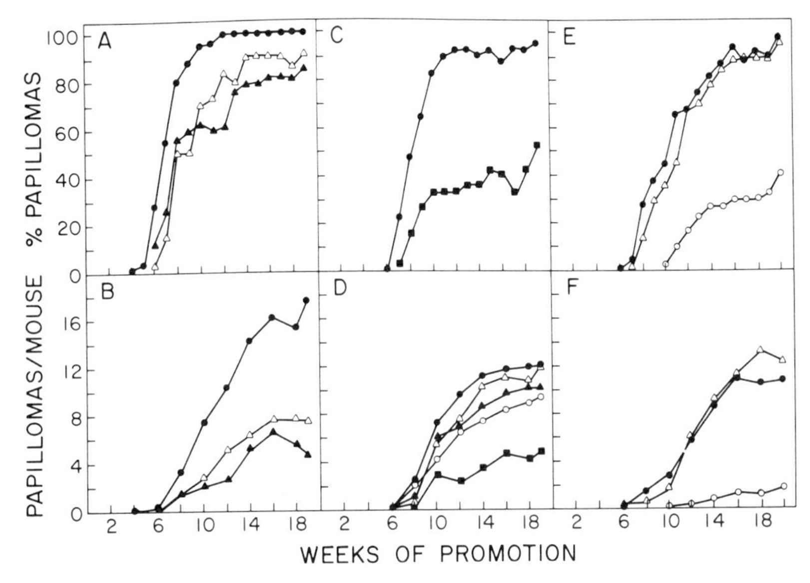

Figure 5 The effect of treatment with retinoic acid on the incidence of mouse skin papillomas promoted with TPA. All mice were initiated with 0.2 μmole (51.2 μg in 0.2 ml acetone) of DMBA; 14 days after initiation, all mice were promoted twice a week (days 1 and 4) with either 8 (Panels E, F) or 17 (Panels, A, B, C, D) nmoles of TPA for the duration of the experiment. There were 30 mice in each treatment group. The incidence of papillomas was observed weekly. The mice did not lose weight during the entire period of the experiment, and the survival varied from 85 to 100%. A,B: mice were treated with either 1.7 (△) or 17 (▲) nmoles of retinoic acid in 0.2 ml acetone 1 h prior to each promotion with TPA. Control mice were pretreated with acetone (●). D: mice were treated with 68 nmoles of retinoic acid 1 h before initiation with DMBA (▲), 68 nmoles of retinoic acid was applied 1 h after treatment with DMBA as well as every day for 7 days following initiation (○), 68 nmoles of retinoic acid was applied every day for 5 days beginning 7 days after initiation with DMBA (△). Treatment of mice concurrently with 68 nmoles of retinoic acid and 0.2 μmole of DMBA did not inhibit the incidence of formation of papillomas. The values of per cent papilloma bearing mice for treatments of retinoic acid either with, before, or after DMBA (▲, ○, △) which did not differ significantly from control (●) are not shown in the Figure. 68 nmoles of retinoic was applied 1 h before each promotion with 17 nmoles of TPA (■). E, F: 17 nmoles of retinoic acid (○) or acetone (●) was applied 1 h before each promotion with 8 nmoles of TPA, 17 nmoles of retinoic acid was applied 24 h after each TPA treatment (△)

initiated with 0.2 μmole of DMBA, and 14 days later the mice were promoted twice a week with 17 nmoles of TPA. Application of 1.7 and 17 nmoles of retinoic acid 1 h prior to each promotion with TPA resulted in a 57 and 75% reduction in the number of papillomas per mouse respectively (Figure 5, Panel B).

The possibility was examined that retinoic acid treatment may inhibit skin papilloma formation by interfering with the initiation process (the possibilities include inhibition of the metabolic activation of the carcinogen, interference with the interaction of the carcinogen metabolites with macromolecules especially DNA, increased error-free repair of DNA or inactivation of initiated cells). Groups of mice were treated with 68 nmoles of retinoic acid either prior to, concurrent with, or after initiation with DMBA; 14 days following initiation the mice were promoted with 17 nmoles of TPA twice weekly. These treatments resulted in minimal effects on the number of papillomas per mouse or the number of mice bearing papillomas. In contrast, application of 68 nmoles of retinoic acid 1 h before each promotion with TPA resulted in a 60% reduction in the number of papillomas per mouse, and only 53% of the mice bore papillomas (Figure 5, Panel C and D). These results suggest that retinoic acid probably exerts its prophylactic effect on skin carcinogenesis by interference with promotion by TPA.

Results presented so far indicate that retinoic acid applied to the skin of the mouse 1 h before TPA treatment suppresses the induction of ODC activity (Figure 2) as well as reducing the formation of skin papillomas (Figure 5, Panels B, D and F). To further test whether ODC induction is relevant to tumour promotion, groups of mice were treated with 34 nmoles of retinoic acid 24 h after each treatment with 8 nmoles of TPA, the time point when maximum ODC induction has passed and ODC levels have completely returned to original control values (Figure 2). The results (Figure 5, Panels E and F) indicate that retinoic acid treatment 24 h after TPA did not affect the incidence of the formation of papillomas, whereas retinoic acid applied 1 h before each promotion produced an 84% reduction in the number of papillomas per mouse and there were only 42% papilloma bearing mice. These findings strongly suggest that induction of ODC by TPA may be a prerequisite for the promotion of skin tumours.

B. Effect of Other Retinoids on TPA-induced ODC Activity and Formation of Skin Papillomas

Numerous studies have shown that natural retinoids are effective in preventing cancers in experimental animals. These studies include prevention of tumours of the gastrointestinal tract, vagina and uterine cervix, bronchus and trachea, as well as prevention of forestomach papillomas, skin papillomas, lung carcinomas and mammary tumours in rats (Sporn and coworkers, 1976b; Sporn, 1977a, 1977b). High doses of retinoic acid, or its natural analogues, are required to be systemically administered to achieve the desired prophylactic or therapeutic effect. The high levels that are required to inhibit tumours, however, lead to appreciable toxic effects (hypervitaminosis syndrome) (Bollag, 1975b). Thus, there is a need for the development of synthetic retinoids that possess desirable prophylactic and/or therapeutic effects without causing toxicity. A number of retinoids have been synthesized containing alterations either in ring structure, in the side chain or in the terminal polar group. Each new retinoid needs to be evaluated for its biological and pharmacological properties. A number of *in vitro* test systems, namely, newborn mouse epidermal cells in culture (Sporn and coworkers, 1975), hamster trachea in organ culture (Clamon and coworkers, 1974; Sporn and coworkers, 1976a), chick embryo metatarsal skin explants (Wilkoff and coworkers, 1976), and mouse prostate in organ culture (Chopra and Wilkoff, 1977) have been developed for testing the biological as well as anti-carcinogenic properties of retinoids. We have evaluated a number of natural vitamin A analogues (retinal, retinol, retinyl acetate and retinyl palmitate) for their ability to inhibit both the induction of ODC activity and the formation of skin papillomas, and a correlation was found between these two effects (Verma and Boutwell, 1977). These results indicate that in addition to the importance of ODC induction for promotion, inhibition of TPA-induced ODC activity by retinoids may be a simple test for their anti-promoting properties. In order to examine the validity of this *in vivo* test, various synthetic retinoids were tested for their ability to inhibit both the induction of ODC activity and the development of skin papillomas. The quantities of retinoids required to inhibit by 50% the ODC induction by TPA are shown in Table 1. The effects of a number of retinoids on the induction of ODC activity, SAMD activity, and the incidence of skin papillomas are shown in Table 2. Since development of skin papillomas requires repeated applications of a tumour promoter to the initiated skin, these biochemical effects were determined 2 days after the 7th application of TPA to mouse skin initiated with DMBA; the interval between applications was 4 days. Furthermore, it should be noted that the degree of induction of ODC by TPA increased with increasing numbers of applications of TPA; and retinoic acid treatment 1 h before each TPA treatment inhibited the induction of ODC activity by 80% (Figure 6).

It is indicative from the results (Table 2) that those retinoids (retinoic acid, 13-*cis*-retinoic acid, TMMP analogue of ethyl retinoate) which inhibited the induction of ODC activity, inhibited formation of skin papillomas as well. Retinoid treatment did not depress TPA-induced SAMD activity. These findings substantiate that retinoids specifically inhibit ODC induction by TPA, and there appears to be a correlation between the ability of retinoids to inhibit the induction of ODC activity and their ability to inhibit the formation of skin papillomas.

The test system presented for evaluating the anti-promoting property of a retinoid does not provide clues about the toxicity of the retinoid, presumably because only a small dose (nmoles) of the retinoid is required to be applied topically to observe its effect on enzyme induction. Furthermore, it is a short-term assay. Present results demonstrate the prophylactic effect of the retinoids

Table 1 Doses of retinoids that, when administered topically, inhibit 50% of TPA-induced mouse epidermal ODC activity

RETINOID	STRUCTURE	Median inhibitory dose (nmoles)	RETINOID	STRUCTURE	Median inhibitory dose (nmoles)
DMECP analogue of retinoic acid	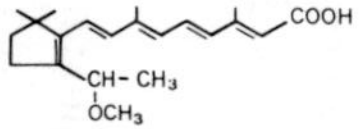	0.09	Trimethylthiophene analogue of ethyl retinoate	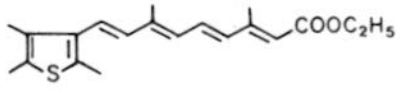	16.4
β-Retinoic acid	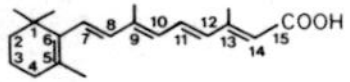	0.12	TMMP thio analogue of retinoic acid	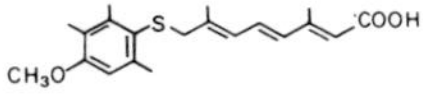	32.0
13-*cis*-Retinal	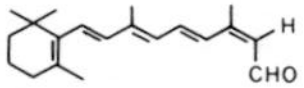	0.14	Lactone of retinoic acid	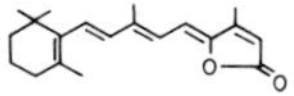	60.0
α-Retinoic acid	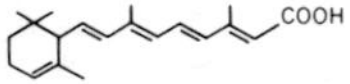	0.20	10-Fluoro-TMMP analogue of 13-*cis*-methyl retinoate	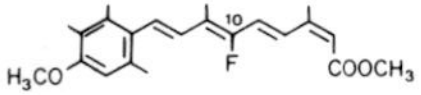	139
8-Fluoro-TMMP analogue of methyl retinoate	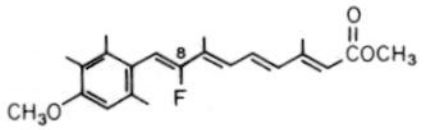	0.21	Phenyl analogue of ethyl retinoate	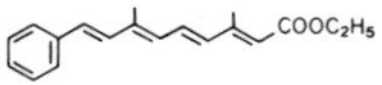	192
13-*cis*-Retinoic acid	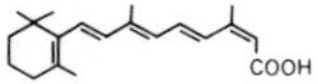	0.24	TMHP analogue of ethyl retinoate	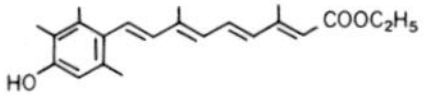	400
5,6-Dihydroretinoic acid	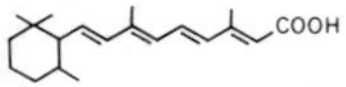	0.43	TMMP analogue of *N*-ethylretinamide	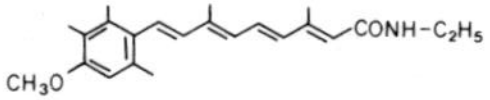	400
DACP analogue of retinoic acid	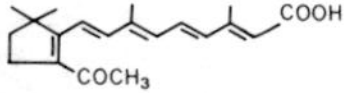	0.54	9-*cis*-10-Fluoro-TMMP analogue of methyl retinoate	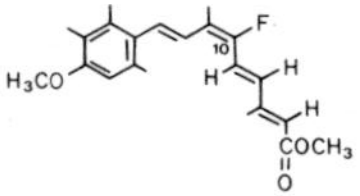	540
12-Fluoro-TMMP analogue of ethyl retinoate	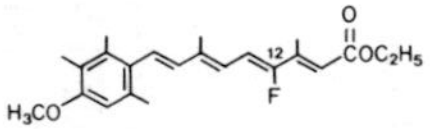	5.00	*N*-(2-Hydroxyethyl) retinamide	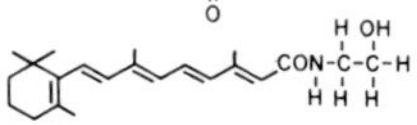	540
10-Fluoro-TMMP analogue of methyl retinoate	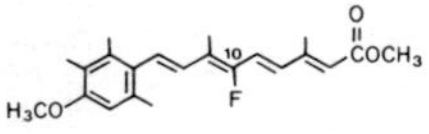	8.90	Furyl analogue of retinoic acid	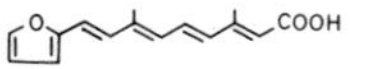	INACTIVE
TMMP analogue of retinoic acid	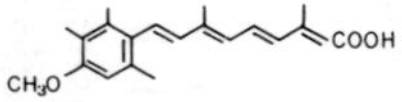	12.8	13-Trifluoromethyl-TMMP analogue of ethyl retinoate	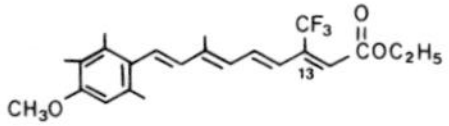	INACTIVE
TMMP analogue of ethyl retinoate	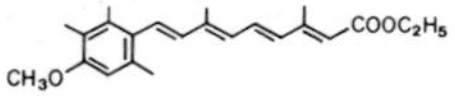	14.0			

Table 2 All mice were initiated with 0.2 μmole of DMBA in 0.2 ml acetone; 14 days after initiation, mice were promoted twice a week with 8 nmoles of TPA. Mice were treated with either 0.2 ml of acetone or a retinoid in 0.2 ml acetone 1 hour before each promotion with 8 nmoles of TPA. The interval between first and second promotion was 3 days. For determination of ODC and SAMD activities, mice were killed 4.5 and 24 hours after the 7th TPA treatment, respectively. In the tumour experiment, there were 30 mice in each treatment group and the incidence of papillomas (Pa) recorded at 20th week after promotion is given in the table

Treatment	Dose (nmoles)	Pa/ mouse	% Pa	ODC activity (nmoles CO_2/30 min/mg protein)	SAMD activity
Acetone	—	9.9	89	3.5 ± 0.4	0.14 ± 0.00
Retinoic acid	34	1.8	54	0.04	0.12 ± 0.00
13-*cis*-Retinoic acid	34	3.7	48	0.5 ± 0.1	0.14 ± 0.01
Retinal	34	7.4	68	1.8 ± 0.3	0.14 ± 0.01
TMMP analogue of ethyl retinoate	140	1.5	38	0.5 ± 0.1	0.14 ± 0.01
TMHP analogue of ethyl retinoate	140	12.9	92	3.3 ± 0.7	0.14 ± 0.01
13-trifluoromethyl TMMP analogue of ethyl retinoate	140	9.6	100	3.8 ± 0.7	0.13 ± 0.01

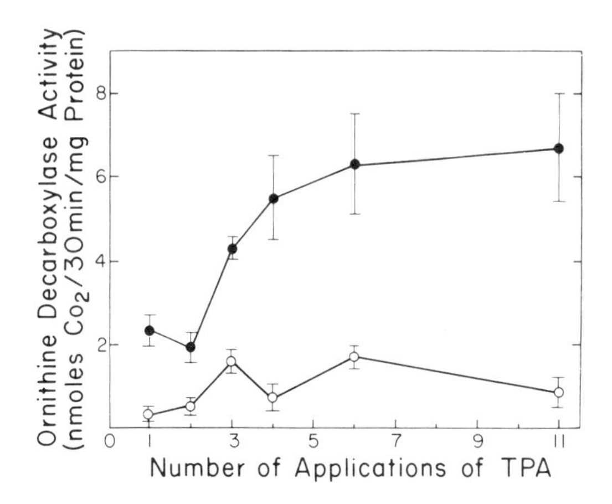

Figure 6 The effect of repeated applications of TPA on induction of epidermal ODC activity, and its inhibition by retinoic acid pretreatments. Groups of mice were initiated with 0.2 μmole of DMBA in 0.2 ml acetone. Fourteen days following initiation mice were either treated with 0.2 ml acetone (●) or 1.7 nmoles of retinoic acid (○) in 0.2 ml acetone 1 h before each treatment with 17 nmoles of TPA on days 1 and 4 of each week. Mice were killed 4.5 h after TPA treatment. ODC activity in soluble epidermal homogenates was determined. Each point is the mean ± s.e. of the determinations carried out in 3 groups of mice; there were 3 mice in each group. ODC activity was not determined after the 11th application of TPA as the mice started bearing papillomas

on skin tumour formation. Therapeutic efficacy of the retinoids on the skin tumour regression was not determined. Bollag (1975b) has shown that systemic administration of retinoids leads to the regression of skin papillomas, and some of the retinoids have a better therapeutic index than β-retinoic acid. We have observed that oral administration of retinoic acid inhibits ODC induction by topically applied TPA (Figure 7) (Verma and coworkers, 1978). It is quite likely that some of the retinoids which were found less active when applied topically may turn out to be active when administered systemically. For example, Bollag (1975a) has reported that systemically administered TMMP analogue of ethyl retinoate is more active than retinoic acid.

The molecular basis for the prophylactic and therapeutic effect of retinoids on carcinogenesis (Bollag, 1971, 1972, 1975a, 1975b; Nettesheim and Williams, 1976; Shamberger, 1971; Sporn and coworkers, 1976b; Sporn, 1977a, 1977b; Sporn and coworkers, 1977) is not clear. Retinoid treatment may modify gene expression. For example, retinyl acetate alters glycoprotein synthesis in cultured epidermal cells (De Luca and Yuspa, 1974), and retinoic acid suppresses production of interferon induced by virus (Blalock and Gifford, 1977). Vitamin A acid is required for the maintenance of normal differentiation of many tissues (Sporn and coworkers, 1976b) and this putative mechanism may possibly form the basis for its control on carcinogenesis at the cellular level. Recently, retinoic acid-binding protein has been detected in many tissues (Ong and Chytil, 1975; Sani and Hill, 1976; Sani and Corbett, 1977). The retinoic acid-binding protein may be involved in the expression

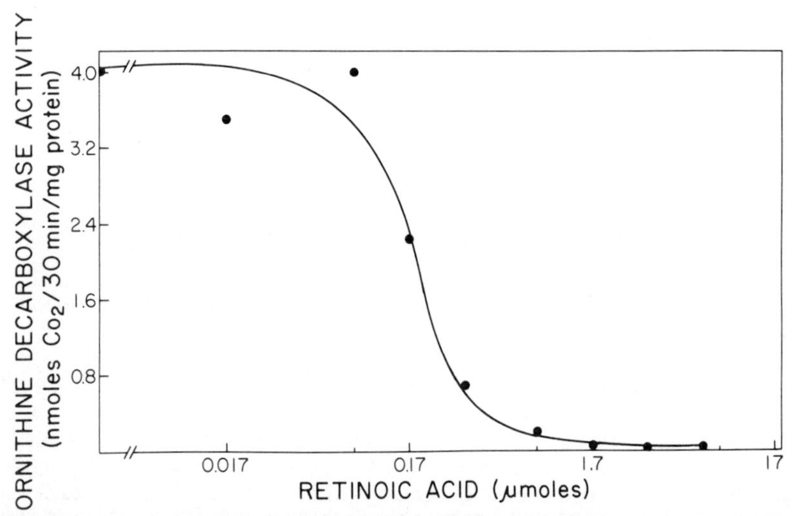

Figure 7 Dose-response curve showing the effect of p.o. administration of retinoic acid on TPA-induced epidermal ODC activity. The mice were given retinoic acid by stomach tube 1 h before topical application of 17 nmoles TPA and were killed 4.5 h after TPA treatment for enzyme assay. Each point represents the mean of triplicate determinations of ODC activity from soluble epidermal extracts prepared from 4 mice

of biological and anti-carcinogenic activities of retinoids. Present results suggest that the mechanisms of the prophylactic effect of retinoic acid on skin tumour formation may involve its ability to inhibit TPA-induced ODC activity.

C. The Role of Prostaglandins in the Induction of ODC Activity by TPA

Prostaglandins are naturally occurring cyclic unsaturated fatty acids which have numerous physiological functions (Pelus and Strausser, 1977; Pike, 1976). A number of prostaglandins in tissues of various species are biosynthesized from their precursor unsaturated fatty acids present in cell membranes in the form of phospholipids (Figure 8). Biosynthesis of prostaglandins E_2 and $F_{2\alpha}$ from the precursor, arachidonic acid, has been shown in skins of various species including mice and human; the prostaglandin synthetase is mostly microsomal (Kingston and Greaves, 1976) and is present in the epidermis (Ziboh, 1973; Wilkinson and Walsh, 1977).

Arachidonic acid is hydrolysed from phospholipids by a membrane-bound enzyme phospholipase A_2. Arachidonic acid is converted into the endoperoxide PGH_2 by an enzymic step that involves cyclooxygenase. PGH_2 serves as an intermediate for the production of prostaglandins, thromboxanes, and prostacyclin (Pace-Asciak, 1977; Sun, Chapman and McGuire, 1977). A number of drugs (e.g. indomethacin, flufenamic acid, acetylsalicyclic acid) are known to inhibit prostaglandin biosynthesis, possibly by irreversibly inactivating the cyclooxygenase enzyme. Indomethacin is the most potent inhibitor of prostaglandin production among these inhibitors (Vane, 1971; Flower, 1974).

Prostaglandins have been implicated as mediators of cutaneous inflammation as well as in various pathological skin conditions (Hinman, 1972; Kingston and Greaves, 1976; Hammarström and coworkers, 1977). A higher than normal level of prostaglandin precursor, arachidonic acid, has been detected in psoriatic skin lesions (Hammarström and coworkers, 1977). In addition, elevated levels of prostaglandins have been detected in various tumours (Jaffe, 1974) as well as in cells transformed with either carcinogens or with SV40 virus (Hong, Wheless and Levine, 1977). Recently, addition of carcinogens (Levine, 1977) or tumour promoting phorbol diesters (Levine and Hassid, 1977) to dog kidney (MDCK) cells cultured in serum-supplemented medium, has been shown to stimulate prostaglandin production. In addition, release of prostaglandin E_2 in cultured peritoneal macrophages was observed as early as 1 h following treatment of cells with a tumour promoter and a correlation was found to exist between the ability of a compound to induce prostaglandin-release *in vitro* and its ability to promote skin tumour formation *in vivo* (Brune and coworkers, 1978). Results from our studies on the mechanism of skin tumour promotion reveal that prostaglandins may play an important role in the induction of ODC activity by TPA, an enzyme involved in mouse skin tumour promotion (Verma and coworkers, 1980).

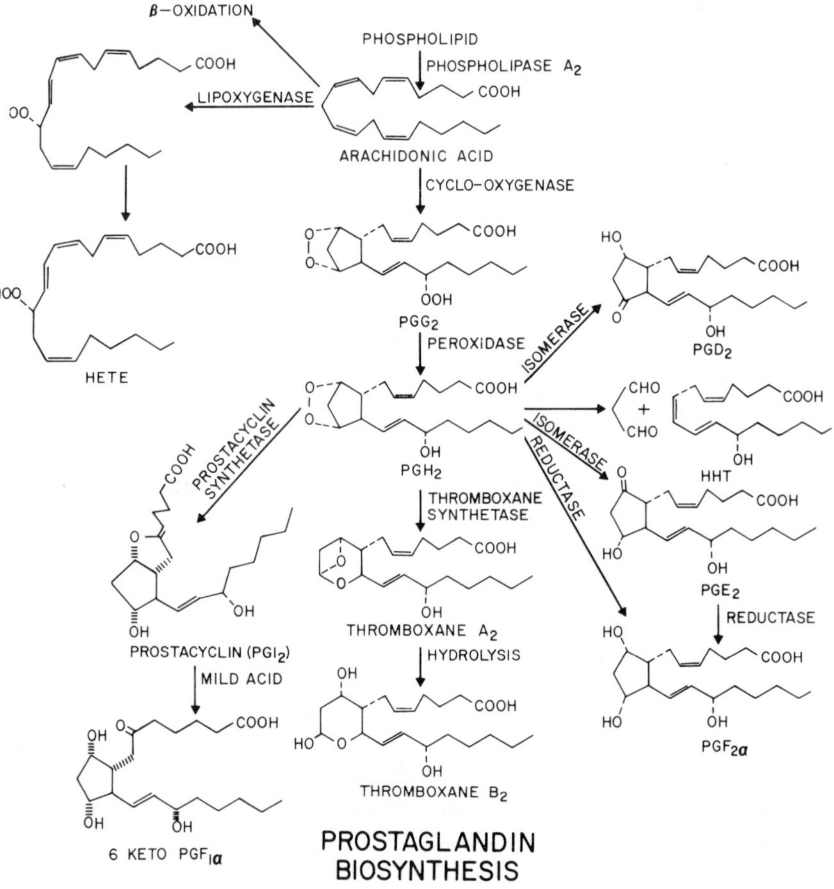

Figure 8 Prostaglandin biosynthesis

As shown in Figure 9, application of 280 nmoles of indomethacin 2 h prior to treatment with 10 nmoles of TPA resulted in an 80% inhibition of the induction of ODC activity and this inhibition was completely overcome when prostaglandin E_2 was applied concurrently with TPA. Application of prostaglandin E_2 alone did not induce epidermal ODC activity. A number of inhibitors of prostaglandin synthesis inhibited ODC induction by TPA: indomethacin > flufenamic acid > acetylsalicylic acid. The inhibition of the induction of ODC activity by these non-steroidal anti-inflammatory drugs was completely counteracted by treatment with either PGE_1 or PGE_2, but not with

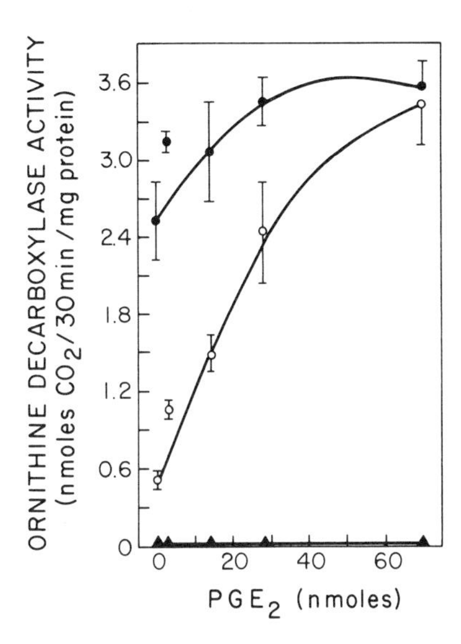

Figure 9 The effect of prostaglandin E_2 on inhibition of TPA-induced epidermal ornithine decarboxylase activity following pretreatment with indomethacin. Groups of mice were treated with either acetone (●) or 280 nmoles of indomethacin in acetone (○) 2 h before treatment with an acetone solution of 10 nmoles of TPA containing various doses of prostaglandin E_2. Control mice were treated with prostaglandin E_2 (▲) 2 h following treatment with acetone. Mice were killed for enzyme assay 4.5 h following final treatments. Each point is the mean ± s.e. of determinations carried out on 3 groups of mice with 3-4 mice per group

$PGF_{1\alpha}$ or $PGF_{2\alpha}$ (Verma, Rice and Boutwell, 1977; Verma and coworkers, 1980). Furthermore, a significant increase in the level of epidermal prostaglandin E_2 was observed as early as 2 h following TPA treatment and this increase can be prevented by treatment of mouse skin with indomethacin (Ashendel and Boutwell, 1979; Verma and coworkers, 1980).

These findings indicate that prostaglandins E_2 or E_1 may mediate or modulate ODC induction by TPA, and the suggestion that prostaglandins may play roles in skin carcinogenesis warrants further experimentation and understanding of the definitive biological role of these compounds in growth and neoplasia.

III. SUMMARY AND CONCLUSIONS

Topical application of vitamin A acid or certain of its analogues 1 h prior to treatment of mouse skin with TPA inhibited the induction of epidermal ODC activity as well as TPA-caused enhanced accumulation of its product, putrescine. In contrast, retinoids did not depress significantly the phorbol ester-induced SAMD activity and accumulation of polyamines spermidine and spermine following TPA treatment.

Application of retinoic acid 1 h before each promotion with TPA suppressed the formation of papillomas. In contrast, application of retinoic acid before, concurrent with, or after initiation with DMBA did not affect incidence of skin papillomas; this suggests that retinoic acid interferes with the promotion step of skin carcinogenesis. Retinoic acid, when applied 24 h after TPA treatment, did not inhibit formation of skin papillomas, this indicates that TPA-induced ODC activity (maximum at 4.5 h and back to control at 24 h after TPA treatment) may be a prerequisite for development of skin papillomas. Furthermore, the retinoids (retinoic acid, TMMP analogue of ethyl retinoate, 13-*cis*-retinoic acid, retinal) that inhibited the induction of ODC activity inhibited the formation of skin papillomas as well. In contrast, the TMHP analogue of ethyl retinoate and the 13-trifluoromethyl-TMMP analogue of ethyl retinoate did not affect either ODC induction by TPA or formation of skin papillomas.

The induction of ODC activity was depressed by prior treatment of mouse skin with indomethacin, acetylsalicylic acid or flufenamic acid, the inhibitors of prostaglandin synthesis. The inhibition of the induction of ODC activity by the prostaglandin synthetase-inhibitors was completely counteracted by treatment with prostaglandin E_1 or E_2 but not with prostaglandin $F_{1\alpha}$ or $F_{2\alpha}$.

The data presented here lends support to the proposal that TPA-induced epidermal ODC activity represents one of the essential biochemical events of the mechanism of skin tumour promotion. Thus, induction of ODC activity by TPA may be a marker for tumour promoters and provides a basis for screening large numbers of chemicals for potential promoting ability. In addition, prostaglandins may mediate or modulate ODC induction by TPA and thus may play an important role in skin tumour promotion.

The findings that the ability of retinoids to inhibit the induction by TPA of ODC activity correlates with their ability to inhibit formation of skin papillomas, suggests that inhibition of the induction of ODC activity by retinoids may be a simple rapid test for the anti-promoting properties of new synthetic retinoids.

ACKNOWLEDGEMENTS

The authors thank Dr. M. B. Sporn, Lung Cancer Branch of the National Cancer Institute, Bethesda, MD and Dr. B. A. Pawson, Hoffmann-LaRoche, Inc., Nutley, NJ for the supply of retinoids. We are indebted to Miss Hope Rice and Mrs. Barbara Shapas for technical assistance. The work was supported by National Institute of Health Program Project Grant CA-07175.

REFERENCES

Ashendel, C. L., and Boutwell, R. K. (1979). *Biochem. Biophysics. Res. Commun.*, **90**, 623–627.
Baird, W. M., Sedgwick, J. A., and Boutwell, R. K. (1971). *Cancer Res.*, **31**, 1434–1439.
Belman, S., and Troll, W. (1978). Hormones, Cyclic Nucleotides, and Prostaglandins. In *Carcinogenesis, A Comprehensive Survey*, (Eds. Thomas J. Slaga, Andrew Sivak, and R. K. Boutwell), Vol. 2, pp. 117–134, Raven Press, New York.
Berenblum, I., and Shubik, P. (1947). *Brit. J. Cancer*, **1**, 379–382.
Blalock, J. and Gifford, G. E. (1977). *Proc. Natl. Acad. Sci. USA*, **74**, 5382–5386.
Bollag, W. (1971). *Experientia*, **27**, 90–92.
Bollag, W. (1972). *European J. Cancer*, **8**, 689–693.
Bollag, W. (1975a). *European J. Cancer*, **11**, *721–724.*
Bollag, W. (1975b). *Chemotherapy*, **21**, 236–247.
Boutwell, R. K. (1964). *Prog. Exp. Tumor Res.*, **4**, 207–250.
Boutwell, R. K. (1974). *CRC Critical Rev. Toxicol.*, **2**, 419–443.
Boutwell, R. K. (1976). *Cancer Res.*, **36**, 2631–2635.
Boutwell, R. K. (1977). The Role of the Induction of Ornithine Decarboxylase Activity in Tumor Promotion. In *Origins of Human Cancer*, Book B, (Eds. H. H. Hiatt, J. D. Watson, and J. A. Winsten), pp. 773–783. Cold Spring Harbor, New York. Cold Spring Harbor Laboratory.
Brune, K., Kälin, H., Schmidt, R., and Hecker, E. (1978). *Cancer Letters*, **4**, 333–342.
Chopra, D. P., and Wilkoff, L. J. (1977). *J. Natl. Cancer Inst.*, **58**, 923–930.
Clamon, G. H., Sporn, M. B., Smith, J. M., and Saffiotti, U. (1974). *Nature*, **250**, 64–66.
Clark-Lewis, I., and Murray, A. W. (1978). *Cancer Res.*, **38**, 494–497.
Colburn, N. H., Lau, S., and Head, R. (1975). *Cancer Res.*, **35**, 3154–3159.
De Luca, L., and Yuspa, S. H. (1974). *Exptl. Cell Res.*, **86**, 106–110.
Flower, R. J. (1974). *Pharm. Rev.*, **26**, 33–67.
Grimm, W., and Marks, F. (1974). *Cancer Res.* **34**, 3128–3134.
Hammarström, S., Hamberg, M., Duell, E. A., Stawiski, M. A., Anderson, T. F., and Voorhees, J. J. (1977). *Science*, **197**, 994–996.
Hecker, E., and Schmidt, R. (1974). In *Progress in the Chemistry of Organic Natural Products.* (Eds. H. Heiz, H. Grisebach and G. W. Kirby). Vol. 31, pp. 377-467. Springer-Verlag, New York.
Hinman, J. W. (1972). *Ann. Rev. Biochem.*, **41**, 161–178.
Hong, S. L., Wheless, C. M., and Levine, L. (1977). *Prostaglandins*, **13**, 271–279.
Jaffe, B. M. (1974). *Prostaglandins*, **6**, 453–461.
Kingston, W. P., and Greaves, M. W. (1976). *Prostaglandins*, **12**, 51–69.
Kreig, L., Kühlmann, I., and Marks, F. (1974). *Cancer Res.*, **34**, 3135–3146.
Levine, L. (1977). *Nature*, **269**, 447–448.
Levine, L., and Hassid, A. (1977). *Biochem. Biophys. Res. Commun.*, **79**, 477–484.
Miller, J. A., and Miller, E. C. (1977). In *Origin of Human Cancers*, Book B., (Eds. H. H. Hiatt, J. D. Watson, and J. A. Winsten), pp. 605–627. Cold Spring Harbor, New York, Cold Spring Harbor Laboratory.
Mottram, J. C. (1944). *J. Pathol. Bacteriol.*, **56**, 181–187.
Mufson, R. A., Simsiman, R. C., and Boutwell, R. K. (1977). *Cancer Res.*, **37**, 665–669.
Murray, A. W., Verma, A. K., and Froscio, M. (1976). Effect of Carcinogens and Tumor Promoters on Epidermal Cyclic Adenosine 3':5'-Monophosphate Metabolism. In *Control Mechanisms in Cancer*, (Eds. W. E. Criss, T. Ono, and J. R. Sabine), pp. 217–229. Raven Press, New York.

Nettesheim, P., and Williams, M. L. (1976). *Intern. J. Cancer*, **17**, 351–357.

O'Brien, T. G. (1976). *Cancer Res.*, **36**, 2644-2653.

O'Brien, T. G., Simsiman, R. C., and Boutwell, R. K. (1975a). *Cancer Res.*, **35**, 1662–1670.

O'Brien, T. G., Simsiman, R. C., and Boutwell, R. K. (1975b). *Cancer Res.*, **35**, 2426–2433.

Ong, D. E., and Chytil, F. (1975). *J. Biol. Chem.*, **250**, 6113–6117.

Pace-Asciak, C. R. (1977). *Prostaglandins*, **13**, 811–817.

Pelus, L. M., and Strausser, H. R. (1977). *Life Sci.*, **20**, 903–914.

Pike, J. E. (1976). *J. Invest. Dermatol.*, **67**, 650–653.

Raineri, R., Simsiman, R. C., and Boutwell, R. K. (1973). *Cancer Res.*, **33**, 134–139.

Rohrschneider, L. R., and Boutwell, R. K. (1973). *Cancer Res.*, **33**, 1945–1952.

Sani, B. P., and Corbett, T. H. (1977). *Cancer Res.*, **37**, 209–213.

Sani, B. P., and Hill, D. L. (1976). *Cancer Res.*, **36**, 409–413.

Shamberger, R. J. (1971). *J. Natl. Cancer Inst.*, **47**, 667–673.

Sims, P., and Grover, P. L. (1974). *Adv. Cancer Res.*, **20**, 165–274.

Sporn, M. B. (1977a). *Nutr. Rev.*, **35**, 65–69.

Sporn, M. B. (1977b). Prevention of Epithelial Cancer by Vitamin A and its Synthetic Analogs (Retinoids). In *Origin of Human Cancers*, Book B., (Eds. H. H. Hiatt, J. D. Watson, and J. A. Winsten) pp. 801-807. Cold Spring Harbor, New York, Cold Spring Harbor Laboratory.

Sporn, M. B., Clamon, G. H., Dunlop, N. M., Newton, D. L., Smith, J. M., and Saffiotti, U. (1975). *Nature*, **253**, 47–50.

Sporn, M. B., Dunlop, N. M., Newton, D. L., and Henderson, W. R. (1976a). *Nature*, **263**, 110–113.

Sporn, M. B., Dunlop, N. M., Newton, D. L., and Smith, J. M. (1976b). *Federation Proc.*, **35**, 1332–1338.

Sporn, M. B., Squire, R. A., Brown, C. C., Smith, J. M., Wenk, M. L., and Springer, S. (1977). *Science*, **195**, 487–489.

Sun, F. F., Chapman, J. P., and McGuire, J. C. (1977). *Prostaglandin*, **14**, 1055–1073.

Tabor, C. W., and Tabor, H. (1976). *Ann. Rev. Biochem.*, **45**, 285–306.

Tabor, H., and Tabor, C. W. (1972). *Advan. Enzymol.*, **36**, 203–268.

Van Duuren, B. L. (1969). *Prog. Exp. Tumor Res.*, **11**, 31–68.

Vane, J. R. (1971). *Nature New Biology*, **231**, 232–235.

Verma, A. K., and Boutwell, R. K. (1977). *Cancer Res.*, **37**, 2196–2201.

Verma, A. K., Froscio, M., and Murray, A. W., (1976a). *Cancer Res.*, **36**, 81–87.

Verma, A. K., Dixon, K. E., Froscio, M., and Murray, A. W. (1976b). *J. Invest. Dermatol.*, **66**, 239–241.

Verma, A. K., Rice, H. M., and Boutwell, R. K. (1977). *Biochem. Biophys. Res. Commun.*, **79**, 1160–1166.

Verma, A. K., Rice, H. M., Shapas, B. G., and Boutwell, R. K. (1978). *Cancer Res.*, **38**, 793–801.

Wilkinson, D. I., and Walsh, J. T. (1977). *J. Invest. Dermatol.*, **68**, 210–214.

Wilkoff, L. J., Peckham, J. C., Dulmadge, E. A., Mowry, R. W., and Chopra, D. P. (1976). *Cancer Res.*, **36**, 964–972.

Ziboh, V. A. (1973). *J. Lipid Res.*, **14**, 377-384.

Chapter 13

Bacterial Amines and Carcinogenesis

ELIZABETH S. BONE

I. INTRODUCTION

The bacterial production of amines has been recognized since the beginning of the present century (Barger and Walpole, 1909; Berthelot, 1913), and the importance of this reaction with respect to human health becomes apparent now when one considers the metabolic activity of the gut bacterial flora.

II. *IN VITRO* AMINE PRODUCTION

A. Decarboxylation

Bacterially produced amines arise mainly from the decarboxylation of amino acids. This results in the formation of the corresponding amine with the liberation of carbon dioxide:

$$RCHNH_2COOH \rightarrow RCH_2NH_2 + CO_2$$

The enzymes responsible for the decarboxylation reaction were investigated systematically by Gale (1946) who demonstrated that each decarboxylase was specific for a single amino acid, and that a third polar group, in addition to the terminal carboxyl group and the amino group, was necessary for decarboxylation. Thus he found amino acid decarboxylases for arginine, ornithine, lysine, histidine, tyrosine, glutamic acid, aspartic acid and tryptophan but not for single monoamino monocarboxylic acids. Later an amino acid decarboxylase active on leucine, isoleucine, valine, nor-valine and α-aminobutyric acid was described (Haughton and King, 1961). This contrasted with the monospecificity and the requirement for a third polar group described by Gale (1946). A decarboxylase enzyme specific for proline has also been identified by Hawksworth (1973). Amino acid decarboxylases are produced optimally *in vitro* at acid pH values and evidently represent an attempt by bacteria to buffer their environment.

Production of amino acid decarboxylases can be used diagnostically to distinguish between genera within the Enterobacteriaceae (Møller, 1954). Initially the determination of the decarboxylases was a research problem but Møller (1955) developed simpler analytical techniques so that the decarboxylase pattern became a useful taxonomic tool in the identification of the Enterobacteriaceae. Since decarboxylases of lysine, arginine, ornithine and glutamic acid are produced by various members of the Enterobacteriaceae and are used diagnostically, they have been screened extensively in these organisms, with the result that the methods for their detection have been greatly improved and are well documented. Screening for decarboxylase activity in other groups of organisms has therefore been concentrated on the decarboxylases of lysine, arginine, ornithine and glutamic acid because it is relatively simple to detect them (or their products). Information on other bacterial decarboxylases is relatively sparse (Table 1).

B. N-Dealkylation

The N-dealkylation of choline by bacteria gives rise to the secondary amine, dimethylamine. Choline is derived from lecithin, a major component of animal fat. Lecithin is converted to choline in the gut by bacterial lecithinase, and this is further metabolized to trimethylamine and then dimethylamine by intestinal bacterial N-dealkylating enzymes (Figure 1). This metabolic sequence has been demonstrated by Asatoor and Simenhoff (1965) and Hawksworth (1973).

Hawksworth (1973) found that clostridial strains were the most active amongst the bacteria isolated from faeces in the N-dealkylation of choline to dimethylamine while there was very little activity amongst the aerobic bacteria. Using a different method Johnson (1977), found both the clostridia and aerobes were very active in the breakdown of choline.

Table 1 Production of amino acid decarboxylases by gut bacteria

Organisms tested	Amino acid decarboxylation										
	Lys.	Arg.	Orn.	Glut. acid	Hist.	Tyr.	Prol.	Trypt.	Leu.	Val.	Norval.
Enterobacteria											
Esch. coli	+	+	+	+	*	*	+		*		
Salmonellae	+	+	+	—							
Shig. sonnei	—	+	+	+							
other Shigellae	—	+	—	+							
P. vulgaris	—	—	—	+							
other Proteus	—	—	+	+					*	*	*
Serratia	+	—	+	—							
Klebsiella	+	—	—	—							
Streptococci											
Strep. faecalis	—	—	—	—	—	+	*				
Group A-F	—	—	—	—	—	*					
Strep. lactis	—	—	—	—	—	—					
Clostridia											
Cl. welchii	—	—	—	+	+	—	+				
Cl. septicum	—	—	+	—	—	—					
Cl. bifermentans	+	+	+				+				
Bacillus spp.	*	*	*	*	*	*		*			
Bacteroides fragilis	*	*	+				+				
Bifidobacteria	*	*	+				+				
Vibriocholera	+		+								
Staphylococci	—	—	—	—	—	—					
Pasteurella	—	—	—	—	—	—					
Pseudomonas	—	—	—	—	—	—					

+ = more than 10 strains tested most of which were +ve. — = more than 10 strains tested; * = individual strains reported +ve, but no screening results available. Reproduced by permission of B. S. Drasar and M. J. Hill.

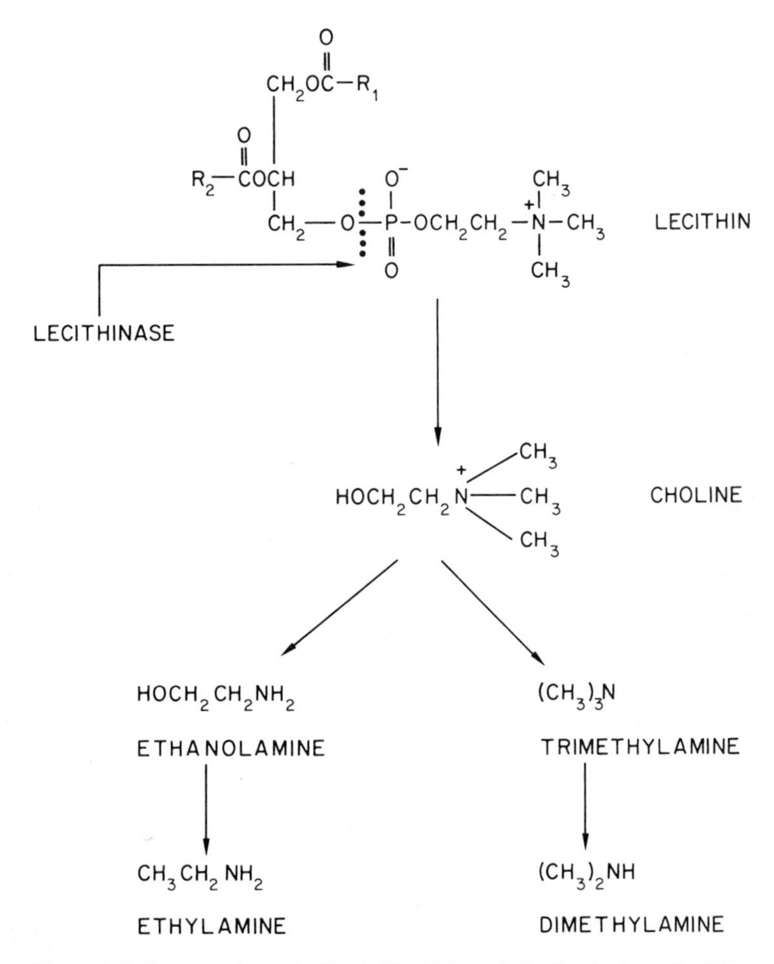

Figure 1 Pathways of metabolism of lecithin and choline by intestinal bacteria

C. Ring Closure

The heterocyclic secondary amines, piperidine and pyrrolidine, are also the products of bacterial metabolism. Following the decarboxylation of lysine and arginine by bacteria, the diamines (cadaverine and putrescine) undergo oxidative deamination followed by ring closure to give the immediate precursors to the cyclic secondary amines, piperidine and pyrrolidine. Hawksworth (1973) showed that the conversion of cadaverine to piperidine, and putrescine to pyrrolidine, was restricted to members of the anaerobic genera, the clostridia, bacteroides and bifidobacteria (Hawksworth, 1973).

III. EVIDENCE FOR *IN VIVO* AMINE PRODUCTION

A. Effects of Antibiotics

A large number of the bacteria able to decarboxylate amino acids *in vitro* were known to flourish in the large intestine but it was not known until 1955 whether they produced amines *in vivo*. In that year evidence for the *in vivo* production of amines came from the work of Melnykowycz and Johansson (1955) who showed that mixed bacterial cultures derived from rat faeces were able to decarboxylate the amino acids histidine, arginine, lysine, glutamic acid and tyrosine, and the effect of the antibiotic chlortetracycline on amino acid decarboxylation was to reduce the number of amines produced by the mixed faecal cultures. They also detected at least seven different amines in rat faeces and found a reduction in the number of amines present in the faeces of rats given food containing chlortetracycline. Irvine, Duthie and Waton (1959) showed that bowel contents from dogs produced histamine when incubated with histidine and found that the antibiotic succinylsulphathiazole diminished this response.

Asatoor and coworkers (1967) showed that oral neomycin reduced the amounts of the primary amine, methylamine, the secondary amines, dimethylamine, pyrrolidine and piperidine and to a lesser extent the tertiary amine, trimethylamine present in stool samples from six normal humans. The almost complete disappearance of these faecal amines during neomycin ingestion indicates that they are largely products of bacterial metabolism.

B. Additional Evidence

Asatoor (1964) incubated a mixed faecal suspension with lysine and found that a large amount of the cyclic secondary amine, piperidine, was being formed. He also found large amounts of the cyclic secondary amine, pyrrolidine, were produced after incubation with ornithine; much smaller amounts being produced after incubation with arginine. This is because arginine, in addition to being decarboxylated to agmatine, is converted by bacteria to citrulline and then to ornithine by arginine dihydrolase, a two enzyme system (Figure 2). Owing to the complexity of this enzyme, the amount of pyrrolidine produced is dependent not only on the ability of the mixed flora to produce it from ornithine but also, and more crucially, on its ability to generate the substrate ornithine from arginine.

Asatoor (1964) confirmed the ability of bacteria to convert an amino acid to a heterocyclic amine using radioisotopes. After incubation of a mixed faecal culture with [^{14}C] lysine, radioactive piperidine was recovered from the incubation mixture.

In determining the pathway for this reaction, Asatoor (1964) noted that the primary products of bacterial action on amino acids are the corresponding primary amines. He therefore used the diamines, putrescine and cadaverine (derived from the bacterial decarboxylation of ornithine and lysine respectively) as substrates in incubation experiments with mixed faecal bacterial. He found that pyrrolidine was produced after incubation with putrescine, and piperidine after incubation with cadaverine. Hawksworth (1973) showed that although the individual reactions in the series leading from the amino acids, arginine

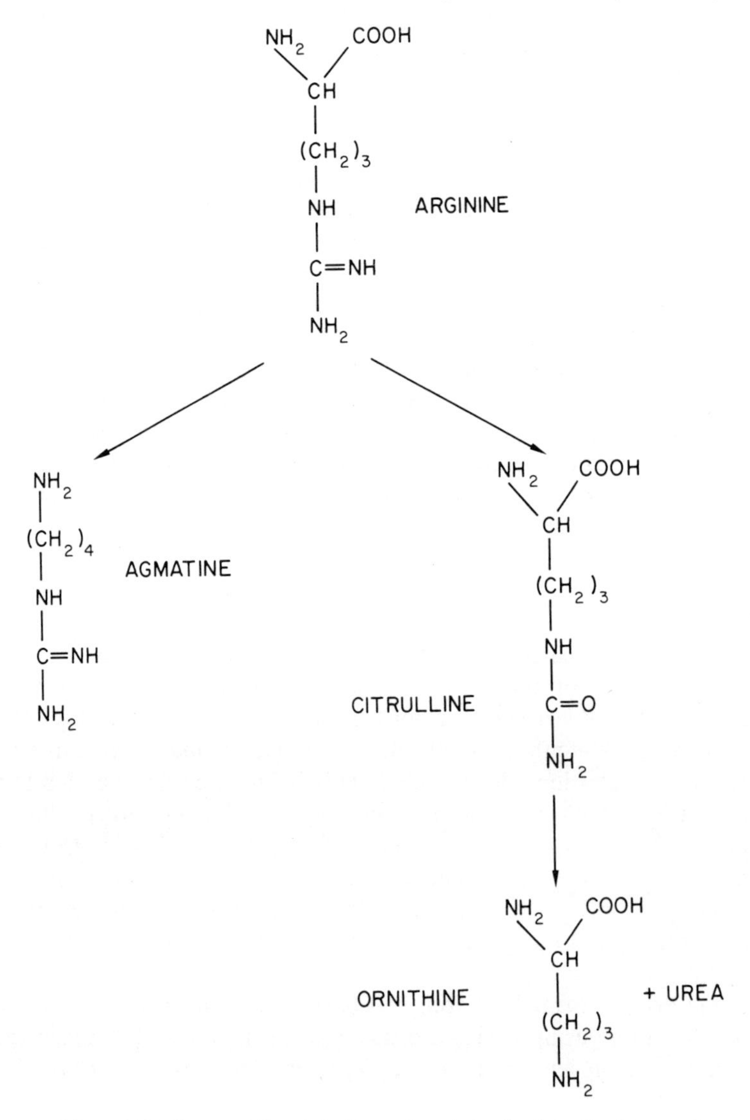

Figure 2 Pathways of metabolism of arginine by intestinal bacteria

and lysine, to their respective cyclic secondary amines could be carried out by strains of a range of bacterial species isolated from the intestine, no single bacterial species tested carried out the whole series. She was able to demonstrate that a simple mixture of selected aerobic and anaerobic bacteria could carry out the complete pathways (Figure 3), thus confirming the work of Asatoor with the mixed faecal suspensions.

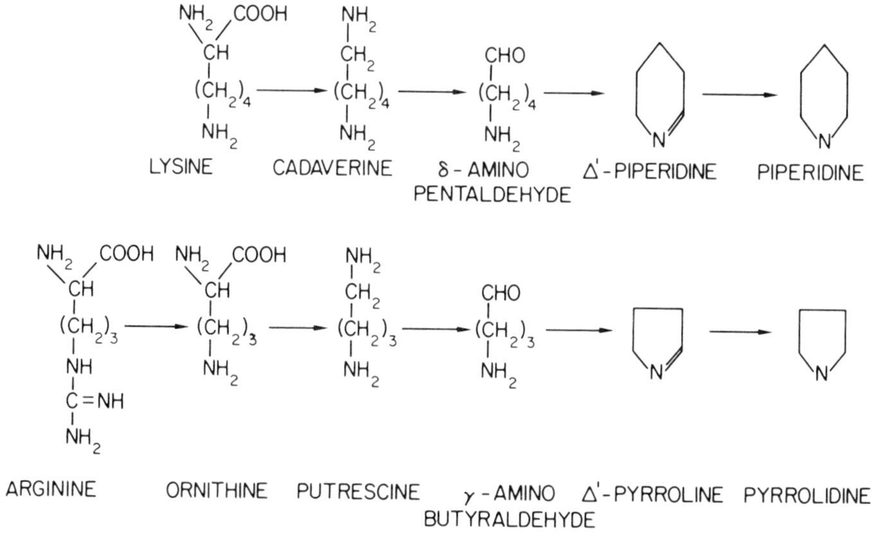

LYSINE → CADAVERINE → δ-AMINO PENTALDEHYDE → Δ'-PIPERIDINE → PIPERIDINE

ARGININE → ORNITHINE → PUTRESCINE → γ-AMINO BUTYRALDEHYDE → Δ'-PYRROLINE → PYRROLIDINE

Figure 3 Pathways of metabolism of lysine and arginine by intestinal bacteria

Asatoor (1968) detected tyramine in faecal extracts from rats and observed that radioactive tyramine was formed when faecal bacteria were incubated with [^{14}C] tyrosine, thus supporting the view that under normal circumstances intestinal bacteria act on dietary tyrosine to form tyramine.

C. Urinary Excretion

A number of amines are normally excreted in the urine and measurement of these in conventional, antibiotic treated and germ free animals have also demonstrated amine production by the gut bacterial flora.

1. Histamine

Histamine is present in the urine both in the free and acetylated form (Roberts and Adam, 1950). Urbach (1949) showed that histamine is acetylated by the gut flora to acetylhistamine which is partially absorbed from the intestinal tract and excreted in the urine as well as in the faeces. Wilson (1954) measured intestinal and urinary histamine in normal rats and rats treated with a range of antibiotics; the antibiotic treatment reduced the amount of histamine excreted by more than 50%. Irvine, Duthie and Waton (1959) found that the increase in the excretion of urinary histamine after ingestion of a meat meal was similar in amount and duration to the rise obtained when the approximate L-histidine content of the meal was given instead of the meal itself. The increases were significantly diminished in both man and dog after antibiotic treatment. These experiments showed that urinary histamine is derived at least in part from the decarboxylation of L-histidine by intestinal bacteria.

2. *Tyramine*

Perry and coworkers (1966) studied four subjects undergoing intestinal chemotherapy and found that *p*-tyramine excretion in the urine of one subject whose bowel was rendered virtually sterile after antibiotic treatment was decreased 15-fold. (The antibiotic treatment consisted of neomycin, 0.5 g 4 times daily, ampicillin, 0.5 g 4 times daily, and nystatin, 500,000 units 4 times daily). They came to the conclusion that urinary tyramine is either largely or entirely of bacterial origin. In contrast, DeQuattro and Sjoerdsma (1967) found that antibacterial chemotherapy (consisting of neomycin, 0.5 g 4 times daily, sulfasuxidine, 2 g 4 times daily, and furazolidine, 0.1 g 4 times daily) in humans failed to lower the urinary level of tyramine, and concluded that tyramine formation by intestinal bacteria is not a major source of urinary tyramine.

Asatoor (1968) found that subcutaneous injection of [^{14}C] tyrosine into rats resulted in excretion of radioactive tyramine and suggested that tyramine is formed by the decarboxylation of tyrosine in tissues, as well as by intestinal bacteria. He found that oral administration of neomycin to normal subjects diminished the excretion of tyramine. This supported the findings of Awapara and coworkers (1964) in that excretion of tyramine diminishes after administration of an antibacterial agent (Sulfasuxidine, 3 g 4 times daily).

3. *Tryptamine and 5-Hydroxytryptamine*

Tryptamine and 5-hydroxytryptamine are both found in human urine (Oates and coworkers, 1960; Korf, 1969) and are the result of the metabolism of tryptophan by both tissue and bacterial enzymes (Weissbach and coworkers, 1959). Little is known of the relative contribution of the gut flora and the liver to the urinary concentration of these two compounds. Perry and coworkers (1966) observed the changes in tryptamine excretion in the urine of four subjects, all undergoing different antibiotic treatments, and suggested that intestinal organisms are the major and possibly the only source of urinary tryptamine in man. They suggested that antibacterial chemotherapy could either decrease or increase tryptamine excretion depending on whether the intestinal flora is eradicated, or merely altered, with secondary overgrowth of organisms producing tryptophan decarboxylase. The antibiotic regime used by DeQuattro and Sjoerdsma (1967), however, failed to lower the urinary level of tryptamine in humans and they concluded that tryptamine is mainly of tissue rather than bacterial origin.

4. Piperidine and Pyrrolidine

A valuable piece of evidence for the production of the heterocyclic amines, piperidine and pyrrolidine, in the intestine emerged from the work of Nordenström (1951) who showed that after oral administration of cadaverine to rabbits there was a large increase in urinary piperidine excretion, whilst after subcutaneous injections of the same diamine there was no increase in piperidine excretion. He therefore concluded that the piperidine in urine is probably derived from cadaverine by bacterial action in the digestive tract. Blau (1961) found that in the normal male the urinary excretion of piperidine fell to zero after a period of fasting, indicating a dietary origin for piperidine. Perry and coworkers (1966) showed antibiotic treatment considerably reduced urinary piperidine levels suggesting that this amine arises from the bacterial decarboxylation of lysine in the bowel, followed by cyclization of the resulting cadaverine to piperidine, either in the intestine or in tissues.

Asatoor and coworkers (1967) found that oral neomycin treatment reduced the urinary output of piperidine and pyrrolidine to negligible amounts, again showing that these amines are derived mainly from bacterial production within the gut. This was confirmed by Hawksworth (1973) who showed piperidine and pyrrolidine were absent from the urines of germ-free animals.

5. Dimethylamine

The secondary amine, dimethylamine, is also present in the urine. Asatoor and Simenhoff (1965) found a 20% reduction in the urinary excretion of dimethylamine in neomycin treated rats. They also showed that ingestion of 10 g choline chloride in man, and 100 mg of choline chloride in rats, more than doubled the amount of dimethylamine excreted in the urine, but when the same amount of choline was administered by i.p. injection in rats there was no increase in the urinary dimethylamine level. They therefore suggested that urinary dimethylamine is derived partly from ingested lecithin and choline by bacterial action in the intestine. Blau (1961) showed that after fasting there was only a 25% decrease in dimethylamine excretion and concluded that the major portion of urinary dimethylamine arises from endogenous sources.

Hawksworth's studies (1973) of germ-free and conventional pigs indicated, however, that at least 50% of the dimethylamine in urine is the result of gut bacterial action on ingested lecithin and choline.

IV. PHYSIOLOGICAL AND PHARMACOLOGICAL EFFECTS OF BACTERIALLY PRODUCED AMINES

Many bacterially produced amines have pronounced physiological or pharmacological activities so that their production *in vivo* could have important consequences.

The amines produced in the gut are absorbed and enter the portal blood system where they are acted on by a monoamine-oxidase produced by the liver which deaminates a large proportion of the monoamines, thereby rendering them non-toxic and reducing their concentration to safe levels before excretion in the urine. The concentration of these potential toxins is high in the portal blood, and low in the systemic venous blood. In any situation where the portal blood can by-pass the liver their concentration increases in the systemic blood and thus can reach toxic levels; this is the generally accepted aetiology of hepatic coma associated with cirrhosis or with porto-systemic anastamosis.

A. Histamine

Acetyl histamine is pharmacologically inactive but histamine itself has pronounced pharmacological activity. Besides its function in immune processes it is probably a neurotransmitter in the mammalian brain. Recently it has been found that histamine may be a crucial factor in the chemistry of depression (Monitor,1978). It is also a pressor substance; injection of small amounts leading to a fall in blood pressure. It dilates blood vessels and increases the permeability of the capillary walls. Smooth muscle in other tissues is, in general, contracted by histamine. Histamine acts directly on the acid secreting cells in the stomach, and stimulates the secretion of acid into the gastric juice. The fact that histamine can stimulate gastric acid secretion, as well as possessing general inflammatory activity in increasing vascular permeability, has led to the suggestion that intestinal histamine is involved in the aetiology of peptic ulcer associated with hepatic cirrhosis (Irvine and coworkers, 1959).

B. Tyramine

Tyramine is a depressor substance, and injection of small amounts results in increased blood pressure.

C. Tryptamine and 5-Hydroxytryptamine

Both tryptamine and 5-hydroxytryptamine are potent pharmacological agents. Among the complex pharmacological actions of 5-hydroxytryptamine is smooth muscle stimulation which may be manifested as vasoconstriction, vasodilation and ileal spasm.

D. Amines Derived from Arginine and Lysine

Agmatine, the product of arginine decarboxylation has an insulin-like activity. The diamines, cadaverine and putrescine, are weak pressor substances, whilst the secondary amine, piperidine, seems to be intimately connected with the neuronal function of the brain. It acts on the synaptic sites in the brain and affects emotional behaviour, sleeping and extrapyramidal function (Kasé and Miyata, 1976).

V. SECONDARY AMINES, N-NITROSAMINES AND GASTRIC CANCER

The secondary amines, dimethylamine, piperidine and pyrrolidine, formed by the bacterial metabolism of choline, lysine and arginine respectively, warrant special attention because of their possible role in human carcinogenesis. Secondary amines, together with nitrite, are precursors of the potent group of carcinogens, the N-nitrosamines. The reaction between secondary amines and nitrite to form nitrosamines is classically acid catalysed. Sander (1968) demonstrated, however, that certain strains of enterobacteria were able to form nitrosamines from aromatic secondary amines and nitrate at physiological pH values. Hawksworth and Hill (1971) and Hill, Hawksworth and Tattersall (1973) confirmed this work, and showed that the reaction involving nitrate is only performed by nitrate-reducing bacteria, but a proportion of the strains which do not reduce nitrate are able to N-nitrosate secondary amines using nitrite as the nitrosating agent. Thus it became apparent that nitrosamines might be formed *in vivo* wherever secondary amine, nitrate or nitrite and bacteria occurred together.

The bacterially produced secondary amines, dimethylamine, piperidine and pyrrolidine will be present in the large bowel and in the urine. Bacteria are present in the large bowel in great numbers, but are normally absent from the urinary bladder. Urinary tract infections are quite common, however, so that two of the three reactants necessary for nitrosamine formation are always present in the large bowel and sometimes in the bladder.

Transit time in the large bowel would allow more than adequate time for nitrosation to occur. Little work has been done, however, on the presence of nitrate in the large intestine. Studies on the fate of ingested nitrate (Hawksworth and Hill, 1971) indicate a rapid absorption of nitrate from the small intestine, and subsequent excretion in the urine, so that little dietary nitrate will reach the large intestine. None the less, it is possible that nitrate is present in colonic secretions, since it has been demonstrated in human saliva (Tannenbaum, Sinskey and Weisman, 1974), and its presence has been inferred in other body secretions e.g. gastric (Ruddell, Blendis and Walters, 1976) and vaginal (Allsobrook and coworkers, 1974) secretions. This being the case it is possible that all three factors necessary for nitrosamine formation could be present in the colon. Evidence for this has come from the work of Varghese and coworkers (1977) who have found that the faeces of normal individuals contain readily detectable amounts of N-nitroso compounds.

Johnson (1977) found that the numbers of bacteria able to decarboxylate arginine, lysine and ornithine were higher in the faeces of colon cancer patients and patients at a high risk of developing colon cancer than in 'normals'. She also found that the clostridia were most active in the breakdown of choline, and the counts of clostridia were higher in cancer and high-risk patients than in normals. Thus it seems likely that there will be more secondary amine present as a possible co-carcinogen in the large bowel of colon cancer and high-risk patients. This being the case the bacterial production of secondary amines in the intestine may well be an important factor in the aetiology of colon cancer.

The three bacterially produced secondary amines, dimethylamine, piperidine and pyrrolidine are present in quite large amounts in the urine. Dimethylamine is the most abundant, 20-30 mg being excreted daily (Asatoor and Simenhoff, 1965). Piperidine and pyrrolidine are excreted at the rate of less than 1.0 mg per day. This secondary amine level is independent of the day-to-day diet and is relatively constant (Bone, 1977).

Dietary nitrate is excreted in the urine where its concentration is dependent on the amount ingested (Hawksworth and Hill, 1971). Thus in the infected urinary bladder the three factors necessary for nitrosamine formation will be present. Hill and Hawksworth (1972) demonstrated the *in vivo* production of nitrosamines in rats with experimental bladder infection. They showed nitrosation was catalysed by bacteria *in vivo* when large amounts of amine and nitrate were fed to the rats; no nitrosamine was detected in the urine of uninfected rats or in infected rats fed only nitrate or only secondary amine. Hicks and coworkers (1977) have demonstrated the presence of nitrosamines in the urine of people with bladder infections. The results indicated that in addition to dimethylnitrosamine other species of nitrosamines, particularly non-volatile compounds, are formed in some infected urines. Hill and Hawksworth (1974) showed that nitrosamines formed *in vivo* in the infected urinary bladder could be absorbed from the bladder and thus contribute to human cancer production.

Bladder infections are more common than generally thought. A survey in a rural general practice in England showed that the incidence of bacteriuria (symptomatic and asymptomatic) was 184 cases per 1,000 patients per year

(Sinclair and Tuxford, 1971). A screening programme among the middle aged (40-64 yrs) female population in a town in Finland showed the total prevalence of urinary tract infection (treated and untreated) to be 7% (Takala, Jousimies and Sieves, 1977).

Urinary tract infections are common in woman of child-bearing age, and the incidence increases with age (Savage, Hajj and Kass, 1967); they are also common in men over the age of 50 (in association with infected prostates). In the bladder there will be large amounts of secondary amines and nitrate; the bacterial counts are high and these organisms will have plenty of time to act on the reactants; since most bladder infections are asymptomatic and not associated with urinary frequency, the incubation time will optimally be overnight. Thus, the conditions for nitrosamine formation — high reactant concentration, large numbers of organisms and long incubation time — are nearly as ideal as could be expected.

Nitrosamine formation occurring in the infected urinary bladder would be expected to produce cancer in men only in the oldest age groups, whereas in women the lower age groups might be affected more often, since bladder infections in young women are fairly common. The amount of nitrosamine formed will be limited by nitrate concentration, since it appears that it must exceėd that of secondary amine for nitrosation to take place (Hawksworth and Hill, 1971).

Therefore, to determine whether nitrosamine formation *in vivo* has any effect on human carcinogenesis it is necessary to identify a population taking in large amounts of dietary nitrate so that a study of the cancer incidence in such a population can be undertaken.

An interesting feature of the carcinogenic N-nitrosamines is their ability to induce a variety of tumours at different sites in experimental animals, depending on their chemical structure, mode of application and dosage of the N-nitroso compound (Magee and Barnes, 1967; Druckrey and coworkers, 1967), so that although man is undoubtedly susceptible to N-nitrosamine carcinogenesis, no predictions can be made on the basis of animal studies concerning the target organ of any N-nitrosamine. Thus any epidemiological survey of nitrosamine carcinogenesis must take into account cancers of all body sites.

Hawksworth (1973) identified a population with a high nitrate intake in Worksop (Nottinghamshire). Here, the high nitrate content of the drinking water supply resulted in a greatly increased intake of nitrate per day and consequently a high urinary concentration of nitrate (Hawksworth and Hill, 1971). Assuming that the incidence of urinary tract infection in Worksop is not markedly different from that in the rest of the country, this should result in an increase in the incidence of nitrosamine-induced cancers and these should be detectable epidemiologically.

Analysis of age adjusted death rates due to all cancers in Worksop and in a number of control towns (selected for their proximity to Worksop, similar major industries and similar social class structure) showed that the number of deaths due to cancer in Worksop was greater than that expected from data from the control towns, and that the increase was greater for women than for men (93% compared with 38%).

The deaths from cancer in Worksop for the years 1958–1971 were then analysed by site, age and sex and compared with the expected numbers (derived from data from the Sheffield cancer registry). This analysis showed that there was an increase in the number of deaths from cancer of the stomach, liver, kidney and oesophagus in both men and women (Table 2). Although the most spectacular proportional increase was in the deaths from primary hepatic carcinoma, this was not taken to be of extreme significance, because the numbers involved were small and the diagnosis of primary hepatic carcinoma is not absolutely reliable. The increase in the number of deaths from cancer of the oesophagus and kidney were not very significant, again because of the small numbers involved. The increased number of deaths from stomach cancer were very significant because of the large numbers involved.

Table 2 Cancer deaths in Worksop (1958–71) compared with those expected

Site	Males			Females		
	Expected*	Observed+	Observed/Expected	Expected*	Observed+	Observed/Expected
Stomach	70	92	1.31	43	83	1.93
Oesophagus	10.4	14	1.34	8.0	10	1.25
Liver	1.8	10	5.56	1.4	8	5.72
Bladder	39	37	0.95	12	12	1.00
Breast				133	119	0.90

*Expected number of deaths calculated from the age adjusted rates for the Sheffield registry and from the age distribution of the population.
+Observed numbers of deaths taken from the records at the Public Health Department, Worksop.
Reproduced by permission of H. K. Lewis & Co. Ltd.

Further analysis of the deaths due to stomach cancer in Worksop showed that the death rate from stomach cancer was more than double that expected from the control town data for men in the oldest age group. In females the number of deaths in the oldest age group was again almost double that expected, but there was also an excess of deaths at the lower age groups (Table 3).

Thus the results of this epidemiological study in Worksop suggest that with high nitrate intake, carcinogenic nitrosamines are formed by bacterial nitrosation of secondary amines in the infected urinary bladder and that these give rise to gastric cancer.

Table 3 Deaths from stomach cancer (1962–71) analysed by age and sex from four control towns with low nitrate and from Worksop

	Deaths per 10,000 living at ages:			
Males	30–44	45–64	65–74	75+
Sheffield	0.3	6.2	21.4	31.0
Wakefield	0.3	4.5	17.7	23.0
Chesterfield	0.5	3.2	18.9	28.8
Doncaster	0.3	4.4	18.7	31.6
Total	0.3	5.3	20.4	29.3
Worksop	0.6	5.3	22.2	49.0
Females				
Sheffield	0.3	2.3	8.8	22.8
Wakefield	0.3	2.6	11.5	15.7
Chesterfield	0.4	3.2	9.8	22.5
Doncaster	0.1	1.6	8.5	21.6
Total	0.3	2.4	9.0	21.8
Worksop	0.6	2.5	15.2	42.0

(Data from Hill, Personal communication)

Several other studies give evidence supporting this hypothesis. Correa, Cuello and Duque (1970) showed that in Narino, a southern province of Columbia, the drinking water supply contains high levels of nitrate and the incidence of stomach cancer in this area is higher than in the rest of Columbia. The work of Armijo and Coulson (1975) in Chile showed that the mortality from gastric cancer is related to the cumulative nitrate exposure, and that there is a 16-year latency period between tumour induction and manifestation. Haenszel and coworkers (1976) noted that in Japan the incidence of gastric cancer is greater in those drinking from rural wells than from piped water supplies; the water in the wells was found to have a higher nitrate concentration.

VI. SUMMARY AND CONCLUSION

The importance of nitrate intake with respect to nitrosamine formation and gastric cancer has been examined and is now well recognised. Much work is at present being carried out in this field. Findings suggest that the part played by bacterially produced secondary amines in nitrosamine formation should not be overlooked.

REFERENCES

Allsobrook, A. J. R., du Plessis, L. S., Hartington, J. S., Nunn, A. J. and Nunn, J. R. (1974). Nitrosamines in the human vaginal vault. In *N-Nitroso Compounds in the Environment*. Scientific Publication No. 9 (Lyon) pp. 197–199.

Armijo, R. and Coulson, A. (1975). *Int. J. Epidemiol.*, **4**, 301–309.

Asatoor, A. M. (1964). Studies on metabolism by intestinal bacteria. In *E. J. King Memorial Symposium. Proc. Assoc. of Clin. Biochem.* pp. 82–87.

Asatoor, A. M. (1968). *Clin. Chim. Acta.*, **22**, 223–229.

Asatoor, A. M., Chamberlain, M. J., Emmerson, B. T., Johnson, J. R., Levi, A. J. and Milne, M. D. (1967). *Clin. Sci.*, **33**, 111–124.

Asatoor, A. M. and Simenhoff, M. L. (1965). *Biochim. Biophys. Acta.*, **3**, 384–392.

Awapara, J., Perry, T. L., Hanly, C. and Peck, E. (1964). *Clin. Chim. Acta.*, **10**, 286–289.

Barger, G. and Walpole, G. S. (1909). *J. Physiol.*, **38**, 343–352.

Berthelot, A. (1913). *C. R. Acad. Sci. (Paris)*, **56**, 641–643.

Blau, K. (1961). *Biochem. J.*, **80**, 193–200.

Bone, E. S. (1977). Amino acid metabolism by the gut bacterial flora. Ph.D. Thesis, University of London.

Correa, P., Guello, C., and Duque, E. (1970). *J. Nat. Cancer Inst.*, **44**, 297–306.

DeQuattro, V. L. and Sjoerdsma, A. (1967). *Clin. Chim. Acta.*, **16**, 227–233.

Druckrey, H., Preussmann, R., Ivankovic, S. and Schmahel, D. (1967). *Z. Krebsforsch.*, **69**, 103–201.

Gale, E. F. (1946). *Adv. in Enzymology*, **6**, 1–32.

Haenszel, W., Kurihara, M., Locke, F. B., Shimuza, K. and Segi, M. (1976). *J. Nat. Cancer Inst.*, **56**, 265–274.

Haughton, B. G. and King, H. K. (1961). *Biochem. J.*, **80**, 268–277.

Hawksworth, G. M. (1973). Metabolic activities of intestinal bacteria. Ph.D. Thesis, University of London.

Hawksworth, G. M. and Hill, M. J. (1971). *Br. J. Cancer*, **25**, 520–526.

Hicks, R. M., Walters, C. L., Elsebai, I., El Aasser, A. B., El Merzabani, M. and Gough, T. A. (1977). *Proc. Roy. Soc. Med.*, **70**, 413–417.

Hill, M. J. and Hawksworth, G. M. (1972). Bacterial production of nitrosamines *in vitro* and *in vivo*. In *N-nitroso Compounds: Analysis and Formation*. IARC Scientified Publication No. 3. (Lyon), pp. 116–121.

Hill, M. J. and Hawksworth, G. M. (1974). Some studies on the production of nitrosamines in the urinary bladder and their subsequent effects. In *N-nitroso Compounds in the Environment*. IARC Scientific Publication No. 9. (Lyon), pp. 220–222.

Hill, M. J., Hawksworth, G. and Tattersall, G. (1973). *Br. J. Cancer*, **28**, 562–567.

Irvine, W. T., Duthie, H. L., Ritchie, H. D. and Waton, N. G. (1959). *Lancet*, **(i)**, 1064–1068.

Irvine, W. T., Duthie, H. L. and Waton, N. G. (1959). *Lancet*, **(i)**, 1061–1064.

Johnson, K. A. (1977). *Medical Lab. Sciences*, **34**, 131–143.

Kasé, Y. and Miyata, T. (1976). *Adv. in Biochem.*, **15**, 5–16.

Korf, J. (1969). *Clin. Chim. Acta*, **23**, 483–487.

Magee, P. N. and Barnes, J. M. (1967). *Adv. Cancer Res.*, **10**, 163–246.

Melnykowycz, J. and Johannsson, K. R. (1955). *J. Exptl. Medicine*, **101**, 507–517.

Møller, V. (1954). *Acta. Path. Microbiol. Scand.*, **35**, 259–277.

Møller, V. (1955). *Acta. Path. Microbiol. Scand.*, **36**, 158–172.

Monitor. (1977). *New Scientist*, **77**. p. 794.

Nordenström, B. E. W. (1951). *Acta. Pharmacol. et. Toxicol.*, **7**, 287–296.
Oates, J., Gillespie, A., Udenfriend, S. and Sjoerdsma, A. (1960). *Science*, **131**, 1890–1891.
Perry, T. L., Hestrin, M., MacDougall, L. and Hansen, S. (1966). *Clin. Chim. Acta*, **14**, 116–123.
Roberts, M. and Adam, H. M. (1950). *Brit. J. Pharmacol.*, **5**, 526–541.
Ruddell, W. S. J., Blendis, L. M. and Walters, C. L. (1976). *Gut*, **17**, 401.
Sander, J. (1968). *Hoppe Seylers Z. Physiol. Chem.*, **349**, 429–432.
Savage, W. E., Hajj, S. N. and Kass, E. H. (1967). *Medicine (Baltimore)* **46**, 385–407.
Sinclair, T. and Tuxford, A. F. (1971). *The Practitioner*, **207**, 81–90.
Takala, J., Jousimies, H. and Sievers, K. (1977). *Acta Med. Scand.*, **202**, 69–73.
Tannenbaum, S. R., Sinskey, A. J. and Weisman, M. (1974). *J. Nat. Cancer Inst.*, **53**, 78–84.
Urbach, K. F. (1949). *Proc. Soc. for Exptl. Biol. and Med.*, **70**, 146–152.
Varghese, A. J., Land, P. C., Furrer, R. and Bruce, W. R. (1977). *Proc. Am. Assoc. Cancer Res.*, **18**, 80.
Weissbach, H., King, W., Sjoerdsma, A. and Udenfriend, S. (1959). *J. Biol. Chem.*, **234**, 81–86.
Wilson, C. W. M. (1954). *J. Physiol.*, **125**, 534–545.

Chapter 14

Potentiation of the Mutagenic Action of Nitrous Acid by Polyamines

MICHAEL L. MURRAY AND PELAYO CORREA

I. INTRODUCTION

Concern over the possible hazard associated with consumption of nitrite with amines has led to intensive examination of the effects of their presence in the environment. The majority of nitroso compounds are carcinogenic, and can be formed under physiological conditions from the reaction of nitrite with amines (reviewed by Crosby and Sawyer, 1976; Chapter 13). Nitrite-amine reactions are highly pH dependent. Within the optimal pH range, nitrite is largely present as nitrous acid, whereas amines are present as the free base (Ridd, 1961; Austin, 1960; Wartheson and coworkers, 1975). Nitrosation of secondary amines and amides produces nitrosamines and nitrosamides. Nitrosamides act as mutagens and/or carcinogens directly, whereas nitrosamines are carcinogenic and mutagenic only after enzymatic activation by organs capable of metabolizing them (Druckrey and coworkers, 1967; Magee and coworkers, 1976).

Polyamines are ubiquitous compounds which have a high affinity for DNA. They are not themselves mutagenic; on the contrary, they have anti-mutagenic properties (Clarke and Shankel, 1975). Polyamines are suitable substrates for nitrosation. Nine nitrosamine derivatives of the polyamine spermidine, including the carcinogen N-nitrosopyrrolidine (Preussmann, Schmähl and Eisenbrand, 1977) have been isolated after reaction with nitrite at an acid pH (Warthesen and coworkers, 1975; Hildrum, 1975; Hildrum, Scanlan and Libbey, 1977; Hildrum and Scanlon, 1977; Hotchkiss, Scanlan and Libbey, 1977). The pathway suggested to explain the variety of nitrosamine products calls for formation of a nitrosamine at the secondary amino group, while the primary amino groups are diazotized, resulting in the formation of carbonium ions. The carbonium ions may rearrange, react with H_2O or Cl^- to form hydroxylated or halogenated derivatives, eliminate a proton to form a propylene or butylene derivative, or cyclize and eliminate an alkyl group to form N-nitrosopyrrolidine, as diagrammed in Figure 1. The assymetry of the spermidine molecule increases the number of possible products with, for example, four possible monohydroxy alcohols (Hildrum, Scanlon and Libbey, 1977).

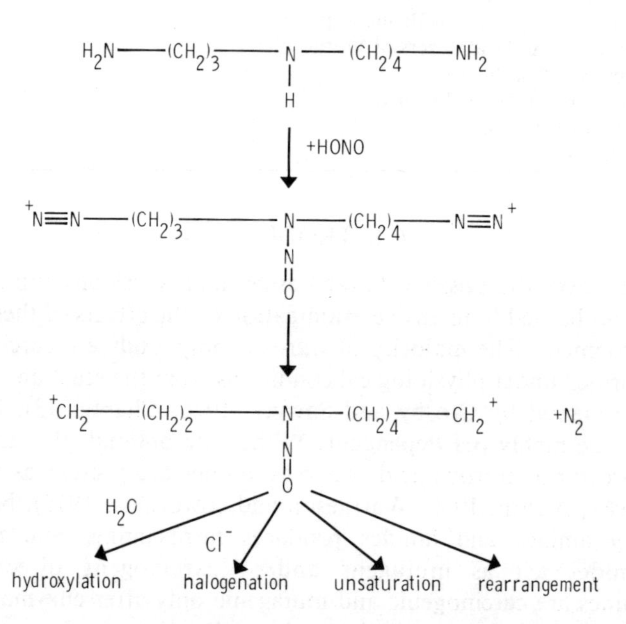

Figure 1 Schematic diagram of the reaction of nitrous acid (HONO) with spermidine

An interesting variation from this scheme has been proposed by Thomas, Hartman and Brown (1978), who point out that it is possible to form a triazene from the nitrosation of polyamines via a bimolecular reaction of nitrite with a diamine or polyamine. The triazene would be in dynamic equilibrium with diazonium ions at the primary amino position (White and coworkers, 1969). The triazene would thus be a reservoir of diazonium ions, the apparent effect of which would be to stabilize the number of carbonium ions. Chemical evidence, however, for the formation of a triazene from a polyamine has not been obtained.

II. DETECTION OF MUTAGENIC ACTIVITY

Several different bacterial systems have been used to assay mutagenic activity of nitrosated polyamines. Using mutant tester strains of *Salmonella typhimurium*, described by Ames, McCann and Yamasaki (1975), various laboratories accumulated data which is directly comparable (McCann and coworkers, 1975). Combining this bacterial mutagenesis indicator with a liver microsomal extract extends the range of sensitivity of the assay system to include compounds which become mutagenic only after enzymatic activation (Ames, McCann and Yamasaki, 1975). The tester strains differ from the wild type principally by their possession of well-defined single site mutations in the genes for the biosynthetic pathway for histidine. As a result of the mutations, they become histidine auxotrophs. Their reversion to histidine independence is scored by the appearance of colonies on medium containing insufficient histidine to support full growth of auxotrophs. The tester strains are designed to cover a spectrum of mutagenic activity, from base substitutions to frameshifts. By analyzing the pattern of reversion of the strains, it is possible to distinguish substances which cause base substitution mutations from those which cause frameshift mutations. Mutagenic activities determined by this test correlate well with carcinogenic activities determined from animal studies (McCann and Ames, 1977; Meselson and Russell, 1977; Purchase and coworkers, 1976).

The use of transforming DNA as a substrate for mutagenesis permits analyses of direct mutagenic action, thus providing information obtained only indirectly in the whole-cell *Salmonella* system (Herriott, 1971).

Lastly, with polyamine auxotrophic mutants of *Escherichia coli*, the effect of alterations in the cellular polyamine pool upon the expression of mutagenic activity by a substance can be measured. Alterations in the quantity and composition of the polyamine pool can be induced by loading starved cells with exogenous polyamines (Cunningham-Rundles and Maas, 1975; Tabor, Tabor and Hafner, 1978; Morris and Jorstad, 1973).

III. POTENTIATION OF MUTAGENESIS BY POLYAMINES

Since nitrosamines are mutagenic only when they have been activated by a microsomal extract, it was surprising to find that polyamines potentiate the mutagenic activity of nitrite without activation by a microsomal extract (Thomas and coworkers, 1978). A solution of sodium nitrite and spermidine or spermine at an acid pH was mutagenic when a small aliquot was mixed with *S. typhimurium* cells suspended in a pH 7.4 buffer, a pH at which little nitrous acid exists. The pH dependence curve for mutagenic activity with the *Salmonella* tester system of sodium nitrite and spermine or spermidine is rather broad with an optimum pH of about 4.2. Spermidine is considerably more effective than spermine in potentiating the mutagenic activity of nitrite (Kokatnur, Murray and Correa, 1978); the diamines putrescine and cadaverine are inactive (Murray and Correa, unpublished results). The amount of mutagenic activity produced varies a great deal with the molar ratio of spermidine to nitrite. When the amount of sodium nitrite was held constant at 0.3 M and the amount of spermidine was varied, mutagenic activity reached a peak at molar ratio of 5:1, respectively, as shown in Figure 2.

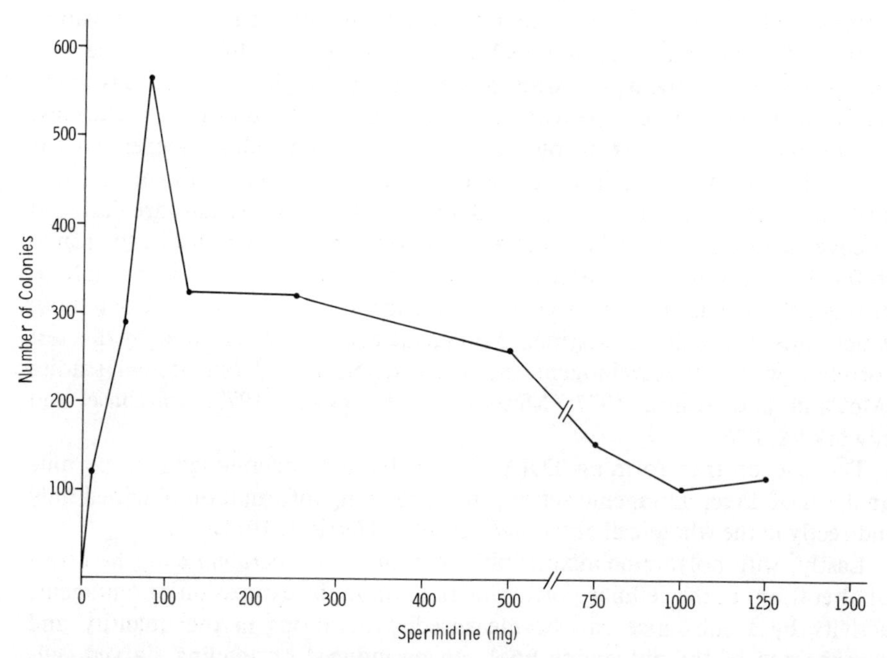

Figure 2 Mutagenic activity obtained when the amount of spermidine was varied and the amount of sodium nitrite was held constant at 0.3 M. The amount of spermidine added is indicated in mg 5 ml^{-1}. Twenty μl of each mixture was assayed for mutagenic activity as described in Kokatnur, Murray and Correa, 1978

P. E. Hartman and his colleagues (1971) used *Haemophilus* transforming DNA to obtain some interesting results implicating polyamines in mutagenesis. They confirmed earlier reports that nitrous acid is highly mutagenic for single stranded DNA but not for double stranded DNA from *Haemophilus influenzae* (Horn and Herriott, 1962; Chevallier and Greth, 1971). Furthermore, they found that the action of nitrous acid on double stranded DNA at pH 4 to 5 was potentiated by the presence of polyamines in the reaction mixture. Spermidine, putrescine, cadaverine, and spermine, in order of increasing activity, all potentiated the mutagenic action of nitrous acid on DNA (Thomas, Mudryj and Hartman, 1976). A secondary amino group is not essential for mutagenic activity, since this is not present in either putrescine or cadaverine. This order of mutagenic activity among the ubiquitous polyamines does not agree with that found by Kokatnur, Murray and Correa (1978). The difference may be due to the effect of competition from endogenous amines present in the whole cell system of Kokatnur, Murray and Correa (1978). In addition, metabolism of added polyamines, most notably by acetylation or oxidation (Dubin and Rosenthal, 1960; Tabor and Tabor, 1970; Tabor and Dobbs, 1970), but also possibly in other ways (Mandel, Ichinotsubo and Mower, 1977), could take place with the whole cell *Salmonella* assay, and not with the transformation assay. The relative mutagenic activities of these polyamines on the pure DNA substrate employed by Thomas, Mudryj and Hartman (1976), should accurately reflect their intrinsic mutagenic capacities. A number of synthetic diamines were also active; mutagenic activity was present as long as the number of carbon atoms separating the amino groups was less than eight. N-monoacetylputrescine, an amide, was active, but no more so than putrescine. The diamide N,N-diacetylputrescine was inactive. Butylamine, a monoamine, was also inactive (Thomas, Mudryj and Hartman, 1977).

Polyamine deficient strains of *E. coli*, starved of polyamines before nitrous acid treatment, were much less sensitive to mutagenesis than cells supplemented with polyamines. The polyamines were effective when added either several hours before harvesting or at the time of nitrous acid treatment (Thomas, Mudryj and Hartman, 1977).

The properties of mutagenic activity detectable from nitrite-treated polyamines differ from those expected of nitrosamines in several ways. The maximum level of mutagenic activity of a polyamine-nitrite mixture is attained at the earliest possible measurement, and remains constant as long as pH of 4 to 5 is maintained (Thomas, Mudryj and Hartman, 1977), in contrast to the steady rise in concentration of nitrosamines with time in such a mixture reported by Hildrum, Scanlan and Libbey (1977). Furthermore, the mutagenic activity generated at pH 4 to 5 is relatively unstable at neutral or alkaline pH, with a half-life of about four min, whereas nitrosamines are more stable (Thomas, Hartman and Brown, 1978).

IV. MECHANISM OF MUTAGENIC ACTION OF POLYAMINES

These observations have led Hartman and colleagues to the conclusion that many mutations induced *in vivo* by nitrous acid are potentiated by small amino compounds. The mutagenic activity of nitrous acid alone on single stranded DNA may not be significant *in vivo*, since most cellular DNA is double stranded. The carbonium ions generated from diazotized polyamines would be effective DNA alkylating agents, and the primary lesions of nitrous acid mutagenesis may be alkylated bases (Pegg, 1977), rather than the more commonly accepted deaminated bases (Zimmermann, 1977). The triazene pathway previously described would provide a mechanism for increasing the effective concentration of carbonium ions at the site of the DNA. Furthermore, O-nitrosated compounds which could be generated from alcohols in the reaction mixture may also act as catalysts for the direct reaction of nitrite with DNA (Davies and McWeeny, 1977). It should be noted that the synthetic triazene, methyl *p*-tolyl triazene, is mutagenic in the *Salmonella* tester system without enzymatic activation, and the carcinogenic product of spermidine nitrosation, N-nitrosopyrrolidine, is not (Thomas, Hartman and Brown, 1978).

The argument that alkylation by diazotized amines is the predominant mechanism for nitrous acid mutagenesis rests heavily on the observation of Thomas, Hartman and Brown (1978) that double stranded DNA is not susceptible to *in vitro* treatment with nitrous acid unless polyamines are present. Not all cellular DNA, however, is double stranded at all times *in vivo*. Henson (1978) has observed the existence of single stranded regions in mammalian DNA, which are not part of replication forks. Activities such as replication, transcription, repair, and recombination all require a transient single stranded state that could be susceptible to direct oxidative deamination by nitrous acid. Furthermore, in transformation of *Bacillus subtilis*, it does not matter whether single or double stranded DNA is treated (Strack, Freese and Freese, 1964). Therefore it is still not clear that *in vivo* mutagenesis by nitrous acid requires the presence of polyamines, although such a mechanism is plausible.

V. REACTION OF NITROUS ACID WITH DNA

The primary effects of nitrous acid on DNA are deamination of guanine, adenine, and cytosine, and the introduction of cross-links (Chapter 13). The rate of deamination of double stranded DNA is much lower than that of single stranded DNA (Litman, 1961; Schuster, 1960: Kotaka and Baldwin, 1964). The effect of the state of DNA on the rate of deamination is reflected in the relative sensitivities of these two forms of DNA to mutagenesis by nitrous acid (Thomas, Mudryj and Hartman, 1976). Cross-linkage of DNA from treatment with nitrous acid, however, occurs along with deamination at a ration of four deaminations to one cross-link (Becker, Zimmerman and Geiduschek, 1964). The cross-links do not appear immediately after nitrous acid treatment, and they continue to form after nitrous acid is removed (Burnotte and Verly, 1971; Verly and Lacroix, 1975). Two kinds of cross-linked nucleosides, which may be the cross-links introduced into the DNA, have been isolated from DNA treated with nitrous acid (Shapiro and coworkers, 1977; Dubelman and Shapiro, 1977). Not all reports are in complete agreement; Luzzati (1962) has noted that *Pneumococcus* and *Bacillus* DNA are relatively immune to cross-linking by nitrous acid, but that *Haemophilus* DNA is quite susceptible to such cross-linking. This result is in direct contradiction to that of Becker, Zimmerman and Geiduschek (1964). If polyamines were present during some of these nitrosation reactions and not others, carbonium ions would have been generated from the diazotized amines which might account for significant differences in the results.

Shepherd and Hopkins (1963) demonstrated clearly that almost all of the cellular polyamines remain associated with phenol-extracted DNA, regardless of the ionic strength of the solution. The DNA-polyamine complex can be dissociated by successive precipitation of the DNA with ethanol in the presence of salts (Bachrach, 1973). These observations have been paid little heed by other investigators in the design of methods for the extraction of DNA. As a consequence, it is likely that DNA preparations differ greatly in their polyamine content, depending on the extraction method. It is impossible to assess the relative purity of the DNA preparations used in most experiments, since the extraction procedures are incompletely described and the problem of polyamine contamination is not usually directly addressed.

VI. GENETIC EFFECTS OF NITROUS ACID

The *in vivo* effects of nitrous acid include those that are 'lethal' and those that are 'mutagenic'. 'Lethal' incidents are those which prevent the production of progeny. Cross-linking of DNA may well have this effect. Phage T2 (Vielmetter and Schuster, 1960) and phage P22 (Prell, 1962) both require approximately four deaminations for each inactivation of phage infectivity. This is precisely the ratio of cross-links to deaminations calculated by Becker, Zimmerman and Geiduschek (1964). Nonn and Bernstein (1977) and Harm (1974) have observed that some nitrous-acid induced lethal lesions in phage T4 are efficiently repaired by a recombination dependent process. Nonn and Bernstein proposed a model for repair involving stimulated recombination at the site of the lesion. Cross-linked lesions formed by psoralen plus light have been found to be repaired by such a stimulated recombination dependent mechanism, the existence of which has been demonstrated through physical evidence (Cole, Levitan and Sinden, 1976).

'Mutagenic' incidents alter the DNA but allow reproduction of the altered form. It is well known that lethal lesions of several types may be repaired by a number of processes which in turn introduce mutations at the site of the lesion (Witkin, 1976). Reactivation of genetically damaged DNA by host cells, which have themselves sustained a low level of genetic damage, is indicative of an inducible repair system (Weigle, 1953; Samson and Cairns, 1977). In error prone repair, increased survival of repaired bacteriophage is at the expense of an increased frequency of mutation. Several different types of mutants which do not possess an error prone inducible repair system ('SOS' repair) have been isolated (Miura and Tomizana, 1968; Defais and coworkers, 1976; Castellazi, George and Butlin, 1972). It should be noted that the SOS repair system is independent of the expression of the *uvrB* gene (Radman and Devoret, 1971), a gene which is deficient in some of the *Salmonella* tester strains described previously. Also notable, *H. influenzae* does not have a repair system analogous to the SOS repair system of *E. coli* (Kimball, Boling and Perdue, 1977).

VII. REPAIR OF NITROUS ACID INDUCED LESIONS

Participation of an inducible error prone repair system in nitrous acid mutagenesis has been indicated by results from two laboratories. The mutation frequency of nitrous acid-damaged bacteriophage λ was increased by a low level of UV irradiation to a host with an SOS^+ phenotype, but not to a host with an SOS^- phenotype. The phages were reactivated in both types of host, but at a reduced level in the SOS^- host (Kerr and Hart, 1972). Moreover, mutagenesis of T4 phage by nitrous acid is dependent upon the activity of polynucleotide ligase (Bernstein and coworkers, 1976), an enzyme which has been implicated in repair activity (Konrad, Modrich and Lehman, 1973). After examination of this evidence, Bernstein and coworkers (1976) proposed a model for nitrous acid mutagenesis via the action of an error prone repair system, rather than the mispairing of bases during replication.

Bases deaminated or otherwise damaged by nitrous acid may be susceptible to excision repair. In particular, uracil produced from deamination of cytosine in *E. coli* DNA is attacked by uracil-DNA glycosylase (Da Roza and coworkers, 1977) and exonuclease V (Gates and Linn, 1977). Mutants deficient in uracil-DNA glycosylase (ung^-) are hypersensitive to the lethal action of nitrous acid (Da Roza and coworkers, 1977). The presence of uracil in DNA is not immediately damaging, however; mutants which were doubly deficient in dUTPase and uracil-DNA glycosylase were viable even though 30% of the thymine residues in DNA in these cells had been replaced by uracil. The DNA containing uracil was fully functional (Warner and Duncan, 1978). Cells containing uracil-substituted DNA have slightly higher mutation frequency than wild type cells. Slow growth is another detrimental effect of the presence of uracil in DNA when nucleases active on uracil sites are not present (Duncan, Rockstroh and Warner, 1978; Warner and Duncan, 1978).

VIII. NITROUS ACID MUTAGENESIS

Oxidative deamination by nitrous acid is thought to be directly mutagenic, rather than lethal. Changes in base-pairing properties as a result of deaminations would give rise to CG→AT and AT→GC transition mutations. For example, oxidative deamination of dAT copolymer by nitrous acid would result in transition mutations such as those resulting from the conversion of adenines to hypoxanthines. The hypoxanthines would pair with cytosines rather than thymidines, eventually resulting in an AT to GC base change. Physical evidence in support of this mechanism has been obtained by Kotaka and Baldwin (1964). *In vitro* replication of dAT copolymer treated with nitrous acid was found to be dependent on the presence of dCTP, and the amount of cytosine incorporated into the newly replicated DNA was found to be proportional to the degree of deamination. This result is consistent with the induction of transition mutations by deamination. The reaction of nitrous acid with the DNA was much more extensive, however, than that sufficient to induce mutations.

Base substitutions which are not transitions are induced by nitrous acid, and these cannot be explained by the deamination reaction (Weigert and Garen, 1965; Prakash and Sherman, 1973; Sherman and Stewart, 1974). Deletion mutations are also induced by nitrous acid (Tessman, 1962; Alper and Ames, 1975; Schwartz and Beckwith, 1969). Among a large number of *S. typhimurium* mutants induced with nitrous acid, 52% were deletions or minus frameshifts (Hartman and coworkers, 1971). In particular, the frequency of large deletions is notably increased by nitrous acid treatment. While the induction of large deletions represents a relatively minor portion of the total mutagenic activity of nitrous acid, this agent is several times more effective than any other tested for their induction (Alper and Ames, 1975). The induction of deletions by nitrous acid is thought to be associated with replication of cross-linked regions of DNA (Iida and Uchida, 1977).

In several types of genetic damage caused by nitrous acid, polyamines evidently do not participate. No potentiation of the induction of frameshift deletions by nitrous acid occurs when they are added (Kokatnur, Murray and Correa, 1978; Thomas, Mudryj and Hartman, 1977). Moreover, inactivation of DNA is not potentiated by polyamines (Thomas, Mudryj and Hartman, 1976, 1977). The primary lesion associated with both of these types of genetic damage is an interesting cross-link. In contrast, polyamines do potentiate base substitution mutations, a type of damage associated with deamination of bases. The most straightforward interpretation of these observations would be that polyamines do not potentiate the induction of interstrand cross-links by nitrous acid.

Mutagenesis by nitrous acid alone also differs from that potentiated by polyamines in the effect of media enrichment. When the minimal medium used in the *Salmonella* assay system was enriched with a complex mixture of nutrients excluding histidine, the number of mutations induced by nitrous acid alone was depressed, whereas the number induced by a spermidine-nitrite mixture was not (Murray and Correa, unpublished results, and Table 1).

IX. POLYAMINES AND ERROR PRONE REPAIR

Reaction of nitrous acid with polyamines may generate mutagenic agents which react directly with DNA, as suggested by Thomas, Mudryj and Hartman (1976, 1977). Nitrite treated polyamines could act as inducers of an error prone repair system. The activity of this repair system on nitrous acid induced lesions would be mutagenic. Variability in the polyamines available for reaction with nitrite and in repair systems of different species could account for differences in the pattern of mutagenesis by nitrous acid noted by various investigators. Thus, with this model, it is possible to explain the dependence of nitrous acid mutagenesis on an inducible repair system and on polynucleotide ligase, potentiation of mutagenesis by polyamines, and the apparent species specificity for mutagenesis of double stranded DNA.

X. INHIBITORS AND CATALYZERS OF MUTAGENICITY

Chemicals which catalyze or inhibit reactions between nitrous acid and amines have the expected effect on the mutagenic activity of polyamine-nitrite mixtures. Ascorbic acid and propyl gallate are effective inhibitors of polyamine-nitrite mutagenesis (Kokatnur and coworkers, 1978 and unpublished results). Addition of ascorbic acid prior to the nitrous acid and polyamine is slightly more effective than addition afterwards (Kokatnur, Murray and Correa, 1978). Ascorbic acid inhibits formation of nitroso compounds (Mirvish and coworkers, 1972) as well as the mutagenic activity of preformed nitroso compounds (Guttenplan, 1977). These results should be interpreted with caution, however, since gallic acid and ascorbic acid will catalyze nitrosation reactions under certain circumstances (Walker, Pignatelli and Castegnaro, 1975; Fiddler and coworkers, 1978). Ascorbic acid actually increased the frequency of adenomas when administered with amines in one animal study (Mirvish and coworkers, 1975).

Thiocyanate is known to catalyze nitrosation reactions (Boyland and Walker, 1973). It also seems to catalyze the production of mutagenic activity from nitrite-treated polyamines. These results, however, are not completely reproducible and should be regarded as only suggestive (Kokatnur, 1978).

XI. MEDICAL IMPLICATIONS

Recent works on the aetiology of cancer has focused on the hypothesis of formation of mutagenic/carcinogenic N-nitroso compounds from nitrite and amines. Many amines have been investigated as possible substrates for reaction with nitrite, but most of the reaction products require activation with a liver microsomal extract. The organotropic carcinogenic activity characteristic of such indirect mutagens makes them unlikely candidates for the aetiology of many types of cancer, particularly gastric cancer (Tannenbaum and coworkers, 1977). The direct mutagenic activity of nitrite-treated polyamines makes them attractive candidates for the aetiology of cancer of the stomach, and possibly of other organs having direct contact with nitrite and polyamines. The pH dependence of the mutagenic activity is compatible with a role in gastric carcinogenesis. In stomachs with chronic atrophic gastritis, the gastric juice is consistently less acid (usually around pH 5) than in people with normal mucosa; these abnormal stomachs are cancer-prone (Correa and coworkers, 1976). Significant mutagenic activity was obtained in the pH range of 3 to 6 with the nitrite polyamine system, and would be expected in stomachs suffering from chronic atrophic gastritis.

The potential mutagenic activity of polyamines is particularly significant because of their ubiquity. These endogenous agents may have significant implications not only for cancer, but also for ageing and teratogenesis. The significance of the wide distribution of the co-mutagens is underscored by the difficulty in avoiding nitrate. The major source of physiological nitrite is nitrate in the diet, which is reduced to nitrite by various members of the bacterial flora of the mouth (Tannenbaum and coworkers, 1974). Approximately 80% of the nitrate intake is from vegetables, and less than 17% from cured meats (White, 1975; 1976). In addition, some nitrite orginates *de novo*, presumably from the metabolic action of bacteria in the lumen on endogenous nitrogenous substrates (Tannenbaum and coworkers, 1978).

Table 1 Comparison of mutant production on minimal medium and on HA* enriched medium

Additions ±	Mutants in minimal medium	Mutants in HA enriched medium
Spermidine, $NaNO_2$	290	304
—— , $NaNO_2$	57	15
Spermidine, ——	13	6
—— , ——	16	6

*HA = Minimal medium supplemented with 1% histidine assay medium (Difco).

±The reaction mixture contained 0.1 M spermidine ·3HCl, and 0.3 M $NaNO_2$, as indicated, and 0.1 M Sorenson's citrate buffer, pH 4.2. Twenty μl of this solution was tested for mutagenic activity on cells suspended in pH 7.4 phosphate buffer, 0.1 M, as described by Kokatnur, Murray and Correa, 1978.

XII. SUMMARY AND CONCLUSIONS

In summary, mixtures of polyamines and sodium nitrite increase the frequency of base substitution mutations. The pH dependence of the production of mutagenic activity is consistent with well known reactions of nitrous acid with primary and secondary amines. Three different bacterial systems have been used to characterize the mutagenic activity, with similar but not identical results.

Mechanisms proposed for the potentiation of nitrous acid mutagenesis by polyamines include the generation of alkylating species from diazotized primary amines, the catalysis of the nitrous acid reaction by O-nitrosated alcohols produced in the reaction mixture, and the induction of an error prone repair system by components of the reaction mixture. These mechanisms are not necessarily mutually exclusive.

It is reasonable to expect that the coordination of these two separate areas of research, nitrous acid mutagenesis and polyamine physiology, will lead to illuminating and productive discoveries. Hopefully, these fundamental findings may in turn find practical medical application in the fields of oncology, gerontology, and teratology.

ACKNOWLEDGEMENTS

This work was supported by the National Cancer Institute, USPHS (Contract N01-CP-53521). We thank P. E. Hartman and S. R. Tannenbaum for sending manuscripts prior to publication. M. L. M. also thanks P. E. Hartman and H. Thomas for stimulating and encouraging discussions, and M. D. Murray for preparation of the manuscript.

REFERENCES

Alper, M. D., and Ames, B. N. (1975). *J. Bacteriol.*, **121**, 259–266.

Ames, B. N., McCann, J., and Yamasaki, E. (1975). *Mutation Res.*, **31**, 347–364.

Austin, A. T. (1960). *Nature, Lond.*, **188**, 1086–1088.

Bachrach, U. (1973). *Function of Naturally Occurring Polyamines*, Academic Press, New York and London.

Becker, E. F., Zimmerman, B. K., and Geiduschek, E. G. (1964), *J. Mol. Biol*, **8**, 377–391.

Bernstein, C., Morgan, D., Gensler, H. L., Schneider, S., and Holmes, G. E. (1976). *Molec. Gen. Genet.*, **148**, 213–220.

Boyland, E., and Walker, S. A. (1973). *Nature, Lond.*, **248**, 601–602.

Burnotte, J., and Verly, W. G. (1971). *J. Biol. Chem.*, **246**, 5914–5918.

Castellazi, M., George, J., and Buttin, G. (1972). *Molec. Gen. Genet.*, **119**, 139–152.

Clarke, C. H., and Shankel, D. M. (1975). *Bacteriol. Rev.*, **39**, 33–53.

Chevallier, M. R., and Greth, M. L. (1971). *Molec. Gen. Genet.*, **110**, 27–30.

Cole, R. S., Levitan, D., and Sinden, R. R. (1976). *J. Mol. Biol.*, **103**, 39-59.
Correa, P., Cuello, C., Duque, E., Burbano, L. C., Garcia, F. T., Bolanos, O., Brown, C., and Haenszel, W. (1976). *J. Natl. Cancer Inst.*, **57**, 1027-1035.
Crosby, N. T., and Sawyer, R. (1976). *Advances in Food Research*, **22**, 1-71.
Cunningham-Rundles, S., and Maas, W. K. (1975). *J. Bacteriol.*, **124**, 791-799.
Da Roza, R., Friedberg, E. C., Duncan, B. K., and Warner, H. R. (1977). *Biochem.*, **16**, 4934-4939.
Davies, R., and McWeeny, D. J. (1977). *Nature, Lond.*, **266**, 657-658.
Defais, M., Fauquet, P., Radman, M., and Errera, M. (1976). *Mol. Gen. Genet.*, **148**, 125-130.
Druckrey, H., Preussman, R., Ivankovic, S., and Schmähl, D. (1967). *Z. Krebsforsch.*, **69**, 103-201.
Dubleman, S. and Shapiro, R., *Nucleic Acid Res.*, **4**, 1815-1827.
Dubin, D. T., and Rosenthal, S. M. (1960). *J. Biol. Chem.*, **235**, 776-782.
Duncan, B. K., Rockstroh, P. A., and Warner, H. R. (1978). *J. Bacteriol.*, **134**, 1039-1045.
Fiddler, W., Pensabene, J. W., Piotrowski, E. G., Phillips, J. G., Keating, J., Mergens, W. J. and Newmark, H. L. (1978). *J. Agric. Food Chem.*, **26**, 653-656.
Gates, F. T., and Linn, S. (1977). *J. Biol. Chem.*, **252**, 1647-1653.
Guttenplan, J. B. (1977). *Nature*, **268**, 368-370.
Harm, W. (1974). *Mutation Research*, **24**, 205-209.
Hartman, P. E., Hartman, Z., Stahl, R. C., and Ames, B. N. (1971). *Advances in Genetics*, **16**, 1-34.
Henson, P. (1978). *J. Mol. Biol.*, **119**, 487-506.
Herriott, R. M. (1971). Effects on DNA: Transforming Principle. In *Chemical Mutagens* (Ed. A. Hollaender), pp. 175-217, Plenum Press, New York.
Hildrum, K. I., and Scanlan, R. A. (1977). *J. Agric. Food Chem.*, **25**, 255-257.
Hildrum, K. I., Scanlan, R. A., and Libbey, L. M. (1975). *J. Agric. Food Chem.*, **23**, 34-37.
Hildrum, K. I., Scanlan, R. A., and Libbey, L. M. (1977). *J. Agric. Food Chem.*, **25**, 252-255.
Horn, E. E., and Herriott, R. M. (1962). *Proc. Nat. Acad. Sci. USA*, **48**, 1409-1416.
Hotchkiss, J. H., Scanlan, R. A., and Libbey, L. M. (1977). *J. Agric. Food Chem.*, **25**, 1183-1189.
Iida, S., and Uchida, H. (1977). *Virology*, **83**, 227-286.
Kerr, T. L., and Hart, M. G. L. (1972). *Mutation Res.*, **15**, 247-258.
Kimball, R. F., Boling, M. E., and Perdue, S. W. (1977). *Mutation Res.*, **44**, 183-196.
Kokatnur, M. G., Murray, M. L., and Correa, P. (1978). *Proc. Soc. Exp. Biol. and Med.*, **158**, 85-88.
Konrad, E. B., Modrich, P. and Lehman, I. R. (1973). *J. Mol. Biol.*, **77**, 519-529.
Kotaka, T., and Baldwin, R. L. (1964). *J. Mol. Biol.*, **9**, 323-339.
Litman, R. M. (1961). *J. Chim. Phys.*, **58**, 997-1004.
Luzzati, D. (1962). *Biochem. and Biophys. Res. Comm.*, **9**, 508-516.
Magee, P. N., Montesano, R., and Preussman, R. (1976). N-nitroso compounds and related carcinogens. In C. E. Searle (Ed.), *Chemical Carcinogens, Monograph 173*, American Chemical Society, New York, p. 491.
Mandel, M., Ichinotsubo, D., and Mower, H. (1977). *Nature, Lond.*, **267**, 248-249.
Meselson, M. and Russell, K. (1977). Comparison of carcinogenic and mutagenic potency. In J. D. Watson, H. H. Hiatt and J. A. Winsten (Eds.), *Origins of Human Cancer*, Cold Spring Harbor Laboratory, Cold Spring Harbor, New York, pp. 1473-1481.

McCann, J. and Ames, B. N. (1977). The *Salmonella*/microsome mutagenicity test: predictive value for animal carcinogenicity. In J. D. Watson, H. H. Hiatt and J. A. Winsten (Eds), *Origins of Human Cancer*, Cold Spring Harbor Laboratory, Cold Spring Harbor, New York, pp. 1431–1450.

McCann, J. E., Choi, E., Yamasaki, E., and Ames, B. N. (1975). *Proc. Nat. Acad. Sci. USA*, **72**, 5135–5139.

Mirvish, S. S., Cardesa, A., Wallcave, L., and Shubik, P. (1975). *J. Natl. Cancer Inst.*, **55**, 633–636.

Mirvish, S. S., Wallcave, L., Eagen, M., and Shubik, P. (1972). *Science*, **177**, 65–68.

Miura, A., and Tomizawa, J. (1968). *Mol. Gen. Genet.*, **103**, 1–10.

Morris, D. R., and Jorstad, C. M. (1973). *J. Bacteriol.*, **113**, 271–277.

Nonn, E. M., and Bernstein, C. (1977). *J. Mol. Biol.*, **116**, 31–47.

Pegg, A. E. (1977). *Advances in Cancer Research*, **25**, 195–269.

Prakash, L., and Sherman, F. (1973). *J. Mol. Biol.*, **79**, 65–82.

Prell, H. (1962). *Z. Vererbungslehre*, **93**, 320–335.

Preussman, R., Schmähl, D., and Eisenbrand, G. (1977). *Z. Krebsforsch.*, **90**, 161–166.

Purchase, I. F. H., Longstaff, E., Ashby, J., Styles, J. A., Anderson, D., Lefevre, P. A., and Westwood, F. R. (1976). *Nature, Lond.*, **264**, 624–627.

Radman, M., and Devoret, R. (1971). *Virology*, **43**, 504–506.

Ridd, J. H. (1961). *Quart. Rev. Chem. Soc., Lond.*, **15**, 418–441.

Samson, L., and Cairns, J. (1977). *Nature, Lond.*, **267**, 281–282.

Schuster, H. (1960). *Biochem. Biophys. Res. Comm.*, **5**, 320–323.

Schwartz, D. O., and Beckwith, J. R. (1969). *Genetics*, **61**, 371–376.

Shapiro, R. Dubelman, S., Feinberg, A. M., Crain, P. F., and McCloskey, J. A. (1977). *J. Am. Chem. Soc.*, **99**, 302–303.

Shepherd, G. R., and Hopkins, P. A. (1963). *Biochem. Biophys. Res. Comm.*, **10**, 103–108.

Sherman, F., and Stewart, J. W. (1974). *Genetics*, **78**, 97–113.

Strack, H. B., Freese, E. B., and Freese, E. (1964). *Mutation Res.*, **1**, 10–21.

Tabor, C. W., and Dobbs, L. G. (1970). *J. Biol. Chem.*, **245**, 2086–2091.

Tabor, C. W., and Tabor, H. (1970). *Biochem. and Biophys. Res. Comm.*, **41**, 232–238.

Tabor, C. W., Tabor, H., and Hafner, E. W. (1978). *J. Biol. Chem.*, **253**, 3671–2676.

Tannenbaum, S. R., Archer, M. C., Wishnok, J. S., Correa, P., Cuello, C., and Haensgel, W. (1977). Nitrate and the etiology of gastric cancer. In J. D. Watson, H. H. Hiatt and J. A. Winsten (Eds), *Origins of Human Cancer*, Cold Spring Harbor Laboratory, Cold Spring Harbor, New York, pp. 1609–1625.

Tannenbaum, S. R., Fett, D., Young, V. R., Land, P. C., and Bruce, W. R. (1978). *Science*, **200**, 1487–1489.

Tannenbaum, S. R., Sinskey, A. J., Weisman, M., and Bishop, W. (1974). *J. Natl. Cancer Inst.*, **53**, 79–84.

Tessman, I. (1962). *J. Mol. Biol.*, **5**, 442–445.

Thomas, H. F., Hartman, P. E., and Brown, D. L. (1978). Direct-acting mutagens resulting from nitrosation of polyamines, lysine, synthetic diamines, and N-mono-acetylputrescine', *Abstracts of the Environmental Mutagen Society*.

Thomas, H. F., Mudryj, M., and Hartman, P. E. (1976). *Abstr. Ann. Mtg. American Society for Microbiology*, H76.

Thomas, H. F., Mudryj, M., and Hartman, P. E. (1977). *Abstr. Ann. Mtg. American Society for Microbiology*, H58.

Verly, W. G., and Lacroix, M. (1975). *Biochim. Biophys. Acta*, **414**, 185–192.

Vielmetter, W., and Schuster, H. (1960). *Biochem. Biophys. Res. Comm.*, **2**, 324–328.

Walker, E. A., Pignatelli, B., and Castegnaro, M. (1975). *Nature, Lond.*, **258**, 176.
Warner, H. R., and Duncan, B. K. (1978). *Nature, Lond.*, **272**, 32–34.
Wartheson, J. J., Scanlon, R. A., Bills, D. D., and Libbey, L. M. (1975). *J. Agric. Food Chem.*, **23**, 898–902.
Weigert, M. G., and Garen, A. (1965). *Nature, Lond.*, **206**, 992–994.
Weigle, J. J. (1953). *Proc. Nat. Acad. Sci. USA*, **39**, 628–636.
White, J. W. (1975). *J. Agric. Food Chem.*, **23**, 886–891.
White, J. W. (1976). *J. Agric. Food Chem.*, **24**, 202.
White, E. H., Maskill, H., Woodcock, D., and Schroeder, M. A. (1969). *Tetrahedron Letters*, **21**, 1713–1716.
Witkin, E. W. (1976). *Bacteriol. Rev.*, **40**, 869–907.
Zimmermann, F. K. (1977). *Mutation Res.*, **39**, 127–148.

Chapter 15

Tumour Promoters and Polyamines

THOMAS G. O'BRIEN

I. INTRODUCTION

The development of many cancers in the human population is probably the result of environmental exposure for long periods of time to various chemicals, some of which may not be carcinogenic themselves, yet can modify the carcinogenic effect of other agents. In addition, a prominent feature of human neoplasia is its progressive nature and apparent multi-stage character (Foulds, 1969). Included among the critical unanswered questions in cancer research are: (1) the relative contributions of environmental carcinogens and co-carcinogens to the total human cancer incidence and, (2) the mechanisms responsible for progression through the various 'stages' in neoplasia.

Perhaps the most widely used experimental system for the study of co-carcinogenesis and the multi-stage nature of cancer has been the dorsal skin of mice (Berenblum and Shubik, 1947). In addition to being susceptible to tumour induction by many carcinogens, such as polycyclic hydrocarbons, this tissue has also been shown to be sensitive to the co-carcinogenic action of chemical irritants such as croton oil. Mottram (1944) reported an important refinement by demonstrating that a carcinogen need only be applied *once*, instead of repetitively, to skin before croton oil treatments were applied in order to produce tumours. Carcinogen treatment alone, or croton oil treatment alone produced few, if any, tumours. This second treatment was termed tumour 'promotion' by Friedewald and Rous (1944), while the single carcinogen application was called 'initiation' (for review see Boutwell, 1964).

Interest in the study of tumour promotion markedly increased when the active principles of croton oil were isolated and characterized as diesters of phorbol largely by the groups of Hecker (1971) and Van Duuren (1969). When pure phorbol esters, both naturally occurring and partially synthesized from phorbol, became available, research on the mode of action of these agents was greatly facilitated. A single topical application of a potent phorbol ester, such as 12-*0*-tetradecanoylphorbol-13-acetate (TPA) evokes numerous biological and biochemical responses (for review see Boutwell, 1974) including irritation, inflammation, hyperplasia, and stimulated incorporation of precursors into epidermal macromolecules. Because of the complex nature of the skin's response to even a single application of an active promoter, it has been difficult to identify essential components of the promotion process.

Since the last comprehensive review of tumour promotion (Boutwell, 1974), studies have focused on a particular response of mouse skin to tumour promoters and its significance in the mechanism of promotion; this is the ability of active promoters to produce a dramatic increase in the activity of ornithine decarboxylase (L-ornithine carboxy-lyase, E.C. 4.1.1.17) (ODC), after topical application to mouse skin (O'Brien, Simsiman and Boutwell, 1975a; Chapter 12). Polyamine levels, especially putrescine, are also markedly elevated in skin after promoter treatment (O'Brien, 1976).

In this article, I will review the studies in mouse skin that suggest an important role for polyamines in tumour promotion and also describe some recent data from several *in vitro* systems that may provide additional insights into the mode of action of promoters.

II. STUDIES IN MOUSE EPIDERMIS

All studies have utilized pure promoting agents, especially the phorbol ester series of promoters, which are by far the most active promoting agents known. The structure of TPA and other phorbol diesters used is shown in Figure 1. Modification of the acyl groups at positions 12 and 13 of the phorbol molecule greatly changes the promoting activity of the various phorbol diesters and the compounds range in activity from non-promoting to excellent and have proven useful in studies on the mechanism of promotion.

The initial observation that tumour promoters stimulate polyamine biosynthesis was made in 1975 (O'Brien, Simsiman and Boutwell, 1975a); a single topical application of 17 nmole TPA to normal mouse skin caused a remarkable 230-fold induction of ODC at 4 h post-treatment (Figure 2). After a very rapid induction, peaking at 4–6 h, the ODC activity declined by 24 h to the very low basal level characteristic of untreated epidermis. S-Adenosyl-methionine decarboxylase (E.C. 4.1.1.50) (SAMD) was also induced by TPA

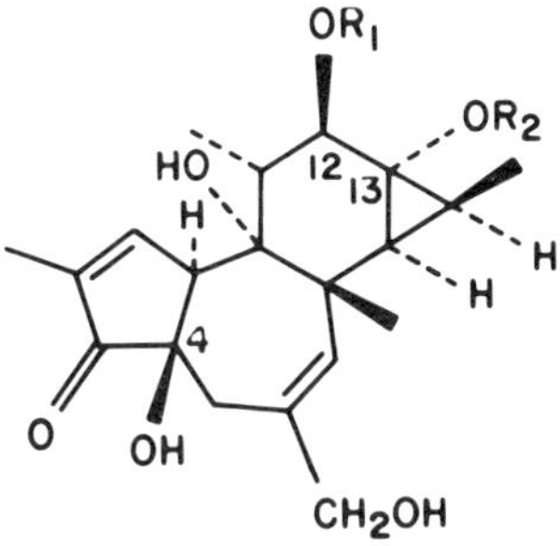

Compound	R1	R2	Promoting Activity
Tetradecanoylphorbol acetate	tetradecanoate	acetate	++++
Phorbol didecanoate	decanoate	decanoate	+++
Phorbol dibenzoate	benzoate	benzoate	+
Phorbol diacetate	acetate	acetate	-
4α-Phorbol didecanoate	decanoate	decanoate	-
Phorbol	H	H	-

Figure 1 The structure of phorbol and the diesters of phorbol used in many studies on promoters. The relative promoting activities of the compounds were taken from Boutwell (1974)

but to a lesser extent (6-7-fold) and with different kinetics (Figure 2). Figure 3 shows the changes in polyamine levels that occur under these treatment conditions. Putrescine concentration was elevated 10-fold at 7 hours after TPA treatment but declined rapidly thereafter probably due, at least in part, to its conversion into spermidine, which showed a biphasic pattern of accumulation. In contrast, the spermine level did not change appreciably.

The idea that the induction of ODC may be a specific response of mouse skin to promoter treatment and perhaps an important component of the mechanism of promotion is supported by the following observations (see O'Brien, Simsiman and Boutwell, 1975a, b):

(1) The ODC induction after a single application of TPA is dose-dependent over the range 0.17 nmole-17.0 nmole. This is practically identical to the dose response of skin to promotion by repeated applications of TPA.

(2) When a series of phorbol esters (see Figure 1) were tested for their ability to induce ODC, non-promoting derivatives did not induce the enzyme and, in general, a good correlation was found between the magnitude of enzyme induction and the relative promoting activity of several active esters.

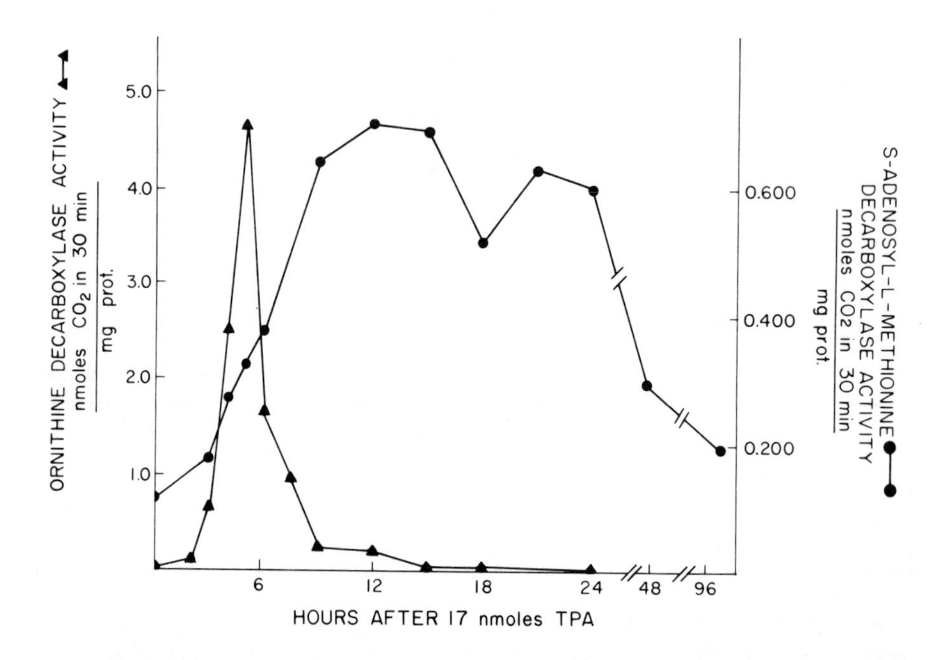

Figure 2 The effect of a single topical application of 17 nmoles TPA on the activities of epidermal ornithine decarboxylase (▲) and S-adenosylmethionine decarboxylase (●). Groups of five mice were treated with either 0.2 ml acetone or TPA dissolved in acetone and killed at the times indicated. Each point represents the mean of triplicate determinations of enzyme activity

(3) Other promoting agents apart from the phorbol esters also induced ODC although at much higher dose levels. These compounds (iodoacetic acid, anthralin, and Tween 60) are much weaker promoters and were also much weaker inducers of ODC than the active phorbol diesters.

(4) When hyperplastic agents that promote weakly or not at all were tested, they either had no effect or were only weakly effective in inducing ODC, suggesting that this enzyme induction is not strictly related to growth stimulation as has been found for many other tissues (Russell and Snyder, 1968; Pegg, Lockwood and Williams-Ashman, 1970; Inoue and coworkers, 1974).

(5) Single carcinogenic doses of several skin carcinogens induced ODC but initiating doses of these same compounds had no effect. High doses of a pure initiator for mouse skin, urethan, were also completely ineffective (O'Brien, Simsiman and Boutwell, 1975b).

(6) Epidermal tumours produced by a two-stage protocol contained high levels of ODC compared to normal epidermis. Benign tumours had moderately

elevated ODC activity, whereas malignant tumours possessed very high levels of this enzyme. Polyamine concentrations in these tumours were also markedly elevated (O'Brien and Boutwell, unpublished observations).

Although promoters, especially the phorbol esters, induced both ODC and SAMD, the evidence for a critical role of ODC, and not SAMD as a 'marker' for promoters, is derived from the following considerations:

(a) Hyperplastic agents for mouse skin, whether they are promoters or not, induce SAMD but only promoting agents induce ODC.

(b) Epidermal tumours obtained by a two-stage procedure did not always have elevated SAMD levels compared to normal epidermis.

(c) In normal, untreated epidermis, the level of ODC is nearly undetectable (0-20 pmoles/30 min/mg protein) while reasonable levels of SAMD exists (100 pmoles/30 min/mg protein). This suggests that in this tissue ODC is the rate-limiting enzyme for polyamine synthesis. Thus, in order to obtain enhanced rates of polyamine synthesis, ODC would have to be induced but the basal (uninduced) level of SAMD might be sufficient to convert the increased amount of putrescine formed by ODC to spermidine and spermine. Hence, it would appear that elevated levels of polyamines could occur in skin only under conditions in which ODC is induced, i.e. after promoter treatment. The levels of polyamines in skin after treatment with promoting agents versus hyperplastic non-promoting agents have not been reported, however, so this assumption must be considered with caution.

The observed correlation between ODC induction and promoter treatment has been confirmed and extended by several workers. Takigawa and coworkers (1977), reported that low doses of ethyl phenylpropionate (EPP), a potent hyperplastic agent with weak promoting activity, induced ODC in whole skin to a greater extent than the high dose of this compound used by O'Brien, Simsiman and Boutwell (1975a,b). Promotion experiments with these doses of EPP, however, were not reported. A recent report demonstrated that wounding, a promoting stimulus for mouse skin, induced ODC but skin massage, a non-promoting treatment, did not (Clark-Lewis and Murray, 1978). Both treatments, however, stimulated cell proliferation in the epidermis.

The above studies established convincingly that treatment of mouse skin with a promoting stimulus (either chemical or physical) causes the induction of ODC and its extent reflects the relative promoting potency of the various treatments. This event does not appear to be related to the stimulated cell proliferation caused by all promoters because non-promoting agents that are capable of stimulating cell proliferation in epidermis have little or no effect on ODC activity. Mouse epidermis, therefore, appears to be one example of a tissue in which the induction of ODC is *not* necessary for subsequent cell or tissue growth.

If ODC activity and polyamines are intrinsically involved in skin tumour promotion, inhibiting or limiting polyamine synthesis after promoter treatment should result in a lower incidence of tumour formation. A few studies along these lines have been reported. Perhaps the most convincing evidence has been obtained with a series of vitamin A analogues (retinoids) (Chapter 2).

In addition, glucocorticoid hormones also have potent anti-promoting activity. Lichti and coworkers (1977) have compared the effect of fluocinolone acetonide (FA), a potent synthetic steroid, on promoter-stimulated ODC activity and DNA synthesis in mouse epidermis. In contrast to the results with retinoids, the steroid (22 nmoles/application did not inhibit, but even slightly enhanced, the ODC induction by TPA (17 nmoles/application) but virtually abolished the stimulation of DNA synthesis by the promoter. This result emphasizes that there is probably a sequence of essential events that occur in mouse epidermis after promoter treatment, and specific inhibition of any of these steps will reduce the efficacy of promotion. Since all promoters are hyperplastic agents for sin (but the converse is not true), the potent anti-proliferative action of steroids such as FA suggest why these compounds are effective anti-promoting agents: one function of an active promoter is to

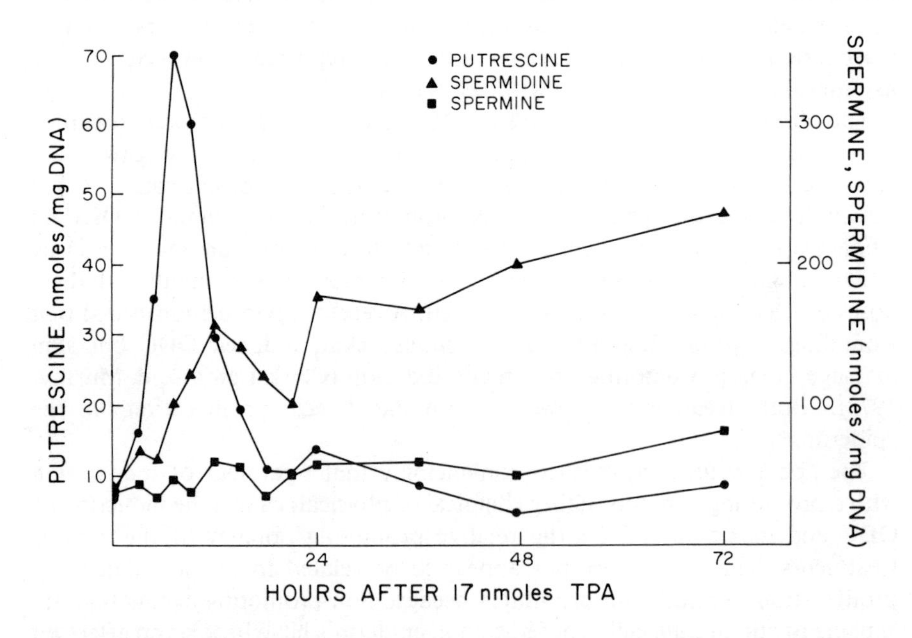

Figure 3 The effect of 17 nmoles TPA on the levels of putrescine (●), spermidine (▲), and spermine (■) in mouse epidermis. Groups of four mice were treated topically with either 0.2 ml acetone or TPA dissolved in acetone and killed at the times indicated

selectively stimulate the continued proliferation of 'initiated' cells, and agents that interfere with proliferation in general would be expected to inhibit promotion. The results of Lichti and coworkers (1977) indicate that the anti-promoting effects of steroids cannot be explained by inhibition of the induction of ODC, although other studies by these authors in epidermal cell cultures suggest that FA might interfere with polyamine *function* in promoter-treated cells thereby inhibiting proliferation (Lichti, Yuspa and Hennings, 1978).

In spite of our increased knowledge of what promoters do and some information on their mechanism of action, there is still much that is *not* known about promoters. For instance, we really know little about how promoters interact with epidermal cells to bring about such pronounced biochemical effects as the induction of ODC. It would also be of interest to know if normal, initiated, and malignant epidermal cells respond differently to promoter treatment and in what ways. Moreover, because of the accumulated circumstantial evidence that polyamines may be critically involved in tumour promotion, detailed knowledge of the control of polyamine biosynthesis in normal, initiated, and malignant cells is clearly an important goal. In order to study these and other questions, several groups in recent years have turned to cell culture systems. The remainder of this article will deal with what has been learned about tumour promoters and polyamine synthesis in several different models systems *in vitro*.

III. TUMOUR PROMOTERS AND POLYAMINE BIOSYNTHESIS IN CELL CULTURE

Cell culture model systems have recently been employed to study the action of promoters *in vitro*. The pioneering work of Sivak and Van Duuren (1967), who showed that the tumour-promoting phorbol esters could enhance the phenotypic expression of known malignant cells in mixed populations of SV40-transformed 3T3 cells and normal 3T3 cells, provided suggestive evidence that these agents could exert effects in cell culture that might be analogous to their actions in mouse skin. Perhaps the most convincing data supporting this view has been provided by Heidelberger and his associates (Mondal, Brankow and Heidelberger, 1976; Mondal and Heidelberger, 1976) who demonstrated a two-stage protocol for the malignant transformation* of a mouse fibroblast cell line, C3H/10T1/2, by the sequential action of non-transforming doses of either polycyclic hydrocarbons or UV light followed by prolonged, continuous treatment with the tumour-promoting phorbol diesters. Other studies have detailed a variety of biological and biochemical effects of

*The term transformation in this article will mean the acquisition by cells in culture of the capacity to produce malignant tumours after inoculation into appropriate animal hosts.

phorbol diesters on various cell culture systems, including effects on growth (Diamond and coworkers, 1974), DNA synthesis (Yuspa and coworkers, 1976a), phospholipid metabolism (Süss, Kreibich and Kinzel, 1971), transport of small molecules (Wenner, Moroney and Porter, 1978), cell differentiation (Cohen and coworkers, 1977; Diamond, O'Brien and Rovera, 1977; Rovera, O'Brien and Diamond, 1977; Yamasaki and coworkers, 1977; Yuspa and Harris, 1974) and prostaglandin synthesis (Levine and Hassid, 1977). I will only be concerned here with what is known of the effects of promoters on polyamine biosynthesis in cell culture systems.

The first demonstration that tumour promoters could induce ODC in a cell culture system was made by Yuspa and coworkers (1976b) who reported a good correlation between the ability of a series of phorbol esters to induce ODC and to stimulate DNA synthesis in primary cultures of mouse epidermal cells. Since in many respects the response of this culture system to promoters mimics the mouse skin system *in vivo*, it may be a useful system for studies of tumour promotion *in vitro*. Like other primary cultures of differentiated tissues, however, the cells fail to grow well and are not capable of subcultivation, which makes this system unsatisfactory for long-term studies.

Another approach to the study of tumour promoters *in vitro* has been to determine whether they elicit biochemical and biological changes in established cell lines. O'Brien and Diamond (1977) demonstrated that tumour promoters induce ODC in normal hamster embryo fibroblasts (HEF) and chemically transformed hamster cell lines. In these initial experiments, TPA was added to the cultures without medium renewal, since fresh medium containing 10% fetal bovine serum is itself a good inducer of ODC. As shown in Figure 4, the induction was much greater in transformed (30-fold) *versus* normal (5-fold) cells. Similar results were obtained with three other chemically or spontaneously transformed cell lines derived from HEF. The experiments reported here were done with the chemically transformed line HE68BP and the spontaneously transformed line HE75S. These data suggest that malignant hamster cells respond differently, at least quantitatively, to promoter treatment than do normal cells. This idea is reinforced by the results of experiments in which each of these cell types was exposed to two inducers of ODC, fresh medium and TPA. Figure 5 illustrates this point in detail: normal cells exposed to fresh medium containing TPA respond with an ODC induction roughly equivalent to the sum of the inductions obtained when each inducer is added separately. In transformed cells, a remarkable synergistic effect of TPA and fresh medium results in a much larger ODC induction than could be expected from the separate inductions due to TPA and fresh medium alone. This ability of TPA and other active promoters to 'potentiate' the serum induction of ODC has been observed in all transformed hamster cell lines tested, but never in normal HEF (O'Brien, unpublished observations). These results suggest that normal and transformed hamster cells exhibit both quantitative and qualitative

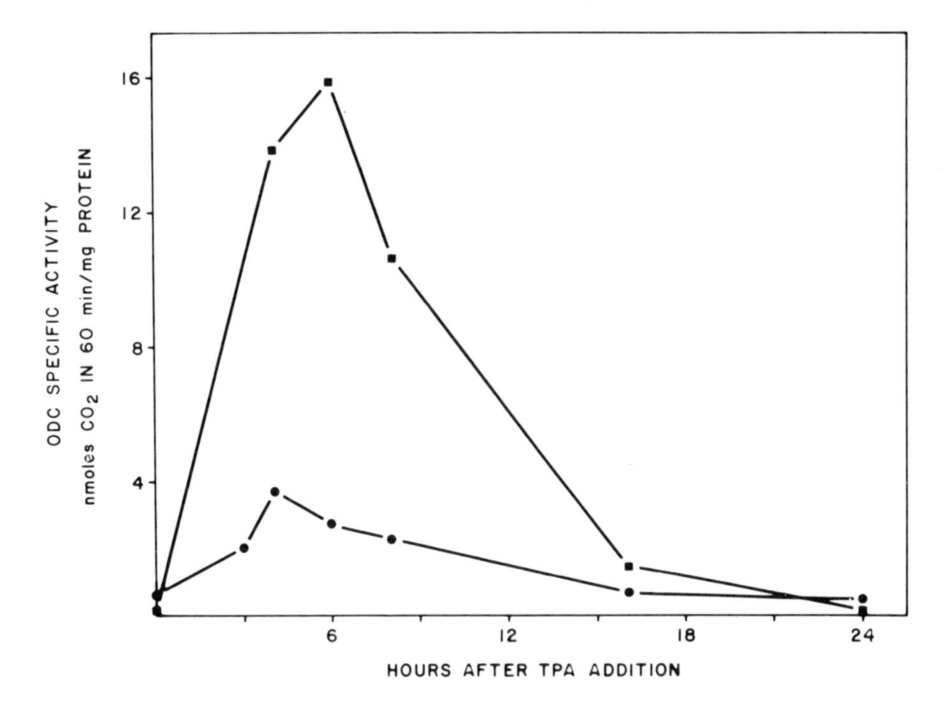

Figure 4 The effect of TPA (1.6×10^{-7} M) on ODC activity in normal (●) and transformed (■) HEF cells. TPA was added without changing the medium to 5-day-old cultures of tertiary HEF or HE68BP cells, and the cells were harvested for enzyme determinations at the indicated times. Each point is the average enzyme activity from two dishes assayed separately

differences in their responses to tumour promoters, at least with respect to ODC activity.

The changes in ODC activity brought about by promoters in transformed HE68BP cells are reflected in increased levels of polyamines as shown in Table 1. TPA added without medium renewal caused a 7.5-fold increase in the intracellular putrescine concentrations 9 h after treatment but only slight changes in spermidine and spermine concentrations. In contrast, no significant changes in polyamine concentrations occurred in normal HEF over a 72 h period after treatment with TPA. It should be pointed out that although these two cell types differ in various aspects of polyamine synthesis, their growth characteristics are similar. The population doubling times (18.5 h for HEF and 19.0 h for HE68BP cells) and maximal cell densities attained ($1.6\text{-}1.8 \times 10^5$ cells/cm^2) are virtually identical.

Recent studies demonstrate that TPA can also induce SAMD in normal and transformed hamster cells in culture. Shown in Figure 6 are the data for

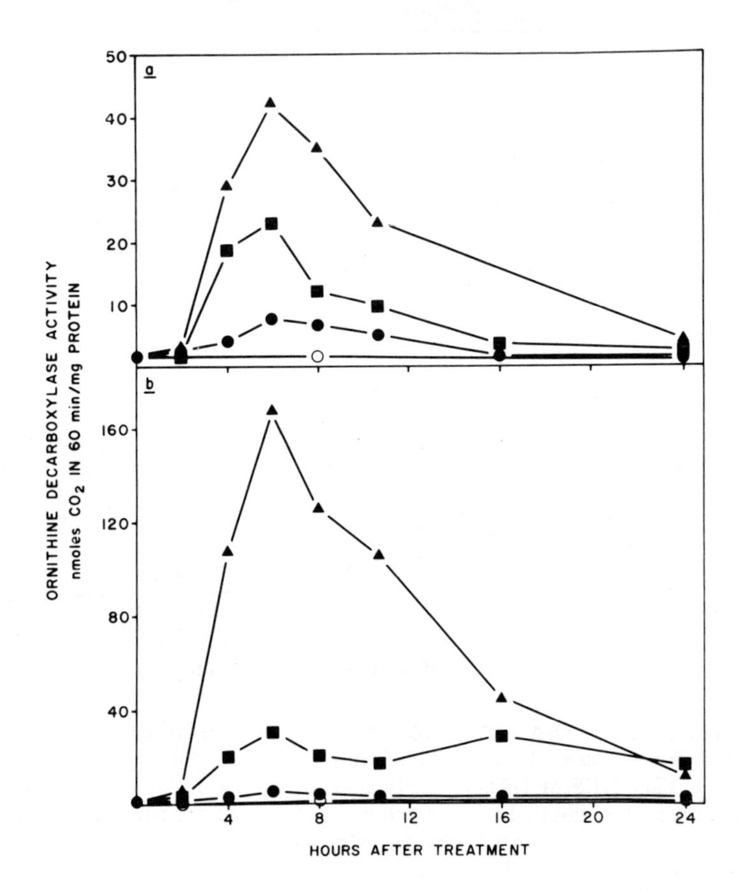

Figure 5 The effect of TPA on ODC activity. Five-day-old confluent cultures of secondary HEF (top) and HE68BP cells (bottom) were: treated with TPA (final concentration 1.6×10^{-7} M) without medium renewal (●); refed with fresh medium (■); refed with fresh medium containing TPA (1.6×10^{-7} M) (▲); or left untreated (○). At the indicated times the cells were harvested and assayed for ODC activity. Each point is the average of two dishes

normal HEF. TPA added without medium renewal caused a 5-fold induction at 8 h after treatment. A medium change was also an inducing stimulus and the combination of fresh medium with TPA appeared to delay somewhat the induction without affecting its magnitude. The changes in SAMD activity in transformed hamster cells after these treatments is shown in Figure 7. TPA alone caused the greatest induction (9-fold), and the combination of fresh medium plus TPA produced a shift in the peak of activity to 24 h instead of the earlier times seen in cells treated with TPA or fresh medium. Comparison of

Table 1 The effect of TPA on polyamine levels in HE68BP cells
TPA in conditioned medium was added to 4-day-old cultures of HE68BP cells to give a final concentration of 0.16 μM. At the indicated times thereafter, the cellular polyamine concentrations were determined as described previously (O'Brien and Diamond, 1977). The values represent the mean ± s.e. of 3 dishes

Time (h) after TPA addition	Polyamine concentration (nmoles/mg DNA)		
	Putrescine	Spermidine	Spermine
Control	28 ± 2	409 ± 29	206 ± 40
4	105 ± 3	458 ± 18	278 ± 14
6	157 ± 14	525 ± 31	225 ± 7
9	211 ± 31	524 ± 61	214 ± 53
16	157 ± 22	544 ± 101	211 ± 11
24	86 ± 10	473 ± 39	164 ± 15

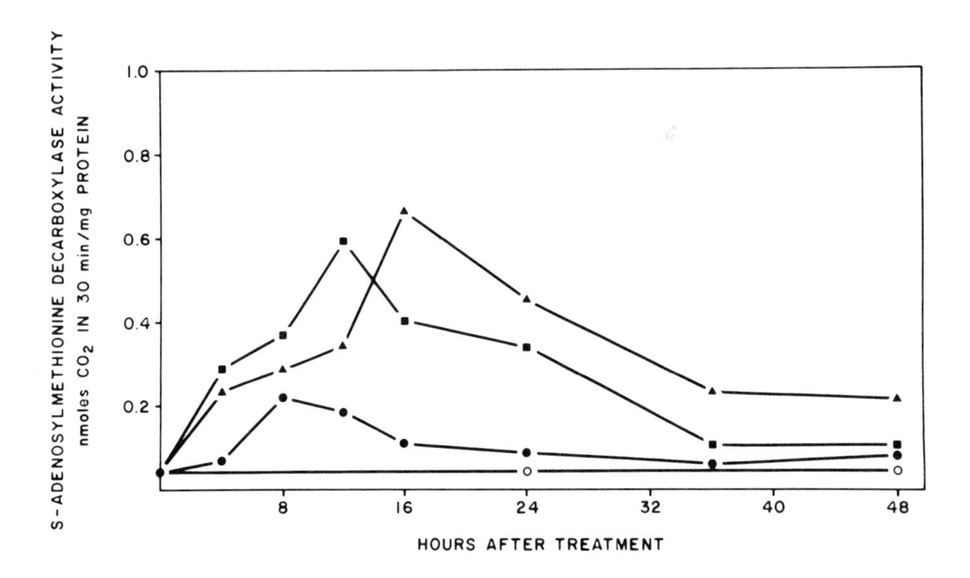

Figure 6 The effect of TPA on SAMD activity. Five-day-old confluent cultures of secondary HEF were: treated with TPA (final concentration 1.6×10^{-7} M) without medium renewal (●); refed with fresh medium (■); refed with fresh medium containing TPA (1.6×10^{-7} M) (▲); or left untreated (○). At the indicated times the cells were harvested and assayed for SAMD activity. Each point is the average of two dishes

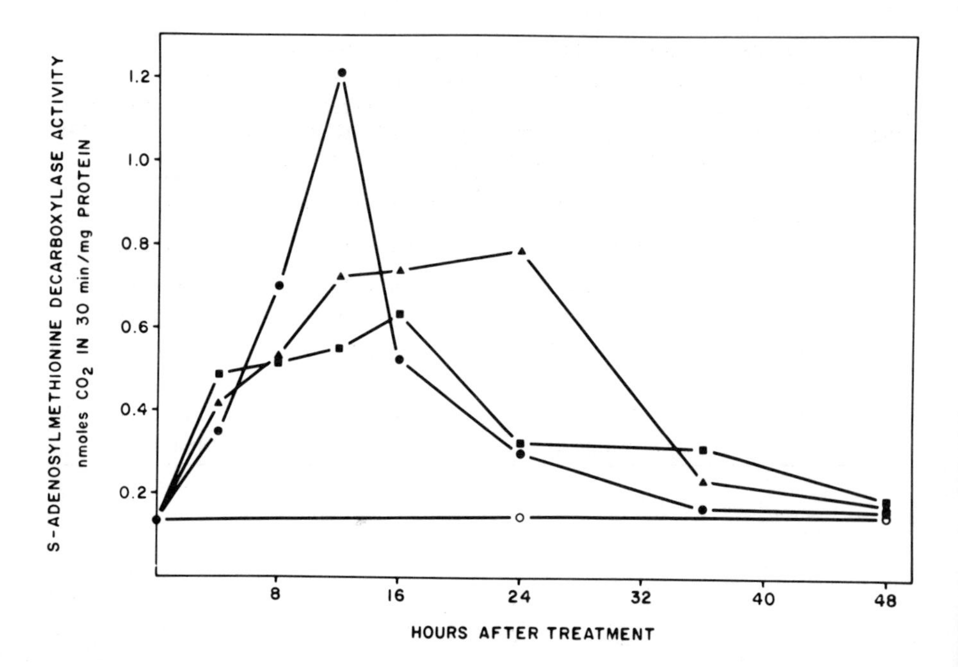

Figure 7 The effect of TPA on SAMD activity in transformed hamster cells. Five-day-old confluent cultures of a spontaneously transformed hamster cell line (HE75S) were treated exactly as described in the legend to Figure 6 (the symbols are the same). Each point is the average SAMD activity of two dishes

the absolute enzyme activities for ODC and SAMD under both basal and induced conditions illustrates a basic difference between most tissues *in vivo* and most cells in culture. Although ODC activity is almost undetectable in several resting animal tissues (brain, skin, liver) and the level of SAMD is much higher than ODC, the reverse is true for these hamster cells in culture and for at least one other cell type in culture (Heby and coworkers, 1975). Under comparable treatment conditions SAMD activity is always much lower by a factor of 10 or 50) than ODC activity, and thus this enzyme (SAMD) may constitute the actual 'rate-limiting' step for spermidine and spermine synthesis in at least some cells in culture.

It is important to note that TPA did *not* potentiate the fresh medium induction of SAMD in these cells as it did for the ODC induction. This interesting specificity in the ability of TPA to selectively potentiate only the expression of certain genes emphasizes that the induction of ODC both *in vivo* and *in vitro* may be one of the important parts of the mechanism of promoter action. Since TPA does not potentiate the ODC induction by fresh medium in normal HEF, our data also suggest that transformed cells in culture (and perhaps pre-neoplastic cells) respond differently to promoter treatment than do normal cells, a finding that also has important implications for the mechanism of promotion *in vivo*.

Because of the close association in animal tissues *in vivo* between a stimulation of polyamine biosynthesis and subsequent tissue growth, it was of interest to examine this relationship in hamster cells *in vitro*, since as we have just shown, TPA can induced both ODC and SAMD and cause polyamine accumulation in these cells. TPA does not, however, stimulate DNA synthesis, as measured by the fraction of cells pulse-labelled with [^{3}H] thymidine in either normal HEF (Figure 8) or in a transformed hamster cell line (Figure 9) in which ODC activity is elevated 20-50-fold after TPA treatment (data not shown). In each cell type, TPA added without medium renewal or in combination with a known 'mitogen' (fresh serum-containing medium) had no effect on the percentage of cells synthesizing DNA at any time up to 48 h after treatment. These results indicate that the ability of promoters to stimulate polyamine biosynthesis is not simply a reflection of growth stimulation by these agents.

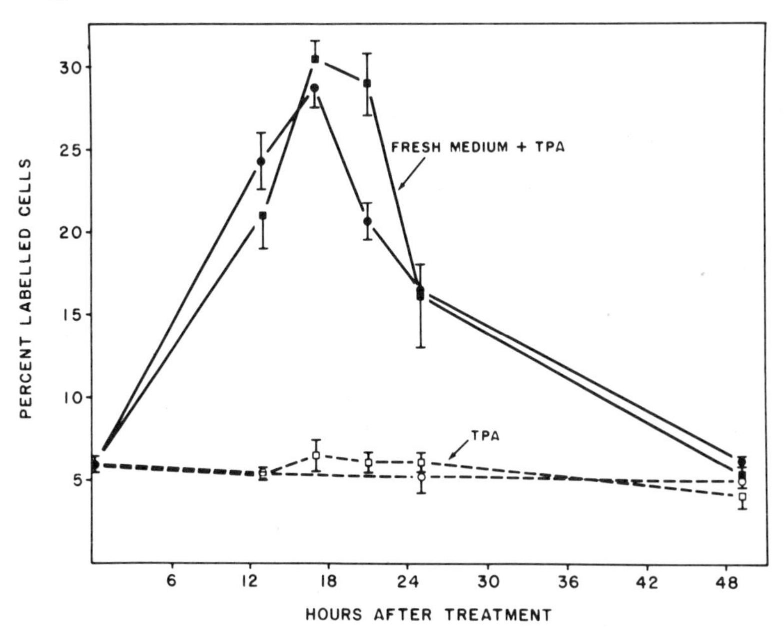

Figure 8 The effect of TPA (1.6 $\times$ 10^{-7} M), fresh medium, and fresh medium containing TPA on the percentage of normal HEF pulse-labelled with tritiated thymidine. Five-day-old cultures of tertiary HE cells grown on coverslips were treated at zero time, and cells were fixed for autoradiography at the indicated times. Tritiated thymidine (final concentration 1 μCi/ml, 0.5 μM) was added to the cultures 60 min before fixation. Each point is the mean ($\pm$ s.e.) percentage of labelled cells (at least 400 cells scored per coverslip) on three to five coverslips in two or more dishes. ○– – – –○, control; ●——●, fresh medium; □– – – –□, TPA; ■——■, fresh medium and TPA

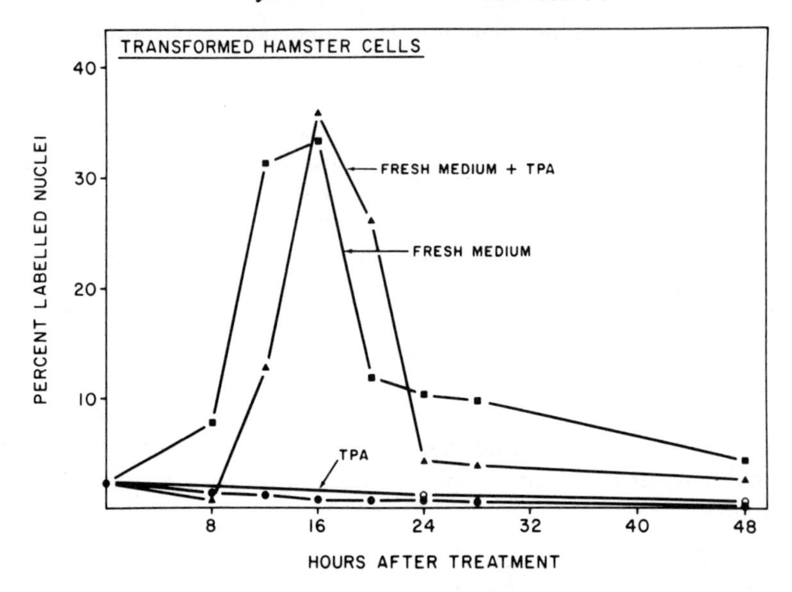

Figure 9 The effect of TPA (1.6×10^{-7} M), fresh medium, and fresh medium containing TPA on the percentage of transformed HEF pulse-labelled with [^{3}H] thymidine. Five-day-old cultures of the HE75S cells grown on coverslips were treated exactly as described in the legend to Figure 8. Each point is the mean ± s.e. of four coverslips from two dishes. ○, control; ■, fresh medium; ●, TPA; ▲, fresh medium plus TPA

In support of this idea, we have recently obtained further evidence that the ODC induction after promoter treatment is not associated with cell division: we found that the promoter TPA does *not* induce ODC in confluent cultures of human fibroblasts, but does stimulate a large percentage (50% or greater) of these cells to enter S phase. Clearly, in these cells an induction of ODC is not necessary for DNA synthesis. Our current thinking is that polyamine accumulation after a growth stimulus is not necessary, since most cells in culture have sufficient levels of polyamines for at least one round of cell division. Therefore, the enhanced rate of polyamine synthesis after promoter treatment may have some other function(s) related to the phenotypic expression of malignancy. These results also point out that the response of various cell types to promoters can be quite different (witness HEF *versus* human fibroblasts) and such differences will probably become more evident as other effects and other cell types are examined.

IV. SUMMARY AND CONCLUSION

The tumour-promoting phorbol diesters induce the two key enzymes involved in polyamine biosynthesis, ODC and SAMD, and cause an actual net accumulation of polyamines both in mouse epidermis *in vivo* and in cell culture systems. These agents, however, cause many other biochemical changes in mouse epidermis and in varous cells *in vitro*, many of which are probably unrelated to tumour promotion *per se*. Several lines of evidence from studies in mouse skin suggest that the induction of ODC and subsequent polyamine accumulation may be an important component of the promotion process. Most of these studies are purely correlative, however, and experiments providing direct evidence for this hypothesis are lacking.

An important finding with promoters in cell culture is that the stimulation of polyamine synthesis by promoters does not appear to be associated with cell proliferation. The most convincing demonstration of this can be seen by comparing the responses of two cell types to promoter treatment, hamster embryo fibroblasts and human fibroblasts. TPA induces ODC in both normal and transformed HEF but does not stimulate DNA synthesis, whereas in human cells, ODC is not induced but TPA does stimulate DNA synthesis.

If the increased polyamine levels in tissues or cells treated with promoters are not involved in cell division, what then is their significance and how might they be involved in mechanism of promotion? We do not have any firm answers, but I would like to speculate here on ways in which polyamines might be important in the development of skin carcinogenesis. The recent observations in cell culture that tumour promoters inhibit terminal differentiation have led to hypotheses about how such a phenomenon could be involved in promotion (Diamond, O'Brien and Rovera, 1977). In a mixed population of stem cells and terminally differentiating cells, elevated polyamine levels could influence whether a cell continues to proliferate or enters a pathway committed to terminal differentiation. By enlarging the fraction of uncommitted 'stem' cells in the population, tumour promoters could increase the probability that an 'initiated' cell would begin to express its malignant phenotype. The intracellular polyamine concentrations may be one of the ways in which a cell capable of either differentiation or continued proliferation makes such a decision.

Another possibility for the role of polyamines in tumour promotion is that polyamine levels in 'excess' of a cell's needs could bind to cell membranes and alter the response to growth-regulating substances, such as chalones (Chapters 20 and 21). In a cell with a defective regulation of polyamine biosynthesis ('initiated' cell?), the resulting continued high levels of polyamines could confer a selective growth advantage and allow such a cell to escape the growth controls still operative in normal, non-initiated cells.

The above discussion has been highly speculative and emphasizes our incomplete knowledge of the function of polyamines in normal versus neoplastic cells as well as proliferating versus differentiating cells. These ideas will provide a framework, however, for designing experiments in which to test some of these hypotheses. For example, it would be expected that normal and initiated cells would respond differently to promoter treatment. Such experiments are impossible in mouse skin but may be feasible in cell culture. Tumour promoters cause a much greater induction of ODC in transformed hamster cells than normal HEF and can markedly potentiate the serum induction of ODC in transformed cells but not in normal cells. This potentiation is at least somewhat specific, since TPA did not potentiate the serum induction of SAMD in these same cells. Related to these findings is the recent report that TPA can greatly potentiate the production of the protease plasminogen activator in Rous sarcoma virus-transformed chick embryo fibroblasts (Weinstein and coworkers, 1978). Clearly, it will be important in future studies to find out in what other ways transformed cells differ from normal cells in their response to promoters and to extend these experiments to premalignant cell populations.

Since many cell culture systems seem to respond to promoter treatment in some ways, such as the induction of ODC (see Chapter 6) that are analogous to the effects of promoters in mouse epidermis, the development of characterization of *in vitro* systems capable of responding biologically to these agents will be a major thrust of future research. In addition to fibroblastic cell types, recently several epithelial cell culture systems in which promoters have been used have been described, notably mouse epidermal cell cultures (Yuspa and coworkers, 1976a) and rat tracheal explants in culture and cell lines derived from carcinogen-exposed trachea (Marchok and coworkers, 1977). The availability of such model systems will greatly facilitate carcinogenesis studies in general and in particular, could establish the generality of tumour promotion by the phorbol diesters in other cell types besides mouse skin.

ACKNOWLEDGEMENTS

This work was supported in part by grants from the U.S. National Institutes of Health. I acknowledge the technical assistance of Ruby Simsiman, Linda Rogers and Douglas Salidik. I thank Drs. R. K. Boutwell and Leila Diamond for providing a stimulating environment in which to pursue these studies.

REFERENCES

Berenblum, I., and Shubik, P. (1947). *Brit. J. Cancer*, **1**, 383–391.
Boutwell, R. K. (1964). *Progr. Exptl. Tumor Res.*, **4**, 207–250.
Boutwell, R. K. (1974). *CRC Crit. Rev. Toxicol.*, **2**, 419–443.
Clark-Lewis, I., and Murray, A. W. (1978). *Cancer Res.*, **38**, 494–497.
Cohen, R., Pacifici, M., Rubinstein, N., Biehl, J., and Holtzer, H. (1977). *Nature*, **266**, 538–540.
Diamond, L., O'Brien, S., Donaldson, C., and Shimizu, Y. (1974). *Intr. J. Cancer*, **13**, 721–730.
Diamond, L., O'Brien, T. G., and Rovera, G. (1977). *Nature, Lond.*, **269**, 247–249.

Foulds, L. (1969). In *Neoplastic Development*, Chapter 3, Academic Press, New York, pp. 41–89.
Friedewald, W. F., and Rous, P. (1944). *J. Exptl. Med.*, **80**, 101–125.
Heby, O., Marton, L. J., Wilson, C. B., and Martinez, H. M. (1975). *J. Cell Physiol.*, **86**, 511–552.
Hecker, E. (1971). *Methods in Cancer Res.*, **6**, 439–484.
Inoue, H., Tanioka, H., Shiba, K., Asada, A., Kato, Y., and Takeda, Y. (1974). *J. Biochem.*, **75**, 679–687.
Levine, L., and Hassid, A. (1977). *Biochem. Biophys. Res. Comm.*, **79**, 477–484.
Lichti, U., Slaga, T. J., Ben, T., Patterson, E., Hennings, H., and Yuspa, S. H. (1977). *Proc. Natl. Acad. Sci. USA*, **74**, 3908–3912.
Lichti, U., Yuspa, S. H., and Hennings, H. (1978). Ornithine and S-adenosylmethionine decarboxylase in mouse epidermal cell cultures treated with tumor promoters. In T. J. Slaga, A. Sivak and R. K. Boutwell, *Mechanisms of Tumor Promotion and Cocarcinogenesis*, Vol. 2, Raven Press, New York, pp. 221–232.
Marchok, A. C., Rhoton, J. C., Griesemer, R. A., and Nettesheim, P. (1977). *Cancer Res.*, **37**, 2254–2260.
Mondal, S., Brankow, D. W., and Heidelberger, C. (1976). *Cancer Res.*, **36**, 2254–2260.
Mondal, S., and Heidelberger, C. (1976). *Nature, Lond.*, **260**, 710–711.
Mottram, J. C. (1944). *J. Pathol. Bacteriol.*, **56**, 181–187.
O'Brien, T. G. (1976). *Cancer Res.*, **36**, 2644–2653.
O'Brien, T. G., and Diamond, L. (1977). *Cancer Res.*, **37**, 3895–3900.
O'Brien, T. G., Simsiman, R. C., and Boutwell, R. K. (1975a). *Cancer Res.*, **35**, 1662–1670.
O'Brien, T. G., Simsiman, R. C., and Boutwell, R. K. (1975b). *Cancer Res.*, **35**, 2426–2433.
Pegg, A. E., Lockwood, D. H., and Williams-Ashman, H. G. (1970). *Biochem. J.*, **117**, 17–31.
Rovera, G., O'Brien, T. G., and Diamond, L. (1977). *Proc. Natl. Acad. Sci. USA*, **74**, 2894–2898.
Russell, D. H. and Snyder, S. H. (1968). *Proc. Natl. Acad. Sci. USA*, **68**, 1420–1428.
Sivak, A., and Van Duuren, B. L. (1967). *Science*, **157**, 1443–1444.
Süss, R., Kreibich, G., and Kinzel, V. (1971). *Eur. J. Cancer*, **8**, 299–304.
Takigawa, M., Inoue, H., Gohda, E., Asada, A., Takeda, Y., and Mobi, Y. (1977). *Exptl. Molecular Pathol.*, **27**, 183–196.
Van Duuren, B. L. (1969). *Progr. exp. Tumor Res.*, **11**, 31–68.
Weinstein, I. B., Wigler, M., Fisher, P. B., Sisskin, E., and Pietropaolo, C. (1978). Cell culture studies on the biologic effects of tumor promoters. In T. J. Slaga, A. Sivak and R. K. Boutwell (Eds), *Mechanisms of Tumor Promotion and Cocarcinogenesis*, Raven Press, New York, pp. 313–333.
Wenner, C. E., Moroney, J., and Porter, C. W. (1978). Early membrane effects of phorbol esters in 3T3 cells. In T. J. Slaga, A. Sivak and R. K. Boutwell (Eds), *Mechanisms of Tumor Promotion and Cocarcinogenesis*, Raven Press, New York, pp. 362–378.
Yamasaki, H., Fibach, E., Nudel, U., Weinstein, I. B., Rifind, R. A., and Marks, P. A. (1977). *Proc. Natl. Acad. Sci. USA*, **74**, 3451–3455.
Yuspa, S. H., and Harris, C. C. (1974). *Exptl. Cell Res.*, **86**, 95–105.
Yuspa, S. H., Ben, T., Patterson, E., Michael, D., Elgjo, K., and Hennings, H. (1976a). *Cancer Res.*, **36**, 4062–4068.
Yuspa, S. H., Lichti, U., Ben, T., Patterson, E., Hennings, H., Slaga, T. J., Colburn, N., and Kelsey, W. (1976b). *Nature, Lond.*, **262**, 402–404.

Chapter 16

Novel Bacterial Polyamines

MARIO DE ROSA, AGATA GAMBACORTA, MARIA CARTENI'-FARINA AND VINCENZO ZAPPIA

I. INTRODUCTION

Putrescine and spermidine represent the common polyamines occurring in bacterial cells (Tabor and Tabor, 1976) while spermine is the most abundant polyamine in animal tissues (Zappia, 1977a). Low amounts of spermine (Stevens and Morrison, 1968; Bachrach, 1973a) 1,3-diaminopropane (Tabor and Tabor, 1964; Weaver and Herbst, 1958), 1,5-diaminopentane and 3-aminopropil-1,5-diaminopentane (Dion and Cohen, 1972) have also been reported, however, in bacterial cells. Some unusual amine derivatives described in prokaryotes are also of interest: 2-hydroxyputrescine has been detected in two *Pseudomonas* strains (Kullnig, 1970; Tobari, 1971; Karrer, Bose and Warren, 1973), glutathionyl-spermidine in *Escherichia coli* (Dubin, 1959; Tabor and Tabor, 1970; Tabor and Tabor, 1975) and edeine A and edeine B, two complex antibiotics containing spermidine or guanylspermidine covalently linked to a pentapeptide, in *Bacillus brevis* Vm 4 (Bachrach, 1973b; Hettinger and Craig, 1968a; Hettinger, Kurylo-Borowska and Craig, 1968b).

More recently two new polyamines, sym-nor-spermidine (1,7-diamino-4-aza-eptane) and sym-nor-spermine-(1,11-diamino-4,8-diaza-undecane), named respectively caldine and thermine (Figure 1), have been identified in a series of genetically different thermophilic bacteria (Oshima, 1975; De Rosa and coworkers, 1976).

NH_2-CH_2-CH_2-CH_2-NH-CH_2-CH_2-CH_2-NH_2

sym-nor-spermidine

(caldine)

NH_2-CH_2-CH_2-CH_2-NH-CH_2-CH_2-CH_2-NH-CH_2-CH_2-CH_2-NH_2

sym-nor-spermine

(thermine)

Figure 1 New polyamines in *Caldariella acidophila*

In this article we summarize the recent developments in the distribution, biosynthesis and physical characterization of these new polycations.

II. DISTRIBUTION AND LEVELS OF NOVEL POLYAMINES

The occurrence of sym-nor-spermine in several extreme thermophilic bacteria as *Thermus thermophilus*, *Thermus flavus* and *Thermus aquaticus* was first reported by Oshima (1975). A roughly equivalent amount of spermine has also been detected in these microorganisms although a quantitative evaluation has not been established. The relationship between the age of cultures and the cellular content of tetra-amines has also been indicated, the higher levels of polyamines being present at an early exponential phase. The relative abundance of these two polycations also appears to be related to the age of cells, since sym-nor-spermine increases in the later stage of growth. The biological occurrence of sym-nor-spermidine, the precursor of sym-nor-spermine, has been proved unequivocally for the first time by De Rosa and coworkers (1976), in *Caldariella acidophila*, an extreme thermoacidophilic microorganism.

Table 1 shows the amount of polyamines present in *C.acidophila* cultures harvested during the exponential-phase and stationary-phase of growth. The overall content of polyamines in this bacterium appears to be the highest reported in comparison with the data for other prokaryotes (Bachrach, 1973a). Besides the presence of sym-nor-spermidine, the results reported in Table 1 differ from those reported by Oshima (1975) in *Thermus* genera two points as follows: (1) the total absence of spermine in *C.acidophila* and (2) the lack of any relation between the overall polyamine content and culture age. It is also

Table 1 Polyamine concentration in *Caldariella acidophila*

Polyamines	Exponential-phase	Stationary-phase
	μmol/g wet cells	
Putrescine	N.D.	N.D.
Spermidine	12.8	8.2
Sym-nor-spermidine	2.0	1.8
Spermine	N.D.	N.D.
Sym-nor-spermine	8.1	15.6

N.D. = not detectable

noteworthy that the decrease of spermidine in the stationary-phase is paralleled by an almost equivalent increase of sym-nor-spermine. The simultaneous presence of spermine, sym-nor-spermidine and sym-nor-spermine in the same microorganism supports the hypothesis that sym-nor-spermidine is presumably the precursor of sym-nor spermine.

III. BIOSYNTHETIC PATHWAY

The biosynthetic pathway of sym-nor-spermidine and sym-nor-spermine has been investigated (De Rosa and coworkers, 1978) in *C.acidophila*, where the absence of spermine simplifies the study on the enzymatic steps involved in the biosynthesis of the new tetra-amines.

The main feature of this biosynthetic pathway supported by isotopic evidence and enzymatic studies, is the unusual role of spermidine as donor of the 1,3-diaminopropane moiety (see Figure 2).

Growing cultures of microorganisms, in separate experiments, were incubated with several labelled precursors and the radioactivity distribution into the three polyamines was analyzed (see Table 2).

The radioactivity of putrescine and spermidine labelled in the tetramethylene moiety was recovered only as spermidine, while the radioactivity of methionine as well as that of spermidine labelled in the propylamine moiety was recovered in sym-nor-spermidine (spermidine), and sym-nor-spermine. If labelled spermine was supplied as precursor, nor radioactivity was detected in the polyamine pool, thereby suggesting the existence of a critical mechanism of membrane permeability.

The radioactivity recovered by using spermidine labelled in the tetramethylene moiety, was specifically localized in the polyamine pool and was significantly lower compared to that recovered when spermidine labelled in the propylamine moiety was used, and this result suggests that the tetramethylene moiety of spermidine is rapidly excreted. The results obtained by the use of spermidine labelled with [^{14}C] in the tetramethylene moiety and [^{3}H] in the propylamine moiety are in agreement with this hypothesis: the percentage

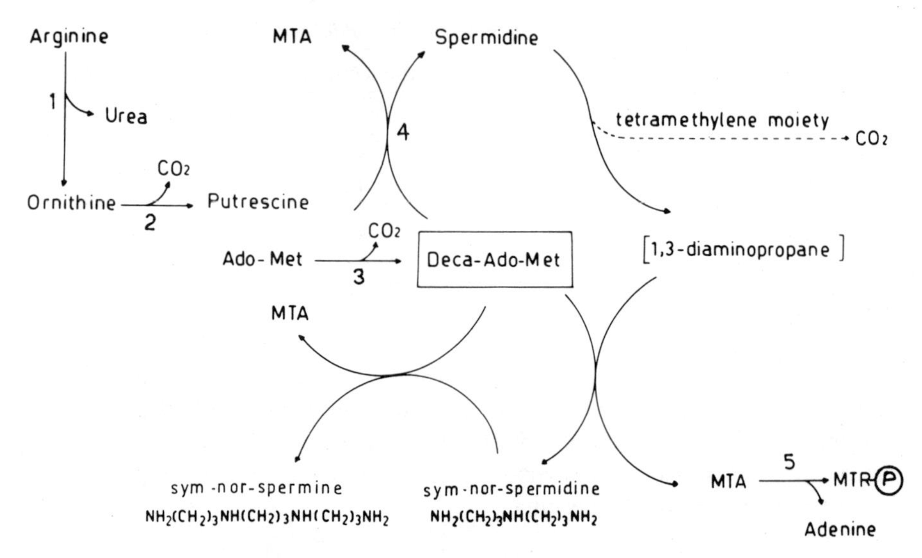

Figure 2 Proposed biosynthetic pathway

of [^{14}C] recovered as [^{14}C]O_2, formed during the microorganism growth was equal to that recovered as [^{3}H] in sym-nor-spermidine plus sym-nor-spermine.

A systematic investigation on the occurrence of γ-aminobutyraldehyde, Δ^1-pyrroline and γ-aminobutyric acid, as probable intermediates of the catabolism of spermidine tetramethylene moiety, has been unsuccessful (De Rosa and coworkers, 1978). The reported results are indicative for a fast turnover of the 4 carbon-chain fragment and also account, in part, for the low cellular recovery of radioactivity in the experiments with labelled putrescine and for the absence of this diamine in the polyamine pool of *C.acidophila*.

The role of 1,3-diaminopropane as 'free' or protein-bound intermediate in sym-nor-spermidine and sym-nor-spermine biosynthesis is a matter of investigation; however, no detectable amounts of this compound have been observed in the polyamine pool.

The biosynthesis of spermidine resembles the pathway occurring in mesophilic microorganisms (Tabor and Tabor, 1976) and mammalian tissues, in that putrescine provides the 1,4-diaminobutane moiety and methionine donates the propylamine group. The intermediary role of S-adenosylmethionine (Ado-Met) and its decarboxylated product (Deca-Ado-Met) was reasonably supported by the detection of both sulphonium compounds (60 nmg^{-1} of wet cells Ado-Met and 5nmg^{-1} of wet cells Deca-Ado-Met). The presence of S-adenosylmethionine at levels comparable with those reported in other microorganisms and eukaryotes (Salvatore and coworkers, 1971; Porcelli and coworkers, 1978) was unexpected in view of the thermal lability of this

Table 2 Biogenesis of polyamines in *Caldariella acidophila*

The labelled precursors, [1,4-(n-^{3}H)] putrescine dihydrochloride (0.35 mCi), L-[G-^{3}H]methionine (0.25 mCi), [tetramethylene-1,4-^{14}C] spermidine trihydrochloride (0.25 mCi),[3-aminopropyl-3-(n-^{3}H)] spermidine trihydrochloride (0.25 mCi) were added to the culture (25 l) at the beginning of the exponential phase. The cells were harvested in the late-exponential-phase (25 h incubation) by continuous flow centrifugation. The yield in bacteria and polyamines was comparable in the four separate experiments.

Precursor	% of recovered radioactivity: in the cells	% of recovered radioactivity: in the polyamine pool[1]	Total radioactivity in the resolved polyamines TFA by GLC and specific radioactivity in brackets (dpm/mmol): sym-nor-spermidine		spermidine		sym-nor-spermine	
[1,4-(n-^{3}H)] putrescine[2] dihydrochloride (3.1×10^{8} dpm; 12.3 Ci/nmol)	2.3	52.9	0	(0)	3.65×10^{6}	(19)	0	(0)
L-[G-^{3}H] methionine (1.0×10^{9} dpm; 250 mCi/mmol	4.0	11.2	0.18×10^{6}	(6)	1.93×10^{6}	(10)	2.05×10^{6}	(17)
[tetramethylene-1,4- ^{14}C] spermidine trihydrochloride (4.0×10^{8} dpm; 122 mCi/mmol)	34.5	97.6	0	(0)	1.30×10^{8}	(630)	0	(0)
[3-aminopropyl-3-(n-^{3}H)] spermidine trihydrochloride (7.2×10^{8} dpm; 423 mCi/mmol)	88.9	99.4	0.43×10^{8}	(1400)	4.34×10^{8}	(2200)	1.23×10^{8}	(1000)

[1]Percentage of radioactivity recovered in the total polyamine pool; 100 refers to the total radioactivity recovered in the whole cells.
[2]No radioactivity was recovered in correspondence of the retention time of putrescine.

molecule (Schlenk, 1977). The protection against the hydrolysis at the elevated growth temperature of the microorganism (87°C) could be ascribed either to an interaction with a specific binding protein (Smith, 1976) or to a masking of the carboxylic group of methionine which is in turn responsible for the intramolecular hydrolytic reaction (Zappia and coworkers, 1977b).

The following enzyme activities related to the biosynthetic pathway have been detected in crude extracts of *C.acidophila*: arginase (1), ornithine decarboxylase (2), S-adenosylmethionine decarboxylase (3),spermidine synthase (4) and 5′-methylthioadenosine (MTA) phosphorylase (5), (for numbering see Figure 2). It is interesting to note that the optimal temperature for activity of these enzymes is close to the growing temperature of the microorganism (87°C); moreover, the enzymic proteins are highly thermostable. For example the activities of purified preparations of MTA phosphorylase did not decrease appreciably by exposure to 100°C for one h (Zappia and coworkers, 1977c).

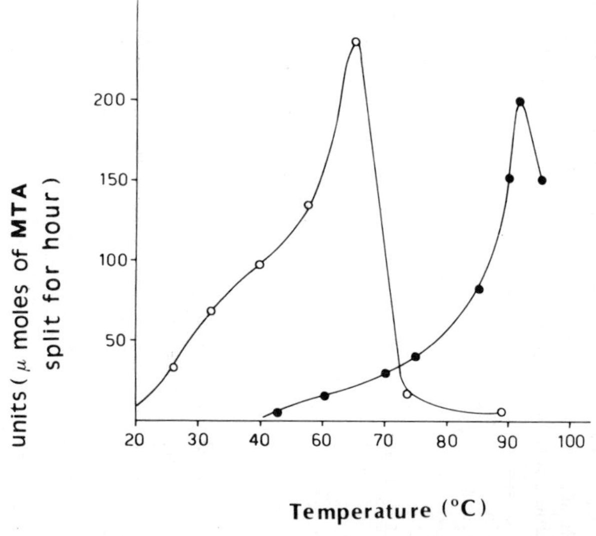

Figure 3 Effect of temperature on reaction rate of methylthioadenosine (MTA)-nucleosidase purified from *Caldariella acidophila* and from human placenta

In Figure 3 the profiles of activity versus temperature of MTA-phosphorylase, purified from *C.acidophila* and human placenta, are compared. One of the major features of the biosynthesis of sym-nor-spermidine and sym-nor-spermine, is the considerable amount of 5′-methylthioadenosine released; in fact, 2 mol of MTA were produced per mol of sym-nor-spermidine and 3 mol of MTA per mol of sym-nor-spermine. The observation that 5′-methylthioadenosine inhibits some, if not all, S-adenosyl-methionine-dependent methyltransferases (Zappia, Zydek-Cwich and Schlenk, 1969), has stimulated the interest about the catabolism of the thioether.

In *C.acidophila*, as well as in other bacterial cells, MTA accumulation is prevented by the high levels of MTA degrading enzymes. Probably any accumulation of this nucleoside could be deleterious to cells, not only for its adverse effect on transmethylation reactions but also because it could deplete the adenine nucleotide pool.

MTA nucleosidase is the main enzyme related to 5′-methylthioadenosine degradation. The enzyme was described for the first time by Shapiro and Mather (1958) in *Aspergillus aerogenes*; later Duerre (1962), isolated from *E.coli* an hydrolytic nucleosidase acting either on S-adenosyl-L-homocysteine or on MTA. More recently the enzyme has been purified 220-fold from *E.coli* (Ferro, Barrett and Shapiro, 1976). The pattern of the reaction catalyzed by the reported enzyme involves the hydrolytic cleavage of the glycosidic bond of MTA followed by the release of methylthioribose (MTR) and adenine. The enzyme responsible for MTA degradation occurring in *C.acidophila* requires inorganic phosphate for its activity, as has been demonstrated for MTA nucleosidases from rat ventral prostate (Pegg and Williams-Ashman, 1969), human placenta (Cacciapuoti, Oliva and Zappia, 1978) and human prostate (Zappia and coworkers, 1978). Purification to homogeneity of the enzyme from *C.acidophila* allowed a prompt elucidation of the cleavage mechanism of MTA. This enzyme catalyzes a phosphorolytic cleavage of 5′-methylthioadenosine to yield adenine and methylthioribose-1-phosphate.

IV. ANALYTICAL METHODS FOR CHARACTERIZATION OF NEW POLYAMINES

A. H.V.Electrophoresis

The two polyamines sym-nor-spermidine and sym-nor-spermine are quite well resolved from the correspondent analogues spermidine and spermine by paper electrophoresis in citrate buffer 5×10^{-2} M pH 3.4 (De Rosa and coworkers, 1976).

B. Gas Liquid Chromatography

The GLC is the most useful tool for the separation of sym-nor-spermidine and sym-nor-spermine from the correspondent homologous spermidine and spermine. The chromatographic characterization of the new polyamines has been performed either with the free bases or with the trifluoracetyl derivatives. In both cases the compounds are eluted in the order of increasing molecular weight with symmetrical and well resolved peaks (see Figure 4).

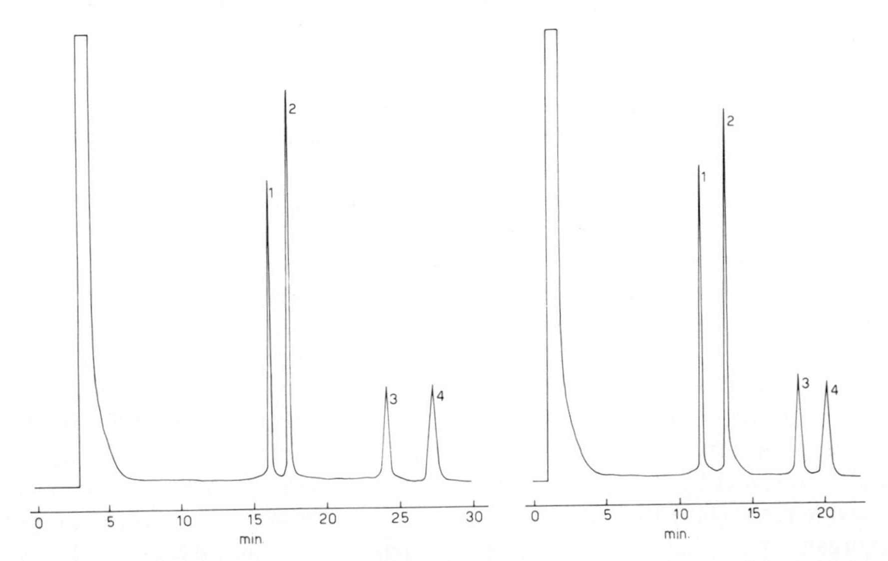

Figure 4 Gas chromatographic profiles of sym-nor-spermidine (1),spermidine (2),sym-nor-spermine (3) and sym-nor-spermine (4) on the left and their trifluoracetylated derivatives on the right. Free polyamines were resolved on Pyrex glass column (2 m × 3 mm I.D.) packed with Carlo Erba glass beads (100–120 mesh) coated with 1% KOH and 0.4% Carbowax 20M (Benninati, Sartori and Argenio-Cerù, 1977); N_2 (20 ml/min) was used as carrier gas with temperature program 10°C/min from 80° to 210°C with 5 min of initial isotherm. Trifluoracetylated derivatives of polyamines were chromatographed on Pyrex glass column (1.5 m × 3 mm I.D.) packed with SE-30 on Chromosorb W silanized (80-100 mesh), N_2 was employed as carrier gas with temperature program of 8°C/min from 120 to 300°C

C. Mass Spectrometry

The TFA polyamines derivatives showed weak but quite visible molecular ions [M^+ 419 for sym-nor-spermidine and 517 for sym-nor-spermine] which differed for 14 mass units from molecular ions of spermidine and spermine, respectively. In Figures 5 and 6 the mass spectra of TFA sym-nor-spermidine and sym-nor-spermine are compared with the corresponding spectra of trifluoroacetylated spermidine and spermine, respectively. All the molecular ions lose CF_3 to form fragments at m/e 350, 364, 503 and 517 (from TFA sym-nor-spermidine, TFA spermidine, TFA sym-nor-spermine and TFA spermine,

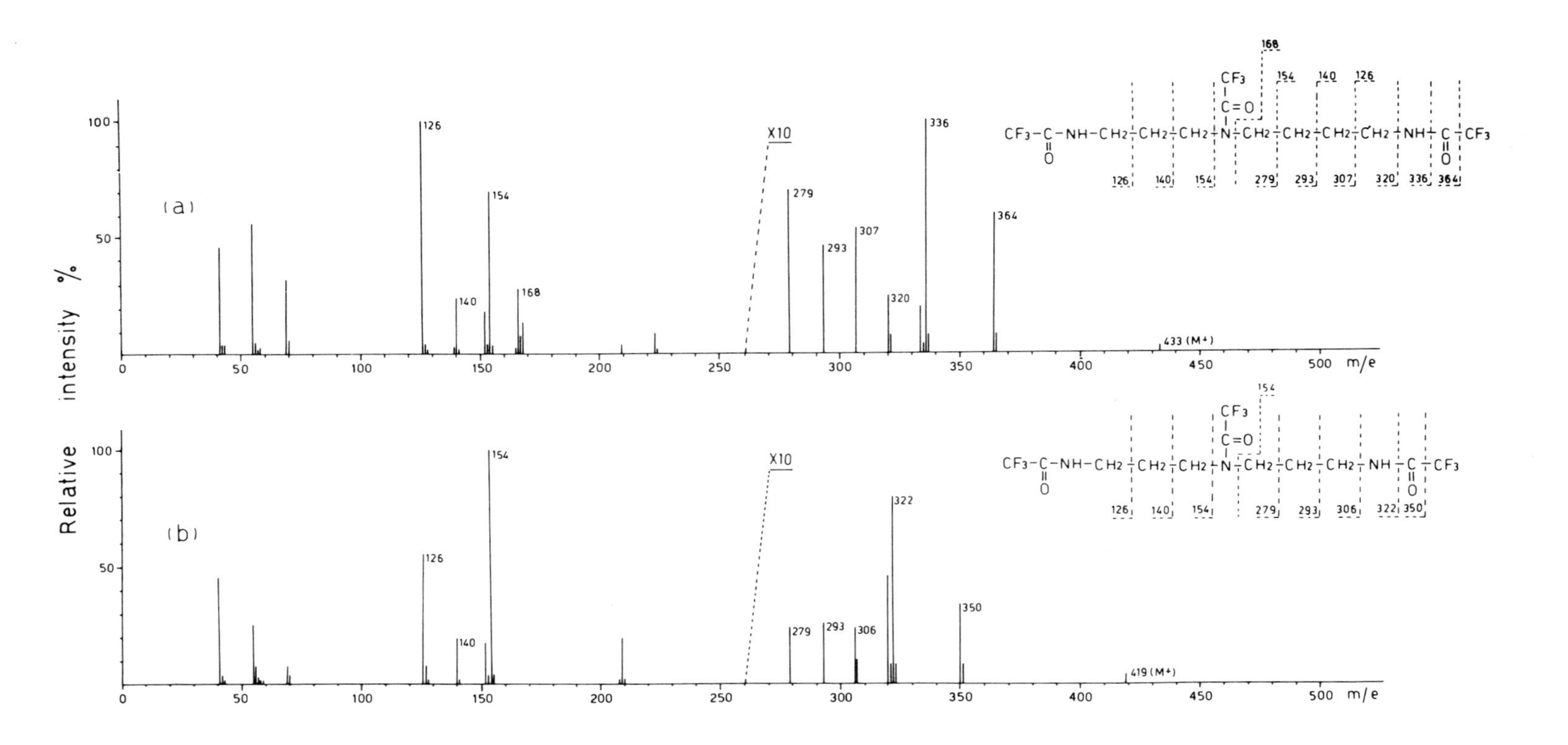

Figure 5 Normalized mass spectra of (a) trifluoracetylated spermidine and (b) sym-nor-spermidine

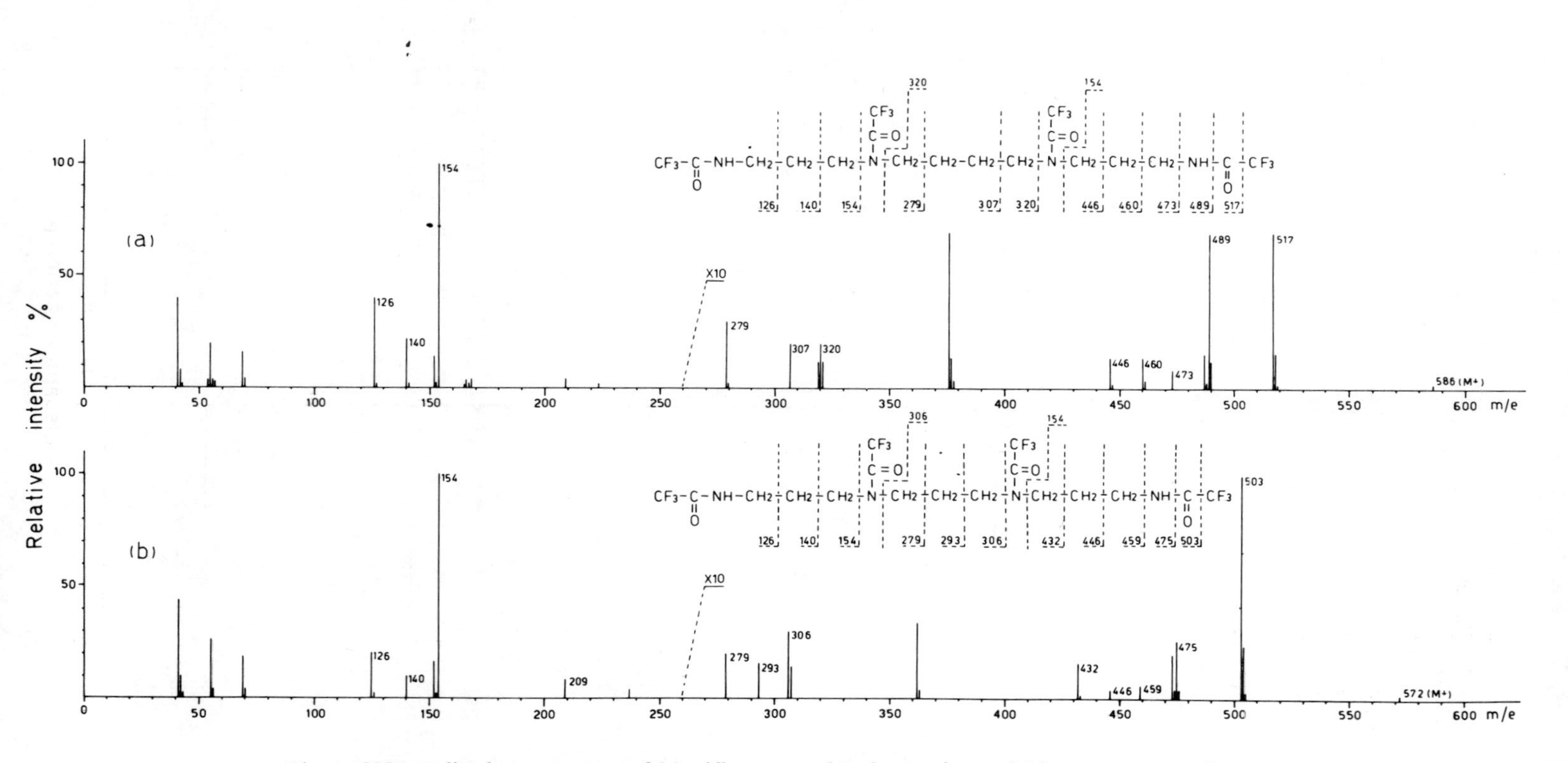

Figure 6 Normalized mass spectra of (a) trifluoroacetylated spermine and (b) sym-nor-spermine

respectively) and CF_3CO to form fragment ions of m/e 322, 336, 475 and 489 (from TFA sym-nor-spermidine, TFA spermidine, TFA sym-nor-spermine and TFA spermine, respectively). All the other fragments were originated by the cleavage of the CH_2-CH and CH-N bonds. The occurrence of the fragment at m/e 154, base peak of TFA sym-nor-spermidine spectrum, was originated by CH_2-N cleavage of iminic group at both ends of the chain. This ion unambiguously localizes the CH_2-N function in the symmetrical position of the molecule and allows its identification as 1,7-diamino-4-aza-heptane. Similarly, the pair ions at m/e 154 and 168 allow identification of spermidine as 1,8-diamino-4-aza-octane. The comparative analysis of the mass spectra of the TFA sym-nor-spermine and TFA spermine unambiguously localizes the iminic functions on the molecules. The fragmentation pattern of the two compounds are identical at low mass values (fragments at m/e 126, 140, 154 and 279), while the sym-nor-spermine fragments at m/e 306, 432, 446, 459, 475 and 503 are systematically shifted 14 mass units in the spectrum of the spermine. This clearly demonstrates that in sym-nor-spermine the two iminic functions are separated by a trimethylene chain, allowing the identification of this molecule as 1,11-diamino-4,8-diaza-undecane.

D. [^{13}C] NMR

The structure of sym-nor-spermidine, sym-nor-spermine and their homologues spermidine and spermine, has also been investigated by [^{13}C] NMR analysis.

The [^{13}C] NMR spectra of the four polyamines, reported in Figure 7, fully confirm the symmetrical structures of the new polyamines from thermophilic microorganisms.

The comparison of the spectra of sym-nor-spermidine and spermidine clearly demonstrate the symmetry of the first molecule; in fact, sym-nor-spermidine shows only three signals integrating for two carbons, while the spectrum of spermidine presents seven separate signals integrating for one carbon. Also in the case of sym-nor-spermine, the [^{13}C] NMR spectra elucidated unequivocally the structure of the molecule. The spectrum of sym-nor-spermine showed five resolved signals as its symmetrical structure requires, one of them integrating for one carbon and the others for two carbons; similarly, the spectrum of spermine showed five separate signals, all integrating for two carbons as required for the symmetrical nature of the molecule.

The assignments are based on (1) chemical shifts rules (Stothers, 1972); (2) comparison with appropriate model compounds, and (3) observed multiplicities and carbon shifts (Koch and coworkers, 1974). Comparison of water solution spectra of the polyamines and their hydrochlorides, indicates that the difference of α, γ and δ-carbon shifts between an amine and its protic salt are not relevant (Table 3). In contrast, the $\Delta\delta$ value for the β-carbon is significantly elevated

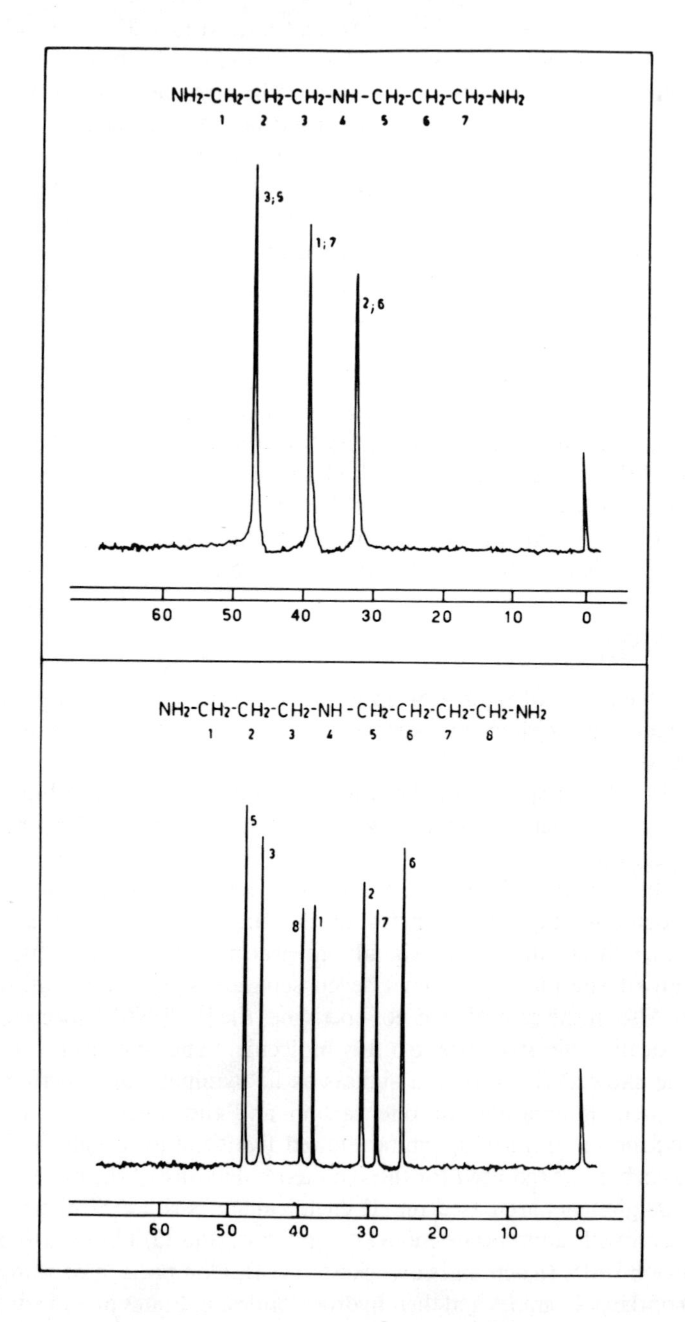

Figure 7 [^{13}C] NMR spectra of polyamines

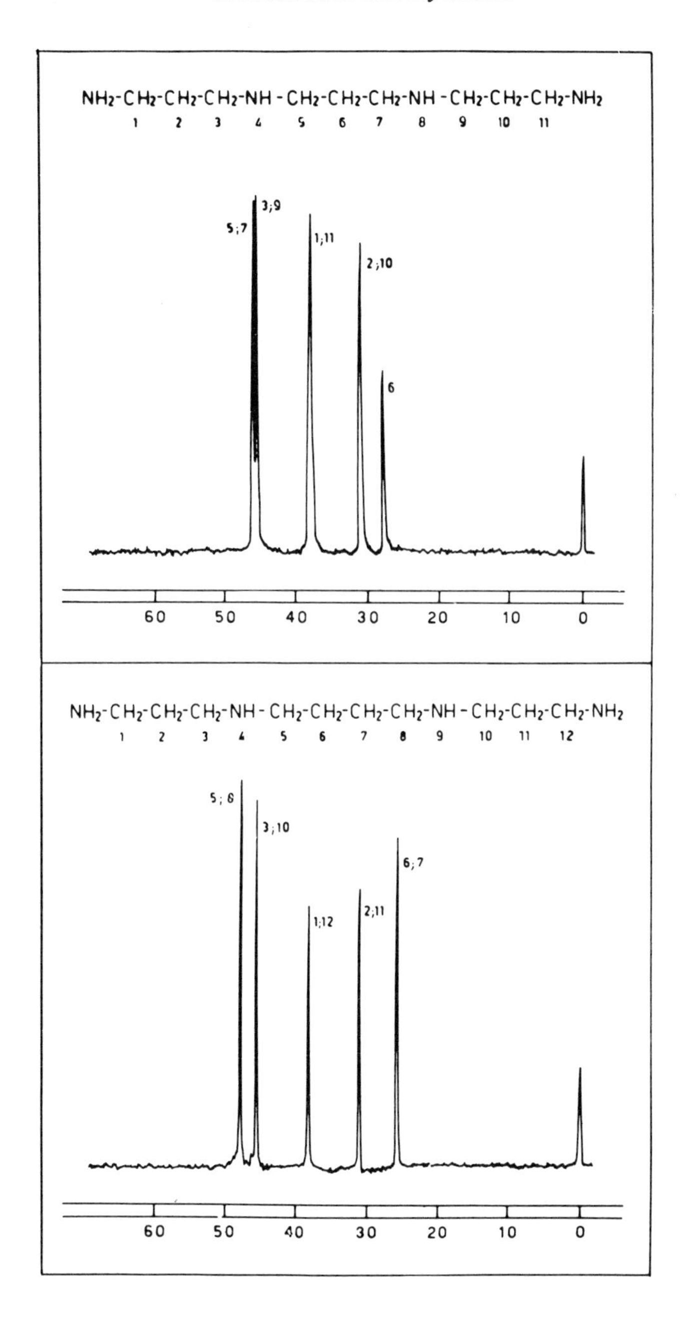
NH2-CH2-CH2-CH2-NH-CH2-CH2-CH2-NH-CH2-CH2-CH2-NH2
1 2 3 4 5 6 7 8 9 10 11
5;7
3;9
1;11
2;10
6
60 50 40 30 20 10 0
NH2-CH2-CH2-CH2-NH-CH2-CH2-CH2-CH2-NH-CH2-CH2-CH2-NH2
1 2 3 4 5 6 7 8 9 10 11 12
5;8
3;10
1;12
2;11
6;7
60 50 40 30 20 10 0

and hence diagnostically valuable in structure analysis (Koch and coworkers, 1974). From the data of Table 3 $\Delta\delta$ values of these polyamines suggest that protonation-induced shifts are additive.

E.[^{1}H]-NMR

Proton NMR spectra of sym-nor-spermidine and sym-nor-spermine are reported in Figure 8. Both spectra are characterized by two complex signals at δ1.6 and 2.7 ascribed to CH_2 and CH_2—N respecitvely and by a sharp singlet at δ1.2 ascribed to N—H. In the spectrum of sym-nor-spermidine, the signal at δ1.6 integrating for 4 protons is, as expected, a pentet with J = 1.1 and the signal at δ2.7, integrating for 8 protons, is a quartet, resulting from the partial overlapping of two equivalent triplets J = 1.0.

In the spectrum of sym-nor-spermine the signal at δ1.6 is a complex signal with 10 lines resulting from a partial overlapping of two pentets J = 1.1. The signal integrates in total for 6 protons and the pentet at low field represents one-third of the total signal; the signal at δ 2.7, integrating for 12 protons, is a quartet resulting from the partial overlapping of two triplets J = 1.0; the triplet at low field represents one-third of the total signal.

The NH signal in both spectra appears as a sharp singlet integrating in the first case for 5 protons and in the second for 6 protons. The absence of coupling between NH and protons on adjacent carbon atoms is justified by the rapid NH exchange of these aliphatic amines.

V. SUMMARY AND CONCLUSION

Polyamine distribution in bacteria appears to be related to the optimal growth temperature of microorganism; in fact, while putrescine and spermidine occur in mesophilic bacteria, spermine is the most abundant component in the moderate thermophile *Bacillus stearothermophilus* (Stevens and Morrison, 1968). Moreover, sym-nor-spermine is also present in the extreme thermophiles, where the concentrations of putrescine and spermidine are lower (Figure 9).

The distribution of the polyamines in the bacteria and the observation that tetra-amines were obligatory for the stimulation *in vitro* of protein synthesis in the extreme thermophile *Thermus thermophilus* HB 8 (Ohno-Iwashita, Oshima and Imanori, 1975), could suggest that polyamines play a role in the thermophilicity and thermostability of these microorganisms. Stevens (1967) demonstrated that sym-nor-spermine increases the 'melting temperature' of DNA from calf thymus and from *A.aerogenes*. In this respect it could be interesting to investigate the existence *in vitro* of a specific stabilizing effect of the new tetra-amine on the homologous DNA according to the DNA-polyamines model proposed by Tsuboi (1964) and Liquori and coworkers (1967).

Table 3 [^{13}C] Chemical shifts and shifts differences of polyamines

	Spermidine		Sym-nor-spermidine		Spermine		Sym-nor-spermine	
	δ	$\Delta\delta$	δ	$\Delta\delta$	δ	$\Delta\delta$	δ	$\Delta\delta$
C-1	38.2	1.9	39.3	2.6	38.0	2.0	38.1	1.5
C-2	31.0	7.8	32.0	9.0	30.8	7.4	31.1	7.8
C-3	45.5	1.4	46.9	2.4	45.3	1.3	45.8	1.5
C-5	48.0	1.4			47.6	1.5	46.1	1.8
C-6	25.7	3.6			25.6	3.6	28.0	5.9
C-7	29.8	6.6						
C-8	40.0	1.4						

δ values are from spectra of water solutions at pH 11 or higher recorded in ppm downfield from TMS; $\Delta\delta = \delta(\text{pH})\ 11) - \delta(\text{pH} < 1)$. The numbering system is indicated in Figure 7.

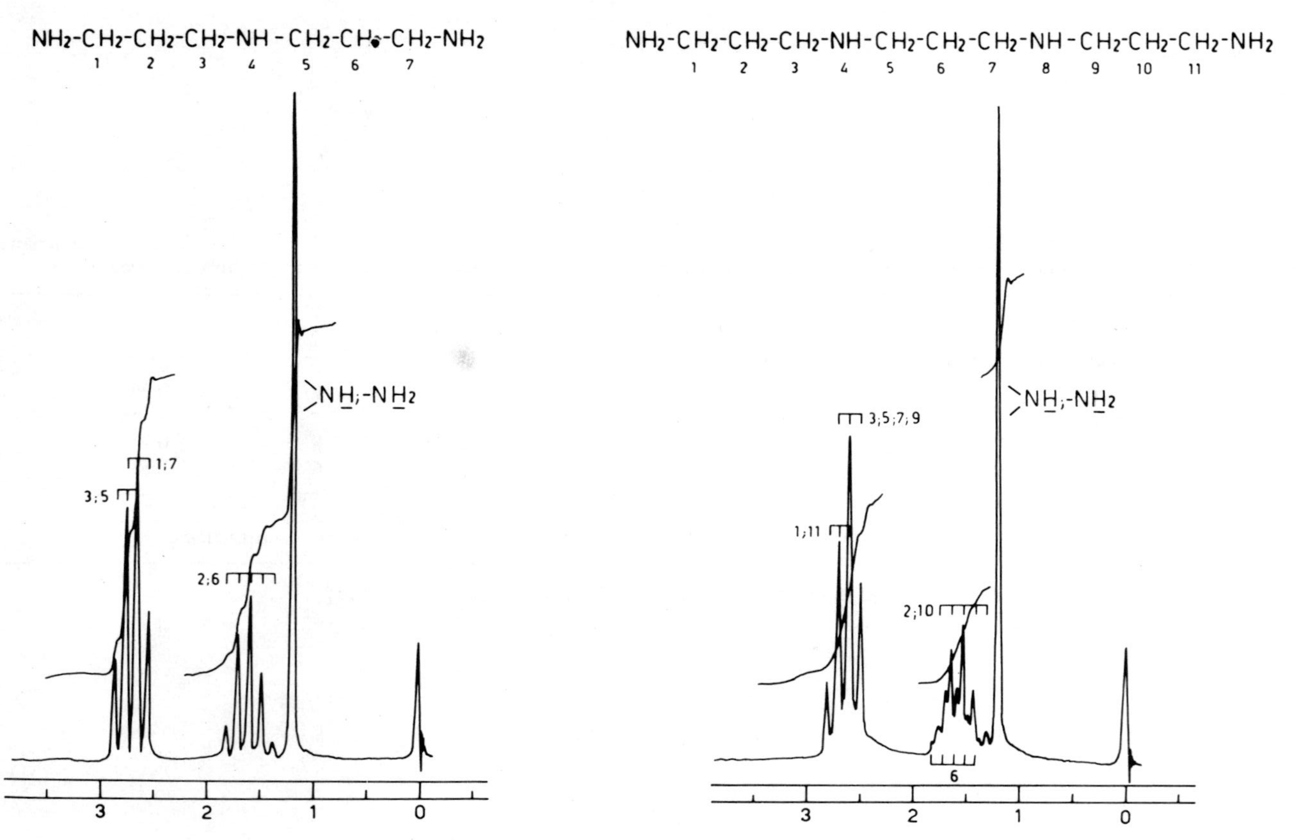

Figure 8 [^{1}H] NMR spectra of new polyamines

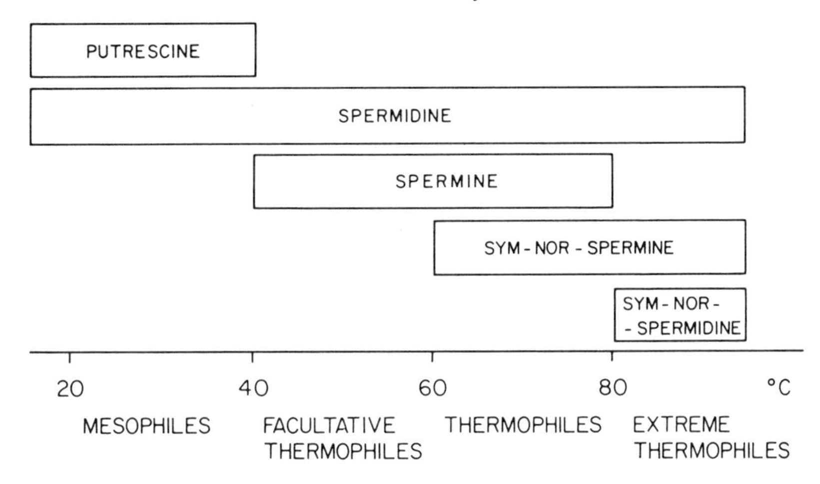

Figure 9 Occurrence of polyamines in bacteria growing at different temperatures

The recent observation on the presence of sym-nor-spermidine and sym-nor-spermine in the shrimp *Penaeus satiferus* (Stillway and Walle, 1977) adds new questions on the physiological role(s) played by these new polycations.

REFERENCES

Bachrach, U. (1973a). Occurrence of polyamines. In *Function of Naturally Occurring Polyamines*, Academic Press, New York and London, pp. 15–17.

Bachrach, U. (1973b). Polyamine and diamine conjugates. In *Function of Naturally Occurring Polyamines* Academic Press, New York and London, p. 32.

Benninati, S., Sartori, C. and Argenio-Cerù, M.P. (1977). *Anal. Biochem.*, **80**, 101–107.

Cacciapuoti, G., Oliva, A. and Zappia, V. (1978). *Int. J. Biochem.*, **9**, 35–41.

De Rosa, M., De Rosa, S., Gambacorta, A., Carteni-Farina, M. and Zappia, V. (1976). *Biochem. Biophys. Res. Comm.*, **69**, 253–261.

De Rosa, M., De Rosa, S., Gambacorta, A., Carteni-Farina, M. and Zappia, V. (1978). *Biochem. J.* (in press).

Dion, A. S. and Cohen, S. S. (1972). *Proc. Natl. Acad. Sci. USA*, **69**, 213–217.

Dubin, D. T. (1959). *Biochem. Biophys. Res. Comm.*, **1**, 262–265.

Duerre, J. A. (1962). *J. Biol. Chem.*, **237**, 3737–3741.

Ferro, A. J., Barrett, A. and Shapiro, S. K. (1976). *Biochem. Biophys. Acta*, **438**, 487–494.

Hettinger, T. P. and Craig, L. C. (1968a). *Biochemistry*, **7**, 4147–4153.

Hettinger, T. P., Kurylo-Borowska, Z. and Craig, L. C. (1968b). *Biochemistry*, **7**, 4153–4160.

Karrer, E., Bose, R. J. and Warren, R. A. J. (1973). *J. Bacteriol.*, **114**, 365–366.

Koch, K. F., Rhoades, J. A., Hagaman, E. W. and Wonkert, E. (1974). *J. Amer. Chem. Soc.*, **96**, 3300–305.

Kullnig, R., Rosano, C. L. and Hurwitz, C. (1970). *Biochem. Biophys. Res. Comm.*, **39**, 1145–1148.

Liquori, A. M., Costantino, L., Crescenzi, V., Elia, V., Gigli, E., Puliti, R., De Santis-Savino M. and Vitagliano, V. (1967). *J. Mol. Biol.*, **24**, 113–122.

Ohno-Iwashita, J., Oshima, T. and Imanori, K. (1975). The effect of polyamines on thermostability of a free cell protein synthesizing system of an extreme thermophile. In H. Zuber (Ed.), *Enzymes and Proteins from Thermophilic Microorganisms*, Birkhauser Verlag, Basel and Stuttgart, pp. 333–345.

Oshima, T. (1975). *Biochem. Biophys. Res. Comm.*, **63**, 1093–1097.

Pegg, A. E., and Williams-Ashman, H. G. (1969). *J. Biol. Chem.*, **244**, 682–693.

Porcelli, M. Della Ragione, F., Giordano, F. and Zappia, V. (1978). *Ital. J. Biochem.* (in press).

Salvatore, F., Utili, R., Zappia, V., and Shapiro, S. K. (1971). *Anal. Biochem.*, **41**, 16–22.

Schlenk, F. (1977). Recent studies of the chemical properties of adenosylmethionine. In F. Salvatore, E. Borek, V. Zappia, H. G. Williams-Ashman and F. Schlenk (Eds), *The Biochemistry of S-Adenosylmethionine*, Columbia Univ. Press, New York, pp. 3–17.

Shapiro, S. K. and Mather, A. N. (1958). *J. Biol. Chem.*, **233**, 631–633.

Smith, J. D. (1976). *Biochem. Biophys. Res. Comm.*, **73**, 7–10.

Stevens, L. and Morrison, M. R. (1968). *Biochem. J.*, **108**, 633–640.

Stevens, L. (1967). *Biochem. J.*, **103**, 811–815.

Stillway, L. W. and Walle, T. (1977). *Biochem. Biophys. Res. Comm.*, **77**, 1103–1107.

Stothers, J. B. (1972). In *Carbon-13 NMR Spectroscopy*, Academic Press, New York and London.

Tabor, H. and Tabor, C. W. (1964). *Pharmacol. Rev.*, **16**, 245–300.

Tabor, C. W. and Tabor, H. (1970). *Biochem. Biophys. Res. Comm.*, **41**, 232–238.

Tabor, H. and Tabor, C. W. (1975). *J. Biol. Chem.*, **250**, 2648–2654.

Tabor, H. and Tabor, C. W. (1976). *Ann. Rev. Biochem.*, **45**, 285–306.

Tobari, J. and Tchen, T. T. (1971). *J. Biol. Chem.*, **246**, 1262–1265.

Tsuboi, M. (1964). *Bull. Chem. Soc. Japan*, **37**, 1514–1522.

Zappia, V., Zydek-Cwich, C. R. and Schlenk, F. (1969). *J. Biol. Chem.*, **244**, 4499–4509.

Zappia, V., Carteni-Farina, M. and Galletti, P. (1977a). Adenosyl-methionine and polyamine biosynthesis in human prostate. In F. Salvatore, E. Borek, V. Zappia, H. G. Williams-Ashman and F. Schlenk, *The Biochemistry of Adenosylmethionine*, Columbia Univ. Press, New York, pp. 471–492.

Zappia, V., Galletti, P., Oliva, A. and De Santis, A. (1977b). *Anal. Biochem.*, **79**, 535–543.

Zappia, V., Carteni-Farina, M., Oliva, A., De Rosa, M., De Rosa, S. and Gambacorta, A. (1977c). *Abstr. 3rd SIB Meet.*, Siena n.C38.

Zappia, V., Oliva, A., Cacciapuoti, G., Galletti, P., Mignucci, G. and Carteni-Farina, M. (1978). *Biochem. J.* (in press).

Chapter 17

Diamine Oxidase

ANNE-CHARLOTTE ANDERSSON, STIG HENNINGSSON
AND ELSA ROSENGREN

I. INTRODUCTION

The presence in mammalian tissue of an enzyme inactivating a diamine was first pointed out by Best (1929). The enzyme was thought to attack principally histamine and for that reason it was named 'histaminase'. Later it was shown by Zeller (1938) that several diamines including putrescine, cadaverine and agmatine are also catabolized by this enzyme. At present, most authors consider histaminase identical with diamine oxidase (DAO). It has been shown that putrescine and cadaverine are the preferred substrates (Zeller, 1965; Bardsley, Hill and Lobley, 1970), so the enzyme is consequently proposed to be called diamine oxidase. This enzyme is widely distributed in animal tissues.

Diamine oxidase (EC 1.4.3.6) catalyzes the oxidation of a diamine yielding in the process aminoaldehyde, hydrogen peroxide and ammonia as products. According to Zeller (1938) the enzyme controls the following transformation:

$$RCH_2NH_2 + O_2 + H_2O \rightarrow RCHO + NH_3 + H_2O_2$$

The main emphasis in the present survey is on the role of DAO in the metabolism of putrescine. Excellent reviews dealing with different aspects of DAO have been published (Zeller, 1963; Buffoni, 1966; Kapeller-Adler, 1970). Diamine oxidase might merely be of importance in the metabolism of endogenously formed cadaverine. Our knowledge of cadaverine function, however, in higher vertebrates is scanty. Until recently nothing had been disclosed on the formation of cadaverine in mammalian tissues. Its formation from lysine was shown to occur in the kidney of mice after the injection of an anabolic steroid (Henningsson, Persson and Rosengren, 1976; Persson, 1977).

II. CATABOLISM OF PUTRESCINE

It has been proposed that putrescine formation is the rate-limiting step in the synthesis of polyamines (Jänne and Raina, 1968; Williams-Ashman, Pegg and Lockwood, 1969). The oxidation of putrescine in animal tissues has been much less investigated. It is well established, however, that when putrescine is incubated in the presence of purified DAO γ-aminobutyraldehyde is formed which almost quantitatively is non-enzymatically transformed into the cyclic compound Δ^1-pyrroline. At steady-state the two compounds exist in equilibrium (Jakoby and Fredericks, 1959). Besides, both *in vitro* and *in vivo* studies have recently shown that tissues are readily able to transfer γ-aminobutyraldehyde to γ-aminobutyric acid (GABA), a reaction which is catalyzed by γ-aminobutyraldehyde dehydrogenase.

GABA has been shown to be a constituent of non-neural tissue (Zachmann, Tocci and Nyhan, 1966; Whelan, Scriver and Mohyuddin, 1969). The kidney is the only non-neural organ in which its concentration is considerable (Lancaster and coworkers, 1973). GABA is alleged to stimulate protein synthesis (Baxter, 1970). This is of particular interest, since several groups of workers have suggested that polyamines are implicated in growth processes and cell proliferation (Chapters 1–10, 22). Thus GABA formed from putrescine may have a similar regulatory function in growth processes.

In vivo studies with [^{14}C] GABA have shown that it enters the tricarboxylic acid cycle and is rapidly degraded to [^{14}C]CO_2 (Roberts, 1956; Wilson, Hill and Koeppe, 1959). Jänne (1967) was first to show that the injection of [^{14}C] putrescine is followed by expiration of large amounts of radioactive CO_2 within a couple of hours. Later it was found that GABA is an intermediate in the catabolism of putrescine to carbon dioxide (Seiler and Eichentopf, 1975; Henningsson and Rosengren, 1976). The catabolism of [^{14}C]putrescine to radioactive GABA and CO_2 was blocked by low concentrations of aminoguanidine, a specific DAO inhibitor. Thus, a main route in the catabolism of putrescine seems to be a multi-step sequence involving several enzymes, the first two steps of the route being catalyzed by DAO and γ-aminobutyraldehyde dehydrogenase.

In our studies, administration of nandrolone which is an anabolic steroid with low androgenic activity, produced a thousand-fold elevation of putrescine formation in the kidneys of gonadectomized mice. The elevated putrescine formation was reflected in a striking increase in the urinary excretion of free putrescine (Henningsson and Rosengren, 1976). The increased formation of putrescine after the administration of nandrolone led us to investigate the catabolic fate of [^{14}C] putrescine, particularly in the kidney when stimulated to growth by the administration of anabolic steroid (Henningsson and Rosengren, 1976). The results of the determinations of [^{14}C] compounds found in the kidneys of control and hormone treated animals 1 h and 24 h after the injection of [^{14}C] putrescine, are presented in Figure 1. From the Figure it is obvious that only a small fraction of the administered radioactivity was recovered in the kidneys. Of the radioactivity observed in the kidneys 1 h after the injection a large part (40%) of it appeared as unchanged [^{14}C] putrescine. After 24 h this portion was reduced to a few per cent. A major part of the radioactivity was found incorporated into spermidine both 1 h and 24 h after the injection. In the kidneys of mice receiving nandrolone the incorporation of radioactivity into spermidine was significantly increased at 1 h and at this time considerable amounts of radioactive carbon were recovered as GABA.

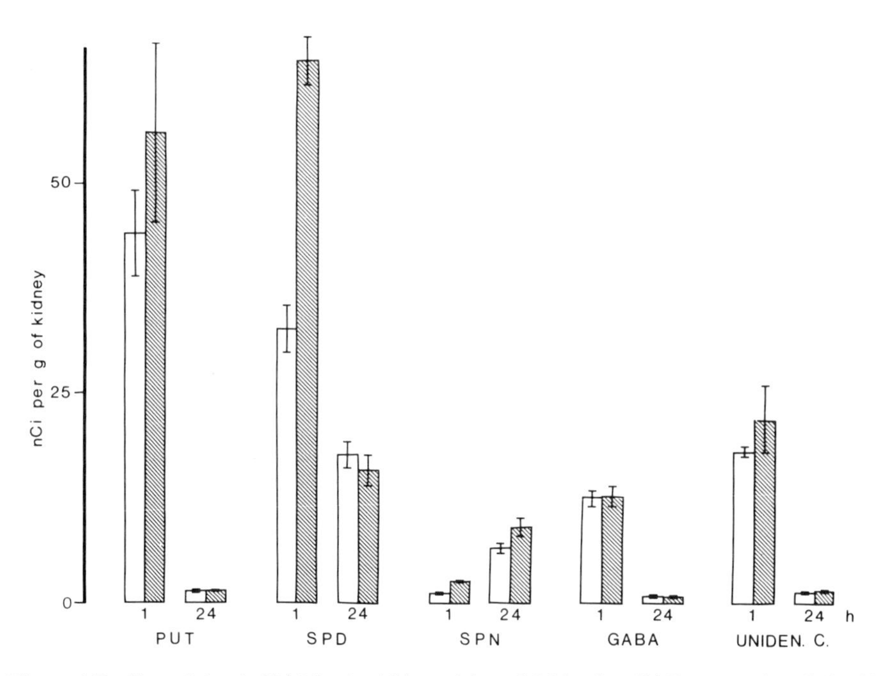

Figure 1 Radioactivity (nCi/g) in the kidney 1 h and 24 h after [^{14}C] putrescine (2.5 μCi s.c.). Open columns: controls; hatched columns: mice injected with nandrolone (0.1 mg daily for 3 days). PUT = putrescine, SPD = spermidine, SPN = spermine, GABA = γ-aminobutyric acid, UNIDEN. C. = unidentified compound. Each column is the mean ± s.e. mean of 4 observations. Total kidney weights: controls 260 mg, nandrolone-treated animals 360 mg; no significant difference in body weight (Henningsson and Rosengren, 1976). Reproduced by permission of the British Pharmacological Society

In our experiments with [^{14}C] putrescine injections to mice, we used the ion-exchange column of an automatic amino acid analyzer in order to separate the catabolic products of putrescine. In addition to the above mentioned radioactive compounds found in the kidneys, there was an unidentified radioactive compound. The unidentified compound appeared before GABA in the eluate and could be separated.

A large proportion of injected putrescine is excreted as CO_2 in the expired air. Figure 2 shows the time course of the occurrence of [^{14}C]CO_2 in the expired air during the 2 h period following the injection of [^{14}C] putrescine. Almost 50% of the injected radioactivity was recovered in the expired air within 2 h as radio-labelled CO_2. Thus, the putrescine was rapidly metabolized. Pretreatment of the animals with aminoguanidine half-an-hour before the [^{14}C] putrescine injection inhibited the expiration of labelled CO_2 almost completely.

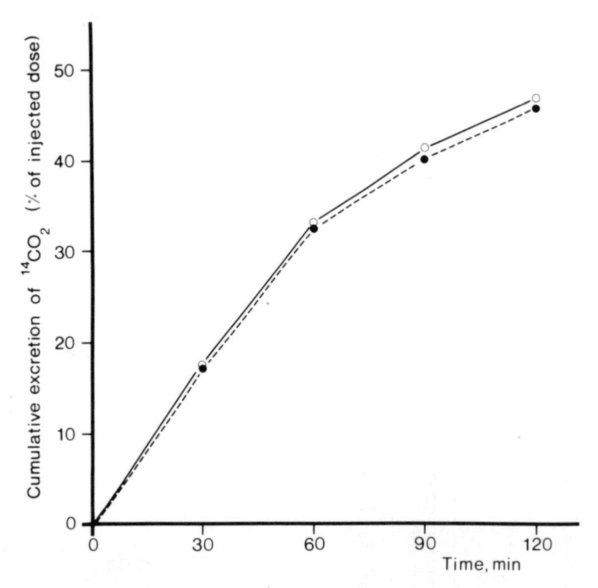

Figure 2 Cumulative radio-labelled CO_2 excretion following injection of [^{14}C] putrescine in mice injected with nandrolone (0.1 mg daily for 3 days; ●) and in controls (○). Mean values of 3 experiments (Henningsson and Rosengren, 1976). Reproduced by permission of the British Pharmacological Society

Furthermore, in the mice in which kidney [^{14}C] putrescine metabolites were determined, urine was collected and its [^{14}C] compounds were examined (Table 1). Significant amounts of the injected radioactivity were incorporated into GABA. Some unidentified carbon-labelled compound or compounds were

Table 1. Radioactivity (nCi/24 h) in urine after injection of [^{14}C]putrescine (2.5 μCi s.c.) in 6 controls and 6 mice injected with nandrolone (0.1 mg daily for 3 days). Mean $\pm$ s.e. mean and the degrees of significance between nandrolone-treated animals and controls are given. (Henningsson and Rosengren, 1976). Reproduced by permission of the British Pharmacological Society

	Total radioactivity	Putrescine	Spermidine	Spermine	GABA	Unidentified compound
Controls	490.8 ± 118.05	300.5 ± 94.33	2.5 ± 0.38	0.4 ± 0.12	17.2 ± 2.65	128.2 ± 13.11
Nandrolone	975.4 ± 167.75*	694.9 ± 126.70*	8.0 ± 1.00**	0.8 ± 0.20	21.8 ± 0.90	178.4 ± 22.23

* <0.05, ** <0.01.

present also in the urine and appeared in the same fractions of the column eluate in which we had observed the unidentified radioactive substance in the kidney extracts. The amount of this substance in the urine was almost 10-fold that of GABA, about 50-fold that of spermidine and near 300-fold that of spermine. The synthesis of radioactive GABA and the unidentified compound from [^{14}C] putrescine were inhibited by aminoguanidine (unpublished). These findings support the view of GABA being a major intermediate in the catabolism of putrescine to CO_2 and show the importance of DAO in this catabolism.

The reported findings on putrescine catabolism became the basis of the development of a new approach to the assay of DAO activity in tissues taking into consideration the main products formed in the incubate, thus applying our *in vivo* findings of putrescine catabolism on *in vitro* measurements (Andersson and coworkers, 1978). Homogenates of guinea-pig liver and human placenta, tissues known to be rich in DAO, were incubated with [^{14}C] putrescine and the metabolites formed were determined. In the incubate of guinea-pig liver, the major metabolites were GABA and some unidentified material; Δ^1-pyrroline and CO_2 were also obtained. In both the maternal and fetal parts of human placenta, radioactive GABA and the unidentified material as well as Δ^1-pyrroline were found. The results indicated that GABA was an important intermediate from putrescine catabolism. The amount of Δ^1-pyrroline formed in guinea-pig liver homogenate was found unrelated to the amount of tissue and the time of incubation; this is in contrast to the production of GABA and the unknown compound(s) which was approximately proportional to the tissue amount and the time of incubation (Figure 3). In both supernatant and homogenate of guinea-pig liver, Δ^1-pyrroline formation represented only a tiny fraction of the total amount of metabolites.

The most commonly used method for *in vitro* measurements of DAO activity is based upon the determination of one product, i.e. Δ^1-pyrroline, formed from [^{14}C] putrescine (Okuyama and Kobayashi, 1961). It would thus appear that this method should not be used indiscriminately to measure the enzyme activity in tissues.

The monoacetyl derivative of putrescine has been found low in concentrations in various tissues, human brain (Perry, Hansen and MacDougall, 1967) and mouse brain, liver, kidney and spleen (Seiler, Al-Therib and Knödgen, 1973). Monoacetyl derivative of putrescine has also been recovered from human urine (Perry, Hansen and MacDougall, 1967). Monoacetylputrescine can be oxidized to N-acetyl-γ-aminobutyrate which is then converted to GABA in the presence of monoamine oxidase. It is necessary to elucidate the significance of the pathway of putrescine to GABA via acetylation and oxidative deamination by monoamine oxidase. Studies based on experiments using inhibitors of DAO and monoamine oxidase indicate that the direct pathway catalyzed by DAO is quantitatively the most important. For this reason the main metabolic pathway of putrescine is purported to be:

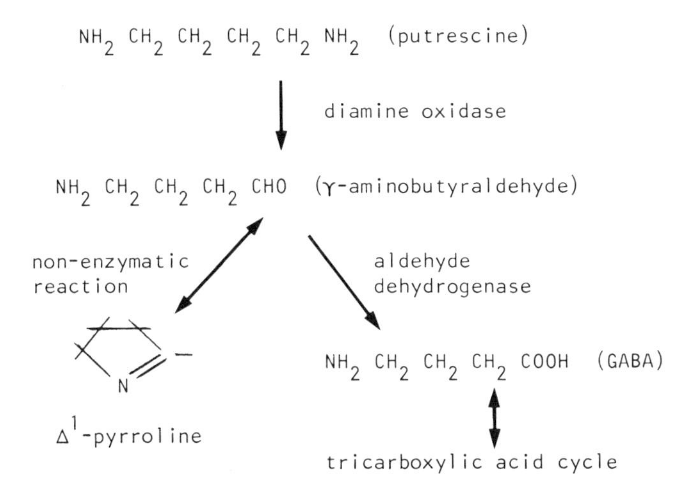

Scheme of putrescine metabolism via DAO

Methodological Considerations

Automated liquid chromatography appears to be very useful in research on diamine metabolism (Henningsson, 1978). We have combined this procedure with the split-stream technique as described by Lou (1973). It was thereby possible to analyze simultaneously radioactive metabolites in physiological fluids or tissue extracts and to determine the non-labelled amine content in the same tissue sample. In addition, it appeared possible to discover and identify hitherto unknown compounds in biological samples. An example of the merit of this technique was the detection and identification of radioactive GABA in the kidney and urine of mice after administration of [^{14}C] putrescine (Henningsson and Rosengren, 1976).

III. SUMMARY AND CONCLUSIONS

With regard to the postulated role of polyamines in the synthesis of macromolecules and cell growth, it is most interesting that even DAO activity is altered in several forms of rapid growth. For example, in human pregnancy plasma DAO activity attains values about 1,000-fold the average for non-pregnant women (Ahlmark, 1944). Large increases are also detected in the placental and serum DAO activity of the pregnant rat (Roberts and Robson, 1953; Kobayashi, 1964). Plasma DAO activity can be raised by injection of progesterone (Swanberg, 1950). DAO activity has also been reported to be

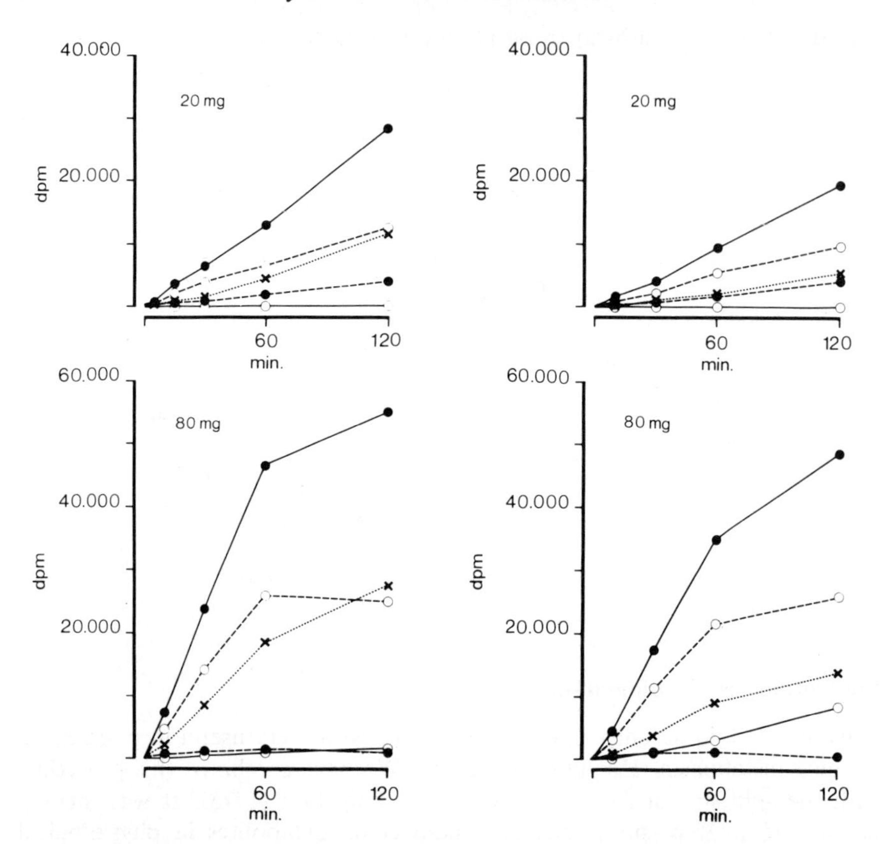

Figure 3 Rate of formation of different metabolites from [^{14}C] putrescine in guinea-pig liver as related to the time of incubation. The metabolites were CO_2 (○———○), Δ^1-pyrroline (●– – – – ●), GABA (○– – – – ○) and unknown compound(s) (× · · · · · ×). The total amount of these metabolites is also given (● ——— ●). (Andersson and coworkers, 1978). Reproduced by permission of *Acta Physiologica Scandinavica*

associated with various types of cancer. Borglin and Willert (1962) showed that the enzyme activity was elevated in plasma of patients with endometrial adenocarcinoma, uterine myosarcoma and granulosa cell carcinoma. Later, Baylin and coworkers (1970) reported a significant elevation of the enzyme activity in medullary carcinoma of the thyroid gland and in the serum of patients with this tumor. More recently elevated DAO activity has been reported in ovarian cancer (Lin, Orcutt and Stolbach, 1975) and in small-cell carcinoma of the lung (Baylin and coworkers, 1975).

Correlations of changes in levels of polyamine synthesis with levels of DAO activity during stages of cell growth might provide clues to the role of the enzyme in certain neoplasms and clarify the possibility of utilizing its measurement for cancer diagnosis and clinical evaluation of tumor progression and regression during therapy.

Ornithine decarboxylase appears to be the key regulatory enzyme in the synthesis of the polyamines (see Chapters 6–10). It has the shortest half-life of any enzyme so far reported. Its activity is rapidly increased, e.g. by growth hormone or partial hepatectomy. The cellular concentration of ornithine is regulated by ornithine-keto acid aminotransferase activity (Räihä and Kekomäki, 1968). An additional point of control in the putrescine level might reside in the activity of DAO.

The pathway of oxidative deamination of putrescine in animal tissues is not as well established as the voluminous literature on DAO might lead one to suppose. Some conclusions may be drawn from the recent work reviewed above. Investigations in rats, mice and guinea-pigs have shown that the catabolism of putrescine is very rapid. Our experiments suggest that the direct pathway via DAO plays the major role in the catabolism of putrescine in the body, although acetylation of putrescine followed by oxidative deamination by monoamine oxidase may be important in some tissues.

Decarboxylation of glutamic acid has up to recently been looked upon as the only reaction responsible for the formation of GABA in mammalian tissue (Baxter, 1970). Our results together with those of Seiler's group show that GABA is an important intermediate in the catabolism of putrescine to carbon dioxide via a direct pathway which does not include glutamic acid as an intermediate (Seiler and Eichentopf, 1975; Henningsson and Rosengren, 1976; Andersson and coworkers, 1978). Very little is known regarding the physiological significance of the formation of GABA from putrescine. It is interesting that GABA stimulates protein synthesis in the brain (Tewari and Baxter, 1969; Sandoval and Tapia, 1975). This is of particular interest with regard to the postulated role of polyamines in growth processes. Thus, GABA and polyamines might have two important properties in common. Both appear to prevent the larger polysomes from breaking up into monomers. Both GABA and polyamines also stimulate aminoacyl-tRNA synthetases (Baxter, 1976). In this context it should be mentioned that in preliminary experiments we have been able to show an increased formation of radioactive GABA from 5-[^{14}C] ornithine via putrescine in the mouse kidney stimulated to growth by the anabolic steroid nandrolone (unpublished).

Determination of the rate of polyamine excretion has been suggested as a test in the early detection of cancer (Russell and coworkers, 1971). Our study showed that after the administration of radioactive putrescine more urinary radioactive carbon was found in the fractions of GABA and in some unidentified compound(s) than in the polyamine fractions, i.e. spermidine and spermine, which constituted a small fraction of the excreted compounds. This demonstrates the relative importance of the different pathways in putrescine metabolism. If it were possible to determine endogenous amounts of the yet unidentified putrescine metabolite the determination of this compound in the urine might be a more sensitive tool for revealing activity in malignancy.

ACKNOWLEDGEMENTS

This work was supported by grants from the Swedish Medical Research Council, the Faculty of Medicine, University of Lund and from the Foundation of John and Augusta Persson.

REFERENCES

Ahlmark, A. (1944). *Acta Physiol. Scand.*, **9**, *suppl. 28*, 1–107.

Andersson, A.-Ch., Henningsson, S., Persson, L., and Rosengren, E. (1978). *Acta Physiol. Scand.*, **102**, 159–166.

Bardsley, W. G., Hill, C. M., and Lobley, R. W. (1970). *Biochem. J.*, **117**, 169–176.

Baxter, C. F. (1970). The nature of γ-aminobutyric acid. In A. Lajtha (Ed.), *Handbook of Neurochemistry*, Vol. 3. Plenum Press, New York, pp. 289–353.

Baxter, C. F. (1976). Effect of GABA on protein metabolism in the nervous system. In E. Roberts, T. N. Chase and D. B. Tower (Eds), *GABA in Nervous System Function*, Raven Press, New York, pp. 89–102.

Baylin, S. B., Abeloff, M. D., Wieman, K. C., Tomford, J. W., Ettinger, D. S. (1975). *New Engl. J. Med.*, **293**, 1286–1290.

Baylin, S. B., Beaven, M. A., Engelman, K., and Sjoerdsma, A. (1970). *New Engl. J. Med.*, **283**, 1239–1244.

Best, C. H. (1929). *J. Physiol.*, **67**, 256–263.

Borglin, N. E., and Willert, B. (1962). *Cancer*, **15**, 271–275.

Buffoni, F. (1966). *Pharmacol. Rev.*, **18**, 1163–1199.

Henningsson, S. (1978). *Agents and Actions*, **8**, 392–393.

Henningsson, S., Persson, L., and Rosengren, E. (1976). *Acta Physiol. Scand.*, **98**, 445–449.

Henningsson, S., and Rosengren, E. (1976). *Br. J. Pharmacol.*, **58**, 401–406.

Jakoby, W. B., and Fredericks, J. (1959). *J. Biol. Chem.*, **234**, 2145–2150.

Jänne, J. (1967). *Acta Physiol. Scand.*, *Suppl.* **300**, 1–71.

Jänne, J., and Raina, A. (1968). *Acta Chem. Scand.*, **22**, 1349–1351.

Kapeller-Adler, R. (1970). *Amine Oxidases and Methods for Their Study*, Wiley, New York.

Kobayashi, Y. (1964). *Nature, Lond.*, **203**, 146–147.

Lancaster, G., Mohyuddin, F., Scriver, C. R. and Whelan, D. T. (1973). *Biochim. Biophys. Acta*, **297**, 229–240.

Lin, C.-W., Orcutt, M. L., and Stolbach, L. L. (1975). *Cancer Res.*, **35**, 2762–2765.

Lou, M. F. (1973). *Analyt. Biochem.*, **55**, 51–56.

Okuyama, T., and Kobayashi, Y. (1961). *Archs Biochem. Biophys.*, **95**, 242–250.

Perry, T. L. Hansen, S., and MacDougall, L. (1967). *J. Neurochem.*, **14**, 775–782.

Persson, L. (1977). *Acta Physiol. Scand.*, **100**, 424–429.

Räihä, N. C. R., and Kekomäki, M. P. (1968). *Biochem. J.*, **108**, 521–525.

Roberts, E. (1956). Formation and utilization of γ-aminobutyric acid. In S. A. Korey and J. I. Nurnberger (Eds), *Progress in Neurobiology: I. Neurochemistry*, Hoeber, New York, pp. 11–25.

Roberts, M., and Robson, J. M. (1953). *J. Physiol.*, **119**, 286–291.

Russell, D. H., Levy, C. C., Schimpff, S. C., and Hawk, I. A. (1971). *Cancer Res.*, **31**, 1555–1558.

Sandoval, M-E., and Tapia, R. (1975). *Brain Res.*, **96**, 279–286.

Seiler, N., Al-Therib, M. J. and Knödgen, B. (1973). *Hoppe-Seyler's Z. Physiol. Chem.*, **354**, 589–590.

Seiler, N., and Eichentopf, B. (1975). *Biochem. J.* **152**, 201–210.

Swanberg, H. (1950). *Acta Physiol. Scand.*, **23**, *Suppl. 79*, 1–69.

Tewari, S., and Baxter, C. F. (1969). *J. Neurochem.*, **16**, 171–180.

Whelan, D. T., Scriver, C. R., and Muhyuddin, F. (1969). *Nature, Lond.*, **224**, 916–917.

Williams-Ashman, H. G., Pegg, A. E., and Lockwood, D. H. (1969). Mechanisms and regulation of polyamine and putrescine biosynthesis in male genital glands and other tissues of mammals. In G. Weber (Ed), *Advances in Enzyme Regul.*, Vol. 7, Pergamon Press, New York, pp. 291–323.

Wilson, W. E., Hill, R. J., and Koeppe, R. E. (1959). *J. Biol. Chem.*, **234**, 347–349.

Zachmann, M., Tocci, P., and Nyhan, W. L. (1966). *J. Biol. Chem.*, **241**, 1355–1358.

Zeller, E. A. (1938). *Helv. Chim. Acta*, **21**, 880–890.

Zeller, E. A. (1963). Diamine oxidases. In P. D. Boyer, H. Lardy and K. Myrbäck (Eds), *The Enzymes*, 2nd edn, Academic Press, New York, pp. 315–335.

Zeller, E. A. (1965). *Fed. Proc.*, **24**, 766–768.

Chapter 18

Polyamine Oxidases

D. M. L. MORGAN

I. INTRODUCTION

The definition and classification of amine oxidases in the literature is somewhat confused. The original division into two groups, mono-, and diamine oxidases, was based on substrate specificity (Zeller, Stern and Wenk, 1940). It soon became apparent (Blaschko and Duthie, 1945; Blaschko, 1952) that some monoamine oxidases are capable of oxidizing long-chain aliphatic diamines which are not attacked by diamine oxidases. These observations, together with many other studies of the substrate and inhibitor specificities of these enzymes (for reviews see Zeller, 1963; Blaschko, 1963; Kapeller-Adler, 1970 and Chapter 17) led to the classification put forward by Zeller (1963) and adopted by Malmström, Andreason and Reinhammer (1975):

> 'Diamine oxidases, in contrast to monoamine oxidases, do not oxidise secondary amines and are not inhibited by the potent monoamine oxidase inhibitor 2-phenylcylopropylamine.
>
> Monoamine oxidases are not inhibited by hydrazine and unsubstituted acylhydrazides while the diamine oxidases are strongly inhibited by hydrazine and semicarbazide.'

Zeller goes on to make the point that 'This clear distinction should not be obscured by arbitrary nomenclature . . . and by quoting data without making this differentiation between the two enzymes.' Unfortunately this advice has not always been followed.

The current *Recommendations of the Commission on Biochemical Nomenclature* on the nomenclature and classification of enzymes (International Union of Biochemistry, 1973) lists two amine oxidases. EC 1.4.3.4 amine: oxygen oxidoreductase (deaminating) (flavin-containing) for which the recommended name is amine oxidase (flavin-containing) and EC 1.4.3.6 amine: oxygen oxidoreductase (deaminating) (pyridoxal-containing) or amine oxidase (pyridoxal-containing). The first is defined as a flavoprotein acting on primary, secondary and tertiary amines; the second as a group of pyridoxal-phosphate-proteins containing copper that oxidize primary monoamines and diamines, including histamine. These enzymes were previously known as monoamine oxidase and diamine oxidase respectively. In both cases the enzyme is considered to act on a $—CH_2—NH_2$ group of the donor and oxygen is the acceptor. EC 1.5.3.3 spermine oxidase is now deleted.

For the purposes of this review amine oxidases able to catalyse the oxidative deamination of the polyamines spermine and spermidine will be classed as *polyamine oxidases*, this, irrespective of whether or not they also act on mono- or diamines.

II. PLASMA POLYAMINE OXIDASES

The presence of an enzyme in sheep and bovine sera which brought about a rapid oxidative deamination of spermine and spermidine was first reported by Hirsch (1953), who estimated the activity of the enzyme by monitoring oxygen uptake and ammonia production. The rate of oxidative attack on these two polyamines was at least ten-fold higher than on any of the other amines tested. The absence of significant oxidation of typical monoamines (tyramine, 1-propylamine) or typical diamines (putrescine, histamine) as shown by manometry, clearly distinguished this enzyme from the previously reported monoamine and diamine oxidases. This enzyme has been purified (Tabor, Tabor and Rosenthal, 1954; Gorkin and coworkers, 1962; Sakamoto, Ogawa and Hayashi, 1963; Unemoto, 1963) and crystallized from bovine plasma (Yamada and Yasunobu, 1962a), and has been shown to act on spermine and spermidine as follows (Tabor, Tabor and Rosenthal, 1954; Tabor, Tabor and Bachrach, 1964):

$NH_2(CH_2)_3NH_2(CH_2)_4NH_2(CH_2)_3NH_2 + 2O_2 \longrightarrow$
spermine
$OCH(CH_2)_2NH(CH_2)_4NH_2(CH_2)_2CHO + 2NH_3 + 2H_2O_2,$
N,N′-*bis*(3-propionaldehyde)-
1,4-diaminobutane

$$\underset{\text{spermidine}}{NH_2(CH_2)_4NH(CH_2)_3NH_2} + O_2 \rightarrow \underset{\text{N-(4-aminobutyl)-3-aminopropionaldehyde}}{NH_2(CH_2)_4NH(CH_2)_2CHO} + NH_3 + H_2O_2.$$

The aldehydes produced were characterized by reduction with sodium borohydride and comparison with the corresponding synthetic amino alcohols.

The enzyme is inhibited by isoniazid (4-pyridinecarboxylic acid hydrazide (1), iproniazid (4-pyridinecarboxylic acid 2-(1-methylethyl) hydrazide (2), cyanide, and carbonyl binding agents such as semicarbazide and hydroxylamine (Tabor, Tabor and Rosenthal, 1954). Other compounds found to be inhibitory, as measured by oxygen consumption, include Benadryl (2-diphenylmethoxy-N,N-dimethylethanamine (3), Dibenamine (N-(2-chloroethyl) dibenzylamine hydrochloride (4), and Pyribenzamine (N,N-dimethyl-N′-(phenymethyl)-N′-2pyridinyl-1,2-ethane diamine, (5). Approximately 90% inhibition was observed at concentrations of 10^{-3}—10^{-5} M.

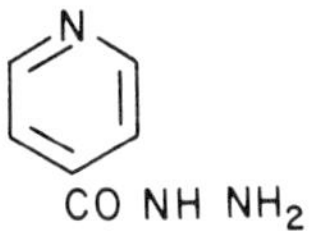

CO NH NH_2

Isoniazid (1)

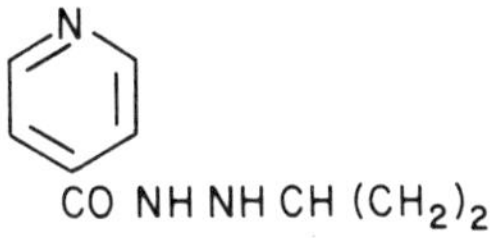

CO NH NH CH $(CH_2)_2$

Iproniazid (2)

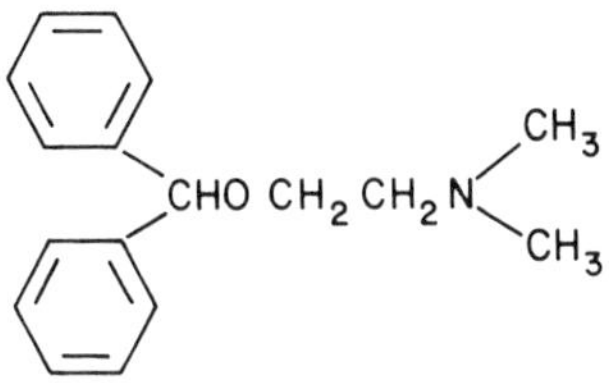

CHO CH_2 CH_2 N(CH_3)(CH_3)

Benadryl (3)

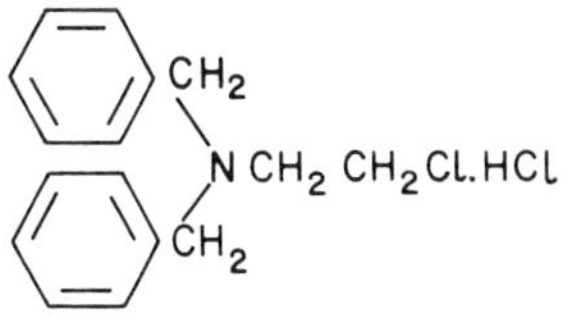

CH_2 / N CH_2 CH_2CL.HCL / CH_2

Dibenamine (4)

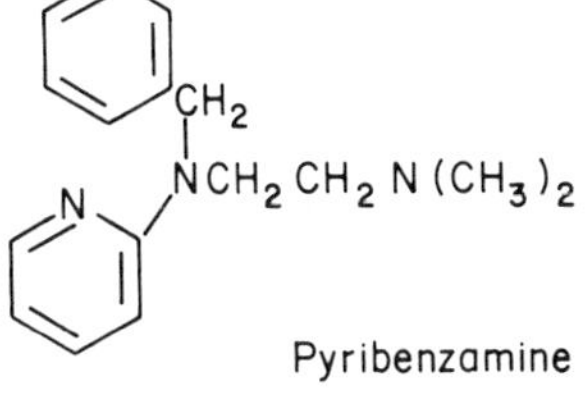

CH_2 / N CH_2 CH_2 N $(CH_3)_2$

Pyribenzamine (5)

The molecular weight of the enzyme is 170,000 daltons (Achee and coworkers, 1968), it has 1-2 atoms of copper per mole, which are essential for activity, and an incompletely defined co-factor that is probably pyridoxal phosphate (Yasunobu and coworkers, 1968; Yasunobu and Smith, 1971; but see also Wanatabe and coworkers, 1972); it does not contain flavin.

The enzyme appears to consist of two subunits of 85,000 daltons linked by a disulphide bridge. The copper is apparently not chelated to the sulphydryl groups, of which there are 2 per mole. Since *p*-chloromercuribenzoate is a non-competitive inhibitor of the enzyme only at high concentrations, the —SH groups do not appear to be essential for enzyme activity (Yasunobu and coworkers, 1968). The enzyme can be inactivated by dialysis against diethyl-dithiocarbamate buffer but is reactivated specifically by subsequent dialysis against cupric copper (Yamada and Yasunobu, 1962b). The enzyme oxidizes primary amino groups of some monoamines as well as those of polyamines (Tabor, Tabor, and Rosenthal, 1954; Yamada and Yasunobu, 1962a). The relative rate of oxidation of spermidine, spermine, benzylamine, heptylamine, amylamine, kyneuramine and butylamine was 10, 10, 6.1, 5.4, 2.5, 1.0 and 0.1. A formal mechanism has been proposed for the reaction (Oi, Inamasu and Yasunobu, 1970).

Following the discovery by Hirsch (1953) of polyamine oxidase in sheep and bovine sera, Bengerst and Blaschko (1957), who also used manometry to measure oxygen consumption, reported the presence of polyamine oxidase activity in goat serum. The distribution of the enzyme in the sera or plasma of other species has been extensively studied by Blaschko and coworkers (for review see Blaschko, 1962) again using manometric techniques. Polyamine oxidase activity was found in the sera of all ruminants examined, but not in the sera of non-ruminants. By the use of a more sensitive radiochemical assay (Chapter 19) we have been able to demonstrate the presence of low polyamine oxidase activity in normal human serum and in rabbit, mouse and guinea-pig serum. McEwan (1956) in a study of human plasma monoamine oxidase noted some activity against spermidine as measured by ammonia production although the activity against monoamines was very much greater.

Buffoni and Blaschko (1964) crystallized an amine oxidase from pig plasma that oxidized benzylamine and histamine and commented on the similarity between their enzyme and that from bovine plasma. Spermine, spermidine, putrescine and cadaverine are not oxidized by the pig plasma enzyme (Buffoni and Blaschko, 1971).

$$NH_2(CH_2)_5NH_2$$

Cadaverine

A polyamine oxidase has also been isolated from human seminal plasma (Höltta and coworkers, 1975). The enzyme has a higher affinity for diamines but polyamines were also degraded at concentrations that can be considered physiological in human semen. The enzyme activity was inhibited by carbonyl reagents such as isoniazid, semicarbazide, and canaline. The activity could not be restored by exogenous pyridoxal-5′-phosphate. An approximate molecular weight of 182,000 was calculated for the enzyme.

III. MICROBIAL POLYAMINE OXIDASES

Oxidation of spermine and spermidine by cultures of *Pseudomonas pycyanoeae* was first reported by Silverman and Evans (1944). Subsequently oxidative degradation of polyamines with cleaving of the substrate molecule has been observed with intact cells of crude extracts of *Mycobacterium smegmatis* (Roulet and Zeller, 1945; Bachrach, Persky and Razin, 1960), and *Pseudomonas aeruginosa* (Razin, Gery and Bachrach, 1959). The oxidative activities for these amines usually increase when the organisms are grown with the amines in the growth medium, suggesting that the relevant enzymes are inducible.

A similar enzyme, apparently specific for polyamines, has been described in the bacterium *Neisseria perflavia* (Weaver and Herbst, 1958a); cyanide, hydroxylamine and semicarbazide inhibit the reaction. *Haemophilus para-influenzae* also degrades spermine and spermidine to 1,3-diaminopropane during growth (Weaver and Herbst, 1958b).

The amine oxidase of *Aspergillus niger*, although similar in many respects to the ruminant enzyme (copper and pyridoxal-containing, inhibited by copper chelators and carbonyl reagents, including iproniazid and isoniazid) has been reported not to oxidize polyamines (Yamada and coworkers, 1965; Yamada, Adachi and Ogata, 1965a; Muraoka and coworkers, 1966). The substrate specificity of the enzyme, however, was determined manometrically and the substrate concentration was not optimal for all the substrates tested. As has been noted (Zeller, 1963), the manometric method may not be sensitive enough to measure adequately low activities occurring in many biological materials.

Serratia marcescens also degrades spermine to 1,3-diaminopropane (Bachrach, 1962a,b) and 4-aminobutyraldehyde,

$$\underset{\text{spermidine}}{NH_2(CH_2)_3NH(CH_2)_4NH_2} + H_2O + O_2 \rightarrow \underset{\text{1,3-diaminopropane}}{NH_2(CH_2)_3NH_2} + \underset{\text{4-aminobutyraldehyde}}{NH_2(CH_2)_3CHO} + H_2O$$

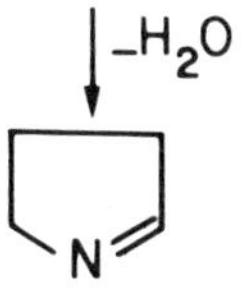

$$\downarrow -H_2O$$

Δ^1-Pyrroline

which later undergoes spontaneous cyclization to Δ^1-pyrroline.

The enzyme, which has been purified to homogeneity (Tabor and Kellog, 1970) has a molecular weight of 76,000 and contains approximately 1 mole of iron-protoprophyrin IX and 1 mole of FAD mole^{-1} of enzyme. The enzyme does not react directly with molecular oxygen but requires an electron acceptor such as ferricyanide, dichlorophenolindophenol or cytochrome c and as such is more properly classified as dehydrogenase (EC 1.4.99.?) although it does not appear as such in the current IUB list. The enzyme is inhibited by *p*-hydroxymercuribenzoate, N-ethylmaleimide, and Quinacrine (N^4-(6-chloro-2-methoxy-9-acridinyl)-N′, N′-diethyl-1,4-pentanediamine) and phenylhydrazine (Tabor and Kellog, 1971). Metal chelators (EDTA, *o*-phenanthroline, cuprizone, cupferron and diethyldithiocarbamate) had no effect on activity. Isoniazid, iproniazid, cyanide, sodium azide, and borohydride reduction did not inhibit the enzyme.

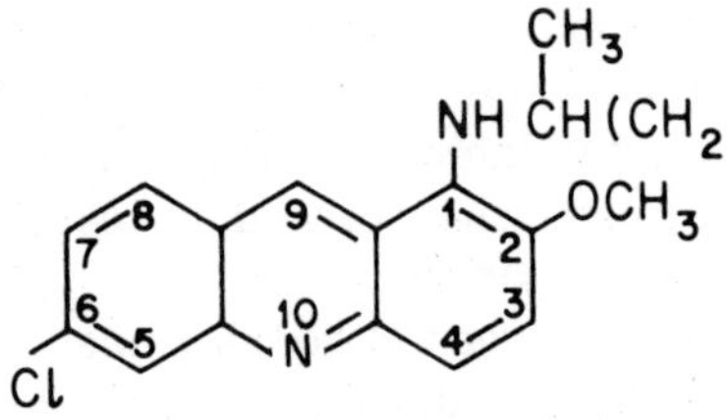

Quinacrine

In addition to spermidine, N^8-acetylspermidine, N,N′-*bis*(3-aminopropyl)-1,3-propanediamine, and N-(3-aminopropyl)-1,3-propanediamine are oxidized. Spermine is oxidized at about one-fifth the rate of spermidine. This enzyme has been used as the basis for an enzymatic assay for spermidine (Bachrach and Oser, 1963).

Spermidine is also oxidized to 1,3-diaminopropane and Δ^1-pyrroline by the putrescine oxidase of *Micrococcus rubens*. The enzyme has been purified to homogeneity (Adachi, Yamada and Ogata, 1966; Yamada, 1971), and has a molecular weight of 80,000. Treatment of the enzyme with ammonium sulphate at pH 4.0 — 4.5 liberates FAD with a corresponding loss of enzyme activity; the treated enzyme is specifically reactivated by FAD (Yamada, Adachi and Ogata, 1965b). The purified enzyme contains 1 mole of FAD mole^{-1} of enzyme. Putrescine (1,4-diaminobutane) is oxidized to 4-amino-butyraldehyde which condenses to Δ^1-pyrroline.

$$NH_2(CH_2)_4NH_2 + O_2 + H_2O \rightarrow NH_2(CH_2)_3CHO + NH_3 + H_2O_2$$

Cadaverine is also oxidized, but spermine and histamine are not attacked.

The enzyme is inhibited by sulphhydryl reagents such as *p*-chloromercuribenzoate, and by cyanide, hydroxylamine and iproniazide, but not by phenylhydrazine, semicarbazide, isoniazid or metal-chelating agents.

When spermidine is oxidized by a *Pseudomonas* species which has been adapted to grow on spermidine as the sole nitrogen source (Padmanabhan and Kim, 1965), the oxidation takes place at the other side of the secondary nitrogen:

$$\text{spermidine} \rightarrow \text{1,4-diaminobutane} + \text{3-aminopropionaldehyde.}$$

Exogenous spermidine and spermine are metabolized rapidly by cultures of *Anacystis nidulans* with the production of 1,3-diaminopropane (Ramakrishna, Guarino and Cohen, 1978). These cultures also appeared able to convert spermidine to a spermine-like molecule.

IV. PLANT POLYAMINE OXIDASES

Pea seedlings (*Pisum sativum*) contain an enzyme which catalyses the oxidative deamination of a wide range of amines containing the $—CH_2NH_2$ group, including putrescine (112) (numbers are relative activity), cadaverine (112), histamine (6), spermine (0.4), spermidine (39), and benzylamine (4.9). The enzyme has been purified about 1,800-fold (Hill, 1971). It contains divalent copper which can be removed with diethyldithiocarbamate, thus inactivating the enzyme; activity can be restored specifically with divalent copper ions. It is strongly inhibited by carbonyl reagents and copper chelators; sulphydryl group reagents do not show significant inhibition. The enzyme acts on spermidine to yield 1-(3-aminopropyl)-2-pyrroline and ammonia. The kinetics of the pea seedling amine oxidase have been studied by Yamasaki, Swindell and Reed (1970).

A similar enzyme has been demonstrated in the leaves of barley (*Hordium vulgare*), maize and oat seedlings (Smith, 1970; 1971; 1972; Smith, Haygarth and Williams, 1976), which acts on both spermine and spermidine to yield 1-(3-aminopropyl)-pyrroline and 1,3-diaminopropane or ammonia respectively.

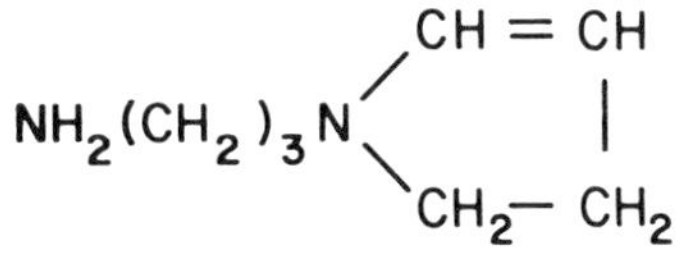

$$NH_2(CH_2)_3N\begin{matrix} / CH = CH \\ \quad\quad\quad | \\ \backslash CH_2 - CH_2 \end{matrix}$$

1(3-aminopropyl)2-pyrroline

V. MAMMALIAN TISSUE POLYAMINE OXIDASES

The diamine oxidase isolated from the mitochondria of hog kidney cortex (Mondovi and coworkers, 1971) is also a copper and pyridoxal (Mondovi and coworkers, 1964; Yamada and coworkers, 1967) containing enzyme acting on histamine and short chain aliphatic diamines, but does not oxidize polyamines and has minimal activity against monoamines.

The mitochondrial monoamine oxidase from beef liver normally oxidizes monoamines such as benzylamine, and selected secondary and tertiary amines, but does not normally act on diamines or polyamines (Yasunobu and Gomes, 1971). Incubation of a highly purified preparation with oxidized oleic acid, however, apparently altered the oxidation state of the sulphydryl groups of the enzyme, and the modified enzyme released ammonia from spermidine phosphate, whereas the untreated enzyme did not (Akopyan, Stesina and Gorkin, 1971). Furthermore, the treated enzyme was completely inhibited by hydroxylamine but not by *trans*-2-phenylcylopropylamine in contrast to the native enzyme where the opposite was the case.

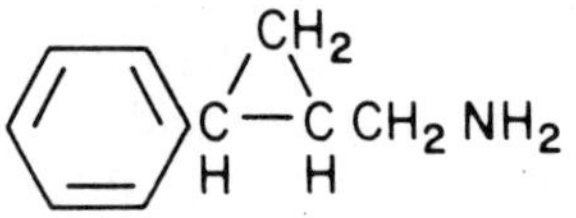

2-Phenylcyclopropylamine

The enzyme appears to contain 1 mole of FAD (or FAD-like component) per mole (assuming a molecular weight of 100,000). A similar enzyme has been isolated from bovine kidney mitochondria (Chuang, Patek and Hellerman, 1974), for which a molecular weight of about 109,000 was calculated, based on titration of the enzyme with the inhibitor Paragyline (N-methyl-N-2-propynyl-benzenemethanamine).

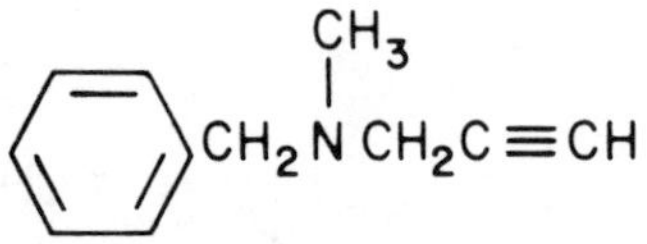

Paragyline

FAD-containing amine oxidases have also been obtained from a number of sources; pig liver mitochondria (Hollunger and Oreland, 1970; Oreland, 1971) where it was shown, by measurement of oxygen uptake, to be inactive against spermine or spermidine; from pig brain mitochondria (Tipton, 1968); and from rat liver mitochondrial outer membranes (Houslay and Tipton, 1974) but these latter two preparations do not appear to have been examined for activity against polyamines. From the sensitivities to clorgyline [N-methyl-N-proparagyl-3-(2,4-dichlorophenoxypropylamine)], benzyl cyanide and 4-cyanophenol, these workers concluded that there were two kinetically different monoamine oxidases in their preparation. They were unable to distinguish between the possibility that they were dealing with a single enzyme species in two different lipid environments, or whether there were two physically separable enzyme species present.

$$Cl\text{-}C_6H_3(Cl)\text{-}O(CH_2)_3N(CH_3)CH_2C\equiv CH.\ HCl$$

Clorgyline

Amine oxidases have also been prepared from various connective tissues. Nakano, Harada and Nagatsu (1974) obtained from the soluble fraction of bovine dental pulp (solely connective tissue), at a 115-fold purification, an enzyme that preferentially oxidized polyamines (spermine and spermidine), benzylamine and peptidyllysine (as lysine-vasopressin). The enzyme, which was inhibited by cuprizone, *p*-chloromercuribenzoate, β-aminopropionitrile, and iproniazid, had a molecular weight of approximately 170,000. The inhibition by iproniazid was reversed by the addition of pyridoxal phosphate.

A similar enzyme, also inhibited by β-aminopropionitrile has been partially characterized by Rucker, Royler and Parker (1969) from chick bone tissue. Amine oxidases have also been observed in bovine and chick aorta (Rucker and O'Dell, 1971; Harris and coworkers, 1974). The chick enzyme, which was active against peptidyllysine substrates, contained 0.14% copper and was inhibited by β-aminopropionitrile, dithiothreitol, and reagents that react with carbonyl groups; was not assayed for activity against polyamines. The material from bovine aorta, however, oxidized spermine and spermidine in addition to benzylamine and some simple aliphatic amines. The bovine aorta enzyme was inhibited by chelating agents, but activity could be partially restored by the addition of cupric ions. Activity was also lost when the aorta enzyme was incubated in the presence of cyanide, hydroxylamine, semicarbazide, isoniazid, and iproniazid. This enzyme also catalysed the oxidation of peptidyllysine when lysine vasopressin was used as substrate, but the bovine plasma enzyme, assayed under the same conditions, was inactive (Rucker and O'Dell, 1971). The occurrence of a spermine oxidase in chick embryo brain has been described by Caldarera and coworkers (1969) who found that the enzyme reaches maximal activity on the 15th day, just before hatching.

More recently Syatkin and Galaev (1977) have demonstrated, by measurements of hydrogen peroxide formation, the oxidation of spermine and spermidine, in addition to tyramine and putrescine by mouse liver homogenates. This may well be due to the presence of more than one enzyme. Höltta (1977), who used a radiochemical assay, has isolated and characterized an oxidase from rat liver peroxisomes which appears to be specific for polyamines. The enzyme, which has been purified 4,000-fold to electrophoretic homogeneity, appears to be a flavoprotein of approximately 60,000 molecular weight, containing FAD as the prosthetic group. It acts on the secondary amino groups of spermine and spermidine to produce 3-aminopropionaldehyde

$$\underset{\text{spermine}}{NH_2(CH_2)_3NH(CH_2)_4NH(CH_2)_3NH_2} + O_2 + H_2O \rightarrow \underset{\text{spermidine}}{NH_2(CH_2)_3NH(CH_2)_4NH_2} + \underset{\text{3-aminopropionaldehyde}}{NH_2(CH_2)_2CHO} + H_2O_2$$

$$\underset{\text{spermidine}}{NH_2(CH_2)_3NH(CH_2)_4NH_2} + O_2 + H_2O \rightarrow \underset{\text{3-aminopropionaldehyde}}{NH_2(CH_2)_2CHO} + \underset{\text{putrescine (1,4-diaminobutane)}}{NH_2(CH_2)_4NH_2} + H_2O_2.$$

A similar cleavage of spermidine and spermine appears to occur in *Pseudomonas* (Padmanabhan and Kim, 1965), and in mycobacteria (Bachrach, Persky and Razin, 1969). The dependence of the reaction on molecular oxygen was tested by adding glucose oxidase (known to have a high affinity for oxygen) and glucose to the reaction mixture. Under these conditions a total inhibition of the polyamine oxidase reaction was observed. Neither glucose nor glucose oxidase alone markedly inhibited the reaction. No other electron acceptors except molecular oxygen have been found. The oxidase seems to contain sulphydryl groups essential for activity, since sulphydryl reagents were markedly inhibitory. The enzyme was also inhibited by carbonyl reagents but not by typical inhibitors of pyridoxal phosphate enzymes such as isoniazid and canaline (1-amino-4-aminooxybutyric acid) or by copper chelators; iron chelators were inhibitory.

$$NH_2\,O\,CH_2CH_2CH(NH_2)COOH$$

Canaline

Polyamine oxidase activity was considerably inhibited (80%) by the flavoprotein inhibitor Quinacrine at 10^{-4} M. The monoamine oxidase inhibitor Paragyline was without effect at the same concentration. Methylglyoxal *bis*(guanylhydrazine), an inhibitor of diamine oxidase (Höltta, Sinervirta and Jänne, 1973) also had no effect at 10^{-4} M.

$$NH_2\underset{\underset{NH}{\|}}{C}NH\;N{=}CH\;\underset{\underset{CH_3}{|}}{C}{=}N\;NH\underset{\underset{NH}{\|}}{C}NH_2$$

Methylglyoxal *bis*(guanylhydrazone)

A unique feature of this enzyme is the stimulatory effect on the reaction velocity of a number of aldehydes. When spermidine was used as the substrate, the activity (—fold) of the reaction was as follows: formaldehyde (2.3), acetaldehyde (1.7), propanal (5.3), butanal (9.0), hexanal (16.3), benzaldehyde (74.3), anisaldehyde (73.3), salicyaldehyde (3.7), *p*-hydroxybenzaldehyde (5.0), *o*-aminobenzaldehyde (3.0), phenylacetaldehyde (11.0), and pyridoxal (22.3). The stimulatory effect of benzaldehyde was much greater with spermidine than with spermine as the substrate. A number of non-aldehydic compounds structurally related to benzaldehyde were without effect on the purified enzyme preparation, showing that the presence of the aldehyde group is necessary for the stimulatory effect. The enzyme is also capable of oxidizing N^1- and N^8-acetylspermidine (Höltta, 1977; Blankenship and Walle, 1978), and this activity is also detectable in the soluble fraction of rat kidney, spleen and brain.

$$CH_3CONH(CH_2)_3NH(CH_2)_4NH_2$$

N^1-Acetylspermidine

$$NH_2(CH_2)_3NH(CH_2)_4NH\;CO\;CH_3$$

N^8-Acetylspermidine

As can be seen from the foregoing account there exist both pyridoxal-containing and flavin-containing amine oxidases capable of the oxidative deamination of the biologically important polyamines spermine and spermidine. The existing data tempts the speculative generalization that the copper and pyridoxal-containing enzymes capable of oxidizing polyamines attack primary amino groups while the flavin-containing enzymes cleave the polyamines at the secondary amino groups with the ultimate formation of 1,4-diaminobutane or 1,3-diaminopropane as one of the products.

VI. BIOLOGICAL ROLE OF POLYAMINE OXIDASES

As first shown for *Mycobacterium tuberculosis* (Hirsch and Dubos, 1952; Fletcher, Epstein and Jewell, 1953), the products formed after oxidation of spermine and spermidine are toxic to a variety of cells (for reviews see Tabor and Tabor, 1964; Bachrach, 1970a,b). These include trypanosomes (Tabor and Rosenthal, 1956), bacteriophages of the T-odd series (but not T-even phages) and RNA phages (Bachrach, Tabor and Tabor, 1963; Bachrach and Leibovici, 1965, 1966; Fukami and coworkers, 1967; Oki and coworkers, 1968, 1969), animal and plant viruses (Katz and coworkers, 1967; Kremzner and Harten, 1970; Bachrach and Don, 1971; Bachrach and coworkers, 1965), bacteria (Bachrach and Persky, 1964; Bachrach, Abzug, and Bekierkunst, 1967; Fukami and coworkers, 1967; Kimes and Morris, 1971a), tumour cells (Alarcon, Foley and Modest, 1961; Israel, Rosenfield and Modest, 1964; Bachrach, 1970a), mammalian spermatozoa (Tabor and Rosenthal, 1956; Pulkkinen and coworkers, 1978), and a variety of mammalian and chick cell lines in culture (Alarcon, Foley and Modest, 1961; Halevy, Fuchs and Mager, 1962; Otsuka, 1971; Higgins and coworkers, 1974; Katsuta and coworkers, 1975). All the above observations were made in the presence of calf serum, or using oxidized polyamines prepared with bovine plasma enzyme, or in the presence of bovine serum albumin (Cohn fraction V), which contains polyamine oxidase as a contaminant (Katsuta and coworkers, 1975; Pulkkinen and coworkers, 1978; Morgan, unpublished). Replacement of calf serum by human or horse serum abolished or markedly reduced these effects (Alarcon, Foley and Modest, 1961; Higgins and coworkers, 1974). Attempts to establish the structure of the active product, oxidized spermine, were hampered by its lability; dioxidized spermine, mono-oxidized spermine and oxidized spermidine have been shown to have half-lives of the order of 42, 137 and 137 min respectively (Kimes and Morris, 1971b).

$$OCH(CH_2)_2NH(CH_2)_4NH(CH_2)_3NH_2$$

Mono-oxidized spermine

$$OCH(CH_2)_2NH(CH_2)_4NH(CH_2)_2CHO$$

Oxidized spermidine

$$OCH(CH_2)_2NH(CH_2)_4NH_2$$

Oxidized spermidine

It was claimed that the product is identical with either β-propylal-γ-butylal imine (Carvajal and Carvajal, 1957) or acrolein (Alarcon, 1964). Although both these compounds are toxic for bacteria and mammalian cells, it was later suggested that they might be artefacts produced during the isolation of the active principle (Bachrach, 1970b). The identification of the products of the enzymic oxidative deamination of spermine and spermidine as N,N′-*bis*(3-

$$OCH(CH_2)_2NH(CH_2)_3CHO$$

β-Propylal-γ-butylal-imine

$$CH_2 = CHCHO$$

Acrolein

propionaldehyde)-1,4-diaminobutane and N-(4-aminobutyl)-3-aminopropionaldehyde respectively (Tabor, Tabor and Rosenthal, 1954; Tabor, Tabor and Bachrach, 1964), together with the demonstration of the toxicity of synthetic N,N′-*bis*(propionaldehyde)-1,4-diaminobutane (Fukami and coworkers, 1967) appeared to provide conclusive evidence for the role of these compounds as the cytotoxic agents. None the less, doubt has been thrown on this conclusion by the later work of Alarcon (1970) and of Kimes and Morris (1971a) who have shown that acrolein is formed in significant amounts from enzymatically oxidized spermine or spermidine, probably by the spontaneous decomposition of the intermediate oxidation products, the unstable aminoaldehydes, at 37°C. Furthermore, it has been shown (Kimes and Morris, 1971b) that these aminoaldehydes can also undergo a condensation process, possibly aldol-type, to yield as yet unidentified products containing carbonyl groups. The observations that synthetic analogues of spermine and spermidine, in the presence of calf serum, demonstrate comparable inhibitory activity against tumour cells to that of the naturally occurring polyamines (Israel, Rosenfield and Modest, 1964); and that synthetic oxidized polyamines analogous to oxidized spermine markedly inactivated phage T5 (Oki and coworkers, 1968) serve to reinforce the view that more work is necessary to determine the molecular structure of the active agent(s).

The mechanism by which oxidized spermine exerts its cytotoxic effects is also unclear. It has been suggested (Bachrach and Leibovici, 1966; Oki and coworkers, 1969) that in phages the oxidized spermine forms a biologically inactive complex with the phage DNA, preventing the separation of the complementary DNA strands, and although it has been shown that in intact cells the ultimate effect is an inhibition of nucleic acid and protein synthesis (Kimes and Morris, 1971a), the mechanism by which the oxidation products enter the cell and whether or not they undergo further modification during this process is not clear.

Although there are a number of amine oxidase activities in various animal tissues and extracts the role of these enzymes in controlling tissue levels of polyamines is unknown. It seems clear that the polyamine oxidase found in the blood of ruminants does not provide a general mechanism for the biodegradation of polyamines in animal tissues for two reasons: firstly, this enzyme is not found in non-ruminants, and secondly the oxidation products formed differ from those of the FAD-containing tissue enzymes. This difference is readily demonstrable by electrophoresis of the reaction products formed by oxidation of [^{14}C] labelled spermine (Morgan, unpublished; see also Chapter 19). The biological importance of the 3-aminopropionaldehyde formed by the rat liver enzyme is also unknown, although significant changes in the amount of this amino aldehyde in the sera of cancer patients have been reported (Quash and Majaraj, 1970; Quash and Taylor, 1970). Elevated levels of an enzyme capable of converting [^{14}C] labelled spermine to a product that co-migrates with spermidine on paper electrophoresis have also been observed in the sera of some patients with hepatitis (Morgan and Allison, unpublished).

Precursors for cross-linking in elastin and collagen are the lysine- and hydroxylysine-derived aldehydes which are formed enzymically in a step believed to be similar to the oxidative deamination of amines by amine oxidases (Traub and Piez, 1971; Feeney, Blankenhorn and Dixon, 1975) followed by a series of spontaneous aldol and aldimide condensations. The reported activity of some polyamine oxidases from connective tissue against peptidyllysine (Rucker and O'Dell, 1971; Harris and coworkers, 1974; Nakano, Hamada and Nagatsu, 1974) suggests the possiblity of a common identity, since it appears that the enzyme responsible for the oxidative deamination step in collagen cross-linking is also copper and perhaps pyridoxal requiring, and is inhibited by β-aminoproprionitrile.

VII. SUMMARY AND CONCLUSIONS

The close involvement of polyamines in cell growth and proliferation (for review see Jänne, Pösö and Raina, 1978) together with the observations that at the end of a period of growth polyamine biosynthesis declines and is followed by a reduction in intracellular polyamine, either by release into the culture medium (Fuller, Gerner, and Russell, 1977; Chapter 23) or by intracellular degradation (Ramakrishna, Guarino, and Cohen, 1978) serves to highlight the lack of knowledge of the catabolism and excretion of polyamines (for review see Raina and Jänne, 1975; Chapters 26 and 27). Though the biosynthetic pathway appears to be relatively well-established (Tabor and Tabor, 1976; but see also Tait, 1976), a possible involvement of polyamine oxidase in the control of cellular polyamine levels cannot be excluded. The role of polyamines in the organization of tRNA structure and activity appears now to be well-established (Cohen, 1978) and in view of the finding that the ultimate cellular effect of oxidized polyamines is an inhibition of nucleic acid and protein synthesis (Kimes and Morris, 1971b; Tabor and Tabor, 1972) it is tempting to speculate that polyamine oxidase could exert a regulatory role by oxidizing the nucleic acid-bound polyamine with the production of a biologically inactive complex (as suggested by Bachrach and Leibovici, 1966) as a first step in a degradative pathway.

The diversity of effects described for polyamines in living organisms (Jänne, Pösö and Raina, 1978), together with the current lack of knowledge of the degradative pathways of these important compounds, leaves the biological role of the polyamine oxidases very much an open question.

ACKNOWLEDGEMENT

I thank Miss Jennifer Seems for help in the preparation of the manuscript.

REFERENCES

Achee, F., Chervenka, C. H., Smith, R. A. and Yasunobu, K. T. (1968). *Biochemistry*, **7**, 4329–4335.
Adachi, O., Yamada, H. and Ogata, K. (1966). *Agr. Biol. Chem. (Tokyo)*, **30**, 1202–1210.
Akopyan, Z. I., Stesina, L. N. and Gorkin, V. Z. (1971). *J. Biol. Chem.*, **246**, 4610–4618.
Alarcon, R. A. (1964). *Arch. Biochem. Biophys.*, **106**, 240–242.
Alarcon, R. A. (1970). *Arch. Biochem. Biophys.*, **137**, 365–372.
Alarcon, R. A., Foley, G. E. and Modest, E. J. (1961). *Arch. Biochem. Biophys.*, **94**, 540–541.
Bachrach, U. (196a). *Nature, Lond.*, **194**, 377–378.
Bachrach, U. (1962b). *J. Biol. Chem.*, **237**, 3443–3448.
Bachrach, U. (1970a). *Ann. N. Y. Acad. Sci.*, **171**, 939–956.
Bachrach, U. (1970b). *Ann. Rev. Microbiol.*, **24**, 109–134.
Bachrach, U., Abzug, S. and Bekierkunst, A. (1967). *Biochim. Biophys. Acta*, **134**, 174–181.
Bachrach, U. and Don, S. (1971). *J. Gen. Virol.*, **11**, 1–9.
Bachrach, U. and Leibovici, J. (1965). *Israel J. Med. Sci.*, **1**, 541–551.
Bachrach, U. and Leibovici, J. (1966). *J. Mol. Biol.*, **19**, 120–132.
Bachrach, U. and Oser, I. S. (1963). *J. Biol. Chem.*, **238**, 2098–2101.
Bachrach, U. and Persky, S. (1964). *J. Gen. Microbiol.*, **37**, 195–204.
Bachrach, U., Persky, S. and Razin, S. (1960). *Biochem. J.*, **76**, 306–310.
Bachrach, U., Rabina, S., Lachenstein, G. and Eilon, G. (1965). *Nature, Lond.*, **208**, 1095–1096.
Bachrach, U., Tabor, C. W. and Tabor, H. (1963). *Biochim. Biophys. Acta*, **78**, 768–770.
Bengerest, B. and Blaschko, H. (1957). *Brit. J. Pharmacol.*, **12**, 513–516.
Blankenship, J. and Walle, T. (1978). *Advan. Polyamine Res.*, **2**, 97–110.
Blaschko, H. (1952). *Pharmacol. Rev.*, **4**, 415–458.
Blaschko, H. (1962). *Advan. Comp. Physiol. Biochem.*, **1**, 67–116.
Blaschko, H. (1963). Amine oxidase. In *The Enzymes*, 2nd edn. vol. 8. (Ed. Boyer, P. D., Lardy, H. and Myrbäck, K.), pp. 337–351. New York and London, Academic Press.
Blaschko, H. and Duthie, R. (1945). *Biochem. J.*, **39**, 478–481.
Buffoni, F., and Blaschko, H. (1964). *Proc. Roy. Soc.* (Lond.), Series B, **161**, 153–158
Buffoni, F. and Blaschko, H. (1971). Amine oxidase (pig plasma). In H. Tabor and C. W. Tabor (Eds), *Methods in Enzymology*, Vol. 17B, Academic Press, New York and London, pp. 682–686.

Caldarera, C. M., Moruzzi, M. D., Rossoni, C. and Barbiroli, B. (1969). *J. Neurochem.*, **16**, 309–316.

Carvajal, G. and Carvajal, E. J. (1957). *Amer. Rev. Resp. Dis.*, **76**, 1094–1096.

Chuang, H. Y. K., Patek, D. R. and Hellerman, L. (1974). *J. Biol. Chem.*, **249**, 2381–2384.

Cohen, S. S. (1978). *Nature, Lond.*, **274**, 209–210.

Feeney, R. E., Blankenhorn, G., and Dixon, H. B. F. (1975). *Advan. Protein Chem.*, **29**, 136–203.

Fletcher, F., Epstein, C. and Jewell, P. I. (1953). *J. Gen. Microbiol.*, **8**, 323–329.

Fukami, H., Tomida, I., Morino, T., Yamada, H., Oki, R., Kawasaki, M. and Ogata, K. (1967). *Biochem. Biophys. Res. Comm.*, **28**, 19–24.

Fuller, D. J. M., Gerner, E. W. and Russell, D. H. (1977). *J. Cell. Physiol.*, **93**, 81–88.

Gorkin, V. Z., Avakyan, A. A., Venevkina, I. V. and Komissarova, N. V. (1962). *Vop. Med. Khim.*, **8**, 638–645; translated in *Fed. Proc.*, **22**, T619–T622 (1963).

Halevy, S., Fuchs, Z. and Mager, J. (1962). *Bull. Res. Council Israel*, **11A**, 52–53.

Harris, E. D., Gonnerman, W. A., Savage, J. E. and O'Dell, B. L. (1974). *Biochim. Biophys. Acta*, **341**, 332–344.

Higgins, M. L., Tillman, M. C., Rupp, J. P. and Leach, F. R. (1974). *J. Cell Physiol.*, **74**, 149–154.

Hill, J. M. (1971). Diamine oxidase (pea seedling). In H. Tabor and C. W. Tabor (Eds), *Methods in Enzymology*, Vol. 17B, Academic Press, New York and London, pp. 730–735.

Hirsch, J. G. (1953). *J. Exper. Med.*. **97**, 345–355.

Hirsch, J. G. and Dubos, R. J. (1952). *J. Exper. Med.*, **95**, 191–208.

Hollenger, G. and Oreland, L. (1970). *Arch. Biochem. Biophys.*, **139**, 320–328.

Höltta, E. (1977). *Biochemistry*, **16**, 91–100.

Höltta, E., Pulkkinen, P., Elfving, K. and Jänne, J. (1975). *Biochem. J.*, **145**, 373–378.

Höltta, E., Sinervirta, R. and Jänne, J. (1973). *Biochem. Biophys. Res. Comm.*, **54**, 350–357.

Houslay, M. D. and Tipton, K. F. (1974). *Biochem. J.*, **139**, 645–652.

International Union of Biochemistry (1973). *Enzyme Nomenclature*. Elsevier Scientific Publishing Co., Amsterdam.

Israel, M., Rosenfield, J. S. and Modest, E. J. (1964). *J. Med. Chem.*, **7**, 710–716.

Jänne, J., Pösö, H. and Raina, A. (1978). *Biochim. Biophys. Acta*, **473**, 241–293.

Kapeller-Adler, R. (1970). *Amine Oxidases and Methods for Their Study*. Wiley-Interscience, New York and London.

Katsuta, H., Takaoka, T., Nose, K. and Nagai, Y. (1975). *Japan J. Exper. Med.*, **45**, 345–354.

Katz, E., Goldblum, T., Bachrach, U. and Goldblum, N. (1967). *Israel J. Med. Sci.*, **3**, 575–577.

Kimes, B. W. and Morris, D. R. (1971a). *Biochem. Biophys. Acta*, **228**, 235–244.

Kimes, B. W. and Morris, D. R. (1971b). *Biochim. Biophys. Acta*, **228**, 223–234.

Kremzner, L. T. and Harter, D. H. (1970). *Biochem. Pharmacol.*, **19**, 2541–2550.

Malmström, B. G., Andréasson, L.-E. and Reinhammer, B. (1975). Copper containing oxidases and superoxide dismutase. In P. D. Boyer (Ed.), *The Enzymes*, 3rd edn., Vol. 12, Academic Press, New York and London, pp. 507–579.

McEwan, C. M. (1965). *J. Biol. Chem.*, **240**, 2003–1010; 2011–2018.

Mondovi, B., Rotilio, G., Agio, A. F. and Santoro, A. S. (1964). *Biochem. J.*, **91**, 408–415.

Mondovi, B., Rotolio, G., Costa, M. T. and Agio, A. F. (1971). Diamine oxidase

(pig kidney). In H. Tabor and C. W. Tabor (Eds), *Methods in Enzymology*, Vol. 17B, Academic Press, New York and London, pp. 735–740.
Muraoka, S., Hoshika, A., Yamasaki, H., Yamada, H. and Adachi, O. (1966). *Biochim. Biophys. Acta*, **122**, 544–546.
Nakano, G., Harada, M. and Nagatsu, T. (1974). *Biochim. Biophys. Acta*, **341**, 366–377.
Nara, S., Igaue, I., Gomes, B. and Yasanobu, K. T. (1966). *Biochem. Biophys. Res. Comm.*, **23**, 324–328.
Oi, S., Inamasu, M. and Yasunobu, K. T. (1970). *Biochemistry*, **9**, 3378–3383.
Oki, T., Kawasaki, H., Ogata, K., Yamada, H., Tomida, I., Morino, T. and Fukami, H. (1968). *Agr. Biol. Chem. (Tokyo)*, **32**, 1349–1354.
Oki, T., Kawasaki, H., Ogata, K., Yamada, H., Tomida, I., Morino, T. and Fukami, H. (1969). *Agr. Biol. Chem. (Tokyo)*, **33**, 994–1000.
Oreland, L. (1971). *Arch. Biochem. Biophys.*, **146**, 410–421.
Otsuka, H. (1971). *J. Cell. Sci.*, **9**, 71–84.
Padmanabhan, R., and Kim, K. (1965). *Biochem. Biophys. Res. Comm.*, **19**, 1–5.
Pulkkinen, P., Piik, K., Poso, P. and Jänne, J. (1978). *Acta Endocrinol.*, **87**, 845–854.
Quash, G. and Maharaj, K. (1970). *Clin. Chim. Acta*, **30**, 13–16.
Quash, G. and Taylor, D. R. (1970). *Clin. Chim. Acta*, **30**, 17–23.
Raina, A., and Jänne, J. (1975). *Med. Biol.* **53**, 121–147.
Ramakrishna, S., Guarino, L. and Cohen, S. S. (1978). *J. Bacteriol.*, **134**, 744–750.
Razin, S., Gery, I. and Bachrach, U. (1959). *Biochem. J.*, **71**, 551–558.
Roulet, F. and Zeller, E. A. (1945). *Helv. Chim. Acta*, **28**, 1326–1342.
Rucker, R. B. and O'Dell, B. L. (1971). *Biochim. Biophys. Acta*, **235**, 32–43.
Rucker, R. B., Royler, J. C. and Parker, H. E. (1969). *Proc. Soc. Exper. Biol. Med.*, **130**, 1150–1155.
Sakamoto, Y., Ogawa, Y. and Hayashi, K. (1963). *J. Biochem. (Tokyo)*, **43**, 292–294.
Silverman, M. and Evans, E. A. (1944). *J. Biol. Chem.*, **154**, 521–534.
Smith, T. A. (1970). *Biochem. Biophys. Res. Comm.*, **41**, 1452–1456.
Smith, T. A. (1971). *Biol. Rev.*, **46**, 201–241.
Smith, T. A. (1972). *Endeavour*, **31**, 22–28.
Smith, T. A., Haygarth, W. L. and Williams, J. F. (1976). *Biochem. Soc. Trans.*, **4**, 76–77.
Syatkin, S. P. and Galaev, Y. V. (1977). *Biochemistry SSR*, **46**, 783–786.
Tabor, C. W. and Kellog, P. D. (1970). *J. Biol. Chem.*, **245**, 5424–5433.
Tabor, C. W. and Kellog, P. D. (1971). Spermidine dehydrogenase *(Serratia marcescens)*: In *Methods in Enzymology*, vol. 17B. (Ed. Tabor, H. and Tabor, C. W.), pp. 746–753. New York and London, Academic Press.
Tabor, C. W. and Rosenthal, S. M. (1956). *J. Pharmacol. Exper. Therap.*, **116**, 139–155.
Tabor, H. and Tabor, C. W. (1964). *Pharmacol. Rev.*, **16**, 245–300.
Tabor, C. W. and Tabor, H. (1976). *Ann. Rev. Biochem.*, **45**, 285–306.
Tabor, C. W., Tabor, H. and Bachrach, U. (1964). *J. Biol. Chem.*, **239**, 2194–2203.
Tabor, H., and Tabor, C. W. (1972). *Adv. Enzymol.* **36**, 203–268.
Tabor, C. W., Tabor, H. and Rosenthal, S. M. (1954). *J. Biol. Chem.*, **208**, 645–661.
Tait, G. H. (1976). *Biochem. Soc. Trans.*, **4**, 610–612.
Tipton, K. F. (1968). *Eur. J. Biochem.*, **4**, 103–107.
Traub, W. and Piez, K. A. (1971). *Advan. Protein Chem.*, **25**, 243–352.
Unemoto, T. (1963). *Chem. Pharm Bull. (Tokyo)*, **11**, 1255–1264.
Weaver, R. H. and Herbst, E. J. (1958a). *J. Biol. Chem.*, **231**, 647–655.

Weaver, R. H. and Herbst, E. J. (1958b). *J. Biol. Chem.*, **231**, 637–646.

Watanabe, K., Smith, R. A., Imanasu, M. and Yasunobu, K. T. (1972). *Advan. Biochem. Psychopharmacol.*, **5**, 107–117.

Yamada, H. (1971). Putrescine oxidase *(Micrococcus rubens)*. In H. Tabor and C. W. Tabor (Eds), *Methods in Enzymology*, Vol. 17B, Academic Press, New York and London, pp. 726–730.

Yamada, H., Kumagai, H., Kawasaki, H., Matsui, H. and Ogata, K. (1967). *Biochem. Biophys. Res. Comm.*, **29**, 723–727.

Yamada, H., Adachi, O., Kumagai, H. and Ogata, K. (1965). *Mem. Res. Inst. Food Sci. Kyoto Univ.*, **26**, 21–23.

Yamada, H., Adachi, O. and Ogata, K. (1965a). *Agr. Biol. Chem. (Tokyo)*, **29**, 117–123; 864–869; 912–917.

Yamada, H., Adachi, O. and Ogata, K. (1965b). *Agr. Biol. Chem. (Tokyo)*, **29**, 1148–1149.

Yamada, H. and Yasunobu, K. T. (1962). *J. Biol. Chem.*, **237**, 1511–1516.

Yamada, H. and Yasunobu, K. T. (1962b). *J. Biol. Chem.*, **237**, 3077–3082.

Yamasaki, E. F., Swindell, R. and Reed, D. J. (1970). *Biochemistry*, **9**, 1206–1210.

Yasunobu, K. T., Achee, F., Chevvenka, C. and Wang, T. M. (1968). Molecular properties of beef plasma amino oxidase. In K. Yamada, N. Katunoma, and H. Wada (Eds), *3rd Symposium on Pyridoxal Enzymes*, Maruzen, Tokyo, pp. 139–142.

Yasunobu, K. T. and Gomes, B. (1971). Mitochondrial amine oxidase (monoamine oxidase) (beef liver). In H. Tabor and C. W. Tabor (Eds), *Methods in Enzymology*, Vol. 17B, Academic Press, New York and London, pp. 709–717.

Yasunobu, K. T. and Smith, R. A. (1971). Amine oxidase (beef plasma). In H. Tabor, and C. W. Tabor (Eds), *Methods in Enzymology*, Vol. 17B, Academic Press, New York and London, pp. 698–704.

Zeller, E. A. (1963). Diamine oxidases. In P. D. Boyer, H. Lardy and K. Myrbäck (Eds), *The Enzymes*, 2nd edn, Vol. 8, Academic Press, New York and London, pp. 313–335.

Zeller, E. A., Stern, R. and Wenk, M. (1940). *Helv. Chim. Acta*, **23**, 3–17.

Chapter 19

Polyamine Oxidase and Macrophage Function

D. M. L. MORGAN, J. FERULGA AND A. C. ALLISON

I. INTRODUCTION

Amine oxidases have been defined (Malström, Andréasson and Reinhammer, 1975) as enzymes which catalyse the oxidative deamination of mono-, di-, and polyamines with the formation of stoichiometric amounts of aldehyde, hydrogen peroxide, and ammonia according to the equation:

$$RCH_2NH_2 + O_2 + H_2O \rightarrow RCHO + H_2O_2 + NH_3.$$

The presence of an amine oxidase in sheep and bovine serum active against spermine and spermidine was first reported by Hirsch (1953). This enzyme has been purified by Tabor, Tabor and Rosenthal (1954), who showed that it was active against mono-, di-, and polyamines; it was later crystallized by Yamada and Yasunobu (1962). The products formed by the oxidation of spermine and spermidine were subsequently isolated (Tabor, Tabor and Bachrach, 1964) and their identity confirmed (Chapter 18). They were, respectively, N,N′-*bis*(3-propionaldehyde)-1,4-diaminobutane and N-(4-aminobutyl)-3-aminopropionaldehyde. Blaschko and Hawes (1959) observed the presence of a spermine oxidase in serum from a number of ruminants, e.g. camel, llama, fallow deer, giraffe, ox, sheep and goat, but the enzyme was not present in sera from non-ruminants, e.g. cat, lion, man, pig; only trace amounts were found in horse and dog sera (Blaschko and coworkers, 1959), although some of these sera were able to oxidize various primary monoamines. With the aid of the more sensitive assay described here, however, we have been able to demonstrate the presence of low polyamine oxidase activity in normal human sera, and have shown that this is elevated in sera from some but not all patients with hepatitis (Morgan and Allison, unpublished).

Table 1 Comparison of the properties of polyamine oxidase isolated from bovine serum and rat liver

	Bovine serum	Rat liver
Mo. wt.:	170,000	60,000
Co-factors:	pyridoxal phosphate	FAD
Metals:	copper 1–2 atoms/mole^{-1}	? Fe
Inhibitors:	carbonyl reagents	carbonyl reagents
	copper chelators	sulphydryl reagents
Activity stimulated by:	?	aldehydes
Substrates:	spermine 10[1]	spermine[2]
	spermine 10	spermidine
	benzylamine 6.1	N′-acetyl-spermidine
	heptylamine 5.4	
	amylamine 2.5	
	butylamine 0.1	

[1]Relative rates of oxidation (Yamada and Yasunobu, 1962)
[2]Höltta (1977)

The particulate amine oxidase of rat liver differs from the bovine serum enzyme in a number of respects (Höltta, 1977; Table 1). It acts on the secondary amino groups of spermine and spermidine to produce 3-amino-propionaldehyde (Chapter 18). Previously, analysis of the metabolism of polyamines has been hindered by inadequate methods (Tabor, Tabor and Rosenthal, 1961). With the commercial availability of isotopically labelled polyamines, however, this is no longer the case.

A. Estimation of Polyamine Oxidase Activity

The following method is essentially that described by Höltta (1977) with minor modifications.

1. *Preparation of Sample*

(a) *Tissues.* Immediately after removal the tissue (liver, spleen, kidney, etc.) was placed in ice-cold Tris buffer (10 mM, pH 7.4) and chilled in an ice bath. Tissue samples and homogenates were stored frozen (−20°C) until assayed. Pieces of tissue were thawed (if necessary), dried with filter paper, weighed, and homogenized in Tris buffer (9 ml/g tissue).

(b) *Cells and Medium from Cultures of Macrophages.* Cell cultures were prepared as described elsewhere (peritoneal macrophages: Davies, Allison, Butterfield and Williams, 1974; Tolnai, 1975; alveolar cells: Myrvik, Leake and Oshima, 1962; Moore and Myrvik, 1970). Medium was removed from the culture vessels; 2.5 ml was placed in a Minocon concentration cell (type CS-15; Amicon Ltd. Membrane cut-off 15,000 MW) and the volume reduced to 0.1 ml. The cell sheet was washed once with phosphate-buffered saline (PBS; Dulbecco's A), then covered with a volume of 0.9% (w/v) saline containing 0.1% Triton X-100 (v/v) equal to that of the culture medium. The cells were detached by scraping with a sterile silicone rubber bung and 2.5 ml of the suspension concentrated as before.

2. *Reagents*

(a) *Incubation Buffer.* Benzaldehyde (10 mmole) was added to 1,000 ml 0.5 M Tris buffer, pH 8.5, containing 0.1% Triton X-100 (v/v) and the mixture stirred until homogenous.

(b) *[^{14}C] labelled Substrate.* [^{14}C]-spermine tetrahydrochloride (N,N′-*bis*-(3-aminopropyl)-[1,4-^{14}C]-tetramethylene-1,4-diamine tetrahydrochloride), 120 mCi/mmole^{-1}, was obtained from the Radiochemical Centre (Amersham, Bucks). Unlabelled spermine (4 mM; 5 ml) was added to 250 μCi (1.26 ml) of [^{14}C]-spermine to give a stock solution of labelled substrate. For use 0.05 ml of this solution was added to 0.4 ml of 4 mM spermine and the volume made up to 5 ml with the incubation buffer.

3. *Method*

Water (0.5 ml) was placed in a 75 × 12 mm dia. polystyrene tube and a 38 × 7 mm dia. tube, also of polystyrene, inserted (the assay is carried out in the smaller tube but the larger tube is easier to handle; the water acts as a cushion and heat exchanger). The tubes, containing buffered substrate (0.2 ml) were placed in a 37°C water bath. After 30 min for temperature equilibration, sample (0.1 ml) was added and the mixture incubated for 60 to 120 min. The reaction was stopped by the addition of 50% trichloracetic acid (w/v; 0.05 ml) and protein sedimented by centrifugation for 15 min at approximately 600 **g**.

Electrophoretic separation of the reaction products (Herbst, Keister and Weaver, 1958) was carried out on Whatman 3MM paper in a Shandon electrophoretic apparatus (Model U77; Shandon Southern Products Ltd., Runcorn, Cheshire). Sheets of paper (30 × 23 cm) were marked with the origin 6 cm from the positive end. The protein-free supernatant (0.02 ml) was applied as 2 to 3 successive streaks 1 cm long, the area being dried with a stream of hot air between applications. Seven samples could be accommodated on the paper, allowing for 3.5 cm margins and 1.5 cm between spots. Spots 1 to 3 and 5 to 7 were assay samples, spot 4 was a mixture of unlabelled markers (spermine, spermidine, putrescine).

Electrophoretic separation of the polyamines was carried out in 5-sulphosalicylate buffer (0.05 M, pH 4.0) by application of a potential difference across the paper of 150 V for 4 h. Fresh buffer was used for each separation, since some decomposition took place in the anode compartment. After electrophoresis the paper was dried in air overnight at room temperature then stained with ninhydrin (0.1% w/v in acetone containing 0.2% v/v anhydrous pyridine). [^{14}C] polyamine (0.005 ml; 800 dpm) was spotted onto each of the unlabelled markers (to determine the quenching effect of the stain) and the polyamine spots were cut out, by reference to the markers, and placed in vials. Toluene (10 ml) containing 2,5-diphenyloxazole (PPO, g l^{-1}) and (*p-bis*[2-(phenyloxazolyl)]-benzene (POPOP, 62.5 mg l^{-1}) was added and the radio activity counted in a liquid scintillation spectrometer for 100 sec or 10,000 counts.

II. MACROPHAGE POLYAMINE OXIDASE

Examination of the polyamine oxidase activity of rabbit liver, spleen, serum; mouse liver, spleen and serum with the assay described above showed identical distributions of radioactivity on paper electrophoresis which differed from that shown by the calf serum enzyme (Figure 1).

Rabbit alveolar macrophages, obtained by bronchial lavage (Myrvik, Leake and Oshima, 1962; Moore and Myrvik, 1970) also contain a polyamine oxidase giving apparently identical reaction products to that of the other tissue enzymes (Figure 1), as do mouse peritoneal macrophages. Activation of macrophages *in vivo* results in an increased content of polyamine oxidase per cell although the specific activity remains constant (Table 2).

As first shown by Hirsch and Dubos (1952), the products formed after oxidation of spermine and spermidine inhibit the growth of a number of cell types including tumour cells, bacteria, trypanosomes, plant and animal viruses and bacteriophages (Tabor and Tabor, 1972). Recently Byrd, Jacobs and Amoss (1977) have reported that micromolar quantities of spermine and spermidine reversibly inhibit transformation of primary cultures of murine spleen cells.

Our observations show that the supernatants of mouse peritoneal adherent cell cultures, activated by lipopolysaccharide, contain demonstrable polyamine oxidase activity (Ferluga, Morgan and Allison, unpublished). Moreover, such supernatants in the presence of spermine inhibit proliferation of mouse thymocytes cultured without serum. They also inhibit the proliferation of P 815 mastocytoma cells. Supernatants of unactivated mouse peritoneal cell cultures have no such activity. Polyamine oxidase activity was also readily

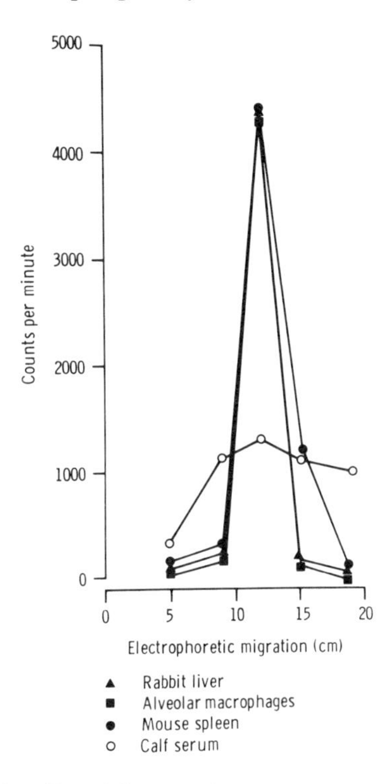

Figure 1 Distribution of radioactivity on electrophoretograms of reaction products formed by the action of polyamine oxidase from rabbit liver, rabbit alveolar macrophages, mouse spleen, and newborn calf serum on [^{14}C] spermine

demonstrable in the spleens of mice injected intravenously with *Corynebacterium parvum*. The supernatants of such spleen cells in the presence of spermine likewise inhibit the proliferation of mitogen stimulated mouse thymocytes. Elevated levels of polyamine oxidase activity have also been found during the rising phase of parasitaemia in spleens of mice injected

Table 2 Polyamine oxidase activity in normal and activated macrophages

	Normal	Activated[1]
Mouse peritoneal	3.3	6.2 pmole min^{-1} 10^6 cells
macrophages	145	147 pmole min^{-1} mg^{-1} protein
Rabbit alveolar	6.0	23 pmole min^{-1} 10^6 cells
macrophages	87	96 pmole min^{-1} mg^{-1} protein

[1]Peritoneal cells collected 3d after intraperitoneal injection of 2 ml proteose-peptone. Alveolar cells obtained 4 days after 5 mg BCG given i.v. to rabbits immunized with BCG in Freund's incomplete adjuvant 3 weeks earlier (Myrvik, Leake and Oshima, 1962).

intraperitoneally with *Babesia microti* (Morgan and Christensen, unpublished).

III. SUMMARY

Our observations suggest that activated adherent peritoneal cells, and spleen cells (presumably macrophages) synthesize and release polyamine oxidase. In the presence of spermine (which is formed and released by actively dividing lymphocytes and tumour cells) the oxidase generates products that inhibit cellular proliferation. This possibly represents a mechanism by which macrophages could regulate proliferation of lymphocytes, tumour cells and other cell types.

REFERENCES

Blaschko, H., Friedman, P. J., Hawes, R. and Nilsson, K. (1959). *J. Physiol.*, **145**, 384–404.

Blaschko, H. and Hawes, R. (1959). *J. Physiol.*, **145**, 124–131.

Byrd, W. J., Jacobs, D. M. and Amoss, M. S. (1977). *Nature, Lond.*, **267**, 621.

Davies, P., Allison, A. C., Butterfield, A. and Williams, S. (1974). *Nature, Lond.*, **251**, 423–425.

Herbst, E. J., Keister, D. L. and Weaver, R. H. (1958). *Arch. Biochem. Biophys.*, **75**, 178–185.

Hirsch, J. G. (1953). *J. Exper. Med.*, **97**, 345–355.

Hirsch, J. G. and Dubos, R. J. (1952). *J. Exper. Med.*, **95**, 191–208.

Höltta, E. (1977). *Biochemistry*, **16**, 91–100.

Malmström, B. G., Andréasson, L.-E. and Reinhammer, B. (1975). Copper containing oxidases and superoxide dismutase. In: *The Enzymes* 3rd edn. vol. 12. Ed. Boyer, P. D. New York and London, Academic Press.

Moore, V. L. and Myrvik, Q. A. (1970). *Infection and Immunity*, **2**, 810–814.

Myrvik, Q. N., Leake, E. S. and Oshima, S. (1962). *J. Immunol.*, **89**, 745–751.

Tabor, H. and Tabor, C. W. (1972). *Advan. Enzymol.*, **36**, 203–268.

Tabor, C. W., Tabor, H. and Bachrach, U. (1964). *J. Biol. Chem.*, **239**, 2194–2203.

Tabor, C. W., Tabor, H. and Rosenthal, S. M. (1954). *J. Biol. Chem.*, **208**, 645–661.

Tabor, H., Tabor, C. W. and Rosenthal, S. M. (1961). *Annu. Rev. Biochem.*, **30**, 579–604.

Tolnai, S. (1975). *Tissue Culture Assoc. Manual*, **1**, 17–19.

Yamada, H. and Yasunobu, K. T. (1962). *J. Biol. Chem.*, **237**, 1511–1516.

Chapter 20

Melanocytes, Chalones and Polyamines

D. L. DEWEY

I. MELANOCYTES AND CHALONES

The melanocytes may be regarded as cells of an organ whose function is the production of melanin and the excretion of the melanin granules into keratinocytes and hair cells. Melanocytes are found between the basal cells of the epidermis where they are distinguishable in stained sections by their melanin content after pre-incubation with dihydroxyphenylalanine (DOPA). Melanocytes have a different embryonic origin to epidermis. The cells contain tyrosinase which first oxidizes tyrosine to DOPA and then converts DOPA through several intermediate stages to the final product, melanin. The transfer of melanin from epidermal melanocytes to keratinocytes and from hair bulb melanocytes to hair cells is through active phagocytosis of the distal portion of the dendritic processes of melanocytes by keratinocytes and hair cells. The cell wall of the phagocytized dendritic process then disappears, and the melanin particulars are dispersed throughout the keratinocyte or hair cell.

Melanin is widespread throughout the animal kingdom where it appears to serve a variety of functions concerned with protection of tissues and gross colouration of animals. Its major function in man would appear to be the protection of the underlying tissue from ultraviolet light (UV). The increase in pigmentation following exposure to UV results from an increase in activity of the melanocytes and not to an increase in the number of cells (Rook and coworkers, 1970).

The number of melanocytes in the basal layers of the skin stays relatively constant between 1,000 and 2,500 cells per square mm depending on the part of the body. They are replaced by cell division as required. In contrast, the melanocytes in the hair bulbs, however, are not replaced, and when the last one in the hair bulb dies, that hair continues to grow but the new growth is without pigment.

Like other cells in the body, melanocytes can lose control of their proliferation rate and grow in excess of requirements, ultimately to the destruction of the host. This is a form of cancer known as melanoma.

Cancer is a state where cells, which should in the normal healthy state be under the proliferative control of their surrounding tissues, have partially lost that control and are multiplying at a faster rate than that needed for normal tissue replacement. It is clear that the defect is in the cell, not the animal, because the cancer cell will continue to multiply when transplanted into other animals. Also, if every tumour cell is removed from a tumour-bearing animal, the animal will be cured and will be no more likely than its litter mates to start another independent tumour. The change is also a genetic one involving the DNA of the cell. This must be so because the daughters of tumour cell divisions are also tumour cells. Thus, we have in cancer a genetic change to a loss of response to a signal from the rest of the tissues of the animal. If that signal is an order not to divide, or in chemical terms an inhibitor of cell division and the signal increases in strength when too many cells divide, then the process is described as *negative feed-back inhibition*.

The search for inhibitors of cell division has been long and frustrating, with numerous pitfalls. No generally accepted control substance has yet been found. Weiss (1952) introduced a theory concerning the specific control of growth which Lenicque (1959) summarized in the following way: 'Protoplasm synthesis of a given organ yields (a) templates for further reproduction, and (b) accessory diffusible compounds capable of inactivating the former. As the latter, accumulating in the common humoral pool, reach critical concentrations, growth ceases. Partial removal of an organ, by reducing the concentration of (b), will automatically entail "compensatory growth" in the rest of the organ.'

Weiss (1952) has included inhibition of growth in a theory of differentiation. He created the conceptions of 'templates' and 'anti-templates', these denoting on the one hand the positive factors of growth, acting as centres of protein synthesis and, on the other hand the negative factors, or inhibitors of growth, which appear as the tissues grow. A balance ensues which permits the tissues and organs to adjust in relation to each other (Weiss and Kavanau, 1957).

Bullough and Laurence (1960) applied the concept of *negative feed-back inhibition* to their work on the control of mitosis in the epidermis. They showed that epidermal extracts could inhibit epidermal mitosis without influencing other cell types. Extracts of tissues other than epidermis did not have the same power to depress epidermal mitotic activity. Bullough and Laurence (1964) proposed that a tissue-specific mitotic inhibitor with the characteristics of that obtained from the epidermis should be called a 'chalone' (Editorial, 1973; Watts, 1974). Operationally, it was suggested that the epidermis contained a tissue-specific inhibitor produced in the distal cell layers of the epidermis. The inhibitor diffused into the basal cell layer to inhibit mitosis; and that, when the distal cells are sufficiently numerous, the production of new basal cells will cease. In this way the epidermal thickness is maintained from day to day, and it is restored after epidermal damage (Bullough, 1973).

Evidence has been presented for the existence of chalones for over a dozen different tissues but not one has yet been completely characterized. Another criticism is that the growth and differentiation of animal organs are too complex to be explained solely in terms of the homeostatic regulation of cell division by chalones. If chalones are physiological controllers of cell division, they do not act by a simple *negative feed-back mechanism* but require the action of a substance to decrease their concentration (Argyris, 1972). A more detailed analysis has shown that chalone systems must be more complicated than was originally thought. Inversen (1973) suggests that instead of a direct proportional feed-back based on the momentary concentration of chalones in or between cells, possibly the speed and direction of alterations in concentration may be the signal to which cells react. The response to a change in chalone concentration, Iversen calls *derivative feed-back*, and the cells' memory for previous chalone concentrations he calls *integral feed-back*.

Of all the chalones previously suggested, Houck and Hennings (1973) considered that only four were sufficiently substantiated to be worth reviewing (epidermal cells, melanocytes, fibroblasts, and lymphocytes). No explanation was given as to why granulocyte chalone was not included (Rytömaa and Kiviniemi, 1968a,b,c, 1969a,b, 1970; Paukovits, 1971).

In this chapter we are primarily concerned with melanocytes, their mitotic rate and inhibitors thereof. Giacometti and Allegra (1967) have shown that after wounding there was a wave of mitosis in melanocytes found among the epidermal cells of the guinea-pig. The time sequence observed for melanocyte mitosis differed from that of epidermal mitosis and thus it is reasonable to assume that melanocytes have their own independent regulation mechanism. Bullough and Laurence (1968) found that their crude extract containing epidermal chalone from pig skin contained a substance that reduced the mitotic rate of melanoma cells. The more highly purified material, however, worked only against epidermal cells and did not have any detectable action against melanocytes. Their assumption was that the crude material contained a melanocyte inhibitor which was removed during purification. The crude pig skin extract containing the anti-melanocyte activity was used in an experiment by Mohr and coworkers (1968) on mice bearing the Harding-Passey melanoma and on Syrian hamsters with amelanotic melanomas with a dramatic result. The tumours first ulcerated and then regressed. The effect was first thought to be due to an effect of the melanocyte chalone, although disappointingly the authors have now advanced an alternative hypothesis based on bacterial contamination (Mohr, Hondius Boldingh and Althoff, 1972).

Adachi, Hu, and Kondo (1971) showed that an extract of B16 mouse melanoma caused complete degeneration of cultured melanoma cells after adding the extract to the culture medium. The active ingredient was thought to be RNA of molecular weight between 20,000 and 50,000 daltons. Dewey (1973) reported a heat-labile extract of Harding-Passey melanoma that was also a potent inhibitor of melanocytes in culture, but the active principle has a much lower molecular weight than that reported by Adachi, Hu, and Kondo (1971). Seiji and coworkers (1974) also used an extract of the Harding-Passey melanoma and obtained two different active fractions, one was of protein in nature in the molecular weight range 30,000 to 50,000 daltons, and the other was a mixture of protein and RNA; both portions were required for activity.

The appeal of a substance which specifically or selectively stops the division of melanocytes is overwhelming to cancer workers. In chemotherapy the aim is to kill cancer cells selectively and thus far has not been achieved. In the special case of melanocyte specificity, if such a drug were given to stop all melanoma and melanocyte division, the maximum adverse effect would be the ultimate loss of pigement producing an albino appearance.

The published work on chalones and inhibitors in melanoma tumours make their characterization a worthwhile task. The presence of a melanocyte inhibitor was confirmed, but unfortunately the specificity was far from absolute. When higher concentrations of the inhibitor became available, these extracts also arrested most of the non-melanocyte cell lines in culture. Nevertheless, it seemed worthwhile to pursue the purification and establish the identity of this potent inhibitor.

II. PURIFICATION OF THE INHIBITOR

Initial work on the isolation and purification of the inhibitor used the rate of change of the increase in cell numbers as a measure of biological activity. Although such a method is simple there are snags, the major defect being the uncertainty of the fate of those cells which do not divide. Assuming an inhibitor is added at a concentration which prevents half of the cells going into division, then, after one cell cycle, the total cell number will be 150% of the starting number instead of 200%, if all cells had divided. Thus, 50% inhibition will only bring the cell count down by 50 in 200, or 25%. If radio-labelled thymidine incorporation is used as a measure of cell division, then 50% inhibition of all cells going through the cell cycle will cause a 50% reduction in the uptake, assuming of course that conditions are adjusted so that the counts are not near background levels. A general obstacle to using the uptake of naturally occurring, but radioactive, precursors to study cellular inhibition is that any test material containing the precursor will dilute the radioactive material. Because the overall specific activity is reduced, the uptake of radioactivity by the cells will be less, and will be falsely recorded as a positive result.

In the case of the partially purified extract tested on melanocytes in culture, identical results were obtained regardless of whether cell counting or radioactive thymidine incorporation was used as the test of inhibitor activity. In general, after the initial purification step, thymidine incorporation was used as the standard assay procedure to obtain the benefit of the increased accuracy as outlined above.

The saline extract of homogenized Harding-Passey melanoma cells contained a dialysable inhibitor. No activity remained inside the dialysis sac after exhaustive dialysis. The active ingredient, originally but erroneously assumed to be a peptide, would pass through a 2,000 MW but not a 500 MW amicon membrane molecular filter. The results, however, were not absolutely clear cut; variations between different experiments using the same membrane filters were considerable.

A major problem in the use of the growth of cells in culture, or their ability to incorporate thymidine, as an end point was that any chemical, such as buffer, added in the purification stage can itself have an effect on the assay. Both phosphate and Hepes buffers reduced the growth rate of the melanoma cell line, even when added to medium at optimal pH. The most practical solution to the problem of removing added substances has been to add the minimum or nothing to the extract, and to use purification methods based on physical rather than chemical systems.

Considerable purification of the active ingredient was obtained by running the extract down a column of Sephadex G25 equilibrated and eluted with distilled water. All biological activity was found in a narrow band well separated from the void volume and most of the other impurities. The question of the degree of purification became an impossible one to answer without knowing the chemical nature of the substance. For example, the activity per unit weight of nitrogen is of no value unless the active ingredient is known to contain nitrogen. The dry weight was equally unhelpful because nearly all of the dry weight in the crude extract was sodium chloride. The percentage recovery of the total activity put on the Sephadex column was also of no help as this value often exceeded 100%. The reason for the enhanced inhibition of the purified material was due to the presence of a stimulator in the crude extract. This stimulator, which made the cells grow faster, was bound onto the Sephadex by some form of absorption and was not eluted until after the low molecular weight material and salts had come through.

For routine use the column was run in isotonic sodium chloride rather than in distilled water. The resolution was not quite so high but the column could be re-used a very large number of times with identical results.

The elution position of the active ingredient was just ahead of vitamin B12, added as a marker, suggesting a molecular weight marginally in excess of 1,357.

Further purification was obtained by collecting the active fraction from the Sephadex column, concentrating the material by freeze drying and fractionating it on a column of Biogel P2, run in distilled water. The inhibitor was eluted in front of the band of salts (added in the sample), and was almost free from all other detectable impurities. A fraction containing most of the biological activity had no detectable, visible or UV absorption.

The final purification stage consisted of electrophoresis in previcon support (Smith, 1968). The ammonium bicarbonate in the buffer was removed by freeze drying and the resultant product was identified as spermidine.

The log of the fraction of thymidine incorporation was proportional to the amount of extract or spermidine added. The slope of this dose effect line on the semi-log plot was a measure of the activity. By plotting the results of a large number of repeat samples in this manner, it was found that the absolute value of activity of the extracts also depended on:

(1) Cell concentration.
(2) Serum concentration.
(3) Batch serum.
(4) Type of cell culture container.
(5) The order in which the constituents were added.

Once the inhibitor had been identified as spermidine, all the foregoing anomalies could easily be explained. The apparent high molecular weight on Sephadex G25 was checked by running a mixture of spermidine and vitamin B12 through the column, and again the spermidine came off just ahead of the vitamin. Thus, the anomalous behaviour was presumably due to the high charge on the spermidine preventing entry into the Sephadex matrix.

Much of the variability in recovery of spermidine could be attributed to the pH of the sample during freeze drying. In acid solution, the spermidine hydrochloride is quite stable but in alkaline solution, such as in ammonium bicarbonate after electrophoresis, the spermidine-free base is volatile and easily lost.

Spermidine and spermine are known to be toxic to cells in the presence of amine oxidase contained in serum (Chapters 19, 22), and this accounts for the variable effect of the container and of different batches of serum. In culture tubes the cells settled to the bottom, whereas much of the spermidine oxidation was probably well away from the cells. More of the toxic products would have combined with components of the serum, or evaporated from the surface in a time shorter than that required to reach the cells by diffusion. In a flat culture flask all the medium was only three or four mm from the cells, and in this situation the cells were most sensitive. When a duplicate of the same flask was incubated on end so that the same number of cells settled over a smaller surface area with a larger distance to the surface of the same overall volume of medium, the cells were more resistant to the same concentration of extract or spermidine.

III. MELANOCYTES AND POLYAMINES

In order to demonstrate that the biological effect of the tumour extract was due only to its polyamine content, an attempt was made to see if the extract would behave like spermidine in reversing the inhibition of methylglyoxal-*bis*-(guanylhydrazone) (methyl-GAG) when added simultaneously (Dewey, 1978). Methyl-GAG inhibits the biosynthesis of spermidine and spermine (Pegg, Corti and William-Ashman, 1973). Both polyamines are involved in the regulation of cellular growth (Chapters 1–4). If the formation of polyamines was prevented by adding methyl-GAG, the cells were held in G_1 (Heby and coworkers, 1977). But if excess spermidine was added with methyl-GAG, then the cells were not held (Mihich, 1963). Once cells have been blocked in G_1 by lack of spermidine, however, they are irreversibly blocked and cannot be induced to proceed through the cell cycle by the addition of spermidine at a later time (Heby and coworkers, 1977).

Using Harding-Passey melanoma cells in culture, it was found that spermidine ultimately reversed the inhibition of methyl-GAG at relatively high concentrations of the latter. The initial experiments at low concentrations of methyl-GAG showed exactly the opposite effect (Figure 1), i.e. methyl-GAG reversed the inhibition by spermidine. Even with concentrations of spermidine up to 50 μM, which alone would inhibit radioactive thymidine uptake by more than 99%, the inhibition was completely abolished by 250 μM methyl-GAG. Very low concentrations of methyl-GAG in the absence of added spermidine slightly increase the amount of thymidine incorporation (Figure 1), possibly due to the effect of the small amount of endogenous polyamines made by the cells and already present in the culture. Growing the cells in very low concentrations of methyl-GAG (1 μM) marginally, but not significantly increased the plating efficiency, which was normally around 20%.

We have been investigating the mechanism of the reverse methyl-GAG effect to find whether the toxic oxidation products of polyamines are inactivated by this compound; or whether methyl-GAG, which is claimed to be highly specific for S-adenosylmethionine decarboxylase (Williams-Ashman and Schenone, 1972; Pegg, Corti and Williams-Ashman, 1973; Heby and Russell, 1974) is in fact also an inhibitor of amine oxidase.

Pure spermidine can be measured in small quantities by gas liquid chromatography; but if the initial sample is serum or urine, the measurement

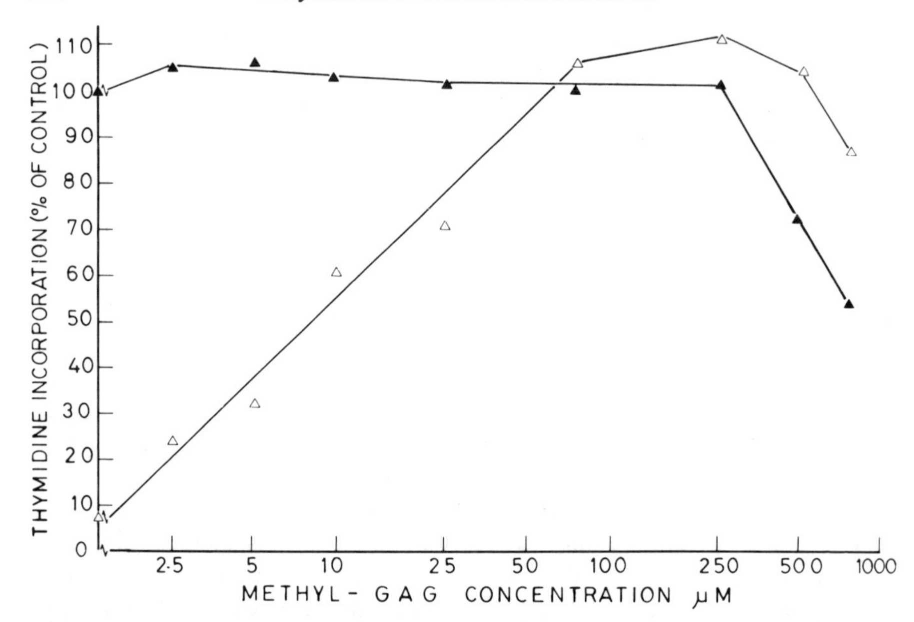

Figure 1 The effect of spermidine and methyl-GAG on the ability of Harding-Passey cells in culture to incorporate tritium-labelled thymidine. ▲ methyl-GAG △ methyl-GAG and 10 μM spermidine

is much more difficult and less reliable (Gehrke and coworkers, 1973; McGregor and coworkers, 1976). Thus, measuring the disappearance of the substrate to study the enzyme reaction kinetics is not a simple matter when the substrate is spermidine in serum. Similarly, measuring the enzyme product presents a different set of problems. The toxic element produced by oxidation of spermidine in animal tissues has been identified as β-aminopropionaldehyde (Quash and Taylor, 1970). On the other hand, acrolein has also been identified as a major product in serum (Alarcon, 1964), and evidence has been produced to show that acrolein is the toxic aldehyde enzymatically produced from spermidine and spermine (Alarcon, 1970).

We have used the spectrophotometric method for the determination of acrolein in combustion gases (Cohen and Altshuller, 1961) in order to measure the acrolein produced from mixtures of bovine serum and spermidine and confirm the results of Alarcon. The technique we used was to continually blow air through a buffered solution of spermidine with serum, or purified enzyme, incubated in a water bath at 37°C.

The volatile acrolein was swept away by the stream of air, which, after passing through an antifoam trap, was bubbled through cold alcohol at —70°C. The acrolein trapped in the alcohol could then be measured directly. Acrolein was produced in high yield from spermidine by the amine oxidase in fetal bovine serum and by commercial purified serum amine oxidase (Miles Laboratories Ltd.), but not from purified diamine oxidase (Sigma Chemical Co). Atmospheric oxygen was used in the oxidation of spermidine under the conditions described above. No acrolein was formed if the air flowing through the system was replaced by nitrogen.

When methyl-GAG was added to the incubation mixture at a final concentration of 100 μM, a low concentration which had no effect on short term cell culture, the oxidation of spermidine to acrolein was inhibited by more than 99%. Partial inhibition of the oxidase was observed at the lower concentrations at which methyl-GAG had no deleterious effects when continuously present in cells in culture over long periods. In control experiments, high concentrations of methyl-GAG could be added without any effect on the recovery of acrolein when spermidine was replaced in the reaction mixture by acrolein and the recovery of acrolein was measured. Thus, it appears that the reason for the reverse effect of low concentrations of methyl-GAG on cell cultures in the presence of spermidine is not an interference with the toxic product but an inhibition of the amine oxidase converting spermidine to acrolein. Thus, methyl-GAG can no longer be claimed to be a specific inhibitor of S-adenosylmethionine decarboxylase, although there is no doubt that this is its main action when preventing cells entering the S phase of the cell cycle.

IV. *IN VIVO* STUDIES

Purified melanoma extract was tested for *in vivo* anti-tumour effects. The total activity from 100 g of mouse tumour, in physiological saline, was injected into each of five mice bearing the Harding-Passey melanoma. Each mouse received the dose as injections into tumour of 0.1 ml twice per day for five days. After 12 days from the start of the treatment, the tumours had completely regressed in two of the mice. One of the five had a delay in the rate of tumour growth, and in the remaining two mice there was no effect. In 15 control tumour-bearing mice from the same batch, injected at the same time but with physiological saline only, there was no regression.

Mohr and coworkers (1968) have already reported that pig skin extract injected into melanoma-bearing mice or hamsters resulted in the complete regression of the tumours. Later the effect was attributed to *Clostridium* contamination in the extract (Mohr, Hondius Boldingh and Althoff, 1972).

The initial experiments of Mohr and coworkers have now been repeated and confirmed (Dewey, Butcher and Galpine, 1978) under conditions where *Clostridium* can be ruled out on the following grounds. Bacterial analysis of the extract, before filtration, in a *Clostridium* selective medium, failed to demonstrate the presence of contamination. The extract was passed through a sterile 0.22 μm filter before injection. The extraction procedure was completed on the day of slaughter of the pig and kept cold, so that even if *Clostridium* had been present, there would not have been time for it to grow and produce toxin. Again, in two out of the five treated mice the tumour regressed and did not reappear, whereas there was no regression in any of the saline treated control mice.

In all the various pig skin extracts tested there was a total of 18 complete regressions among 65 mice. Among the controls, treated with saline or other inert substances, there was a total of one regression among 94 mice.

Now we know that the mechanism of inhibition of the tumour extract in cell cultures is through the action of a combination of amine oxidase and spermidine, and it is, therefore, logical to see if similar effects can be obtained *in vivo* with mixtures of polyamines and enzymes. Spermidine itself is highly toxic to mice, but by the use of sublethal combinations of polyamines and enzymes interesting results have been obtained. In preliminary experiments, complete remissions of Harding-Passey melanoma have been achieved in 100% of the mice (Dewey, Butcher and Galpine, unpublished), but complete absence of tumour recurrence is not yet possible. Analogues of spermidine and other polyamines which can be degraded by similar enzymes offer a new dimension in cancer chemotherapy (Israel and Modest, 1971).

V. CHALONES AND POLYAMINES

The role of polyamines on cells is two-fold, in neither role can they in any way be construed as being a chalone. The polyamines spermidine and spermine are essential requirements for cell division; if their formation is blocked by inhibiting any of the enzymes involved in their formation, then the cell will not enter the S phase of the cell cycle. But inhibition of cell division by removing polyamines is no more a control system than is preventing cell division by withholding or interfering with the supply of any other essential ingredient. There is no reason to suppose that the tissues may not normally control division by withholding one vital chemical, but if it is done this way, the chemical is unlikely to be a polyamine because in the absence of spermidine, the cell does not appear to stop cycling completely but goes into unbalanced growth from which it cannot be revived by added spermidine.

The other action of polyamines is their degradation in the presence of serum amine oxidase to acrolein, which will inhibit cells at very low concentrations, and this inhibition is reversible (Willmer and Wallersteiner, 1939). At higher concentrations acrolein will kill cells, a property that would not be expected of a biological cell regulator. But even in the case of chalones, there is no fundamental reason why a physiologically excessive concentration should not be lethal. Moreover, there is no great specificity in either polyamines or acrolein.

The attempt to find a negative feed-back substance controlling the rate of cell division has not yet been successful. It could be that control substances are impermeable to the normal cell surface but pass between adjacent cells through low resistance junctions and never leave the cell complex. Such a substance extracted from tissues could not be tested by conventional methods as it would not get into the cells.

VI. SUMMARY

In summary it can be said that although we have learnt much about substances that affect cell division, we are still totally ignorant of the fundamental mechanism that makes the decision which cell will divide and which cell will not. Evidence for chalones as the control mechanism falls far short of proof. Polyamines, although involved with cell multiplication, probably play little or no part in the decision making before the initial steps for a new round of cell replication has begun.

ACKNOWLEDGEMENT

I thank the Cancer Research Campaign for support.

REFERENCES

Adachi, K., Hu, F., Kondo, S. (1971). *Biochem. Biophys. Res. Commun.*, **45**, 742–746.
Alarcon, R. A. (1964). *Arch. Biochem. Biophys.*, **106**, 240–242.
Alarcon, R. A. (1970). *Arch. Biochem. Biophys.*, **137**, 365–372.
Argyris, T. S. (1972). *Am. Zoologist*, **12**, 137–149.
Bullough, W. S. (1973). *Nat. Cancer Inst. Monogr.*, **38**, 5–15.
Bullough, W. S., and Laurence, E. B. (1960). *Proc. Royal Soc. B.*, **151**, 517–536.
Bullough, W. S., and Laurence, E. B. (1964). *Expt. Cell Res.*, **33**, 176–194.
Bullough, W. S., and Laurence, E. B. (1968). *Europ. J. Cancer*, **4**, 607–615.
Cohen, I. R., and Altshuller, A. P. (1961). *Analyt. Chem.*, **33**, 726–733.
Dewey, D. L. (1973). *Nat. Cancer Inst. Monogr.*, **38**, 213–216.
Dewey, D. L. (1978). *Cancer Lett.*, **4**, 77–84.
Dewey, D. L., Butcher, F., and Galpine, A. R. Tumour regression factor in pig skin (In preparation).
Editorial (1973). *The Lancet*, **(ii)**, 1248–1249.
Gehrke, C. W., Kuo, K. C., Zumwalt, R. W., and Waalkes, T. P. (1973). The determination of polyamines in urine by gas-liquid chromatography. In D. H. Russell (Ed), *Polyamines in Normal and Neoplastic Growth*, Raven Press, New York, pp. 343–353.
Giacometti, L., and Allegra, F. (1967). *Ad. Biol. Skin*, **8**, 89–95.
Heby, O., Marton, L. J., Wilson, C. B., and Gray, J. W. (1977). *Europ. J. Cancer*, **13**, 1009–1017.
Heby, O., and Russell, D. H. (1974). *Cancer Res.*, **34**, 886–892.
Houck, J. C., and Hennings, H. (1973). *FEBS Lett.*, **32**, 1–8.

Israel, M., and Modest, E. J. (1971). *J. Med. Chem.*, **14**, 1042–1047.
Iversen, O. H. (1973). *Nat. Cancer Inst. Monogr.*, **38**, 225–230.
Lenicque, P. (1959). *Acta Zoologica*, **XL**, 1-202.
McGregor, R. F., Sharon, M. S., Atkinson, M., and Johnson, D. E. (1976). *Prep. Biochem.*, **6**, 403–419.
Mihich, E. (1963). *Cancer Res.*, **23**, 1375–1389.
Mohr, U., Althoff, J., Kinzel, V., Süss, R., and Volm, M. (1968). *Nature, Lond.*, **220**, 137–139.
Mohr, U., Hondius Boldingh, W., and Althoff, J. (1972). *Cancer Res.*, **32**, 1117–1121.
Paukovits, W. R. (1971). *Cell Tiss. Kinet.*, **4**, 539–547.
Pegg, A. E., Corti, A., and Williams-Ashman, H. G. (1973). *Biochem. Biophys. Res. Commun.*, **52**, 696–701.
Quash, G., and Taylor, D. R. (1970). *Clin. Chim. Acta*, **30**, 17–23.
A. Rook, D. S. Wilkinson, and F. J. G. Ebling (1968) (Eds), *Textbook of Dermatology*, Vol. 2, Blackwell Scientific Publications.
Rytömaa, T., and Kiviniemi, K. (1968a). *Europ. J. Cancer*, **4**, 595–606.
Rytömaa, T., and Kiviniemi, K. (1968b). *Cell Tiss. Kinet.*, **1**, 329–340.
Rytömaa, T., and Kiviniemi, K. (1968c). *Cell Tiss. Kinet.*, **1**, 341–350.
Rytömaa, T., and Kiviniemi, K. (1969a). *Nature, Lond.*, **222**, 995–996.
Rytömaa, T., and Kiviniemi, K. (1969b). *Cell Tiss. Kinet.*, **2**, 263–268.
Rytömaa, T., and Kiviniemi, K. (1970). *Europ. J. Cancer*, **6**, 401–410.
Seiji, M., Nakano, H., Akiba, H., and Kato, T. (1974). *J. Invest. Dermatol.*, **62**, 11–19.
Smith, I. (1968) (Ed) *Chromatographic and Electrophoretic Techniques*, Academic Press, London and New York.
Watts, G. (1974). *World Medicine*, **9**, 38–45.
Weiss, P. (1952). *Science*, **115**, 487–488.
Weiss, P., and Kavanau, J. L. (1957). *J. Gen. Physiol.*, **41**, 1–47.
Williams-Ashman, H. G., and Schenone, A. (1972). *Biochem. Biophys. Res. Commun.*, **46**, 288–295.
Willmer, E. N., and Wallersteiner, K. (1939). *J. Physiol.*, **96**, 16P-17P.

Chapter 21

Association of Chalones with Polyamines: Evidence for a Chalone from JB-1 Ascites Tumour Cells as a Spermine Complex

NIELS M. BARFOD

I. INTRODUCTION

It has been assumed for some time that cell proliferation is controlled by endogenous, tissue-specific inhibitors of cell proliferation (Weiss and Kavanau, 1957). Such factors are described as 'chalones' (for review articles, see Houck (1976)). Two types of chalones acting on the G1 and G2 phases, respectively, of the cell cycle of epidermal cells (Elgjo, 1973) and of intestinal cells (Brugal and Pelmont, 1975) have been described.

The chalone concept deals with the growth regulation of normal tissues, but there is some evidence that malignant tumours may synthesize and respond to the chalone of their tissue of origin (Rytömaa and Kiviniemi, 1968).

In a series of studies it has been shown that cell-free ascites fluid from the JB-1 ascites tumour (a hypotetraploid murine plasmacytoma) in the plateau phase of growth, contains substances which arrest JB-1 cells at G_1-S border and in the G_2 phase (Bichel 1970, 1971, 1972, 1973). These substances seem to conform to the definition of chalones in that they induce a cell specific and reversible inhibition of the cell cycle of JB-1 cells (Bichel 1971, 1972). Since more than 80% of the non-cycling cells in the plateau tumours are in G_1 and these cells are responsible for most of the recycling given the proper conditions, we have concentrated on the purification of the G_1-chalone.

The present paper describes the partial purification of this G_1-chalone. The behaviour of the G_1-inhibitory activity using the purification techniques indicates that the active compound consists of a complex between the polyamine spermine and an as yet unidentified acidic carrier. The involvement of polyamines in chalone effects is a very recent observation (Allen and coworkers, 1977; Barfod and Marcker, 1977).

II. PREPARATION OF CHALONE EXTRACT

As starting material for purification of the G_1-chalone we use cell-free ascitic fluid from old JB-1 ascites tumours (Barfod and Bichel, 1976). It is also possible to extract the inhibitory activity by incubating JB-1 tumour cells in Hank's balanced medium at 37°C for 4 h. Cell-free ascitic fluid and JB-1 conditioned Hank's medium are ultra-filtrated through a XM50 ultrafilter which has a cut-off level of 50,000 D. The G_1-inhibitory activity is found in the filtrate.

III. TEST SYSTEMS

All cell cycle perturbations are analysed by measurement of the single cell DNA content by means of flow cytometry (FCM).

A. Estimation of Cell Cycle Compartments from FCM Data

An easy hand calculation method of decomposing DNA histograms into G_1, S and (G_2 + M) compartments (Reddy and coworkers, 1973) described in the legend to Figure 1, is used.

Several investigators have suggested substituting computer techniques for DNA histogram analysis. None of these techniques, however, has yet been proved to be superior in application. The reason for this may be the instrumental instability of the rather complex equipment used in FCM. Such instability can be reduced by appropriate adjustment but very sensitive mathematical methods of analysis of the histograms may lead to erroneous conclusions, if the data are not sufficiently accurate. Consequently a more rough estimation of the measurements may be sufficient. Indeed, the values of the G_1, S and (G_2+M) compartments in the JB-1 ascites tumour calculated by this method have shown a good fit to the values obtained from growth curves and labelled mitosis analysis (Dombernowsky and Bichel, 1975).

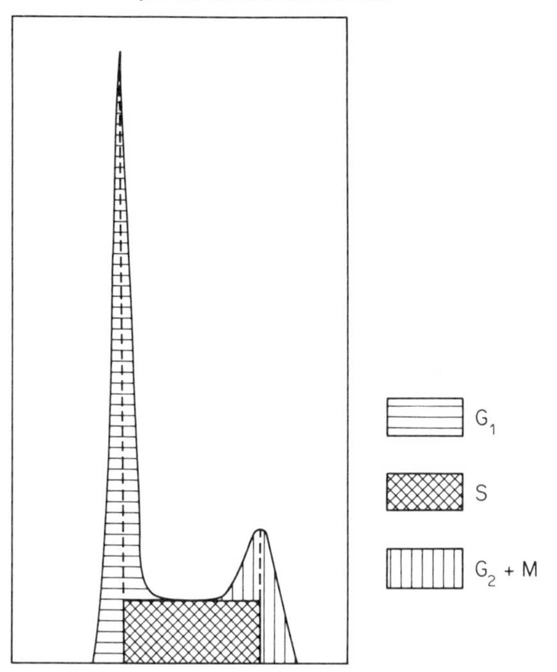

Figure 1 Abscissa: channel number (rel. fluorescence intensity); ordinate: cells/channel. Flow cytometric determination of DNA content of JB-1 tumour cells. Two peaks are seen in the DNA histogram; the first represents cells with G_1 DNA content, the second cells with (G_2 + M) DNA content, while cells in the S phase give relative fluorescence values between the G_1 and (G_2 + M) maxima. The percentage of cells in the S phase may be estimated as the rectangle between two vertical lines through the G_1 and the (G_2 + M) peaks and a horizontal line through the trough between G_1 and (G_2 + M). If a vertical line is drawn halfway between the G_1 and the (G_2 + M) values, the area to the left of this line represents G_1 + (S/2) cells, and that to the right (G_2 + M) + (S/2) cells, from which the amount of cells with G_1 and (G_2 + M)DNA content can be calculated.
Reproduced by permission of Academic Press Inc. New York

B. *In Vitro* Test System

JB-1 cells established *in vitro* are synchronized by starvation. In starved cultures 85% of the cells are in G_1 and when cells from these cultures are explanted to new flasks and fresh medium, a synchronous wave of cells will traverse the cell cycle as seen from the DNA distributions in Figure 2. Figure 3 shows the calculated values of G_1, S and (G_2 + M). During the first 8 h, no major changes in the DNA distributions are observed. From 8 to 22 h, a wave of cells leave the G_1 phase (the percentage of cells with G_1 DNA content decreases from 90 to 2%) and passes through the S phase (the percentage of cells with

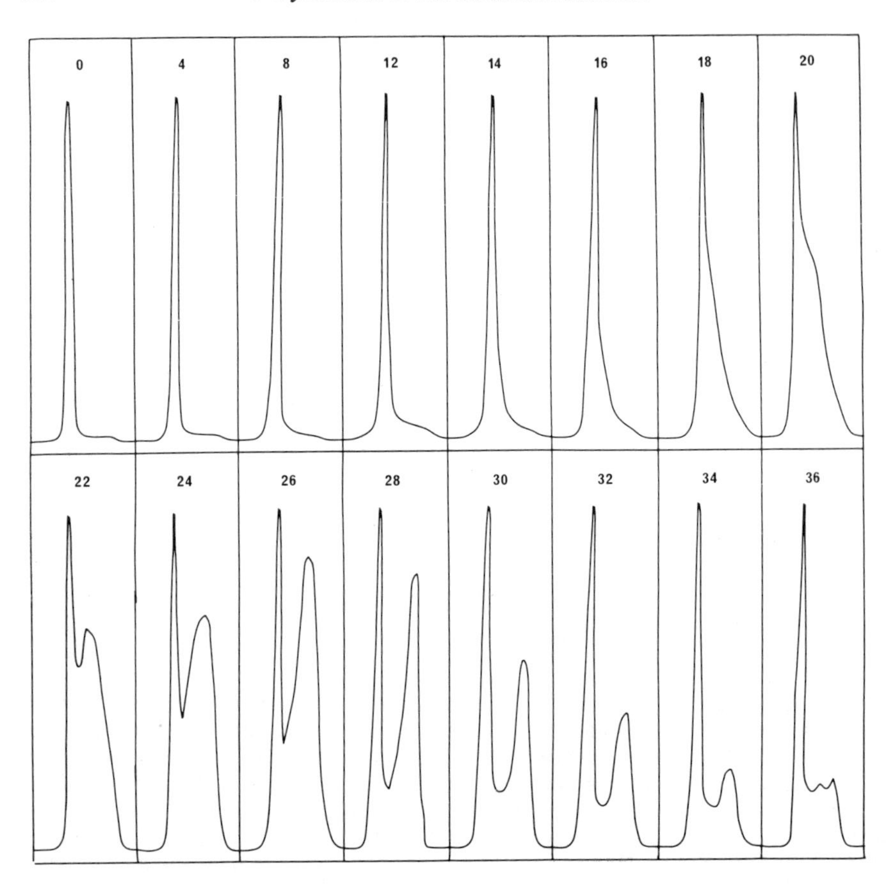

Figure 2 Abscissa, ordinate, see Figure 1. DNA distribution of synchronized JB-1 cells subcultivated at 0 h from stock cultures and harvested during the next 36 h at the times indicated. Reproduced by permission of Academic Press Inc., New York

S DNA content increases from 10 to 90%). During the next 6 h (22–28 h), the cohort of cell leaves the S phase and goes into the G_2 phase (the percentage of cells with (G_2 + M) DNA content increases from 6 to 30%). At the same time cells travel into the G_1 phase again. From 28 to 36 h, the cells leave the G_2 phase (decrease in the percentage of cells with (G_2 + M) DNA content). At 36 h, a new cycle starts as the number of cells with S DNA now increases, while the cells with G_1 DNA are decreasing. The degree of synchronization has now diminished, probably due to variations in the cycle time of individual cells. At approximately 48 h, complete desynchronization of the cells has occurred.

The G_1 inhibitory activity present in the XM50 filtrate of cell-free ascites or

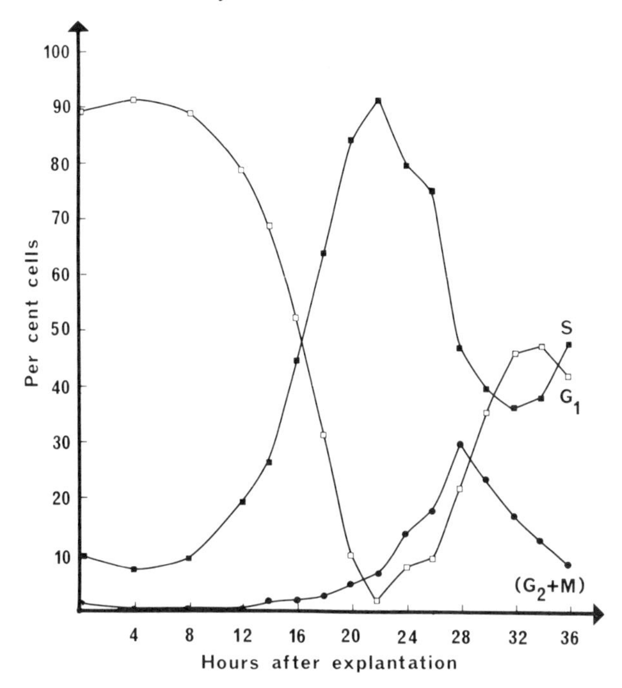

Figure 3 Abscissa: time after subcultivation (hours); ordinate: percentage of cells with G_1 (□ — □); S (■ — ■); and (G_2 + M) DNA content (● — ●). Percentage of G_1, S and (G_2 + M) cells calculated from the DNA distributions in Figure 3. Reproduced by permission of Academic Press Inc., New York

of cell conditioned medium is estimated as an inhibitory effect of the synchronous wave travelling from G_1 into the S phase. Figure 4 represents a dose response experiment measured at 18 h after explantation and shows the gradual inhibition of the influx of cells with G_1 DNA content into the S phase. The calculated values are indicated in the table to Figure 4.

The reversibility of action on the proliferation of the cells is routinely tested by consecutive measurements of chalone treated and untreated cells as we have reported previously (Bichel, Barfod and Jakobsen, 1975; Barfod, 1977).

C. *In Vivo* Test System

The purification and isolation of chalones require relatively simple and rapid tests for screening a great number of fractions. It has become apparent, however, that more reliable test systems are urgently needed for assaying the properties of chalones (Maurer, Weiss and Laerum, 1975; Thornley and Laurence, 1975).

Thus, owing to difficulties with the assessment of cell specificity using the *in*

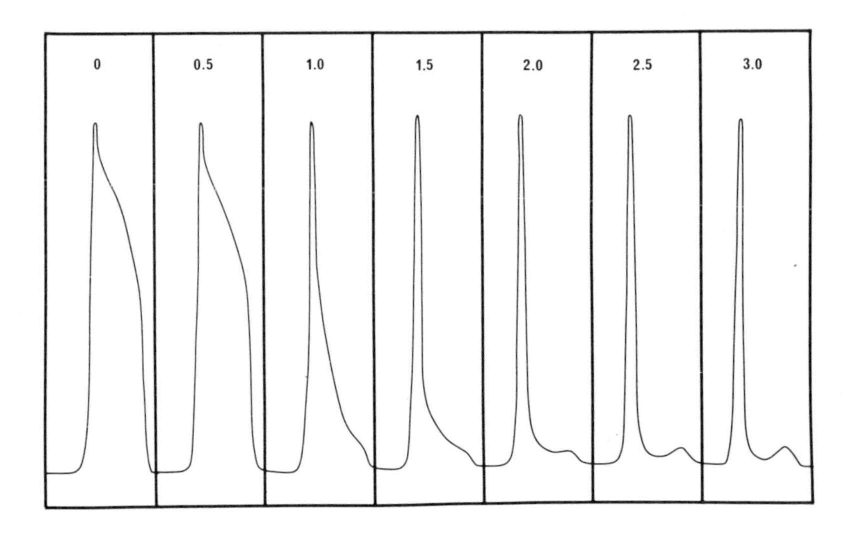

ml of XM50 filtrate	0	0.5	1.0	1.5	2.0	2.5	3.0
% cells in S phase	89	84	53	28	15	8	11
% inhibition	0	6	40	69	82	90	88

Figure 4 Abscissa, ordinate, see Figure 1. Effect of 0.5, 1.0, 1.5, 2.0, 2.5 and 3.0 ml of XM50 filtrate on progression of cells from G_1 to S phase *in vitro* measured at 18 h after subcultivation

vitro system described earlier in the text, we recently developed an *in vivo* test system based on flow-cytometric analysis of the regenerative growth after percutaneous aspiration of JB-1 and $L1A_2$ ascites tumour derived from a plasmacytoma and a bronchogenic sarcoma, respectively, in their plateau phase of growth (Bichel and Barfod, 1977).

Principally the *in vivo* test system is very similar to the *in vitro* test system. In the plateau phase of growth approximately 45 to 60% of the cells are in G_1 phase. After percutaneous aspiration an initiation of regenerative growth is induced. After 8 to 10 h a considerable influx of cells with G_1 DNA content into the S phase is observed in both tumours.

In the JB-1 tumour, these initial regenerative changes can be reversibly blocked by injections of cell-free plateau JB-1 ascitic fluid or an XM50 ultrafiltrate of this ascites as seen in Figure 5 to the left. In contrast, no delay in the regenerative changes is observed in the $L1A_2$ tumour after treatment with JB-1 ascites or the ultrafiltrate (Figure 5 to the right).

This *in vivo* assay has recently been improved by using plateau phase tumour cells retransplanted to a new host (Barfod, 1980). Using this technique

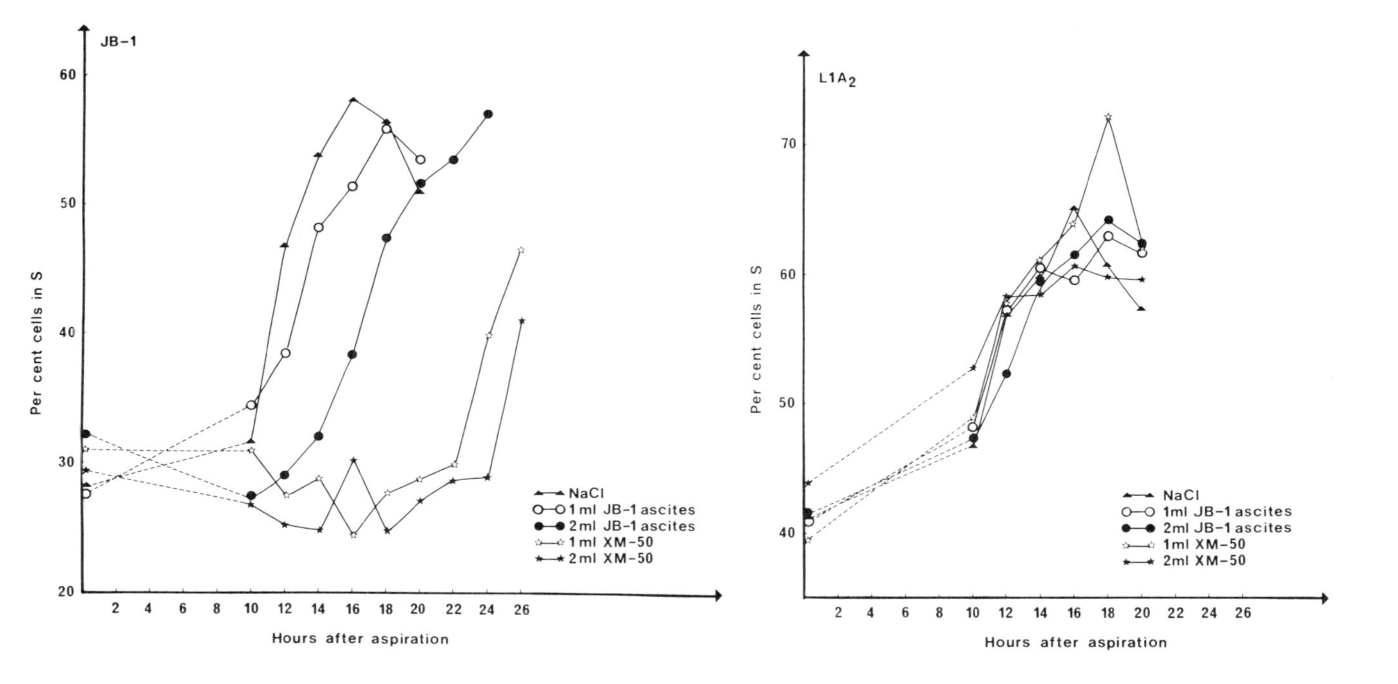

Figure 5 Effect of cell-free JB-1 ascites and a XM50 ultrafiltrate of this on the regenerative growth of JB-1 and L1A$_2$ ascites tumours. Percentages of cells with S DNA content as a function of time after aspiration at 0 h of plateau JB-1 tumours (left part) or L1A$_2$ tumours (right part) and i.p. injections at 1, 2, 3 and 4 h after aspiration of 1 ml of isotonic saline, 1 ml of cell-free JB-1 ascitic fluid, 2 ml of cell-free JB-1 ascitic fluid, 1 ml of XM50 ultrafiltrate of JB-1 ascitic fluid, 2 ml of XM50 ultrafiltrate of JB-1 ascitic fluid. Each point represents mean of four mice. Reproduced by permission of the journal of *Cell and Tissue Kinetics*

an entry of G_1 cells into S phase similar to that of regenerating cells is observed. The new technique considerably reduced the variability in the experimental results. Consequently, fewer mice are required for the *in vivo* assays.

IV. PURIFICATION

A. Molecular Weight Estimation by Ultrafiltration

In order to characterize the G_1-chalone further with regard to molecular weight, we studied the partition of the G_1-chalone activity with ultrafiltration membranes with a lower cut-off level of molecular weight than the XM50 membrane used in the first purification step. Using the PM10 membrane, the activity is found in the filtrate. In most experiments with UM2 the activity is located in the retinate, though in some experiments the activity is equally distributed in retinate and filtrate. Using UM05, the main activity is always found in the retinate. From these experiments an estimated molecular weight of the factor would be more than 500 and less than 1,000.

B. Sephadex Chromatography of G_1-Chalone Activity

The elution of G_1-chalone activity from the G-50 sephadex column as seen in Figure 6 suggests a molecular weight lower than 1,450 if the active molecule is a peptide. Owing to the apparent low molecular weight of the inhibitor we tried G-15 sephadex. Figure 7 shows the elution behaviour of freeze-dried XM50 filtrate on G-15 sephadex. Concentration by freeze-drying was preferred because of great loss of activity by the UM05 ultrafiltration concentration. Two peaks of activity are obtained. The elution constant (Kav = 0.27) of peak 1 again suggests a molecular weight of 500 – 1,000 if the inhibitor is a peptide. Peak 2 contains substances with molecules less than 200. Although peak 2 might be expected to originate from unspecific inhibition by high concentration of salts since this peak coelutes with the conductivity peak, we went into further experiments with not only peak 1 but also peak 2, since it has been reported that the granulocytic chalone elutes in the same way as peak 2 (Paukovits, 1976). XM50 ultrafiltrate concentrated by UM05 ultrafiltration usually did not exhibit as much activity in the region between [^{14}C]-methionine and [^{3}H]-thymidine (in the salt region) as freeze-dried ultrafiltrate. In a few experiments an additional inhibitory peak was observed with the same elution constant as [^{14}C]-methionine.

C. Assay of Tissue Specificity *in Vivo*

Testing of the two peaks from the G-15 column *in vivo* on regenerating ascites tumours showed that both peaks inhibit the entry of JB-1 cells from G_1

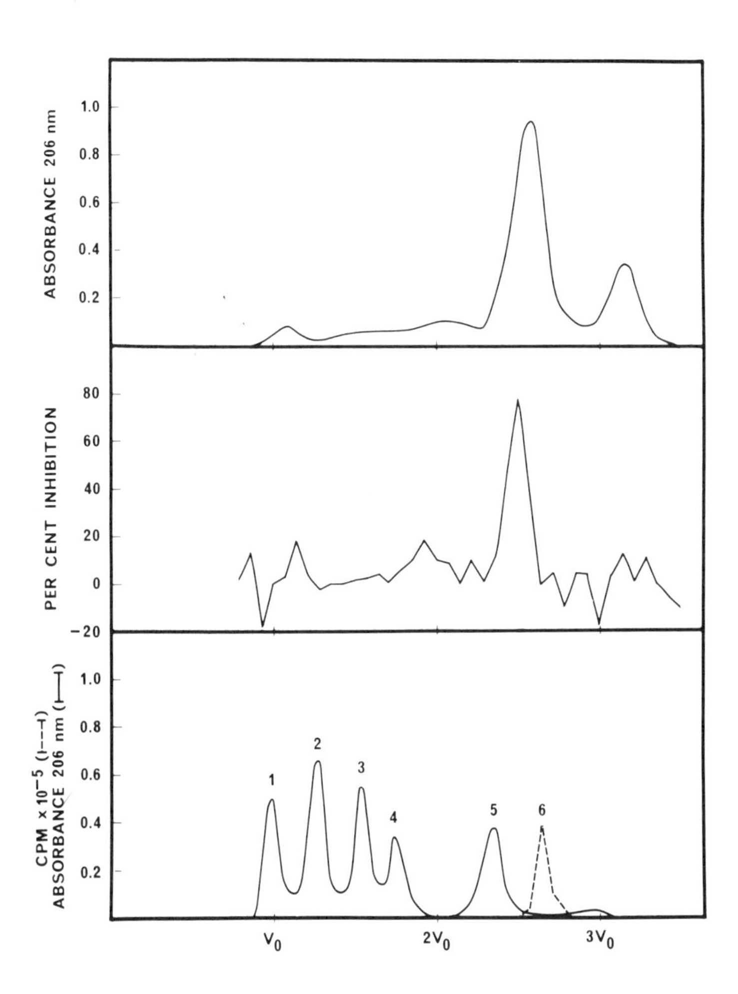

Figure 6 Molecular weight estimation of G_1-chalone activity using G-50 sephadex chromatography. The activity of XM50 filtrate was concentrated five-fold using UM05 ultrafiltration before application to the column. *Upper part*. Absorbance at 206 nm measured continuously in a flow cell. *Middle part*. Inhibitory activity expressed as per cent inhibition in comparison with control. *Lower part*. Molecular weight markers. Following markers were used (molecular weight in brackets). (1) Blue dextran (2,000,000), (2) ovalbumin (45,000), (3) chymotrypsinogen A (25,000), (4) cytochrome c (12,400), (5) bacitracin (1,450) and (6) [^{14}C]-methionine (149.2). 0.2 mg of each marker and 5 μCi of the radioactive marker were dissolved in 2.0 ml of glass-distilled water before application to the column. The absorbance at 206 nm was measured as mentioned above and radioactivity was counted in a Beckmann scintillation counter

into S phase, while only peak 2 inhibits the $L1A_2$ cells (Figure 8). Hence, peak 1 specifically inhibits the JB-1 cells and thus contains the G_1-chalone activity.

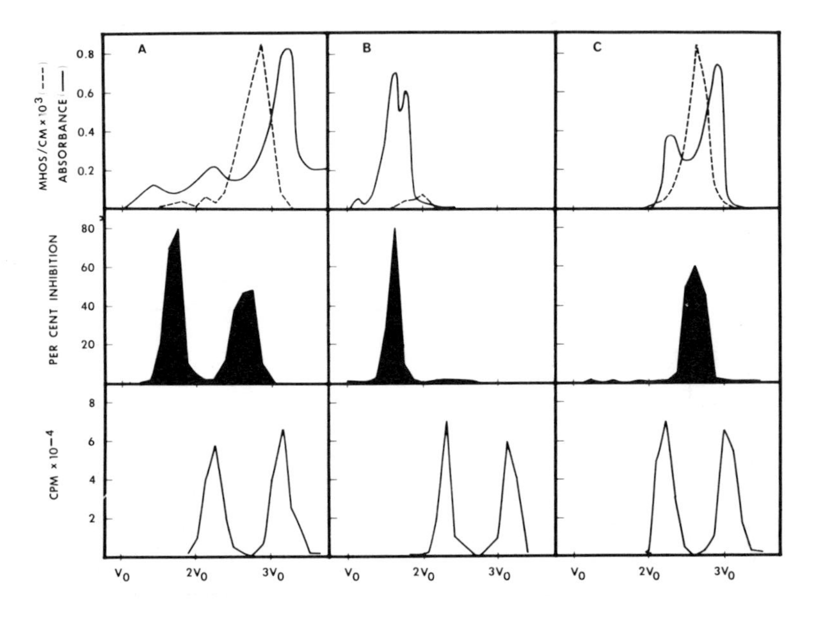

Figure 7 Preparative G-15 sephadex chromatography of G_1-inhibitory activity. *Left column*, elution of freeze-dried XM50 filtrate. *Middle column*, rechromatography of the first inhibitory peak after freeze-drying and redissolving in water. *Right column*, rechromatography of the second inhibitory peak processed as the first peak. *Abscissa*, multiples of the void volume as determined with blue dextran. *Upper ordinate*, absorbance at 254 nm (left curve), 206 nm (middle and right curve) and specific conductance in mhos/cm. *Middle ordinate*, inhibitory activity. *Lower ordinate*, elution of [^{14}C]-methionine (first peak) and [^{3}H]-thymidine (second peak) cochromatographed with the chalone extracts. In spite of a larger molecular weight of thymidine than methionine, the latter is more retarded on the sephadex column due to the presence of an extended aromatic π-electron system. (These internal markers have been used for the calibration of the column)

D. High Voltage Paper Electrophoresis Analysis of Peak 1 from G-15 Sephadex

After rechromatography of the peak it was concentrated and analysed by high voltage electrophoresis at pH 6.5. The spots were eluted from the electrophoresis strip and assayed (Figure 9). The main inhibitory activity was located to some very basic quick moving spots. The spots with the highest activity migrated nearly identically with the polyamine spermine at pH 6.5 (Figure 9) and at pH 2.1 and by paper chromatography. Rechromatography on G-15 of this spot resulted in a decreased molecular weight of the inhibitory peak as seen in Figure 10 (right part). Obviously the position of the peak is now

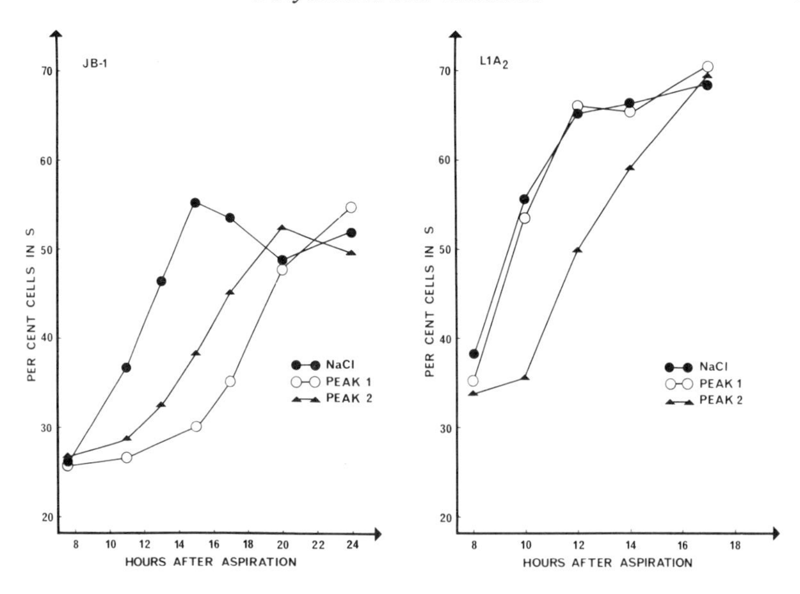

Figure 8 Specificity test of peak 1 and 2 from G-15 sephadex column (see Figure 7). The peaks were pooled apart and the fractions to be tested were made isotonic with NaCl. Percentages of cells with S DNA content as a function of time after aspiration at 0 h of plateau JB-1 tumours *(left part)* or $L1A_2$ tumours *(right part)* and intraperitoneal injections at 1, 2 and 3 h after aspiration of 0.5 ml of isotonic saline, 0.5 ml of peak 1 and 0.5 ml of peak 2. Each point represents mean of five mice

identical with that of [^{14}C]-spermine. Presumably, the spermine-like compound isolated by electrophoresis has been dissociated from the inhibitor.

Previous experiments had shown that the G_1-inhibitory activity (Figure 11 left column) was lost after passing the ultrafiltrate through columns of cation exchangers (e.g. CM-sephadex and phosphocellulose) (Figure 11 middle column). Since elution of activity from such columns turned out to be impossible, it was tentatively presumed that a co-factor for the inhibitor, but not the inhibitor itself was bound to the column. We checked Mg^{2+} and Ca^{2+} and found in some experiments Ca^{2+} but not Mg^{2+} was able to reconstitute the inhibitory activity. The reproducibility was too low to be of any use, however.

From the information obtained with high voltage electrophoresis we tried to add spermine to the eluate from CM-sephadex.

Chromatography on G-15 showed that spermine was able to reconstitute the activity as seen in Figure 11 right column. A peak of activity is observed in the correct position on the column. The second peak originates from an excess of added spermine which has growth inhibitory activity by itself as mentioned below. It is seen from the elution pattern of [^{14}C]-spermine that spermine is able to combine with the inhibitor suggesting non-covalent bonds between the two compounds.

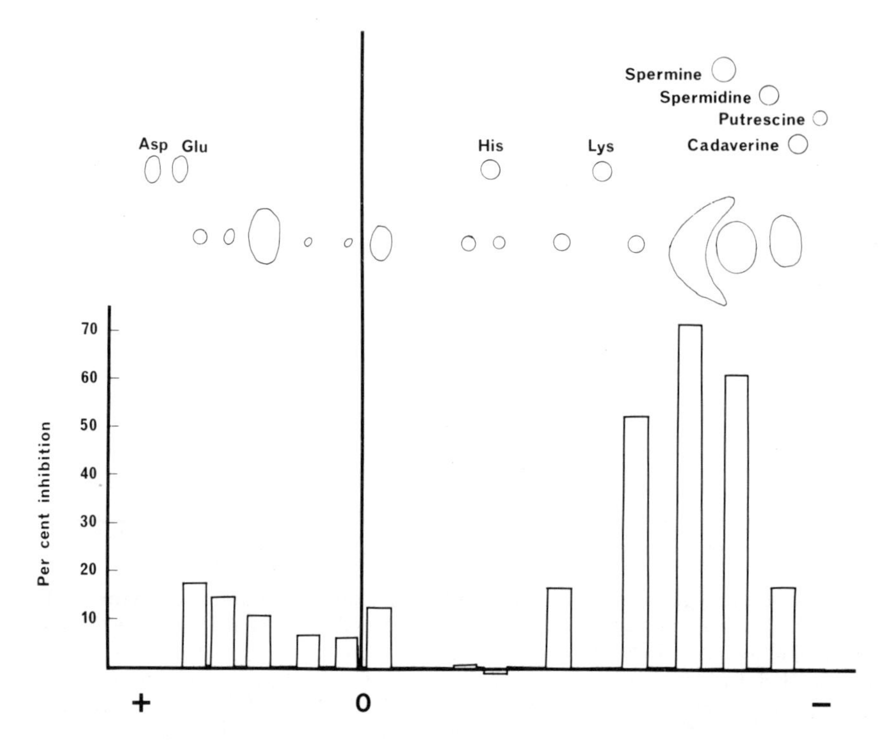

Figure 9 High voltage paper electrophoresis at pH 6.5 of peak 1 from G-15 sephadex (see Figure 7). Paper Whatman 3 MM, 30 volts cm^{-1} for 30 min. The guide strip was stained with ninhydrin

The intact G_1-inhibitor is able to bind [^{14}C] labelled spermine, indicating an exchange reaction with unlabelled native inhibitor-bound spermine as seen in Figure 12.

Since (1) the carrier for spermine is able to pass through the CM-sephadex column and (2) spermine is a very basic substance, we presume that the inhibitor is of acidic or at least neutral nature.

E. Heat Effect

We now had an explanation of earlier experiments concerning the stability of the G_1-inhibitory activity to heat. These experiments had shown that the activity was remarkably stable to heat. Chromatography of the heated inhibitor (70°C for 30 min) showed that the untreated inhibitor (left column in Figure 13) splits into two peaks (right column in Figure 13), the first representing intact inhibitor and the second spermine generated from the inhibitor.

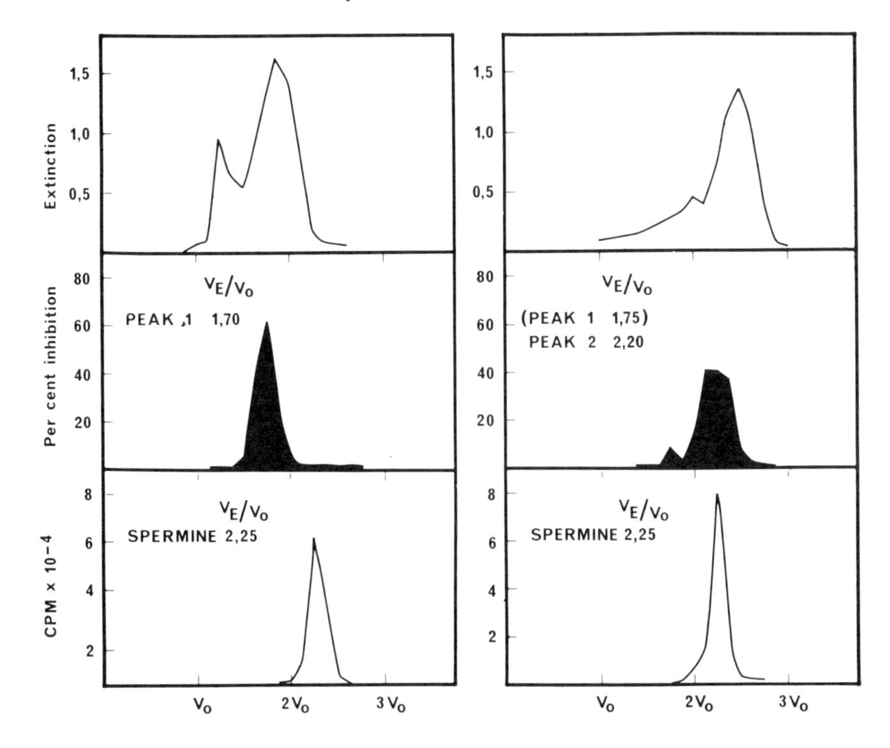

Figure 10 G-15 sephadex chromatography of G_1-inhibitor before *(left column)* and after *(right column)* high voltage electrophoresis. Abscissa, multiples of the void volume (V_o). *Upper ordinate*, absorbance at 254 nm (left curve) and 206 nm (right curve). *Middle ordinate*, inhibitory activity. *Lower abscissa*, elution of [^{14}C]-spermine. V_E = elution volume

V. ELIMINATION OF SPERMINE EFFECT

Since spermine by itself has an inhibitory activity, we had to investigate this effect in our test systems.

It is known that calf serum which is used in the growth medium of our *in vitro* test system contains polyamine oxidases which convert polyamines including spermine (and probably the spermine-containing inhibitor), to a general toxic aldehyde (Chapters 19, 20 and 22). This assumption was verified by adding pure spermine to the *in vitro* test system. Unspecific growth inhibition was observed in several cell lines besides JB-1 at concentrations from 0.5 to 10 μg ml^{-1} (depending on the batch of calf serum used). Modification of the culture medium by substituting calf serum with horse serum (which is known to be without polyamine oxidase activity) eliminated the spermine effect totally (Figure 14). By using cell-free ascites from JB-1 ascites tumour as serum component, the spermine effect is lowered considerably but it is still possible

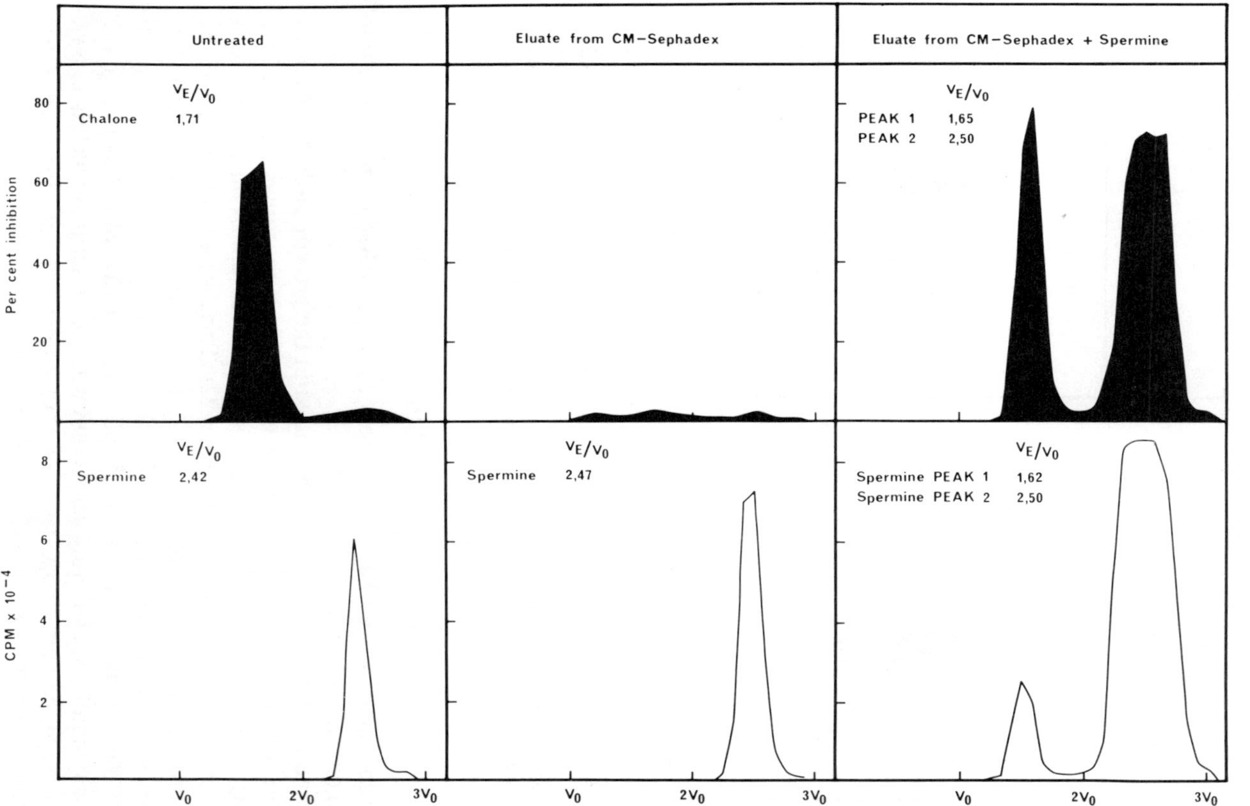

Figure 11 G-15 sephadex chromatography of untreated G_1-inhibitor *(left column)* and after passage through CM-sephadex *(middle column). Right column*, effect of spermine on CM-sephadex eluate. Spermine was incubated with CM-sephadex eluate for 2 h prior to chromatography. *Abscissa* as in Figure 10. *Upper ordinate*, inhibitory activity. *Lower ordinate*, elution of [^{14}C]-spermine

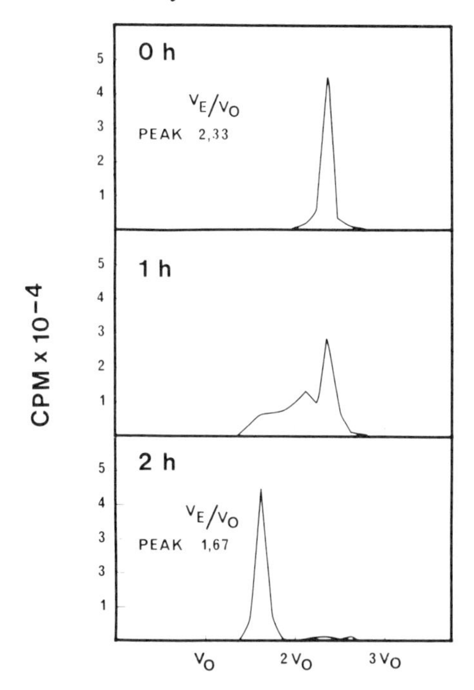

Figure 12 Binding of [^{14}C]-spermine to G_1-inhibitor. Labelled spermine was incubated 0 h *(upper curve)*, 1 h *(middle curve)* and 2 h *(lower curve)* before application to G-15 sephadex column. *Abscissa:* as in Figure 10. *Ordinate:* elution of [^{14}C]-spermine

to measure the cell specific chalone-like activity as seen in Figure 15. The dose response curves in Figure 15 show that JB-1 mouse cell-free ascites give much more cell specific G_1-inhibitory response than calf serum, while horse serum is not usable owing to lack of response of the cells to the inhibitor. Figure 16 shows the G-15 sephadex elution of G_1-inhibitory activity using the modified *in vitro* system (by using JB-1 cell-free ascites as serum component). It is seen that the main peak of inhibitory activity appears to be cell specific when using $L1A_2$ cells as specificity control.

VI. TEST OF THE CARRIER OF SPERMINE

On the basis of the experience gained from the chromatography studies of the elution of the G_1-inhibitory material, a test of possible carriers for spermine was developed as seen in Figure 17. The acidic and neutral spots were eluted from the electrophoresis strip and after addition of spermine they were assayed for the correct elution on G-15, cell-specific activity and binding of [^{14}C]-spermine. Using these criteria we have identified different acidic

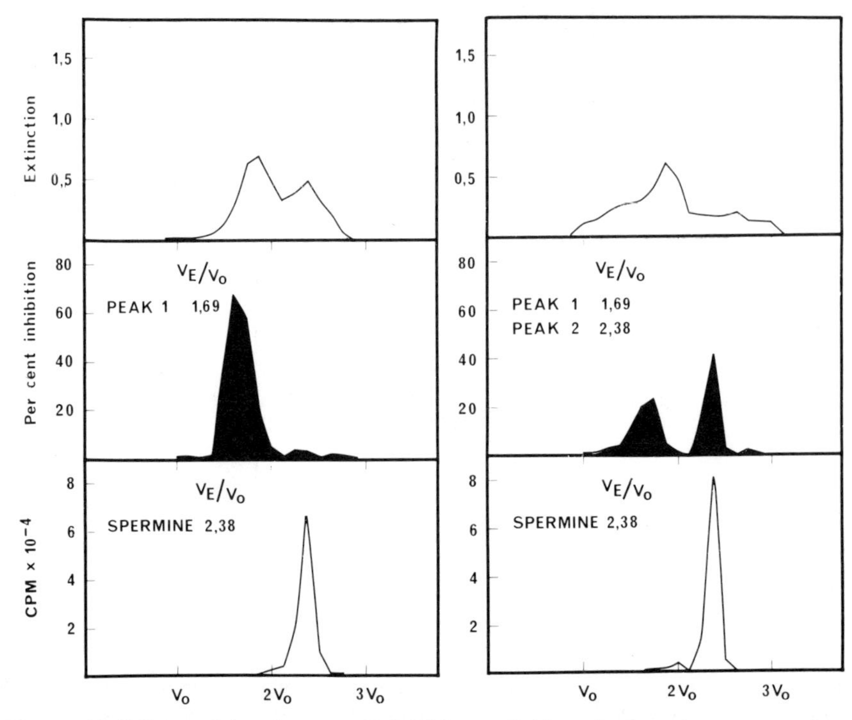

Figure 13 Effect of heating on G_1-inhibitor. G-15 sephadex chromatography of untreated inhibitor *(left column)* and heated inhibitor *(right column)*. Ordinate and abscissa as in Figure 10. The absorbance was monitored at 254 nm

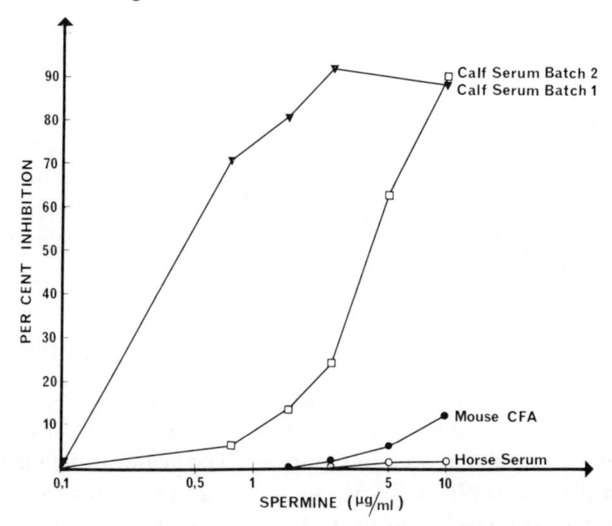

Figure 14 Effect of different sera on the inhibitory effect of spermine on cell proliferation. (Mouse CFA = cell free ascites from JB-1 tumour)

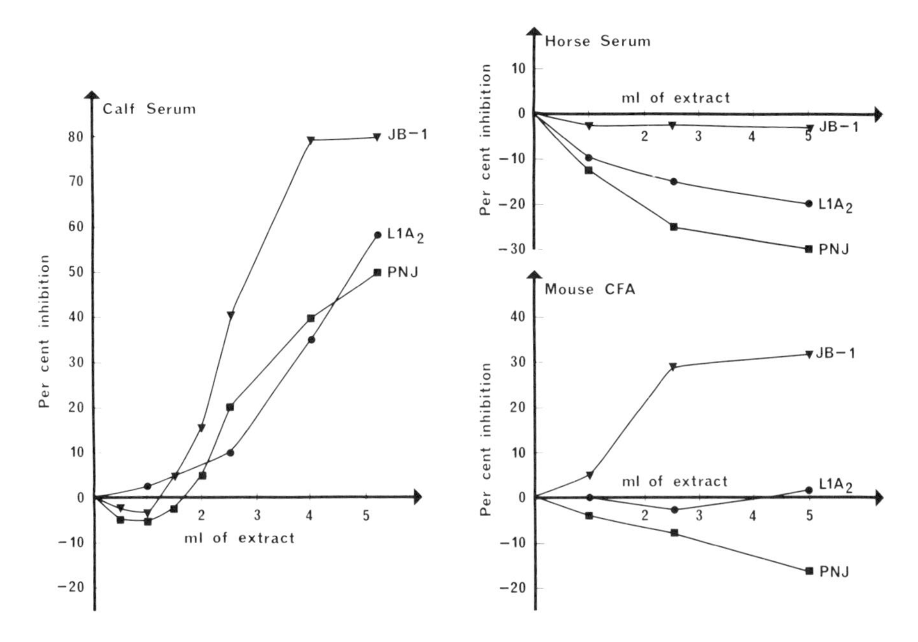

Figure 15 Dose response curves of G_1-inhibitor on JB-1 (plasmacytoma) $L1A_2$ (bronchogenic sarcoma) and PNJ (mammary carcinoma) cells in calf serum, horse serum and mouse CFA

compounds with the ability to make complexes with spermine and correct G-15 elution behaviour as listed in Table 1. Most of these complexes, however, did not exhibit a high degree of specificity. The most cell-specific complex was that of spermine together with (1) a mixed disulphide of glutathione and cysteine and (2), a partially identified spot with high content of glutamic acid (possibly a peptide containing two or more glutamic acid residues). We have synthesized the mixed disulphide peptide (Eriksson and Eriksson, 1967) and tested it after complex formation with spermine *in vitro* and *in vivo*. The complex had no effect in either test systems. One explanation of the discrepancy between the effect of the purified mixed disulphide and the synthetic one is that the active substance migrates electrophoretically like the glutathione-cysteine disulphide and is hidden behind this spot. This is a possibility, since the amino acid analysis from the spot containing the mixed disulphide contained approximately 10% background of other amino acids and other unknown substances.

VII. POSSIBLE MECHANISM OF ACTION OF THE SPERMINE COMPLEX

In order to estimate the influence of the complex formation of spermine and

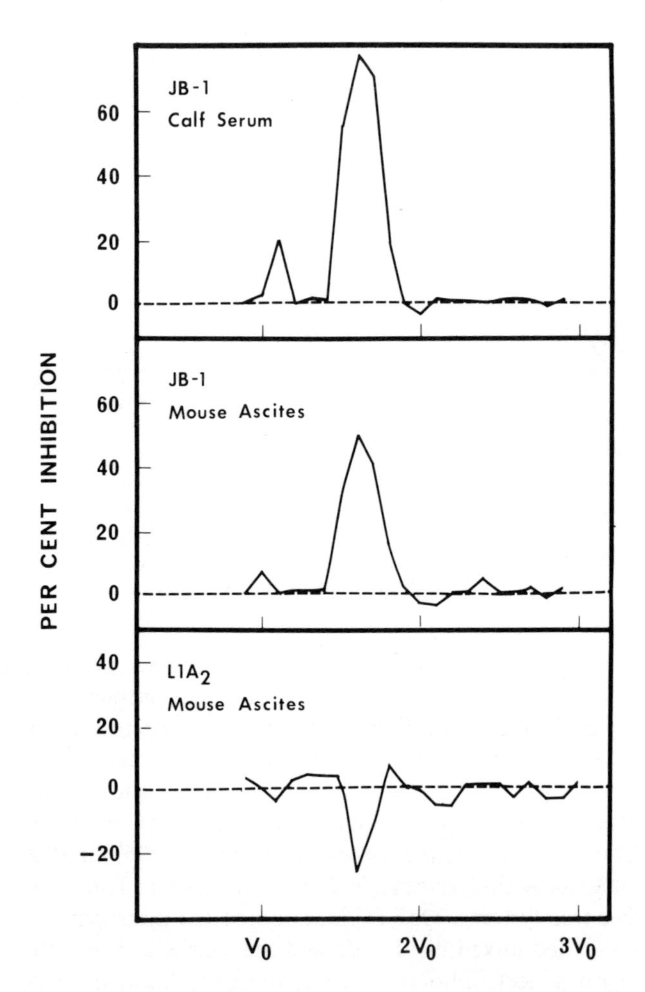

Figure 16 Specificity test of G_1-inhibitor fractionated on G-15 sephadex using JB-1 and $L1A_2$ cultured in mouse CFA. Abscissa as in Figure 10. *Ordinates:* inhibitory activity

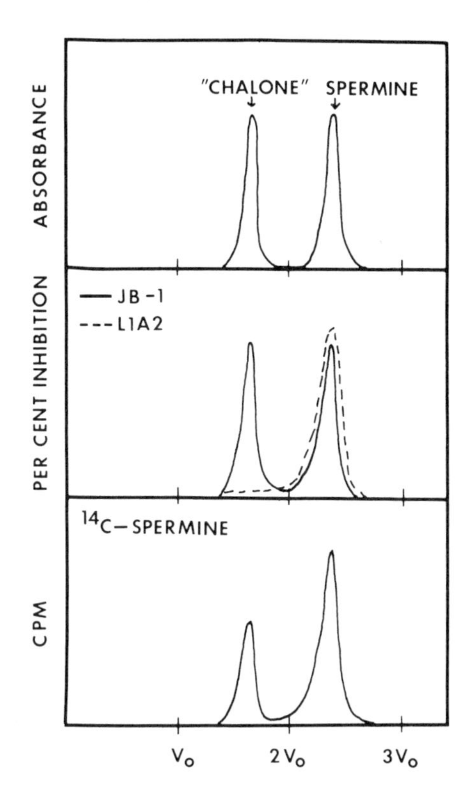

Figure 17 Test of cell specific carrier (chalone) of spermine. The carrier must exhibit the right chalone elution on G-15 sephadex *(upper curve)*, act cell specific *(middle curve)* and be able to bind spermine *(lower curve)*

Table 1 Identified ninhydrin positive compounds capable of complex formation with spermine

Asp aspartic acid
Glu glutamic acid
Spot with high content of Glu + an unidentified ninhydrin positive substance
Glu-Cys-Gly
Cys
Glutathione-cysteine mixed disulphide
Ser- P Serine phosphate
$NH_2CH_2CH_2O$ P Ethanolamine phosphate

the unknown carrier of the chalone effect, we have compared the doses of spermine which exhibit a growth inhibitory activity by themselves and chalone extract needed to obtain an inhibitory effect on JB-1 cells in the different test systems.

From Table 2 it appears that a dose of spermine 20 times higher *in vivo* than *in vitro* is required to get the same effect, whereas only 2 to 3 times more chalone extract is needed *in vivo* than *in vitro*.

Table 2 Dose of spermine and chalone extract to obtain a significant inhibitory effect on JB-1 cells

		Spermine	Chalone extract
	Calf serum	2–10 μg	0.5–1.5 ml
In vitro	Horse serum	No effect	No effect
	Mouse ascites	50–100 μg	1–2.5 ml
In vivo		> 2,000 μg	3–6 ml

Dose factor for obtaining the same effect *in vivo* as *in vitro*

	in vitro (Using mouse ascites serum)	*in vivo*
Spermine	1	20–40
Chalone extract	1	2–3

Thus, there is a discrepancy between the dose relationship *in vivo* and *in vitro* of spermine and chalone extract. From this we conclude that (1) spermine by itself does not account for the main inhibitory activity in the chalone extract as measured *in vivo*, and (2) the measurement of the true chalone effect may be drowned *in vitro* by an unspecific artificial inhibitory effect of spermine.

At present, it is presumably in the *in vitro* system using calf serum that the main part of inhibitory activity is due to an oxidation-product of spermine or spermine complexes. By substituting calf serum with mouse ascites, the main part of the activity is due to chalone-like substances, although there are still some artificial effects of spermine, whilst the *in vivo* system measures solely nearly true chalone activity.

How does spermine or the spermine complex act in the G_1-inhibition? It is known that spermine has high affinity for DNA. Hence, a direct binding to DNA of spermine or oxidized spermine might in some way result in growth inhibition. A few other conceivable mechanisms of action are listed in Table 3, although we have no experimental data to select any of them.

Recently, evidence has been presented that the lymphocyte chalone could be mistaken for a spermine complex (Allen and coworkers, 1977). It was presumed that 'as the complex displayed some tissue-specificity, it is possible

Table 3 Possible mechanisms of action of spermine complex (SpX) as an inhibitor of proliferation

Role of X
(1) Co-factor for the oxidation of spermine
(2) Carrier for spermine or oxidized spermine into the cell
(3) Determining the specificity of the inhibitory effect of spermine or oxidized spermine

that an unidentified tissue-specific material acts as a carrier for spermine' (Allen and coworkers, 1977).

A preliminary report on the isolation of an inhibitor specific for mammary cells, indicates that a low molecular basic compound, possibly a polyamine, may be involved (Gonzalez and Verly, 1976). Finally, Houck has characterized fibroblast chalone and lymphocyte chalone as strongly cationic polypeptides which may contain polyamines (Houck and coworkers, 1976).

VIII. SUMMARY AND CONCLUSION

Evidence is beginning to accumulate for the assumption that polyamines may be involved in chalone-regulation of growth. As has appeared from this presentation, it is very important to distinguish between the true specific chalone activity and more unspecific artificial inhibitory activity. We have identified one factor, namely spermine, which may act disturbingly in chalone *in vitro* analysis.

Since it is possible to measure only cell-specific inhibitory effect on JB-1 cells when culturing them in JB-1 cell-free ascites as serum component (both *in vivo* and *in vitro*), there may be other tumour cell-derived factors present which play a role in the chalone effect. We have only a few indications of such other possible factors at present, but we have the feeling that the chalone effect involves a multi-component system. We have identified one of these components, namely spermine, and there are reasons to hope that future research will shed light on the role of the other component(s) acting in growth regulation.

REFERENCES

Allen, J. C., Smith, C. J., Curry, M. C., and Gaugas, J. M. (1977). *Nature*, **267**, 623–625.

Barfod, N. M. (1977). *Exp. Cell Res.*, **110**, 225–236.

Barfod, N. M. (1980). *Cell Tiss. Kinet.* In press.

Barfod, N. M., and Bichel, P. (1976). *Virchows Arch. B Cell Path.*, **21**, 249–258.

Barfod, N. M., and Marcker, K. (1977). Partial purification and characterization of a G_1 chalone from JB-1 plasmacytoma cells. *International Symposium on Molecular Control of Proliferation in Eukaryotic Cells*, April 1977, Dobogókö, Hungary.

Bichel, P. (1970). *Europ. J. Cancer*, **6**, 291–296.
Bichel, P. (1971). *Europ. J. Cancer*, **7**, 349–355.
Bichel, P. (1972). *Europ. J. Cancer*, **8**, 167–173.
Bichel, P. (1973). *Europ. J. Cancer*, **9**, 133–138.
Bichel, P., and Barfod, N. M. (1977). *Cell Tiss. Kinet.*, **10**, 183–193.
Bichel, P., Barfod, N. M., and Jakobsen, A. (1975). *Virchows Arch. B Cell Path.*, **19**, 127–133.
Brugal, G., and Pelmont, J. (1975). *Cell Tiss. Kinet.*, **8**, 171–187.
Dombernowsky, P., and Bichel, P. (1975). *Acta Path. Microbiol. Scand. Sect. A*, **83**, 222–228.
Elgjo, K. (1973). *Natl. Cancer Inst. Monogr.*, **38**, 71–76.
Eriksson, B., and Eriksson, S. A. (1967). *Acta Chem. Scand.*, **21**, 1304–1312.
Gonzalez, R., and Verly, W. G. (1976). *Proc. Natl. Acad. Sci. USA*, **73**, 2196–2200.
Houck, J. C. (1976). *Chalones*. North-Holland/American Elsevier Publishing Company, New York.
Houck, J. C. and Hennings, H. (1973) *Febs. Lett.*, **38**, 1–8.
Houck, J. C., Kanagalingam, K., Hunt, C., Attallah, A., and Chung, A. (1976). *Science*, **196**, 896–897.
Maurer, H. R., Weiss, G., and Laerum, O. D. (1975). *Virchows Arch. B. Cell Path.*, **20**, 229–238.
Paukovits, W. R. (1976). *In vitro* biological and chemical properties of the granulocytic chalone. In J. C. Houck (Ed.), *Chalones*, North-Holland/American Elsevier Pub. Co., New York, pp. 311–330.
Reddy, S. B., Erbe, W., Linden, W. A., Landen, H., and Baigent, C. (1973). *Biophysik*, **10**, 45–50.
Rytomaa, R., and Kiviniemi, K. (1968). *Europ. J. Cancer*, **4**, 595–606.
Thornley, A. L., and Laurence, E. B. (1975). *Int. J. Biochem.*, **6**, 313–320.
Weiss, P., and Kavanau, J. L. (1957). *J. Gen. Physiol.*, **41**, 1–47.

Chapter 22

Biogenic Diamines and Polyamines in Support and in Inhibition of Lymphocyte Proliferation

JOSEPH M. GAUGAS

I. INTRODUCTION

Polyamines and their diamine precursor, putrescine, are regarded by many cell-biologists as vital in either obscure or unidentified biochemical and electrochemical mechanisms which culminate in cell proliferation. The polyamines, spermidine and spermine, possess a relatively simple molecular structure. They are extremely stable and are able to survive tissue necrosis (Cohen, 1977; Bachrach, 1973). Circumstantial evidence suggests that polyamines support a multiplicity of regulatory mechanisms within the lymphocyte (Chapters 1–4).

Polyamines are abundant in physiological fluids, presumably because they are readily secreted by tissues (Melvin and Keir, 1977). Polyamines are also released from dead cells (Heby and Andersson, 1978). As a consequence of degradation by specific oxidative enzymes, polyamines possibly produce extracellular phenomena including pathological immunosuppression and tissue damage (Chapter 19). Degraded polyamines undergo increased urinary excretion during malignancy and especially in leukaemia where the tumour burden is great (Chapter 26). Hence the role of polyamines in the proliferation of normal and malignant lymphoid cells, as well as for immunosuppression, is of paramount interest.

Firstly, it is just conceivable that spermine^{4+} ions react both intracellularly and extracellularly as nothing more than a cation. This is so even if they are essential for proliferative processes, either together with or independent of calcium (Ca^{2+}) and magnesium (Mg^{2+}) ions. The spermine ion would probably react electrochemically with a greater efficiency than the metal ions on a comparative molar basis. The influx of cations, preponderantly Ca^{2+}, through the intercellular junctions of the plasma-membrane is established as initiatory in lymphocyte transformation. Transformation is characterized by increased protein, increased ribonucleic acid, and *de novo* deoxyribonucleic acid biosynthesis. Transformation is induced *in vitro* by mitogens which bind to the surface membrane of the cell and includes the plant lectins (e.g. phytohaemagglutinin PHA, and concanavalin A ConA), bacterial endotoxin (e.g. lipopolysaccharide LPS) or indeed antigens. Recently, Boynton and coworkers (1977) made the important discovery that neoplastic cells are unlike normal cells in showing ion-influx lesions and are able to proliferate without Ca^{2+}. This invites an emergency or substitute role for free polyamine cations. These could be produced and then may be stored as an inactive complex which would be non-secretable by leukaemia cells. A high binding affinity of polyamines with certain plasma components argues against such a function, (Amman and Werle, 1957; Rosenblum and Russell, 1977). Influx of extracellular Ca^{2+} itself could affect intracellular polyamine reactivity and thereby modify cell metabolism (Durham, 1978).

Secondly, and crucially since reactivity appears specific to polyamines, they may invoke their aliphatic form and electrochemical charge in order to interact with DNA for which a high binding affinity is also shown. Polyamines protect DNA against denaturation (reviewed by Tabor and Tabor, 1976), and may simply hold intact the configuration of the DNA molecule during formation and processing. Chain length of the polyamine molecule is probably a critical factor. Polyamines may interact with RNA (Kay and Pegg, 1972), regulate nucleoside phosphorylation or nucleic acid polymerase. All these are prereplicative conditions for mitosis. Within the mammalian cell, polyamine and diamine biosynthesis is bi phasic occurring during the mid- to late-G_1 phase and G_2-phase of its growth-division cycle as shown in Figure 1. In the former phase polyamine biosynthesis is most likely to be a prerequisite for DNA replication which occurs in the S phase and preceded actual division of the cell or the M phase (Friedman, Bellantone and Cannellakis, 1972; Russell and Stambrook, 1975; Heby and coworkers, 1976; Fuller, Gerver and Russell, 1977).

Thirdly, cells show transient changes in activity of adenyl and guanyl cyclases, cyclic AMP and cyclic GMP (Theoharides and Canellakis, 1975;

Russell and Stambrook, 1975; Hibasami and coworkers, 1977 and Chapter 8), phosphodiesterase (Byus and Russell, 1974), division-protein synthesis and phosphorylation, prostaglandin and in microtubules (depolymerized by Ca^{2+}) which are much favoured in studies of proliferative processes in lymphocytes, macrophages, granulocytes and their stem-cells. These changes could modulate diamine-polyamine function or indeed initiate their synthesis or activation later in the cell cycle. Diamine-polyamine biosynthesis should therefore be evoked by immunologically induced lymphocyte proliferation.

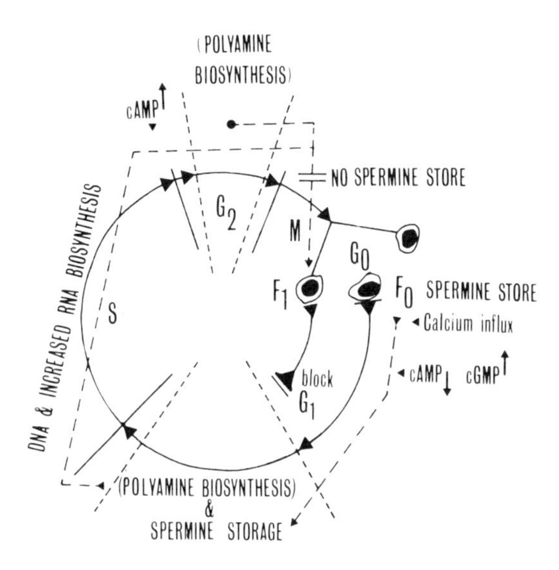

Figure 1 Schematic representation of some events in the cell growth-division cycle of a lymphocyte (F_0 generation) treated in G_0, or early G_1 with a specific inhibitor of diamine-polyamine biosynthesis. The cell traverses G_1, S (stored spermine), G_2, M and divides, but its daughter cell (F_1) is arrested at some undetermined point in G_1 (presumably at the G_1/S border, or at expected onset of diamine-polyamine biosynthesis). Of course, the F_0 cell would not in fact synthesize diamine-polyamine, so the stages of such biosynthesis (G_1 and G_2) is shown for only illustrative purposes representing a normal untreated cell

Since spermine is the end-product of diamine-polyamine biosynthesis, as shown in Figure 2, it might be the sole physiologically functional agent in the cell. An experimental disturbance of the endogenous equilibrium between putrescine, spermidine and spermine, brought about by treatment of cells with commercially available reagents (inhibitors of diamine and polyamine biosynthetic enzymes), or oxidized polyamines, provides a strategem for unravelling the mysteries of the precision of polyamine function.

II. DIAMINES AND POLYAMINES IN SUPPORT OF LYMPHOCYTE PROLIFERATION

In culture, lymphocytes produce a full complement of diamine-polyamine which appears necessary for transformation and proliferation. Low levels of polyamine synthesizing enzymes are found in unstimulated or quiescent lymphocytes which are in the G_0 phase, but optimal mitogen-induced transformation is associated with increased endogenous levels of diamine-polyamine (Kay and Lindsay, 1973; Fillingame, Jorstad and Morris, 1975; Morris, Jorstad and Seyfried, 1977). Polyamine biosynthesis or activation therefore ensues from plasma-membrane situated events. Polyamines by themselves when added to cultured lymphocytes cannot induce a transformation. A remotely possible explanation is active exclusion by the cells.

A. Putrescine Biosynthesis

The biosynthetic pathway for diamine-polyamine in mammalian cells is briefly oulined in Figure 2 and has been reviewed by Tabor and Tabor (1976). Putrescine is derived from L-ornithine along the pathway towards spermidine. Surprisingly, treatment of cultured lymphocytes with the non-biogenic analogue α-methylornithine (α-MO) showed no significant inhibition of PHA-

Table 1 α-Methylornithine and methyl-GAG inhibition of lymphocyte transformation and proliferation

	Inhibitory-dose$_{50}$	
	PHA-treated cells	Growth Bri8 cells
α-MO	Non-inhib.	6.0 mmole*
Methyl-GAG	150-350 μmole**	0.3 μmole**

*reversed by putrescine (160 μmole)
**reversed by spermine (10 μmole)
concentrations expressed as per litre

induced transformation (Table 1), but in contrast it markedly inhibited the proliferation of cultured leukaemia cells by assertion not on the parent cells (F_0 generation) present at the onset of incubation but only on the daughter cells and their progeny (F_1-F_n generations). This was concluded from the analysis of cell-counts and the growth-curve pattern which is presented in Figure 3. In fact, the number of cells in culture approximately doubled before the arrest of proliferation became noticeable. Thus an inhibition was not asserted in mitogen-treated lymphocytes simply because they transform by passing

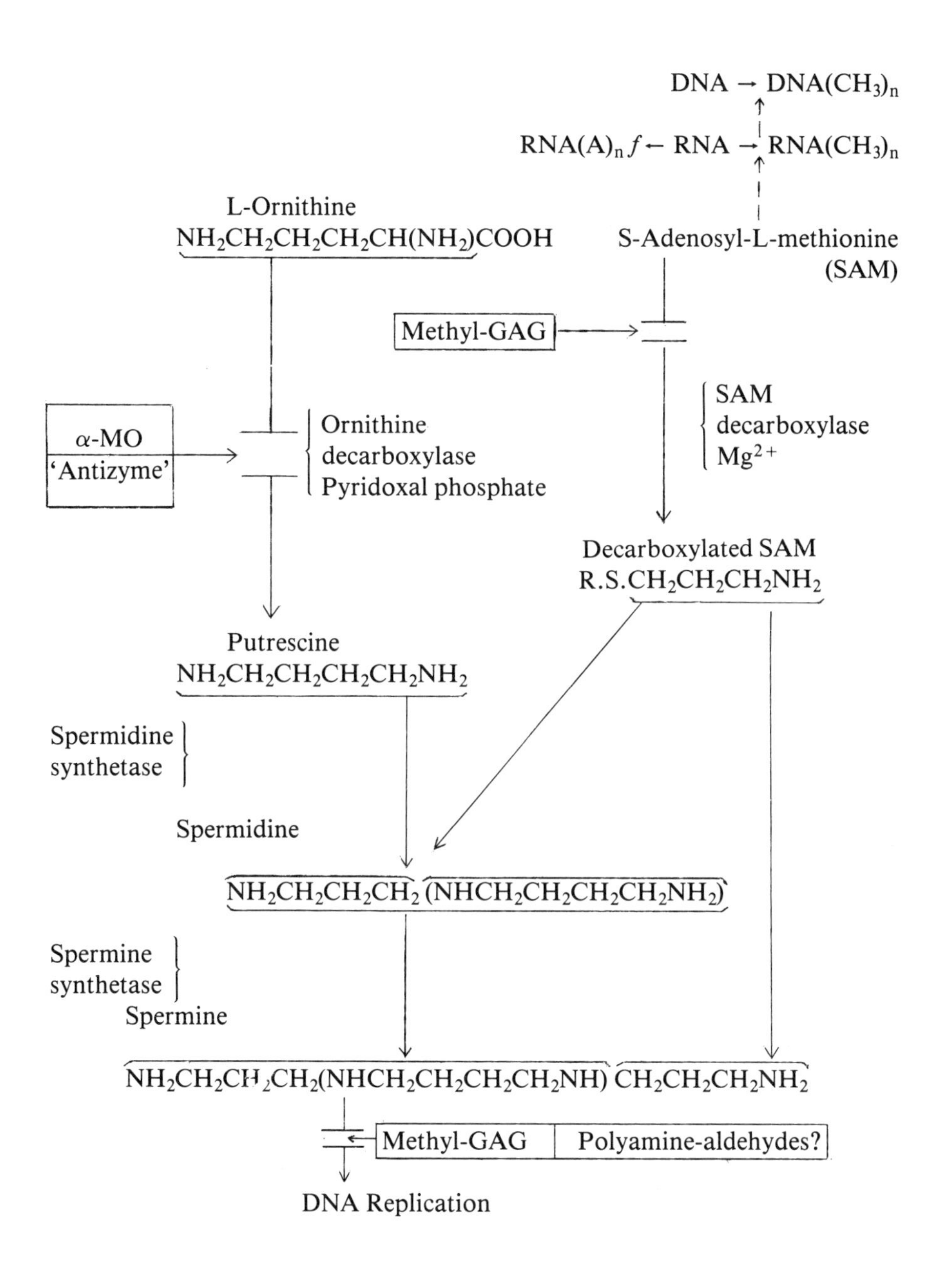

Figure 2 Pathway of diamine-polyamine biosynthesis in mammalian cells, showing points of inhibition by pharmacologically active agents

through the G_1 and S phases but fail to proliferate. The inhibitory effect could be readily prevented by adding putrescine (160 μmole litre^{-1}) or biogenic polyamine (20 μmole). Because α-MO (10 nmole) failed to inhibit during the S phase, taken together the foregoing findings strongly suggest that the block in the cell cycle occurs in the G_1 phase, although a G_2 phase block cannot be positively discounted as additional. Both phases are coincident with diamine-polyamine biosynthesis.

According to Morris, Jorstad and Seyfried (1977) mitogen-treated bovine lymphocytes show elevated contents of putrescine, spermine and spermidine. When they added α-MO (5 mmole litre^{-1}) the accumulation of putrescine was reduced by approximately 75%, but interestingly the spermine level remained stationary. Evidently, such cells with greatly reduced putrescine and spermidine levels, but normal spermine levels, would transform without restriction and then proliferate for *only* a limited number of generations. If putrescine biosynthesis was maximally suppressed, then presumably proliferation for only a *single* subsequent generation could be expected. This evidence undoubtedly demonstrates a storage phenomenon for a factor seemingly essential for proliferative mechanisms. A sufficient reservoir of spermine or spermine-conjugate is apparently transferred along with the cytoplasm from a progenitor cell to its daughter cells for their division. This finding perhaps gives insight into nature's design for diamine-polyamine biosynthesis in the G_2 phase of the cell cycle. Hence theoretically, at least, it should be possible to evoke a transformation in cells without eliciting polyamine biosynthesis. The α-MO suppresses putrescine biosynthesis by interference with ornithine decarboxylase reactivity (Newton and Abdel-Monem, 1977).

The diamines, cadaverine and 1,6-diaminohexane have been reported as *in vivo* inhibitors of ornithine decarboxylase (ODC) in transplanted Ehrlich ascites carcinoma cells in mice. Whether feed-back by diamines is part of the cell's normal physiological regulation of polyamine biosynthesis remains to be fully ascertained (Kallio and coworkers, 1977).

B. Polyamine Biosynthesis

Methylglyoxal-*bis*[guanylhydrazone] (methyl-GAG) is an anti-leukaemia drug and has been reviewed by Mihich (1975). It reduces cell formation of both spermidine and spermine from decarboxylated S-adenosyl-L-methionine (SAMD) and putrescine. This occurs without the drug influencing putrescine biosynthesis (Williams-Ashman and Schenone, 1972). Methyl-GAG was found by Newton and Abdel-Monem (1977) to greatly increase putrescine levels, and to decrease spermidine and spermine levels, in L1210 murine leukaemia cells when they are cultured for two generations (F_0-F_1). They concluded that putrescine and 50% of the endogenous spermidine content which is normally

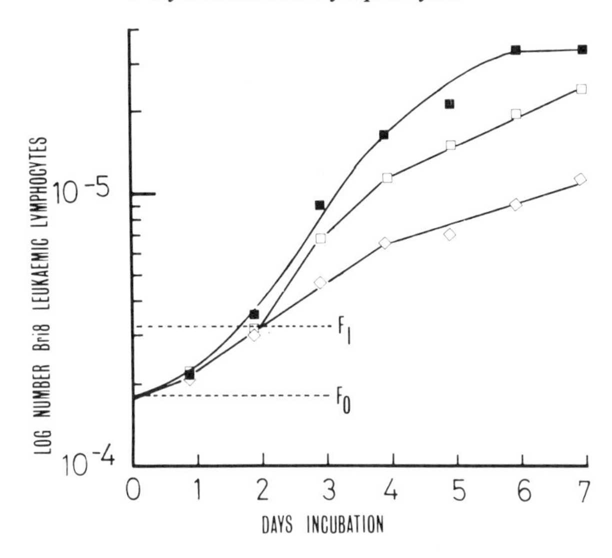

Figure 3 Growth-curve of Bri8 human leukaemic lymphocytes; ■ untreated controls, □ 5 mmole α-MO, ◇ 10 mmole α-MO. Arrest of proliferation was shown in the F_1 generation of cells. Each point represents the mean cell-count (automatic Coulter-Counter) of duplicate cultures. Results statistically self-evident. Concentrations per litre

present were not essential for DNA replication, strongly suggesting that spermine was essential.

When methyl-GAG is added to PHA-treated lymphocytes they fail to transform (Otani and coworkers, 1974) except at exceedingly high concentration, and this finding has now been confirmed (see Table 1). Methyl-GAG failed to inhibit if added to lymphocytes actually synthesizing DNA in the S phase, so must arrest in the G_1 phase of the cell cycle (Gaugas and Chu, 1979). Since it has been established previously that lymphocytes carry an essential reservoir of spermine which could be sufficient for transformation, the circumstance is seemingly paradoxical unless methyl-GAG does not, in fact, react by suppression of SAM decarboxylase in this particular case. The drug probably inhibits by competing with endogenous spermine because its reactivity can be reversed when spermine (10 μmole) is added. Alternatively, it is just feasible that the drug could react as a Ca^{2+} chelator, similar to triethyltetra-amine tetrahydrochloride, with $spermine^{4+}$ ion substitution then responsible for reversibility of proliferation inhibition.

Curiously, whereas about 150-350 μmole per litre was methyl-GAG's ID_{50} (dose giving 50% inhibition) from a range of experiments using mitogen-treated rat thymic lymphocytes, in marked contrast only an ID_{50} of about *one thousandfold less* was required for cultured Bri8 lymphocytes which must be

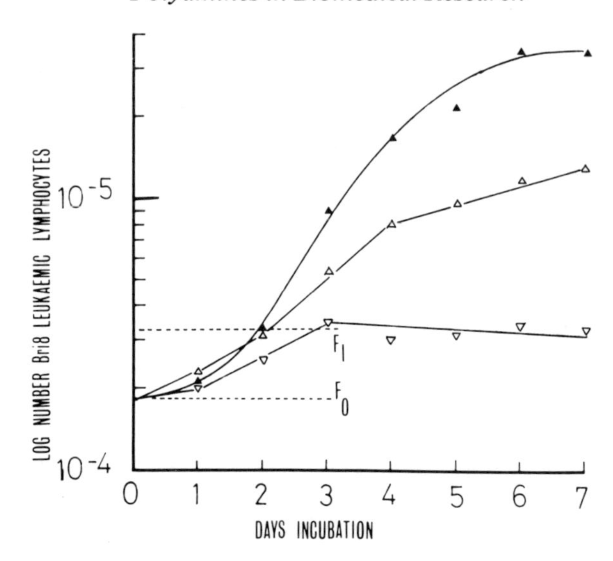

Figure 4 Growth-curve of Bri8 human leukaemic lymphocytes; ▲ untreated controls, △ 0.3 μmole methyl-GAG, ∇ 1.2 μmole methyl-GAG. Arrest probably occurred in the F_1 generation of cells. Each point represents the mean cell-count of duplicate cultures. Concentrations per litre

regarded as exceedingly low. At a somewhat higher concentration (1.2 μmole) complete arrest of lymphocyte proliferation took place, as shown in Figure 4. Once again, the arrest could be ablated by exogenous spermine (10 μmole). The inhibitory system, like that attributed to α-MO, was asserted not on cells in the F_0 generation, but on their progeny (F_1). In this special case methyl-GAG reactivity is undoubtedly expressed by decrease in SAM decarboxylase (SAMD) participation during polyamine biosynthesis in the F_0 generation of cells which results in sequential depletion of cytoplasmic spermine throughout the successive F_1-F_n generations. This occurs, provided the drug concentration in the culture medium was sub-optimal for producing arrest in the F_1 generation cells. After 48 h incubation, but no longer, in methyl-GAG (1.2 μmole) cells in the F_1 generation were washed free of the drug and shown in the presence of putrescine (160 μmole) to recover some degree of proliferative capability. Hence the drug could have a cytostatic action which is limited to F_1 generation cells.

Findings therefore suggest that two separate pharmacological properties for methyl-GAG can be ascribed involving, (1) direct competition with endogenous spermine in its role fulfilling DNA replication, and (2) a considerably more potent reduction on an equimolar basis, of SAMD reactivity, each system independently responsible for complete inhibition of lymphocyte proliferation.

Because of the potent anti-proliferative reactivity of methyl-GAG, studies were extended in order to determine whether bone-marrow haemopoietic stem-cells show a differing susceptibility, which could be relevant to leukaemia therapy. Freshly collected murine bone-marrow cells when placed in culture have the convenient property of completing their S phase but not entering a new cycle. Such cells are not susceptible to inhibition by methyl-GAG (2 mmole). In order to obtain proliferation of bone-marrow cells, the refined colony-culture technique devised by Cline, Warner and Metcalf (1972) was employed. Here, the formation of *in vitro* macrophage/granulocyte colonies from a minority of bone-marrow cells (neonatal mouse donors) seeded into semi-solid LGT agarose (Miles Ltd.) in medium supplemented with 25% v/v mouse-embryo-fibroblast-conditioned medium, can be assessed. It was found that the inhibitory dose range was roughly equal to that for the growth of Bri8 cells which has already been presented. Arrest of colony formation was reversed by addition of spermine (10 μmole), provided that the amine oxidase in the essential fetal-calf serum nutrient was first inactivated. Administration of methyl-GAG to mice was also effective against *in vivo* bone-marrow colony or macroscopic nodule formation in the spleens of recipient mice. Such colonies were produced following intravenous injection of isogeneic and viable bone-marrow cells into the pre-irradiated animals (9 Gy whole-body irradiation). From the 5th day, groups of prepared mice were injected subcutaneously at 8-h intervals with the drug, spermine or a combination of both. After elapse of 10 days all the mice were sacrificed, their spleens surgically removed, the nodules which had formed were counted. Results are summarized in Table 2. Surprisingly, in comparison to *in vitro* potency, methyl-GAG was effective only at a relatively massive regimen of 40 mg kg^{-1} body weight, just below lethal dosage, but nevertheless it completely obviated splenic nodule formation of myeloid and erythroid colonies. Moreover, nodule formation was restored by the simultaneous administration of spermine with methyl-GAG (Gaugas and Chu, 1980). Taken together the foregoing findings could well be decisive in understanding the overall modes of pharmacological action of methyl-GAG. Consequently, they could demonstrate an essential role for spermine in the proliferation of lymphoid cells both *in vivo* and *in vitro*. Methyl-GAG had a relatively poor effect *in vivo*, presumably because of its conjugation with certain tissue components.

C. Polyamines, Protein and RNA Synthesis

Whether active polyamines, in particular spermine, are (Kay and Pegg, 1972) or are not (Morris, Jorstad and Seyfried, 1977) directly involved somehow in the regulatory mechanisms for increased protein or RNA synthesis by transforming lymphocytes, remains problematical. Both authors examined mitogen-treated lymphocytes which transform in the F_0 generation,

Table 2 Inhibition of haemopoietic splenic nodule development in mice by administration of methyl-GAG, and its reversal by the simultaneous administration of spermine

Treatment	Number of nodules per spleen 20 mg kg^{-1}	40 mg kg^{-1}
Saline controls	12.3 ± 1.5	—
Spermine	13.1 ± 1.4	—
Methyl-GAG	5.7 ± 1.0	0.0
Methyl-GAG + Spermine	12.9 ± 5.4	16.4 ± 4.6

(Groups 16 CBA/Ht mice, Mean value ± s.d.). Results statistically self-evident.

but with the benefit of more recently acquired knowledge of spermine storage by cells, polyamine participation in protein (polypeptide chain) biosynthesis should be considered in cells first experimentally depleted of such polyamine. Nevertheless, selective inhibition of ribosomal RNA synthesis was found to inhibit activity of polyamine biosynthetic enzymes, especially ornithine decarboxylase in mitogen stimulated lymphocytes (Kay and Lindsay, 1973).

D. Summary and Conclusions

In summary, diamine-polyamine biosynthesis and activation are essential processes in the preparation for proliferation and DNA replication in lymphocytes and bone-marrow cells. It is likely that spermine is the sole essential physiologically active agent which correlates at least to DNA replication. A revised plan for the broadly accepted cell's growth-division cycle is proposed to encompass both putrescine and polyamine biosynthesis in the G_1 and G_2 phases, in addition to transportation of stored spermine from the progenitor to daughter cells.

Much further detailed investigation is called for in order to confirm or deny any subsidiary roles for diamine or polyamine as has been cursorily outlined in the Introduction.

III. OXIDIZED DIAMINES AND POLYAMINES IN THE SUPPRESSION OF LYMPHOCYTE TRANSFORMATION AND PROLIFERATION

1. *Polyamine Oxidation*

Enzymic deamination of biogenic polyamines under different environmental conditions for cell culture to those employed in the hitherto mentioned investigations are required in order to elicit the *suppression* of normal and tumour cell mitogenesis (e.g. Boyland, 1941; Alarcon, Foley and Modest, 1961; Halevy, Fuchs and Mager, 1962; Bachrach, Abzug and Bikierkunst,

Spermine

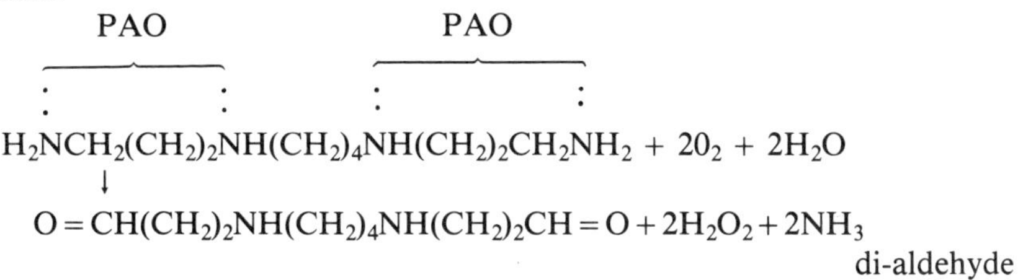

Spermidine

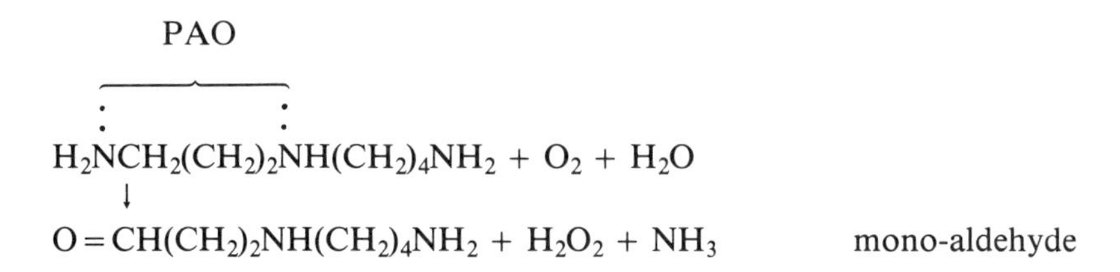

Figure 5 Bovine serum PAO oxidative deamination of biogenic polyamines to produce complex aldehydes with further possible autodegradation (β-elimination) to the simple cytotoxic aldehyde acrolein (or other carbonyls)

1967; Higgins and coworkers, 1969; Bachrach and Ben-Joseph, 1973; Dewey, 1978). Serum is required containing polyamine oxidase (PAO; EC.1.4.3.4. reviewed in Chapter 18) which deaminates terminal amine groups on spermine and spermidine molecules to produce respectively the corresponding amino-dialdehyde and aminomonoaldehyde as shown in Figure 5. These could further autodegrade to some degree to produce putrescine and cytotoxic acrolein ($O=CHCH=CH_2$; Alarcon, 1964; Tabor, Tabor and Bachrach, 1964), or other more complex carbonyl compounds consisting of conjugated oligamines (Kimes and Morris, 1971; Gaugas and Dewey, 1979). It has been elegantly demonstrated by Israel and Modest (1971) that the substrate requirement for plasma PAO is terminal $H_2N(CH_2)_nN$ —, where n = at least 3. The products of PAO-polyamine interaction potently inhibited the transformation of lymphocytes (Byrd, Jacobs and Amoss, 1977) but unexpectedly were effective at sub-cytotoxic concentrations. Thermine, $NH_2(CH_2)_3NH$-$(CH_2)_3NH(CH_2)_3NH_2$, is a bacterial polyamine which had been described by Oshima (1975) and was also found to be a substrate for PAO. Products of PAO-thermine interaction potently inhibited lymphocyte proliferation (Table 3).

Though abundant in ruminant sera the PAO exists in only trace amounts in normal human, rodent or horse sera (Kapeller-Adler, 1970), not enough in fact being present to bio-assay using the culture parameters employed.

The rat liver, at least, has recently been shown to incorporate a novel PAO which has been purified and found to convert both spermine and spermidine to 3-aminopropionaldehyde ($H_2NCH_2CH_2CHO$), but whether it potently inhibits cell proliferation and could cause liver damage has not been ascertained (Hölttä, 1977). Quash and Taylor (1970) also present evidence that bovine plasma PAO can convert spermidine to the aminopropionaldehyde and malondialdehyde ($CHOCH_2CHO$) which should be inhibitory.

In cultured lymphocytes the entry of PAO into cells should be precluded by its macromolecular size unless endocytosed. Extracellular interaction of PAO with polyamines is therefore necessary in order to produce the aldehydes which appear responsible for inhibition of transformation. Fetal-calf serum, FCS, contains a lower level of PAO than adult bovine sera, but at 10% v/v in culture medium FCS is nevertheless supraoptimal for the conversion of biogenic polyamines to inhibitors (Table 3). Because of the surprising observation of lymphocytostasis induced by the inhibitor — rather than the expected cytotoxic effect promised for acrolein — it was thought possible that PAO reactivity need not be implicated in the production of the inhibitor (Byrd, Jacobs and Amoss, 1977). Such challenge was refuted by showing that as little as 0.5–2.0 μg ml^{-1} of purified bovine serum PAO (Miles Ltd.), in human serum as the medium supplement, sufficed for the optimal conversion of polyamines to inhibitors of PHA-induced lymphocyte transformation, but failed to account for the lack of lymphocytotoxicity. None the less Byrd, Jacobs and Amoss' most important discovery of the absence of cytotoxicity is therefore worthy of extensive further investigation.

Using the biochemical assay described by Bachrach and Reches (1966) it was found that 4-h pre-incubation of purified PAO with a polyamine substrate was required for maximal conversion to aldehydes. Thus when testing S-phase lymphocytes pre-incubation of the FCS or PAO and substrate mixture was necessary, and numerous such experiments resulted in a variable degree of inhibition of DNA replication in the range of only 0–35% (10.0–20,000 μmole of spermine) but without correlation to the concentration of polyamine, that is, after attainment of maximal plateau inhibition. In contrast, simple aldehydes such as acrolein and benzaldehyde arrested S-phase cells (see Table 3).

The inhibitory phenomenon of PAO-polyamine interaction on PHA-induced transformation of lymphocytes was demonstrably transient since 24–36 h pre-incubation (37°C) of the mixture in tissue culture medium before adding lymphocytes and mitogen together, failed to end in appreciable suppression of transformability (10 μmole spermine). Thus the inhibitory product must be labile at least in the presence of serum enriched medium with a half-life of 2.3 h (Gaugas and Dewey, 1979). Acrolein cannot be credited with chemical lability, though it is possibly lost from the culture medium through evaporation (bp 53°C). Furthermore acrolein might well be deactivated by glutathione and other low molecular weight thiols abundant in the serum

Table 3 Polyamine interaction with fetal-calf serum, or PAO, in the suppression of lymphocyte transformation or proliferation

	Inhibitory dose 50% (μmole litre^{-1})				
	PHA-treated cells		Thymic cells	LPS-treated cells	Proliferation of Bri8 or L5187Y leukaemia cells
	G_1-S	S	S**	G_1-S	F_0-F_n generations
Polyamines:					
Spermine	6.0	N.I.	N.I.	2.5	2.0
Spermidine	36.0	N.I.	N.I.	—	12.0
1,12-Diaminododecane*	300.00	—	300.0	—	<80.0
Thermine*	4.0	—	—	—	20.0
Diamines:					
Putrescine	>160.0	—	—	N.I.	N.I. (>160)
Cadaverine	>160.0	—	—	—	80.0
1,3-Diaminopropane*	N.I.	—	—	—	N.I.
Aldehydes:					
Benzaldehyde*	>500.0	—	300.0	—	—
Acrolein	15.0	5.0	5.0	—	5.0

*non-biogenic (or bacterial)
** 4-hours pre-incubation before adding lymphocytes
N.I. = Non-inhibitory (ID_{50} >640 μmole litre^{-1})

additive, since leukaemia cells *in vivo* were found to be damaged by only rapid direct contact with acrolein (Motycka and Lacko, 1966). Should some acrolein be produced during the incubation period of lymphocyte cultures containing FCS or PAO and polyamine, it could well be severely limited and then undergo rapid deactivation. When 3-methyl-2-benzothiazolinone hydrazide which conjugates with aminoaldehydes (Bachrach and Reches, 1966) was added (250 μmole) to Bri8 lymphocyte cultures it failed, however, to adjust the degree of inhibition of proliferation attributable to FCS-polyamine products.

Although B lymphocytes (LPS-stimulated mouse splenic lymphocytes were more susceptible than T lymphocytes (PHA-stimulated rat thymic lymphocytes) to the inhibition caused by FCS-polyamine interaction, it is explicable as an artefact due to lability of the inhibitor. In fact, T cell transformation takes longer (64-72 h incubation period) than for B lymphocytes (48 h incubation) so during G_1 these latter cells face a greater relative exposure to the transient inhibitor.

Successive proliferation of Bri8 or L5187Y leukaemic lymphocytes was also inhibited by FCS or purified PAO interaction with polyamines. The results are summarized in Table 3. In both cases, the dose giving 50% inhibition was similar to that for B lymphocyte LPS-induced transformation. Synchronized cell cultures showed that treatment with a mixture of FCS and spermine in the G_1 phase resulted in a failure of entry into the S phase, as shown in Figure 6. In contrast, similarly treated cells selected for culture whilst in the S phase failed to be arrested.

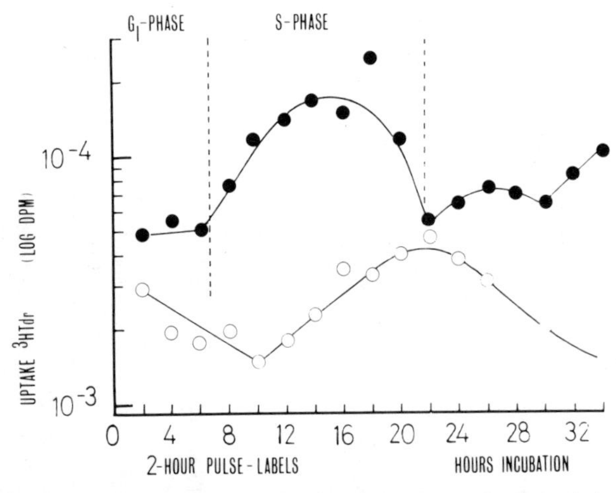

Figure 6 Synchronous-growth of F_0 generation Bri8 human leukaemic lymphocytes (●) 40,000 cells ml^{-1} approximately 66.0% synchrony. Cells treated with FCS-spermine (10 μmole) (○) fail to enter the S phase from G_1. Each point represents the mean (uptake [^{3}H]TdR) of a triplicate-culture, the s.d. being within 10% of the mean. In contrast, cells collected at the S phase at the beginning of incubation fail to show arrest by such treatment (not shown). Concentrations per litre

The oxidative reactivity of PAO on polyamine substrates (10 μmole spermine) could be ablated by the addition of 0.5–250 μmole of 3-hydroxy-benzyloxyamine (Sandev Ltd.) and thus allows for proliferation to continue unheeded. (The enzymic antagonist also inhibits lymphocyte proliferation in its own right as an inhibitor of L-aromatic amino acid decarboxylases in the cell, but only at concentrations greater than 250 μmole.)

Presensitized murine splenic B lymphocytes *in vitro* secretion of antibodies against sheep erythrocytes was not suppressed by the interaction of spermine with FCS, even at 2 mmole (unpublished data). Since massive amounts of aldehyde would probably be produced at such a high level of spermine it seems likely that some plasma-membrane damage should have occurred. It is therefore possible that when exposed to PAO-polyamine interaction *in vivo* presensitized B lymphocytes (G_1 phase) could continue to secrete immunoglobulin. In contrast, T lymphocytes would fail to transform and thus be unable to exert an immunological response. This assumes that such interaction can occur *in vivo* under pathological conditions.

Evidence is therefore amassing that unlike the situation with treatment by inhibitors of diamine-polyamine biosynthesis, the inhibitory reactivity of PAO transformed polyamines occurs in the F_0 generation of cultured lymphocytes. Inhibition is non-cytotoxic and apparently asserted in the G_1 phase of the cell cycle. The effect is probably not due to the polyamine derived aldehydes repressing diamine-polyamine biosynthetic enzymes within the cell. If so, no arrest of cells in the F_0 generation would be forthcoming because of the cell's spermine reservoir. On the other hand, the aminoaldehydes could possibly account for an inhibitory property by competing with endogenous polyamines — instead of merely performing like simple aldehydes in being cytotoxic for S phase cells — in their role supporting DNA replication. If spermine is the sole physiologically functional agent for support of proliferation, then perhaps spermidine when oxidized cannot be an inhibitor. It is, however, known that oxidized spermidine does inhibit. None the less, spermidine has an ID_{50} which is six-fold higher than spermine when interacting optimally with PAO, and could generate a three-fold higher concentration of acrolein in culture to account possibly for its inhibitory phenomenon. Moreover there is some evidence that 50% of the normal level of endogenous spermidine is required for proliferation (Newton and Abdel-Monem, 1977). As a precautionary note, participation of acrolein or other unidentified carbonyl compounds to some unascertained degree in the inhibitory system cannot justly be completely discounted. As pointed out by Kimes and Morris (1971), the instability of oxidized polyamines must be considered when interpreting their physiological effects.

2. *Diamine Oxidation*

Diamine oxidase catalyses the oxidation of putrescine, cadaverine and histamine with the production of aminoaldehydes (reviewed in Chapter 17). Diamine oxidase, DAO, is plentiful in such tissues as liver, kidney, placenta and platelets. The aminoaldehyde formed from histamine is probably imidazoleacetaldehyde (NH.NHCH_2CHO), whilst putrescine and cadaverine substrates degrade to aminobutyraldehyde ($H_2NCH_2CH_2CH_2CHO$) which is then almost quantitatively reduced to $^1\Delta$-pyrroline: α-aminobutyric acid (GABA, $H_2NCH_2CH_2CH_2CO_2H$) is an intermediary metabolite (Andersson and coworkers, 1978). Quash and Taylor (1970) have reported formation of 3-aminopropionaldehyde by hog kidney DAO.

Hence DAO degrades substrates to aldehydes, and aldehydes in general are cytotoxic and have been studied as potential anti-cancer drugs (reviewed by Schauenstein, Esterbauer and Zollner, 1977). The possibility of conversion of biogenic diamines to inhibitors of lymphocyte proliferation was therefore investigated. Results of hog kidney DAO (Sigma Ltd.) interaction in this respect is presented in Figure 7. Whereas both cadaverine and putrescine were converted to inhibitors, histamine or indeed polyamine was not. Moreover

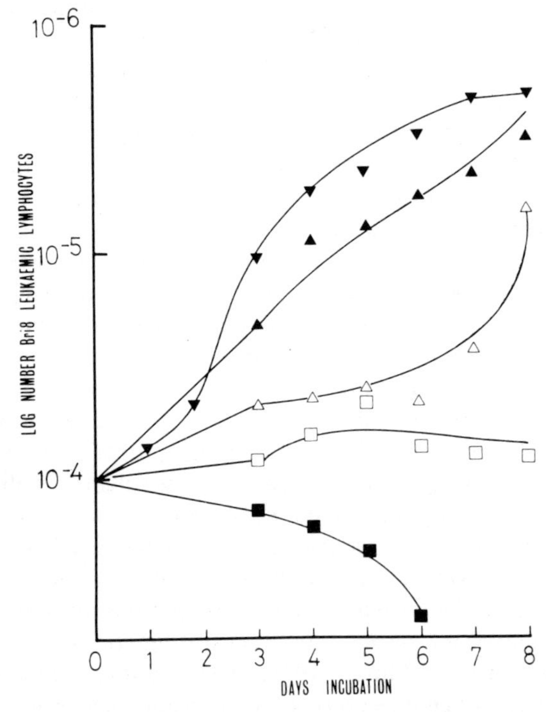

Figure 7 Growth-curves for Bri8 leukaemic lymphocytes in medium (enriched with normal human serum) containing 60 μg ml^{-1} (optimal) of a hog kidney DAO preparation; ▼ untreated controls, ▲ 320 μmole spermine, △ 160 μmole cadaverine, □ 160 μmole putrescine, ■ 320 μmole cadaverine. Each point represents the mean cell-count of duplicate cultures. Concentrations per litre

GABA, commercial imidazole acetic acid HCl, and 4-aminobutyraldehyde diethyl acetal (640 μmole) were all virtually ineffective at inhibition of lymphocyte proliferation. The likelihood is that unidentified intermediary metabolites as reported by Andersson and coworkers (1978) or 3-amino-propionaldehyde are responsible for the observed potent inhibition.

The inhibitory system of diamine oxidation on Bri8 lymphycyte proliferation was reversed by the addition of aminoguanidine (250 μmole). Aminoguanidine is a well known suppressor of DAO and thus confirms that the enzyme was indeed responsible for the conversion of diamine substrates to inhibitors.

3. *Polyamine and Diamine Oxidation in Human Pregnancy*

In order to strengthen the weak arguments for the participation of diamine and polyamine and their oxidative enzymes in immunosuppression or immunoregulation, their increased serum levels should accord with increased levels of amine oxidases. Survival of the human fetoplacental unit in the maternal immunologically hostile environment is an enigma for immunologists, so it was examined in this respect. The fetoplacental unit which is bathed directly in maternal blood, somehow survives despite being immunologically incompatible with the mother because of paternally inherited histocompatibility antigens (reviewed by Beer and Billingham, 1971). Preliminary investigations were carried out in order to determine whether pregnancy sera which had been collected at different stages of gestation (without correlation to parity) were able to convert spermine to an inhibitor of PHA-induced lymphocyte transformation (Gaugas and Curzen, 1978). Potent inhibition was demonstrated. Within the assay limits, inhibition was first apparent at about 15 weeks' gestation and thereafter was maintained maximally until about the 26th week when reactivity usually, but not invariably, ceased (unpublished data). One difficulty encountered in testing a limited number of sera covering 26-weeks gestation, was that even in the absence of polyamine the degree of transformation was exceedingly low, or usually non-existent. Pregnancy sera (16–25 weeks) was at least equal to FCS in its ability to convert spermine.

A rise in serum diamine oxidase (DAO) level occurs on about the 8th week of gestation in humans, becoming maximal by the 20th week and then probably plateaus until parturition (Weingold, 1968; Kapeller-Adler, 1970). Similar rises have been shown for rats (Kobayashi, 1964), but not for marmosets (Hampton and Parmlee, 1969). Southren and coworkers (1966) have produced some evidence for a repressor of amine oxidase in pregnancy sera, so perhaps we should refer to enzyme activity rather than rises in serum level. Not surprisingly, human pregnancy sera were found to interact with putrescine to suppress lymphocyte transformation (Gaugas and Curzen, 1978). The origin of the PAO is unknown, but likely to be a product of the placenta or maternal liver. The former would be most curious since the organ is of fetal origin yet human fetal-cord serum DAO and PAO activity was negligible. Placental extracts include substances that strongly reduce DAO effects on cadaverine and putrescine, but not on histamine (Kapeller-Adler, 1970). For inhibition of lymphocyte proliferation, DAO but not PAO activity was discernible in human and in bovine amniotic fluid (10% v/v in culture medium supplemented with human serum). Pregnancy associated PAO could also convert polyamines to putrescine (Figure 5) which in turn could be converted by DAO to inhibitory substances. This is a situation which requires clarification. Acrolein could also possibly be formed, but there is no evidence for this cytotoxic agent ever existing in the body. Thus diamine and polyamines could be implicated together with respective amine oxidases in the immunoregulation of human pregnancy. The inhibitory system is unlikely to be a prime mechanism, since it does not appear until relatively late in pregnancy. Furthermore, it is not invariably maintained much beyond the second-trimester for PAO but not for DAO activity, unless still reacting locally in the gravid uterus.

A. Conclusions

The products, probably complex aminoaldehydes, or complex carbonyls, which elicit potent and non-cytotoxic suppression of both lymphocyte transformation and proliferation, obtained by the interaction of DAO and PAO with their respective substrates could possibly be implicated in clinical disorders, such as pre-eclampsia which features reversible kidney damage, spontaneous abortion, degenerative diseases (e.g. kidney, liver and brain), ageing; and especially when immunosuppression occurs as in advanced malignancy, and some infectious diseases (bacterial, viral and protozoan). This system of suppression of lymphocyte function, as well as possibly other tissue damage, could therefore be of considerable clinical significance.

ACKNOWLEDGEMENT

This work was supported by the Cancer Research Campaign.

REFERENCES

Alarcon, R. A., Foley, G. E. and Modest, E. J. (1961). *Arch. Biochem. Biophys.*, **94**, 540–541.

Alarcon, R. A. (1964). *Arch. Biochem. Biophys.*, **106**, 240–242.

Allen, J. C., Smith, C. J., Curry, M. C., and Gaugas, J. M. (1977). *Nature, Lond.*, **267**, 623–625.

Amman, R., and Werle, E. (1957). *Klin. Wschr.*, **35**, 22–26.

Andersson, A-C., Henningsson, S., Persson, L., and Rosengren, E. (1978). *Acta Physiol. Scand.*, **102**, 159–166.

Bachrach, U., and Reches, B. (1966). *Analyt. Chem.*, **17**, 38–48.

Bachrach, U., Abzug, S., and Bekierkunst, A. (1967). *Biochem. Biophys. Acta*, **134**, 174–181.

Bachrach, U. (1973). *Function of Naturally Occurring Polyamines*, Academic Press, New York and London.

Bachrach, U., and Ben-Joseph, M. (1973). Tumour cells, polyamines and polyamine derivatives. In D. H. Russell (Ed), *Polymines in Normal and Neoplastic Growth*, Raven Press, New York, pp. 15–26.

Beer, A. E., and Billingham R. E. (1971). *Advances Immunol.*, **14**, 1–8.

Boyland, E. (1941). *Biochem. J.*, **35**, 1283–1288.

Boynton, A. L., Whitfield, J. F., McManus, J. P., Rixon, R. H., Swierenga, S. H. H., and Walker, P. R. (1977). The interaction of calcium, cyclic nucleotides, microtubules, prostaglandins and polyamines in the positivie control of proliferative development. *International Symposium in Molecular Control of Proliferation in Eukaryotic Cells.* April 1977, Dogobókó, Hungary.

Byrd, W. J., Jacobs, D. M., and Amoss, M. S. (1977). *Nature, Lond.*, **267**, 621–623.

Byus, C. V., and Russell, D. H. (1974). *life Sci.*, **15**, 1991–1997.

Cline, M. J., Warner, N. L., and Metcalf, D. (1972). *Blood*, **39**, 326–333.

Cohen, S. S. (1977). *Cancer Res.*, **37**, 939–942.

Dewey, D. L. (1978). *Cancer Lett.*, **4**, 77–84.

Durham, A. (1978). *New Scientist*, March 30, 1978, 860–862.

Fillingame, R. H., Jorstad, C. M., and Morris, D. R. (1975). *Proc. Nat. Acad. Sci. USA*, **72**, 4042–4045.

Friedman, S. J., Bellantone, R. A., and Cannellakis, E. S. (1972). *Biochem. Biophys. Acta*, **261**, 188–193.

Fuller, D. J. M., Gerver, E. W., and Russell, D. H. (1977). *Cellular Physiol.*, **93**, 81–86.

Gaugas, J. M., and Chu, A. M. (1980). Methylglyoxal *bis*(guanylhydrazone) in *in vitro* and *in vivo* inhibition of murine bone-marrow colony formation. (In preparation)

Gaugas, J. M., and Curzen, P. (1978). *Lancet*, **(i)**, 18–20.

Gaugas, J. M., and Dewey, D. L. (1979). *Br. J. Cancer.* **39**, 548–557.

Halevy, S., Fuchs, Z., and Mager, J. (1962). Bull. Res. Council Israel, **11A**, 52–53.

Hampton, J. K. and Parmlee, M. L. (1969). *Comp. Biochem. Physiol.*, **30**, 367–370.

Heby, O., and Andersson, G. (1978). *Acta Path. Microbiol. Scand.*, **86 (Sect. A)**, 17–20.

Heby, O., Gray, J. W., Lindl, P. A., Marton, L. J., and Wilson, C. B. (1976). *Biochem. Biophys. Research Commun.*, **71**, 99–105.

Hibasami, H., Tanaka, M., Nagai, J., and Tadoa, I. (1977). *AJEBAK*, **55**, 379–383.

Higgins, L. M., Tillman, M. C., Rupp, J. P., and Leach, F. R. (1969). *J. Cell Physiol.*, **74**, 149–154.

Hölttä, E. (1977). *Biochemistry*, **16**, 91–100.

Israel, M., and Modest, E. J. (1971). *J. Med. Chem.*, **14**, 1042–1047.

Kallio, A., Pösö, H., Guha, S. K., and Jänne, J. (1977). *Biochem. J.*, **166**, 89–94.

Kapeller-Adler, R. (1970). In *Amine Oxidases and Methods for Their Study*, Wiley, New York.

Kay, J. E., and Lindsay, V. J. (1973). *Exp. Cell Res.*, **77**, 428–436.

Kay, J. E., and Pegg, A. E. (1972). *FEBS Lett.*, **29**, 301–304.

Kimes, B. W., and Morris, D. R. (1971). *Biochim. Biophys. Acta*, **228**, 223–234.

Kobayashi, Y. (1964). *Nature, Lond.*, **203**, 146–147.

Melvin, M. A. L., and Keir, H. M. (1978). *Biochem. J.*, **174**, 321–326.

Mihich, E. (1975). *Bis*-guanylhydrazones. In A. C. Sartorelli and D. G. Johns (Eds), *Handbook of Experimental Pharmacology*, Vol. 38, Pt. 2, Springer-Verlag, Berlin, pp. 766–788.

Morris, D. R., Jorstad, C. M., and Seyfried, C. E. (1977). *Cancer Res.*, **37**, 3169–3172.
Motycka, K. L., and Lacko, L. (1966). *Z. Krebsforsch.*, **68**, 195–199.
Newton, N. E., and Abdel-Monem, M. M. (1977). *J. Med. Chem.*, **20**, 249–253.
Oshima, T. (1975). *Biopphys. Research Commun.*, **63**, 1093–1098.
Otani, S., Yasuhiro, M., Isao, M., and Sieji, M. (1974). *Molecular Biol. Reports*, **1**, 431–436.
Quash, G., and Taylor, P. (1970). *Clin. Chim. Acta*, **30**, 17–23.
Rosenblum, M. G., and Russell, D. H. (1977). *Cancer Res.*, **37**, 47–51.
Russell, D. H., and Stambrook, P. J. (1975). *Proc. Nat. Acad. Sci. USA.*, **72**, 1482–1486.
Schauenstein, E., Esterbauer, H., and Zollner, H. (1977). In *Aldehydes in Biological Systems*, Pion, London.
Southren, A. L., Kobayashi, Y., Carmody, N-C., and Weingold, A. B. (1966). *Amer. J. Obstet. Gynecol.*, **96**, 502–510.
Tabor, C. W., and Tabor, H. (1976). *Ann. Rev. Biochem., 1976*, 285–306.
Tabor, C., Tabor, H., and Bachrach, U. (1964). *J. Biol. Chem.*, **239**, 2194–2203.
Theoharides, T. C., and Canellakis, Z. N. (1975). *Nature, Lond.*, **225**, 7333–734.
Weingold, A. B. (1968). *Clin. Obstet. Gynec.*, **11**, 1081–1087.
Williams-Ashman, H. G., and Shenone, A. (1972). *Biochem. Biophys. Res. Commun.*, **46**, 288–295.

Chapter 23

Excretion of Polyamines from Mammalian Cells in Culture

MAUREEN A. L. MELVIN AND HAMISH M. KEIR

I. INTRODUCTION

Little is known about the origin or state of the polyamines found in physiological fluids. Polyamines are present in increased amounts in physiological fluids of tumour-bearing animals and of patients with diagnosable cancer (Bachrach, 1976; Russell, 1977). Therapy leading to remission of cancer further increases the urinary content of polyamines for several days after treatment and this is followed by a decline as tumour growth is arrested.

It is difficult to assess what proportion of the polyamines present in physiological fluids originates from tumour tissue and what proportion originates from non-cancerous tissue (Russell, 1977). One approach to this problem was to apply local X-irradiation to the tumour tissue of rats carrying hepatomas (Russell and coworkers, 1976). The amounts of spermidine and putrescine in the serum increased after local irradiation of the tumour. Whereas this was paralleled by a decrease in the spermidine content of the tumour, the polyamine concentrations in the host liver remained unchanged. These results suggest that the increased amounts of polyamines detected in the serum originated from the tumour itself.

A variety of factors complicate studies of polyamine excretion in the intact animal. A different approach is to study the fate of intracellular polyamines when the growth rate of cells in culture has been inhibited. In this case, the cells whose growth rate is being altered can be studied directly without interference from other cells and tissues. We have used BHK-21/C13 cells (Syrian baby hamster kidney fibroblasts, McPherson and Stoker, 1962) growing in monolayer cultures to study the metabolism of polyamines when the cell growth rate is changed.

Little is known about how cells can decrease their polyamine content. We have used a variety of factors that affect cell growth rate and have studied their effects on the metabolism and fate of intracellular polyamines (Melvin and Keir, 1977; 1978a; 1978b; Melvin and coworkers, 1978). The effects of increasing cell density, deprivation of serum, starvation for isoleucine, and exposure to anti-tumour agents, on the intracellular polyamine content and on the fate of intracellular polyamines, are remarkably similar in each case. The results indicate that the control of intracellular polyamine levels is not confined to the pathway of biosynthesis of polyamines, but is also mediated by excretion of polyamines, particularly spermidine, from the cells.

II. METHODS AND MATERIALS

A. Cell Culture

BHK-21/C13 cells were grown in monolayer cultures as described elsewhere (Melvin and Keir, 1978a), using horse serum in the culture medium since the serum of other species (e.g. calf), which contains polyamine oxidase activity, is unsuitable (see Chapter 22).

B. Estimation of Cell Growth

Cell number was determined using a haemocytometer. Protein, RNA and DNA were extracted from cells and determined quantitatively as described previously (Munro and Fleck, 1966; Melvin and Keir, 1978a). Putrescine, spermidine and spermine, extracted in 0.2 M perchloric acid, were converted to their dansylated derivatives and separated by thin layer chromatography, the chromatograms being developed twice in ethylacetate:cyclohexane (2:3, v/v). The amines were determined quantitatively by measurement of the fluorescence of their dansylated derivatives (Herbst and Dion, 1970).

C. Radioactive Labelling of Intracellular Polyamines and their Determination in Cells and Medium

Intracellular polyamines were labelled using exogenous radioactive putrescine, spermidine or spermine, as described previously (Melvin and Keir, 1978a), and were dispensed at a density of approximately 0.8×10^6 cells per dish into 90 mm diameter vented plastic dishes.

At the time of harvesting of experimental cultures, the medium was decanted and centrifuged to remove cell debris. Portions of the medium were (1) assayed directly for radioactivity, (2) extracted with ice-cold 0.2 M perchloric acid and the polyamines dansylated, separated by thin layer chromatography, and assayed for radioactivity, or (3) washed with ice-cold trichloroacetic acid, final concentration 4% (v/v), followed by anhydrous ether, and the extract hydrolysed with 6 M HCl at 110°C for 16 h then dried to powder form (Gerner and Russell, 1977). Samples were redissolved in 0.2 M perchloric acid, treated with dansyl chloride, and the products separated and analysed for radioactivity.

The cell sheet was washed with ice-cold phosphate-buffered saline (Melvin and Keir, 1978a) to remove any radioactivity that had adsorbed to the cell surface, the cells were scraped from the dishes in the same buffer, and the polyamines were extracted either with (1) ice-cold 0.2 M perchloric acid, or (2) ice-cold trichloroacetic acid followed by hydrolysis with HCl; the polyamines were then dansylated and analysed as described above.

Radioactivity was determined by liquid scintillation spectrometry (Melvin and Keir, 1978b).

D. Assay for Ornithine Decarboxylase (ODC) Activity

ODC (EC 4.1.1.17) activity was estimated using a modification of the method of Russell and Snyder (1968), as described by Howard and coworkers, 1974.

1. *Materials*

1,4(*n*)-[^{3}H]putrescine dihydrochloride (16-25 Ci mmole^{-1}), [^{14}C]spermidine trihydrochloride (122 mCi mmole^{-1}), [^{14}C]spermine tetrahydrochloride (115 mCi mmole^{-1}) and D,L-1-[^{14}C] ornithine hydrochloride (58 mCi mmole^{-1}) were purchased from the Radiochemical Centre, U.K.; 6-mercaptoguanosine (6-thioguanosine) from Sigma Ltd. and methylglyoxal *bis*(guanylhydrazone) dihydrochloride (methyl-GAG) from Aldrich Co.

III. INHIBITION OF CELL GROWTH

A. Deprivation of Serum

The growth of BHK-21/C13 cells is arrested when the serum concentration in the culture medium is lowered abruptly from 10% to 1%(v/v), or less, and after 2–3 days the cell population is at rest in the G1 phase of the cell cycle (Bürk, 1970; Howard and coworkers, 1974). One of the earliest changes observed in BHK-21/C13 cells after removal of serum is a decrease in the activity of ODC, the enzyme which catalyses the first step in the pathway of biosynthesis of putresçine, spermidine and spermine. ODC activity shows a precipitous decline after removal of serum, only about 10% of the initial activity remaining 2 h later (Table 1). This contrasts with the effect of replacing the medium with fresh medium containing 10%(v/v) serum, which causes a temporary increase in the activity of ODC 4–6 h later.

Table 1 Ornithine decarboxylase activity in BHK-21/C13 cells after transfer to fresh medium containing either 10% or 1%(v/v) serum

Time after transfer (h)	Ornithine decarboxylase activity (pmoles of CO_2 released/h/mg of protein)	
	10% serum	1% serum
2	0.438	0.045
6	0.569	0.026
24	0.150	0.020

Cell proliferation and the accumulation of protein, RNA and DNA are inhibited by depriving BHK-21/C13 cells of serum (Table 2). The cell content of polyamines decreases in serum-deprived cells compared to control cultures in 10% serum medium, and this is attributable mainly to a decrease in the intracellular content of spermidine while the content of spermine changes less dramatically (Table 2). This is reflected in a fall in the molar ratio of spermidine/spermine in serum-deprived cells, which can usually be detected within 6 h of removal of serum.

Intracellular polyamines can be labelled by incubating non-confluent monolayers of cells in medium containing radioactive polyamine (Table 3). The conversion of putrescine to spermidine and of spermidine to spermine occurs quite rapidly in BHK-21/C13 cells. Cells whose polyamines had been labelled using [^{3}H] putrescine were transferred to fresh medium containing either 10% or 1% (v/v) horse serum, and at various times thereafter the cells and the medium in which they were incubated were harvested, extracted with 0.2 M perchloric acid, and the radioactivity in the extract of each fraction determined. In all cultures, labelled material was lost from the cells into the culture medium (Figure 1). Much more radioactivity was lost from cells incubated in medium containing 1% serum than from cells in 10% serum medium.

The nature of the labelled material in BHK-21/C13 cells incubated in 1% serum medium was investigated and compared with the labelled material lost from the cells into the culture medium. Polyamines were extracted from samples of cells and medium, using 0.2 M perchloric acid, and their radioactivity was determined. Whereas the radioactivity in the cells was distributed between spermine and spermidine (the proportion found in

Table 2 Effect of deprivation of serum on BHK-21/C13 cell growth and polyamine content

Culture	Cell no. ($\times 10^{-6}$)	Protein (mg)	RNA (μg)	DNA (μg)	Polyamine content (nmoles/mg of protein)			
					Spermine	Spermidine	Putrescine	Spermidine/ spermine
Initial culture	1.6	0.55	54	17	11.6	15.3	2.2	1.32
24 h in 10% serum medium	6.0	1.64	180	63	9.0	12.2	1.7	1.35
24 h in 1% serum medium	2.1	0.84	80	28	8.8	7.1	1.2	0.81

The values given for cell number, protein, RNA and DNA represent the average content per dish, 4 dishes being assayed in triplicate for RNA, DNA and protein, and 2 dishes being used for each determination of cell number. Polyamines were extracted from 4 dishes of cells, and the values given are the mean values from triplicate assays.

Table 3 Uptake of exogenous polyamines into BHK-21/C13 cells, and intracellular interconversions of the polyamines

Exogenous polyamine	Time (h)	cpm per dish	Radioactivity (%) Origin	Spermine	Spermidine	Putrescine
[³H] Putrescine	0	60106	37	7	40	16
	4	49390	30	14	54	2
	22	57530	36	31	32	1
[¹⁴C] Spermidine	0	88630	25	10	64	1
	4	74600	32	13	54	1
	22	93115	35	30	34	1
[¹⁴C] Spermine	0	78120	37	60	3	0
	22	100730	33	56	11	0

Cells were incubated for 2 h in medium containing 10%(v/v) horse serum and either [³H]putrescine (0.2 μCi/ml), [¹⁴C]-spermidine (0.15 μCi/ml) or [¹⁴C]spermine (0.15 μCi/ml), before being transferred to non-radioactive medium containing 10% horse serum. The values given are the average for 2 dishes of cells, each analysed in duplicate.

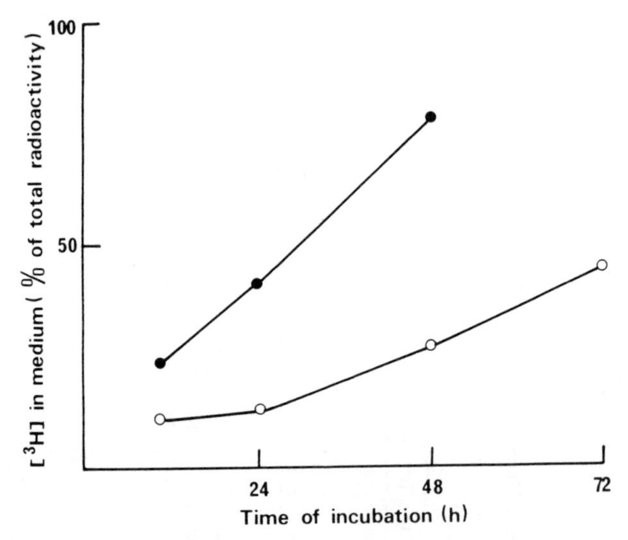

Figure 1 [³H]-Labelled material lost from BHK-21/C13 cells incubated in medium containing either 10% or 1%(v/v) serum

Cells were labelled by incubating them for 24 h, in a 2.24 litre bottle, in medium containing 10%(v/v) horse serum and [³H]putrescine dihydrochloride (0.6 μCi/ml). The cells were dispensed into 90 mm diameter dishes (1.0×10^6 cells per dish), and incubated in non-radioactive medium for 20 h, to allow equilibration of the label with intracellular polyamine pools, then the medium was replaced by fresh medium containing either 10% or 1% horse serum. The initial radioactivity in the cells, at 0 h, was 18.9×10^3 cpm per dish. Values shown represent the mean of 4 separate determinations of radioactivity. Actual values obtained varied by less than 5% of the mean values. ○———○, 10% serum medium; ●———●, 1% serum medium

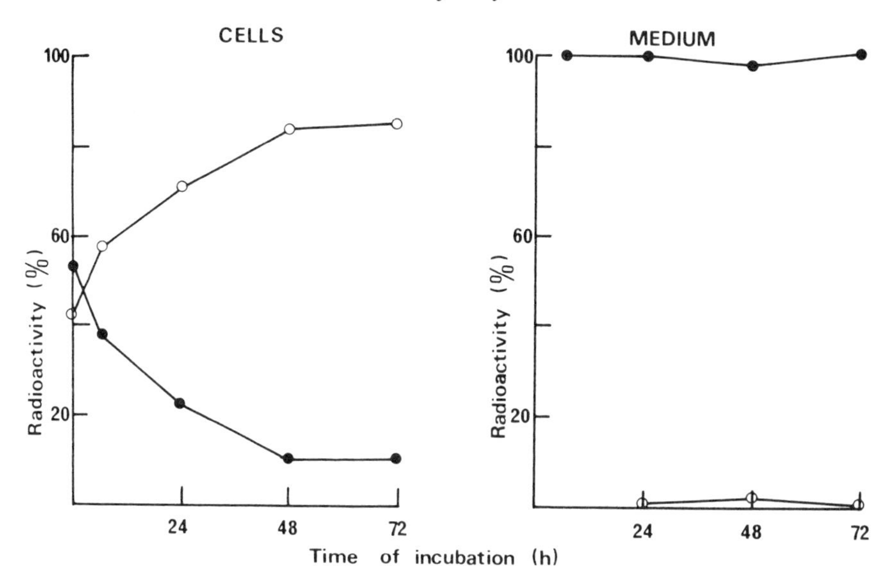

Figure 2 Distribution of [^{3}H]-labelled material in cells and medium when BHK-21/C13 cells were incubated in medium containing 1%(v/v) serum

Cells whose polyamines had been labelled with [^{3}H]putrescine (1.0 μCi/ml of medium, for 24 h), were transferred to medium containing 1%(v/v) horse serum, and the cells and their incubation medium harvested at various times thereafter. Polyamines were extracted using 0.2 M perchloric acid and their radioactivity determined. The initial radioactivity in the cells, at 0 h, was 50.0 × 10^3 cpm per dish. ●——●, radioactivity in spermidine; ○——○, radioactivity in spermine

spermine increasing with time) at all times the radioactivity in the culture medium was located almost exclusively in spermidine (Figure 2). This suggests that in response to inhibition of cell growth rate, caused by depriving cells of serum, the cells specifically excreted spermidine into the culture medium. Excretion of spermidine may be an important mechanism whereby BHK-21/C13 cells dispose of polyamines present in excess of their requirements for growth. Since the conversion of putrescine to spermidine by these cells is very rapid, and since the conversion of spermine to spermidine is known to take place in these cells (Table 3), putrescine and spermine could be disposed of by excretion after they have been converted to spermidine.

Loss of spermidine from BHK-21/C13 cells in these experiments was not due simply to cell death and lysis, as has been suggested as an explanation for the increased amounts of spermidine found in the extracellular fluids of animals and patients responding to therapy for arrest of tumour growth (Russell, 1977). Cell monolayers were examined microscopically 24 and 48 h after their transfer to 1% serum medium, and there was no evidence of extensive cell death or of detachment of cells from the monolayer. Moreover, if loss of polyamines from these cells had been due to cell death and lysis, a substantial amount of radioactivity would have been found in spermine as well as in spermidine in the medium. Following readdition of serum, to 10%(v/v), to BHK-21/C13 cells which had been in low-serum medium for several days, cell growth recommenced (Bürk, 1970; Howard and coworkers, 1974; Melvin and Keir, 1978a). Within 24 h of serum 'step-up', the accumulation of polyamines, particularly spermidine, was considerably increased, and the molar ratio of spermidine/spermine in the cells increased significantly (Melvin and Keir, 1978a). Hence the cells in the serum-starved cultures were not dead, but 'resting'.

B. Effect of Increasing Cell Density

When BHK-21/C13 cells were incubated in medium containing 10%(v/v) serum for several days, the cell growth rate decreased (Table 4). Accumulation of protein, RNA and DNA was inhibited and there was concomitant depletion of the cell content of polyamines, particularly spermidine, so that the molar ratio of spermidine/spermine decreased (Table 4). At the same time, release of polyamines from the cells into the culture medium started to increase (Figure 1). The radioactively-labelled polyamines present in [^{3}H]putrescine-labelled cells and in their incubation medium were examined and it was found that at all times most of the label in the medium was confined in the spermidine, even though the corresponding cells contained more radioactivity in spermine than in spermidine (Table 5). Thus spermidine was excreted specifically from the cells, and increased excretion of spermidine occurred as cell growth was inhibited. These results, which are very similar to those obtained when the cell growth rate was inhibited by deprivation of serum, suggest that specific release of spermidine into the culture medium may be a general response of BHK-21/C13 cells to inhibition of their growth rate.

The cells in the density-inhibited cultures were not dead. They could be stimulated to recommence growth by harvesting them and dispensing them at lower cell density into culture dishes.

C. Deprivation of Isoleucine

It has been suggested that the cell membrane is the primary site of action for the inhibition of cell growth which occurs after deprivation of serum (Rubin and Koide, 1973) or as cell density increases (Rubin, 1971; Gray, Cullum and Griffin, 1976). Excretion of spermidine from cells whose growth has been inhibited by either of these means could be due to alterations in the cell surface membrane, leading to alterations in cell membrane permeability. Alternatively, it might be that in each of these cases the cells are deprived of some nutrients

Table 4 Effect of increasing cell density on BHK-21/C13 cell growth and polyamine content

Time (h)	Cell no. ($\times 10^{-6}$)	Protein (mg)	RNA (μg)	DNA (μg)	Polyamine content (nmoles/mg of protein)			Spermidine/ spermine
					Spermine	Spermidine	Putrescine	
0	1.6	0.55	54	17	11.6	15.3	2.2	1.32
24	6.0	1.64	180	63	9.0	12.2	1.7	1.35
46	8.4	2.66	283	97	9.9	9.0	1.5	0.91
72	8.3	3.20	302	111	8.6	5.9	1.5	0.69

The values given for cell number, protein, RNA and DNA represent the average content per dish, 4 dishes being assayed in triplicate for RNA, DNA and protein, and 2 dishes being used for each determination of cell number. Polyamines were extracted from 4 dishes of cells, and the values given are the mean values from triplicate assays.

Table 5 Distribution of radioactivity in [^{3}H]putrescine-labelled BHK-21/C13 cells and in their incubation medium as cell density increases

Sample	cpm in spermine + spermidine + putrescine	Radioactivity (%) Spermine	Spermidine	Putrescine
Initial cells	47513	45	55	0
(a) Cells				
24h	38827		not determined	
48h	34157	73	26	1
96h	19703	82	17	1
(b) Medium				
24h	8686	16	83	1
48h	13356	1	93	6
96h	27810	20	79	1

Cells were labelled by incubation in medium containing 10%(v/v) horse serum and [^{3}H]putrescine (1.0 μCi/ml) for 5 h. They were then transferred to non-radioactive medium for 16 h to 'chase' the label into intracellular polyamine pools. Cultures were placed in fresh 10% serum medium, and left for the times shown before the cells and medium were harvested and the polyamines extracted with ice-cold 0.2M perchloric acid. The values presented are the means of at least duplicate determinations.

essential for cell growth and proliferation (Holley, 1972; 1975), for example certain amino acids (Ley and Tobey, 1970). When BHK-21/C13 cells in monolayer cultures were transferred from complete medium to medium deficient in isoleucine, their growth rate was arrested within a few hours. Accumulation of protein was inhibited severely, and accumulation of polyamines was prevented (Table 6). The molar ratio of spermidine/spermine decreased in isoleucine-starved cells, and more polyamines were released from [^{3}H]putrescine-labelled cells into the culture medium when the medium contained no isoleucine than when it was complete. The labelled polyamine released from isoleucine-starved cells was predominantly spermidine.

BHK-21/C13 cells deprived of isoleucine were not dead; they could be stimulated to recommence growth by transferring them to medium containing isoleucine.

D. Effects of Anti-tumour Agents

The major aim of our studies has been to use mammalian cells in culture to gain some insight into the effects of anti-tumour agents on the growth of tumour tissue *in vivo*, in particular to understand their effects on the concen-

Table 6. Effect of isoleucine deprivation on the growth of BHK-21/C13 cells and on loss of polyamines into the culture medium

Culture	Protein (mg)	Polyamine content (nmoles)				Radioactivity lost (% of total)
		Spermine	Spermidine	Putrescine	Spermidine /spermine	
Initial culture	0.55	6.4	8.4	1.2	1.32	—
10% serum medium	1.64	14.8	20.0	2.8	1.35	14
Isoleucine-free medium	0.44	4.8	3.9	0.4	0.81	27
1% serum medium	0.84	7.4	6.0	1.0	0.81	40

The values are presented as content per dish, 4 dishes being assayed in triplicate for protein and for polyamines. Cells prelabelled with [^{3}H]putrescine (0.6 μCi/ml of medium, for 24 h), were transferred to fresh medium as indicated, for 24 h. The horse serum was dialysed before use (Ley and Tobey, 1970). The initial radioactivity in the cells was 18.9×10^3 cpm per dish.

trations of polyamines in tumour tissue itself and in physiological fluids (Cohen, 1977). Accordingly, we have examined the effects of two known antitumour agents on the growth of BHK-21/C13 cells and on the fate of intracellular polyamines, (1) 6-thioguanosine, the ribonucleoside of 6-thioguanine, an agent clinically useful in the treatment of acute leukaemia (Gee, Yu and Clarkson, 1969; Clarkson and coworkers, 1975), and (2) methyl-GAG, an analogue of spermidine and spermine, which is used in the treatment of acute leukaemia (Levin and coworkers, 1964), sometimes in combination with 6-thiopurine (Boiron and coworkers, 1965). The clinical effects of these two drugs are different (Levin, Brittin and Freireich, 1963), probably because of the different mechanisms of action of the drugs on the cells: 6-thioguanine interferes with the nucleic acid metabolism of cells (Paterson and Tidd, 1975; Carrico and Sartorelli, 1977a; 1977b), whereas methyl-GAG has potent effects on the pathway of biosynthesis of the polyamines (Williams-Ashman and Schenone, 1972; Heby, Sauter and Russell, 1973).

BHK-21/C13 cells growing in monolayer cultures were transferred to fresh medium containing either various concentrations of 6-thioguanosine, or 20 μM methyl-GAG, (Melvin, Melvin and Keir, 1978; Melvin and Keir, 1978b). Cell proliferation and the accumulation of protein, RNA and DNA were inhibited by each drug, and the accumulation of polyamines was prevented (Table 7). In cells exposed to 6-thioguanosine the concentrations of putrescine, spermidine and spermine were all decreased, the effect being most marked for spermidine, so that the spermidine/spermine molar ratio decreased significantly compared to control cells (no drug). Cells were labelled using [^{3}H]putrescine, then they were transferred to fresh medium containing various concentrations of 6-thioguanosine. Thereafter, the cells and the medium in which they were incubated were harvested at regular intervals and the radioactivity determined. Cells incubated in the presence of 6-thioguanosine lost more radioactivity into the culture medium than did cells incubated in the absence of the drug (Figure 3). Maximum release of radioactivity from the cells corresponded with maximal inhibition of cell growth by the drug (Melvin, Melvin and Keir, 1978). Although cells which had been incubated in the presence of 6-thioguanosine could not be stimulated to recommence growth by transferring them to medium containing no drug, there was no evidence of extensive cellular destruction in cell cultures exposed to 30 μM 6-thioguanosine for 48 h, as determined by microscopic examination and impermeability of the cells to trypan blue dye. Impermeability indicates cell viability.

Cells whose polyamines had been labelled using [^{3}H] putrescine were incubated in the presence of 6-thioguanosine for 46 h, then the cells and the medium in which they were incubated were extracted with 0.2 M perchloric acid and the extracts analysed for radioactivity (Table 8a). [^{3}H]-Radio-labelled spermine and spermidine were both present in significant amounts in extracts from cells, whereas the labelled polyamine in the medium was predominantly

Table 7 Effect of anti-tumour agents on BHK-21/C13 cells

	Initial culture	No drug	10 μM 6-thioguanosine	20 μM methyl-GAG
Cell number ($\times 10^{-6}$)	1.7	5.5	3.2	3.8
Protein (mg)	0.33	1.63	0.84	1.00
RNA (μg)	50	274	103	137
DNA (μg)	18	92	35	49
Putrescine (nmoles)	1.0	5.2	1.9	7.6
Spermidine (nmoles)	6.1	27.2	10.0	8.7
Spermine (nmoles)	2.7	12.4	5.6	4.2
Spermidine/spermine	2.24	2.19	1.78	2.06

Initial cell cultures were transferred to fresh medium containing either no drug, 10 μM 6-thioguanosine, or 20 μM methyl-GAG, and the cells were harvested 24 h later. The results are presented as content per dish, 4 dishes being assayed in triplicate for protein, RNA, DNA and polyamines. 2 dishes were used for each determination of cell number.

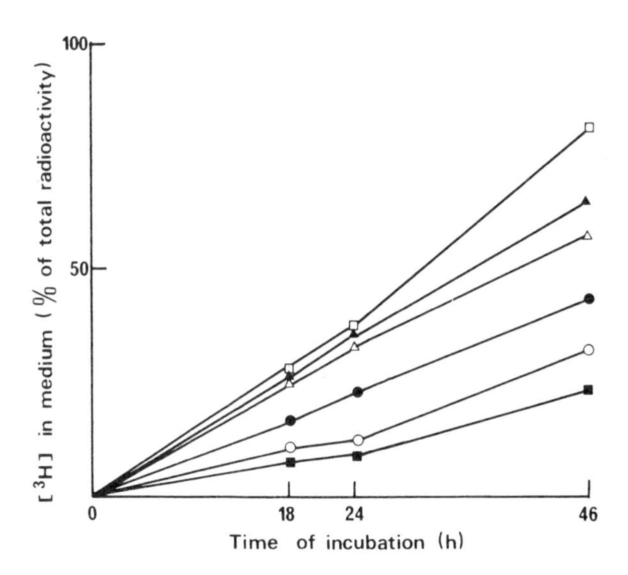

Figure 3 Effect of 6-thioguanosine on release of polyamines from BHK-21/C13 cells into the culture medium.

The radioactivity in the cells at 0 h (100% value) was 47.5 $\times$ 10^3 cpm per dish. Values represent the average of 3 separate determinations. All samples, from cells and medium, were assayed for radioactivity using 10 ml of Triton $\times$-100/toluene scintillation fluid, containing 0.1 ml of 0.2 M perchloric acid and 1.0 ml of medium, so that the efficiency of counting was the same for all samples. Concentrations of 6-thioguanosine used were ■———■, 0 μM, ○———○, 1 μM; ●———●, 3 μM; △———△, 10 μM; ▲———▲, 30 μM; □———□, 100 μM

Table 8 Distribution of radioactivity in [^{3}H]putrescine-labelled BHK-21/C13 cells and in their incubation medium, after incubation in the presence of 6-thioguanosine or methyl-GAG

		Radioactivity (%)			
Drug	Sample	Origin	Spermine	Spermidine	Putrescine
(a) Perchloric acid extracts					
None		16	62	21	1
6-Thioguanosine	Cells	27	36	36	1
Methyl-GAG		31	36	32	1
None		50	3	37	10
6-Thioguanosine	Medium	41	5	45	10
Methyl-GAG		64	1	27	8
(b) Trichloroacetic acid extracts, hydrolysed with 6 M HCl					
None		11	66	22	1
6-Thioguanosine	Cells	11	46	41	2
Methyl-GAG		10	50	39	1
None		11	13	63	13
6-Thioguanosine	Medium	14	9	70	7
Methyl-GAG		10	5	80	5

Cells were labelled for 20 h in medium containing 10% serum and [^{3}H]putrescine (1.0 μCi/ml). They were incubated overnight in non-radioactive medium, then transferred to fresh medium containing no drug, 10 μM 6-thioguanosine, or 20 μM methyl-GAG.

After 46 h, the radioactivity in the medium was 15.3×10^3 cpm in the absence of drug, 28.0×10^3 cpm in the presence of 6-thioguanosine, and 5.9×10^3 cpm in the presence of methyl-GAG, all per dish. The initial radioactivity in the cells was 47.5×10^3 cpm per dish, on average.

spermidine. This result is similar to that obtained when BHK-21/C13 cells were deprived of serum (Figure 2). Loss of polyamines from cells exposed to 6-thioguanosine was not simply a consequence of cell death and lysis, otherwise more radioactivity would have been found in spermine in the medium.

When BHK-21/C13 cells were incubated in medium containing 20 μM methyl-GAG for 24 h, the accumulation of both spermidine and spermine was inhibited to about the same extent, so that there was little change in the spermidine/spermine molar ratio compared with cells incubated in the absence of drug (Table 7). The accumulation of putrescine actually increased. This is consistent with the known mode of action of methyl-GAG on cells (Williams-Ashman and Schenone, 1972), in that synthesis of putrescine continues in the presence of the drug though that of both spermidine and spermine is prevented. When BHK-21/C13 cells, whose polyamines had been labelled by prior incubation in the presence of [^{3}H]putrescine, were transferred to fresh medium containing no drug or 20 μM methyl-GAG, even less radioactivity was

Table 9 Effect of methyl-GAG on loss of polyamines from BHK-21/C13 cells into the culture medium

Drug	Time (hours)	Radioactivity in medium (cpm)	(per cent of total)
None	18	4,791	9.1
None	24	5,848	9.9
20 μM Methyl-GAG	18	3,234	5.8
20 μM Methyl-GAG	24	3,397	5.8

The results are presented as radioactivity per dish, each value being the mean of three separate determinations. All values obtained varied by less than 10 per cent of the mean values. All samples, from cells and medium, were assayed for radioactivity in 10 ml of Triton X-100/toluene-based scintillation fluid, containing 0,1 ml of 0.2 M perchloric acid and 1.0 ml of medium, so that the efficiency of counting was the same for all samples.

lost from cells exposed to the drug than from control cells (Table 9). The radioactive polyamine lost from cells exposed to the drug was predominantly spermidine, even though the radioactivity inside the cells was present in both spermidine and spermine (Table 8a).

The differences we have observed between the effects of methyl-GAG and the other inhibitors of cell growth on the fate of intracellular polyamines in BHK-21/C13 cells probably reflects the different modes of action of the inhibitors on the cells. In cells whose growth rate has been inhibited by methyl-GAG, inhibition of biosynthesis and accumulation of polyamines might result in insufficient spermidine and spermine being available for the continuation of cell growth and proliferation (Heby and coworkers, 1977). In order to conserve intracellular polyamines, and hence maintain cell growth, the mechanism of transport of spermidine out of the cells might be inhibited. In fact, methyl-GAG itself may act as a competitive inhibitor of the transport of spermidine out of cells, as it is known to compete with the transport of spermidine into cells (Dave and Caballes, 1973; Clark and Fuller, 1975).

Excretion of polyamines from cells deprived of serum or an essential amino acid, cells approaching critical cell density, or cells exposed to 6-thioguanosine, may be a specific response of the cells to inhibition to growth rate, and may be an important mechanism whereby cells dispose of polyamines present in excess of their requirements. When Chinese hamster ovary (CHO) cells in monolayer cultures were subjected to heat-shock treatment, there was depletion of the intracellular content of spermidine and spermine and increase in the extracellular levels of these polyamines (Gerner and Russell, 1977). Similarly, when mouse mammary gland explants, normally cultured in the presence of insulin and prolactin, were transferred to medium that contained no hormones, there was marked excretion of spermidine into the culture medium (Kano and Oka, 1976). Hence excretion of polyamines in response to inhibition of cell growth rate probably occurs throughout mammalian cell systems. The alteration of cellular polyamine content through both polyamine biosynthesis and excretion would provide a system for the fine control of cellular polyamine levels and perhaps be involved with the control of cell growth processes.

IV. CONJUGATION OF POLYAMINES

Not all of the radioactivity in BHK-21/C13 cells labelled with exogenous radioactive polyamines could be accounted for by free putrescine, spermidine and spermine; a proportion was found in material which migrated no further than 1.0 cm from the origin of the chromatogram (Table 3). Cells whose polyamines had been labelled using [^{3}H]putrescine were transferred to non-radioactive medium containing 10%(v/v) serum, with or without 6-thioguanosine or methyl-GAG, and the labelled material in the cells and in their incubation medium was extracted with 0.2 M perchloric acid. Much of the radioactivity, in extracts of both cells and medium, was located in material which remained near the origin of the chromatogram (Table 8a). Samples of cells and medium were extracted with trichloracetic acid, and the extracts hydrolysed with 6 M HCl. For all samples, hydrolysis decreased the amount of radioactivity found at the origin of chromatogram, and increased the radioactivity found in free polyamines (Table 8b). Thus the non-migrating radioactive material consisted mainly of polyamines, in a 'conjugated' form (Rosenblum and Russell, 1977; Chapter 25). Polyamines are present, therefore, at least in part, in a conjugated form in BHK-21/C13 cells, and can be released as conjugated derivatives from the cells into the culture medium.

When BHK-21/C13 cells whose polyamines had been labelled with [^{3}H]putrescine were extracted with trichloroacetic acid, and the extracts hydrolysed with HCl, the radioactivity released by hydrolysis was distributed between spermine and spermidine (Table 8b). In contrast, the radioactivity released by hydrolysis of samples of medium in which the cells had been incubated was found predominantly in spermidine. Thus the radioactively-labelled material released from BHK-21/C13 cells into the culture medium when the growth rate was inhibited was almost exclusively spermidine, both free and as a conjugated form. This supports the hypothesis that spermidine is excreted specifically from the cells concomitantly with inhibition of cell growth. The nature of the conjugated derivatives was not investigated further.

When radioactively-labelled polyamines were injected into rats, the label appeared shortly thereafter in the urine and plasma, both as free polyamines and as conjugated derivatives from which free polyamines could be released by hydrolysis (Rosenblum and Russell, 1977; Chapter 25). The conjugation process apparently took place in the liver prior to excretion of the polyamines into the extracellular fluid. The amounts of free polyamines present in physiological fluids are generally very low, and only after extensive hydrolysis of samples can polyamines be detected by standard analytical techniques. N-monoacetylspermidine has been detected as a normal constituent of human urine (Nakajima, Zack and Wolfgram, 1969; Tsuji, Nakajima and Sano, 1975), and it is present in increased amounts in the urine of some cancer patients (Denton and coworkers, 1973). Hence conjugation of polyamines may be a process commonly occurring in mammalian systems, and perhaps is a prerequisite for excretion of polyamines.

V. SUMMARY AND CONCLUSION

Polyamines are secreted, perhaps by an active process or less likely by simple diffusion through the plasma membrane of cultured cells. We have found consistently that in BHK-21/C13 cells whose growth rate has been inhibited, by a variety of means, the proportion of the intracellular ([^{3}H]-labelled) polyamines present in a conjugated form increases compared to control cells (Table 8a). Conjugation of polyamines may be a means by which cells store polyamines. If polyamines are required for cell growth they could be released from the conjugated form into the cell, whereas if growth is inhibited

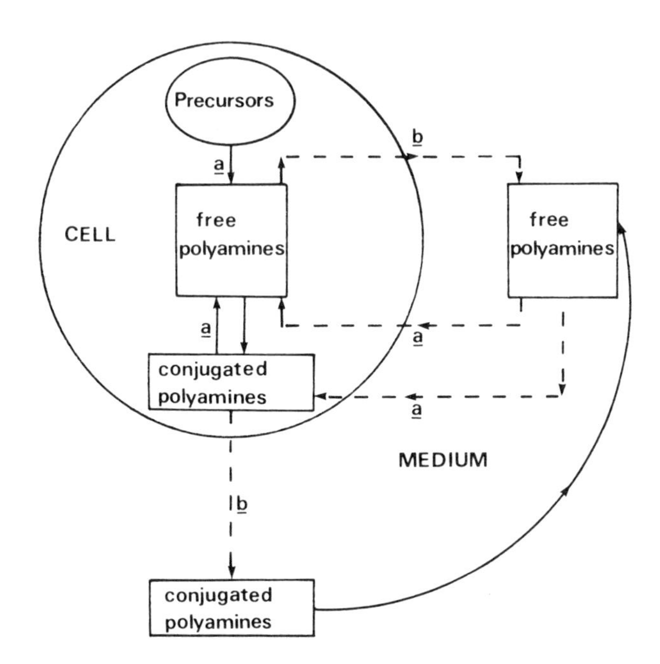

Figure 4 Model for control of intracellular polyamine concentrations
(a) occurs during cell growth. (b) occurs during inhibition of cell growth

polyamines would be released from the cell into the culture medium, at least partly as their conjugated form (Figure 4). Conjugation of polyamines may be yet another means by which mammalian cells can control their intracellular polyamine content.

ACKNOWLEGEMENTS

We thank Mrs Alison Blair for technical assistance, Dr. William T. Melvin for his helpful criticism, and the Medical Research Council and the Faculty of Medicine, University of Aberdeen, for financial support.

REFERENCES

Bachrach, U. (1976). *Ital. J. Biochem.*, **25**, 77–93.

Boiron, M., Jacquillat, C., Weil, M., and Bernard, J. (1965). *Cancer Chemotherapy Reports*, **45**, 69–73.

Bürk, R. R. (1970). *Exp. Cell Res.*, **63**, 309–316.

Carrico, C. K. and Sartorelli, A. C. (1977) (a). *Cancer Res.*, **37**, 1868–1875.

Carrico, C. K. and Sartorelli, A. C. (1977) (b). *Cancer Res.*, **37**, 1876–1882.

Clark, J. L. and Fuller, J. L. (1975). *Biochemistry*, **14**, 4403–4409.

Clarkson, B. D., Dowling, M. D., Gee, T. S., Cunningham, I. B., and Burchenal, J. H. (1975). *Cancer*, **36**, 775–795.

Cohen, S. S. (1977). *Cancer Res.*, **37**, 939–942.

Dave, C., and Caballes, L. (1973). *Fed. Proc. Fed. Amer. Soc. Exp. Biol.*, **32**, 736 (Abstr. 2940).

Denton, M. D., Glazer, H. S., Walle, T., Zellner, D. C., and Smith, F. G. (1973). Clinical application of new methods of polyamine analysis. In D. H. Russell (Ed.), *Polyamines in Normal and Neoplastic Growth*, Raven Press, New York, pp. 373–380.

Gee, T. S., Yu, K-P, and Clarkson, B. D. (1969). *Cancer*, **23**, 1019–1032.

Gerner, E. W., and Russell, D. H. (1977). *Cancer Res.*, **37**, 482–489.

Gray, P. N., Cullum, M. E., and Griffin, M. J. (1976). *J. Cell. Physiol.*, **89**, 225–234.

Heby, O., Marton, L. J., Wilson, C. B., and Gray, J. W. (1977). *Europ. J. Cancer*, **13**, 1009–1017.

Heby, O., Sauter, S., and Russell, D. H. (1973). *Biochem. J.*, **136**, 1121–1124.

Herbst, E. J., and Dion, A. S. (1970). *Fed. Proc. Fed. Amer. Soc. Exp. Biol.*, **29**, 1563–1567.

Holley, R. (1972). *Proc. Nat. Acad. Sci. USA*, **69**, 2840–2841.

Holley, R. (1975). *Nature, Lond.*, **258**, 487–490.

Howard, D. K., Hay, J., Melvin, W. T., and Durham, J. P. (1974). *Exp. Cell Res.*, **86**, 31–42.

Kano, K., and Oka, T. (1976). *J. Biol. Chem.*, **251**, 2795–2800.

Levin, R. H., Brittin, G. M., and Freireich, E. J. (1963). *Blood*, **21**, 689–697.

Levin, R. H., Henderson, E., Karon, M., and Freireich, E. J. (1964). *Clin. Pharm. Therap.*, **6**, 31–42.

Ley, K. D., and Tobey, R. A. (1970). *J. Cell Biol.*, **47**, 453–459.

McPherson, I. A., and Stoker, M. G. P. (1962). *Virology*, **16**, 147–152.

Melvin, M. A. L., and Keir, H. M. (1977). *Biochem. Soc. Trans.*, **5**, 711–712.

Melvin, M. A. L., and Keir, H. M. (1978a). *Exp. Cell Res.*, **111**, 231–236.

Melvin, M. A. L., and Keir, H. M. (1978b). *Biochem. J.*, **174**, 349–352.
Melvin, M. A. L., Melvin, W. T., and Keir, H. M. (1978). *Cancer Res.*, **38**, 3055–3058.
Munro, H. N., and Fleck, A. (1966). The determination of nucleic acids. In D. Glick (Ed.), *Methods of Biochemical Analysis*, Interscience, New York, pp. 159–160.
Nakajima, T., Zack, J. F., and Wolfgram, F. (1969). *Biochim. Biophys. Acta*, **184**, 651–652.
Paterson, A. R. P., and Tidd, D. M. (1975). 6-Thiopurines. In A. C. Sartorelli and D. G. Johns (Eds), *Antineoplastic and Immunosuppressive Agents*, II, Springer-Verlag, Berlin, Heidelberg, New York, pp. 384–403.
Rosenblum, M. G., and Russell, D. H. (1977). *Cancer Res.*, **37**, 47–51.
Rubin, H. (1971). *J. Cell Biol.*, **51**, 686–702.
Rubin, H., and Koide, T. (1973). *J. Cell. Physiol.*, **81**, 387–396.
Russell, D. H. (1977). *Clin. Chem.*, **23**, 22–27.
Russell, D. H., Looney, W. B., Kovacs, C. J., Hopkins, H. A., Dattilo, J. W., and Morris, H. P. (1976). *Cancer Res.*, **36**, 420–423.
Russell, D. H., and Snyder, S. H. (1968). *Proc. Nat. Acad. Sci. USA*, **60**, 1420–1427.
Tsuji, M., Nakajima, T., and Sano, I. (1975). *Clin. Chim. Acta*, **59**, 161–167.
Williams-Ashman, H. G., and Schenone, A. (1972). *Biochem. Biophys. Res. Commun.*, **46**, 288–295.

Chapter 24

Mathematical Models of Polyamines in Growth Processes and in Extracellular Fluids

K. J. HIMMELSTEIN AND M. G. ROSENBLUM

I. INTRODUCTION

The role of polyamines has been studied during the past decade in processes such as the growth of tissue and in excellular fluids to predict the success or failure of cancer chemotherapy (see Chapter 26). Polyamines have been shown to accumulate in tissues with high rates of RNA and protein synthesis. The detailed study of the pathways by which polyamines influence intracellular events has been reviewed in detail (e.g. Russell, 1973; Chapters 1–11). The vast amount of data and information generated by these studies requires a systematic method of understanding the overall growth process.

As a result of this need for systematic organization several mathematical models have been developed to study the nature of the growth process which include putrescine as the growth factor (Pohjanpelto and Rania, 1972) to stimulate growth of groups of cells (Bronk, Dienes and Johnson, 1970; Weiss and Kavanau, 1957).

The observation of increased polyamine levels in rapidly proliferating tissues and biological fluids can be useful both to further understand the growth of neoplastic tissue and also to play the role of biological markers of the presence of disease and response to therapy (Russell and coworkers, 1975). In both of these applications it is necessary to systematically understand the interrelationships between polyamine presence and the state of the neoplastic disease (Russell and coworkers, 1976). Mathematical models have been developed to aid in the quantification of the relationship between tumour cell mass and quantity of polyamines. One type of model predicts or simulates the relationship between tumour cell number and polyamine concentration in the tumour (Woo and Simon, 1973; Woo and Enagonio, 1977). These models will be considered below.

Several studies have shown (Russell and coworkers, 1974) a relationship between polyamine concentration, extracellular fluids and tumour cell mass. It would be desirable to be able to relate tumour cell number to the concentration of the polyamines in biological fluids from which it is easy to take samples so that the presence of polyamines could be used as indicators of the disease's response to therapy. The use of polyamines as markers of neoplastic disease and response to therapy requires an understanding of the pharmacokinetics of polyamines so that the relationship between concentrations of polyamines in extracellular fluids and concentrations of polyamines at the tumour site can be related. Without proper understanding of this relationship it would be impossible to state the relationship between tumour response to therapy and polyamine concentrations due to the possible complicated systematic disposition of polyamines in biological fluids. Thus, a third area of systematic mathematical representation of polyamines, the development of pharmacokinetic models for polyamine disposition, will be considered.

This chapter explores the role of mathematical models in the development of the understanding of the systematic way in which polyamines influence the growth process in both normal and neoplastic tissues. It will also explore those mathematical models which have been developed to describe the pharmacokinetics of polyamines with special emphasis on the application of polyamines as biological markers of tumour response to therapy.

II. MATHEMATICAL MODELS OF POLYAMINES IN THE GROWTH PROCESS

Prior to the time that polyamines were implicated as factors in the biochemical events leading to growth, speculation existed on the regulation of growth processes by means of the presence of key compounds which influence the growth process by acting as stimuli or inhibitors, i.e. messengers between cells in the growing tissue. Two models which use the idea of cooperative regulatory molecules include the works of Weiss and Kavanau (1957), and the work of Bronk, Dienes, and Johnson (1970). Given that putrescine might be a factor that promotes growth in both normal and neoplastic tissues, the co-operative control model of Bronk is of particular interest.

In their model the presence of a key regulatory compound is hypothesized. If a compound is present inside the cell above a critical concentration the cell is then stimulated to divide to form several daughter cells. The key substance is produced by the cell at a constant rate, may leak from the cell, and be adsorbed by other cells proportional to the concentration gradient across the cell wall. The substance may be lost from the medium by an unspecified first order mechanism. The system is then governed by three ordinary differential equations (from Bronk):

$$\frac{dp}{dt} = \alpha(x)\varrho \tag{1}$$

$$\frac{dX}{dt} = K - LX + MY - \alpha X \tag{2}$$

$$\frac{dY}{dt} = L\varrho X - M\varrho Y - \lambda Y \tag{3}$$

where ϱ is the number of cells per ml of medium, α the cell proliferation rate, X is concentration of the key substance inside the cell, K is the production rate per cell, L and M are the directional profusion rate constants across the cell wall, Y is the concentration of the key substance in the medium and λ is the loss or denaturation rate constant of the key substance from the medium.

The three equations are solved numerically by Bronk, Dienes and Johnson (1970) to explore parametrically the response of the cell system to various initial concentrations of cells. The result of one typical parametric study is shown in Figure 1. For a chosen set of parameters the initial cell concentrations are varied. The initial concentration of the key compound in the medium is zero. In the given parametric study initial population density is varied approximately over a factor of five. The population varies from static to one growing in rapid exponential growth, depending on initial cell concentration. At the beginning of the simulation a short spurt of growth occurs, since the concentration of the key substance in the cells is above the critical level. As soon as significant diffusion of the compound into the medium occurs, the growth rate changes depending upon the initial cell concentration. The results clearly demonstrate that a critical population density is necessary if cooperative regulation of the growth process is to occur between cells.

Figure 2 shows the intercellular concentration of the key substance as a function of time in the same simulation as Figure 1. It shows that if the density is not high enough, an inhibition of the growth process may occur should a significant loss of the key substance take place.

Not included in the Bronk model is the possibility of variable key substance

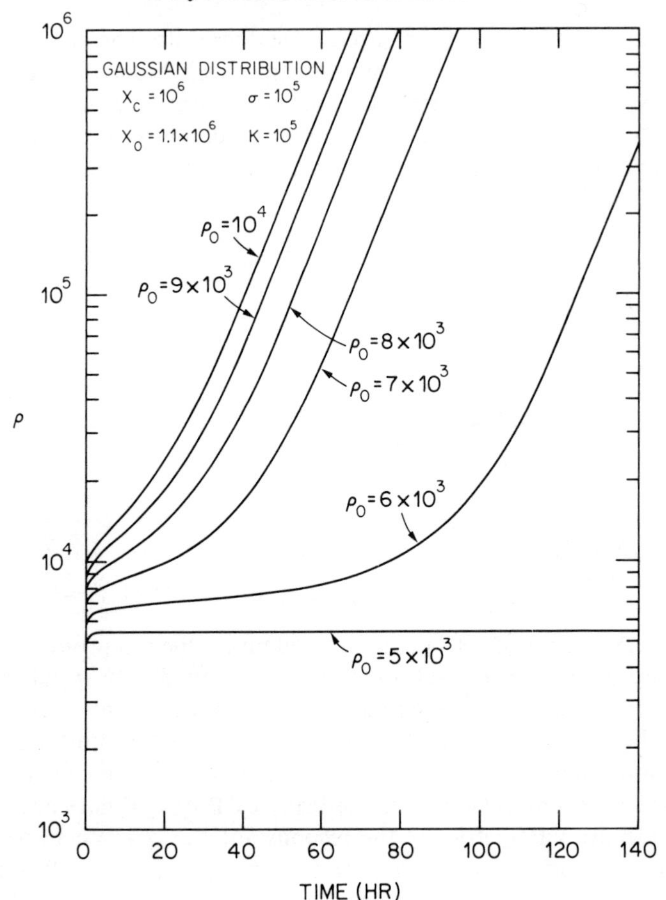

Figure 1 Cell concentration as a function of initial cell concentration. (Reproduced by permission of the *Biophysical Journal*)

production rates or inactivation rates. The models are incapable of predicting either plateau growth or inhibition of growth due to the presence of other compounds. Thus this model can simulate the onset of cellular growth due to growth promoting substances, but lacks the ability to simulate steady state or other growth states which require inhibitory control.

Woo and Simon (1973) have developed a mathematical model of Ehrlich ascites tumours which includes compounds which act as growth stimuli or as inhibitors: a schematic of the general model proposed by these authors is shown in Figure 3. The model contains two tumour cell populations: proliferating tumour cells and non-proliferating cells. Transfer between the two groups of cell occurs at mitosis when a certain fraction of the dividing cells

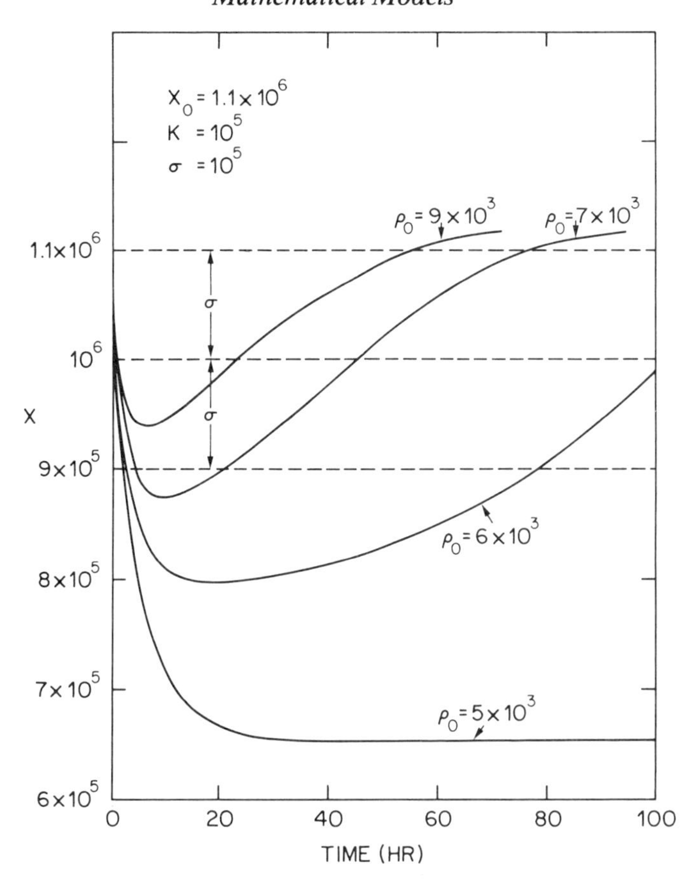

Figure 2 Intracellular concentration of control substance as a function of time. (Reproduced by permission of the *Biophysical Journal*)

is transferred irreversibly from the proliferating cell to the non-proliferating cell compartment. Some cell loss (death) occurs by unspecified means from both tumour compartments. Control of growth in the proliferating tumour cell compartment is a combination of inhibitory feed-back loops which is dependent on cell concentration in both the proliferating and non-proliferating cell compartments. The proliferating tumour cell compartment, the non-proliferating cell compartment, and the rate of production of proliferating cells all exert influence on the production of polyamines, which in turn can be lost from the system by some unspecified mechanism. The polyamines (specifically putrescine) have a stimulatory effect on the production of tumour cells.

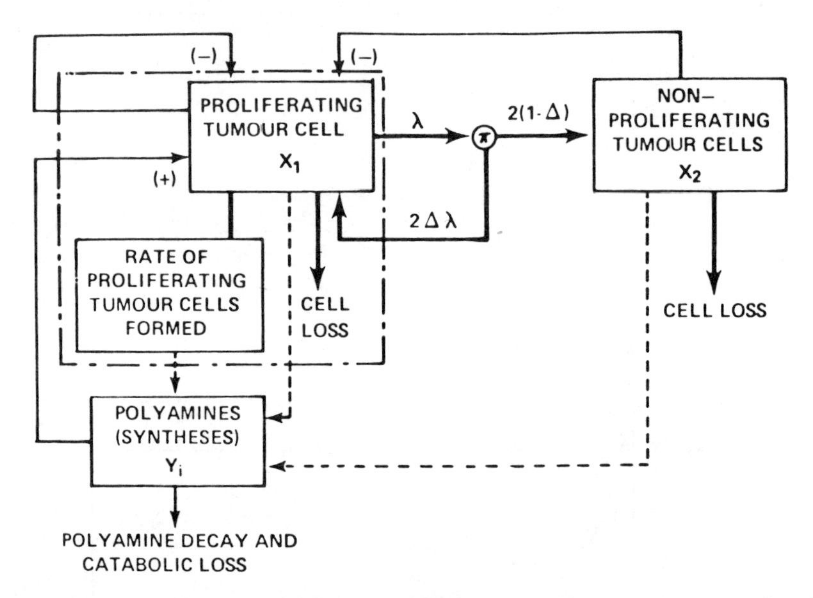

Figure 3 Woo and Simon model for spermidine concentration as a function of the cell cycle. (Reproduced by permission of Raven Press)

The differential equations which comprise the Woo and Simon model for cellular growth are given by:

$$\frac{dx}{dt} \quad 1 = \lambda\,(2\delta - 1)\,\varrho\left(\frac{v1}{x1}\right)\alpha\,(x_1, x_2)\,x_1 - \beta_1 x_1 \tag{4}$$

$$\frac{dx_2}{dt} = \lambda_2\,(1 - \delta)\,\varrho\left(\frac{y1}{x1}\right)\alpha\,(x_1, x_2) - \beta_2 x_2 \tag{5}$$

$$\frac{dyi}{dt} = (K_1 i - \theta)\,\lambda\varrho\left(\frac{y1}{x1}\right)\alpha\,(x_1, x_2)\,x_1 +$$

$$\lambda\,K_{2i}\,\{1 - \varrho\left(\frac{y1}{x1}\right)\alpha\,(x_1, x_2)\,x_1\;1/2\} - \beta_{3i}\,y_i$$

$$i = 1, 2, 3, \tag{6}$$

where x_1 and x_2 are the number of proliferating and non-proliferating tumour cells, λ is the inverse of the generation time for the proliferating cells, δ is the fraction of the daughter cells returning to the proliferating cell compartment, β_1 and β_2 are the loss rate constants for the proliferating and non-proliferating cells, ϱ is the stimulatory control factor exhibited by the concentration of the polyamine, α is the inhibitory control factor for self-regulation of growth by the proliferating and non-proliferating tissues, y is the concentration of the polyamine, k is the production rate of the polyamine by the proliferating cell and is the amount of polyamines consumed or distributed during mitosis.

The control factor, ϱ, is a function of concentration of polyamines relative to the total number of cells present and is assumed to take on one of two forms. ϱ is equal to zero if the relative concentration y/x is less than some critical value and is equal to one if the relative concentration is higher than the critical value. Alternatively, the authors assume that ϱ can take a sigmoid shape as predicted by a Gompertz function. While the authors postulate that putrescine could very well be the polyamine which stimulates growth, they offer no experimental evidence other than that which is already present in the literature (Tabor and Tabor, 1964, 1972; Herbst and Tangway, 1971; Williams-Ashman, Pegg and Lockwood, 1969; Williams-Ashman and coworkers, 1972; Russell and Levy, 1971; and Pohjanpelto and Raina, 1972). The inhibitory factor, α, in the above equations is calculated by assuming Gompertzian growth of the total tumour or tissues. The feed-back loops merely control by this arbitrarily assumed form; no attempt is made to include a mechanism of this inhibitory action. It has been hypothesized (Rose, 1958; and Burns, 1969) that self-inhibition may be due to the build-up of products of the tumour cells, may be simply mediated by the presence of a number of cells themselves, or may be dependent on some critical tumour cell mass whose size is mediated by the surrounding environment. In view of the rather complicated and elaborate attempt to equate polyamine concentrations with a stimulatory effect on the growth of tissue, it seems that this model could be extended or developed further if the inhibitory effect were also more realistically modelled. Realistic inhibition of tissue growth may add additional complexity to the model that is either not warranted or simply makes the model so complex as to be of little practical use. The function, β, is estimated by the potential doubling time data versus actual doubling time for tissues observed. No attempt has been made to discern whether the polyamine concentration is either intra- or extracellular. The developed models are then used to stimulate the growth of Ehrlich ascites tumour.

The various kinetic parameters, such as the retardation parameter α, are estimated from the Gompertz equation fitted to experimental data on Ehrlich ascites tumour cells (EATC) from several authors (Laird, 1964; Andersson and Agrell, 1972). The various other parameters are estimated to explore the results by the model, in light of experimental results. The kinetic parameters are varied so that an accurate representation of the growth of EATC is given by the model as shown in Figure 4. Concentration of the stimulatory substance

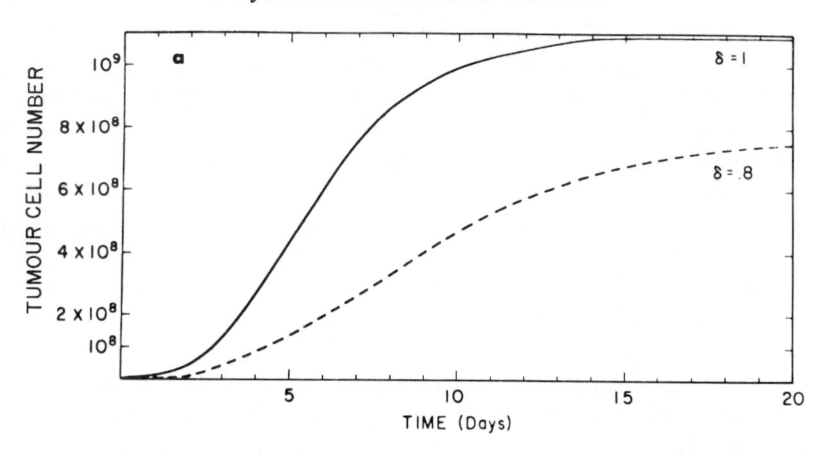

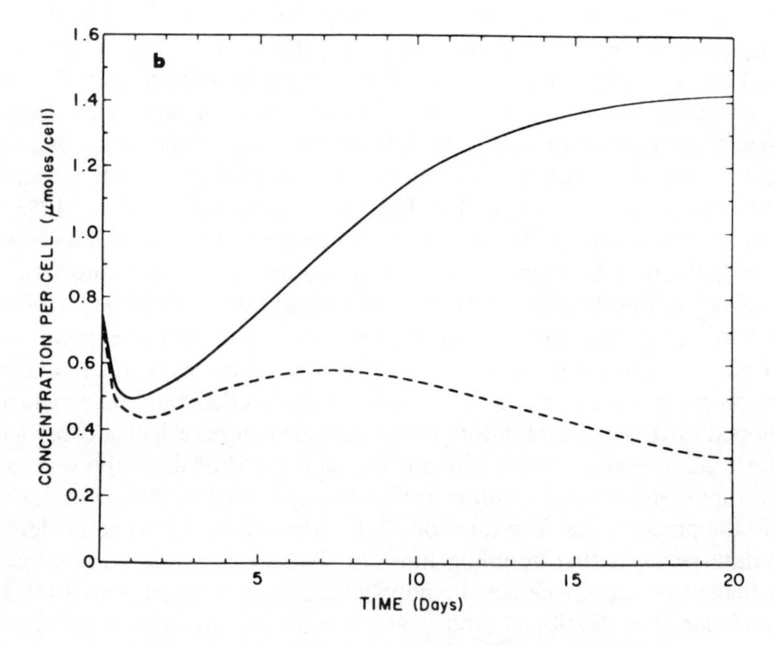

Figure 4 Cell number predicted by Woo and Simon model for tumour cell growth. (Reproduced by permission of Raven Press)

per cell is then presented for several different sets of kinetic parameters in Figure 5. The predicted concentrations are not compared to experimental data, however. This particular model presents a framework from which one might

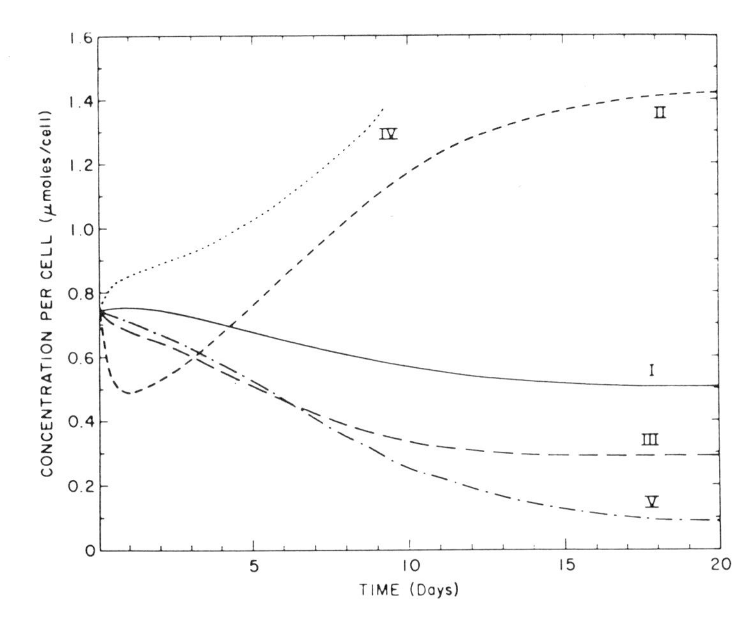

Figure 5 Polyamine concentration from Woo and Simon tumour growth model. (Reproduced by permission of Raven Press)

be able to predict the concentration of putrescine or other polyamines in tumour cells as a function of the state of tumour development. Unfortunately the model is not realistic in that it does not discern between polyamine concentration in cells and in the surrounding medium, and does not include realistic negative feed-back control.

III. THE RELATION OF TUMOUR CELL CYCLE TO BIOLOGICAL MARKERS

Aside from the biochemical role in the growth process, the presence of polyamines could potentially be used as markers or evidence of tissue or tumour cell growth. The intracellular concentration of polyamines in proliferating tissues depends on the stage of the cell cycle (Fuller, Gerner and Russell, 1977); thus, one can use the concentration of polyamines to indicate the phase of the cell cycle and so be used as a replacement for labelled mitosis experiments to study the cell cycle kinetics of a population. Alternatively, in non-synchronized populations, polyamine production is a measure of the proliferating fraction of the cells. A mathematical model which incorporates these observations has been developed by Woo and Enagonio (1977). The general approach taken by these authors is to employ an existing discrete-time cell cycle model (Kim, Bahrami, and Woo, 1974), adding to it an additional expression for the marker content distribution which is viewed as a linear function of the proliferating and non-proliferating cells. Their model is described by the equations:

$$X(k + 1) = \phi_{pp}(K + 1, K) \times (K) + \phi_{pm}(K + 1, K)\, q(K) \quad (7)$$

$$Q(K + 1) = \phi_{mp}(K + 1, K) \times (K) + \phi_{mm}(K + 1, K)\, q(K) \quad (8)$$

$$M(K) = V_m x(K) + V_m q(K) \quad (9)$$

where X is the proliferating cell state function, Q is the non-proliferating cell state function, K represents time (discrete time change Δt when time progresses from K to $K + 1$), and ϕ represents transfer functions between the various cell state functions when going from one time increment to the next. M is the marker concentration, and V_m represents the transformation between marker content and cell age state functions.

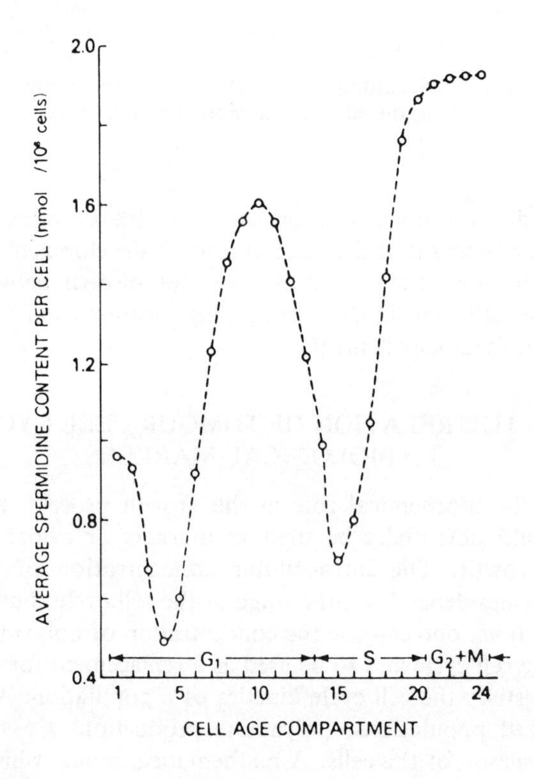

Figure 6 Cell growth predicted by model of Woo and Enagonio. (Reproduced by permission of the American Association for Clinical Chemistry)

The transfer functions are considered to be Gompertz functions. The model is then used to calculate the marker concentrations of a synchronized population using experimental data on Chinese hamster ovarian (CHO) cells (Heby, Marton and Gray, 1976). The discrete time cell cycle model is capable of predicting DNA concentrations for these cells from the experimental data. The DNA and spermidine content of the cells are shown in Figures 6 and 7.

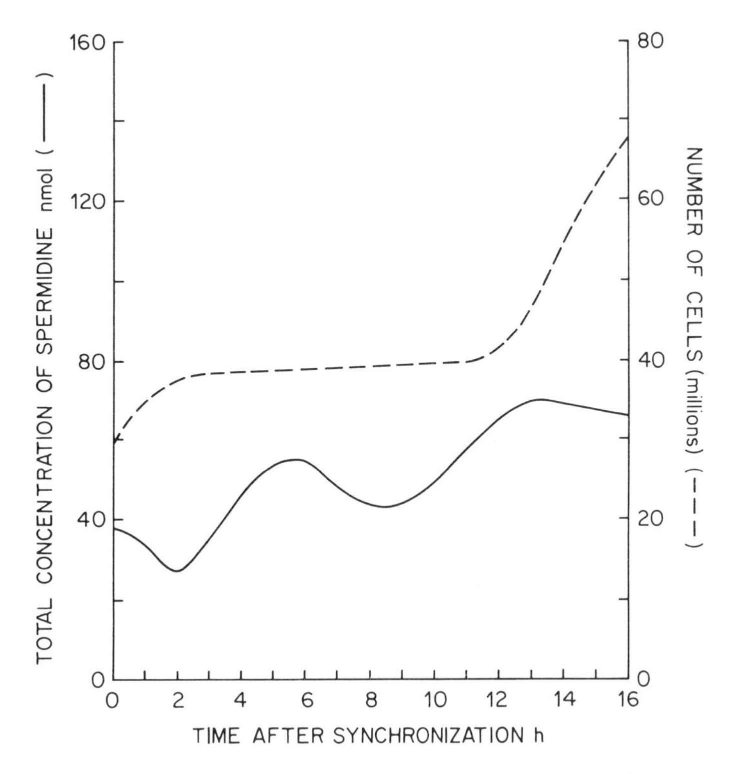

Figure 7 Spermidine concentration as a function of cell cycle as predicted by the model of Woo and Enagonio. (Reproduced by permission of the American Association for Clinical Chemistry)

The marker functions (V_m) are chosen in such a way that they predict spermidine content over the course of a cell cycle. This model owes its success to the degree of flexibility provided by the transition matrices. The model is not extendable in its present form to other cell types unless the matrices are modified; yet qualitatively the model may predict the concentration of spermidine as a function of the cell cycle for other systems. The models could be modified to predict the concentrations of other polyamines such as putrescine in similar fashion. The models can also be integrated as shown in Figure 8 to provide an estimate of the total intracellular concentration of polyamine that might be expected in the cell population. The model has the flexibility to predict a wide range of cellular response, and biomarker concentrations. It is undetermined whether or not the spermidine production matrices which depend on proliferating and non-proliferating cells will predict qualitatively the production of spermidine in other cell lines, once the cell cycle kinetics of that cell line were established in the model.

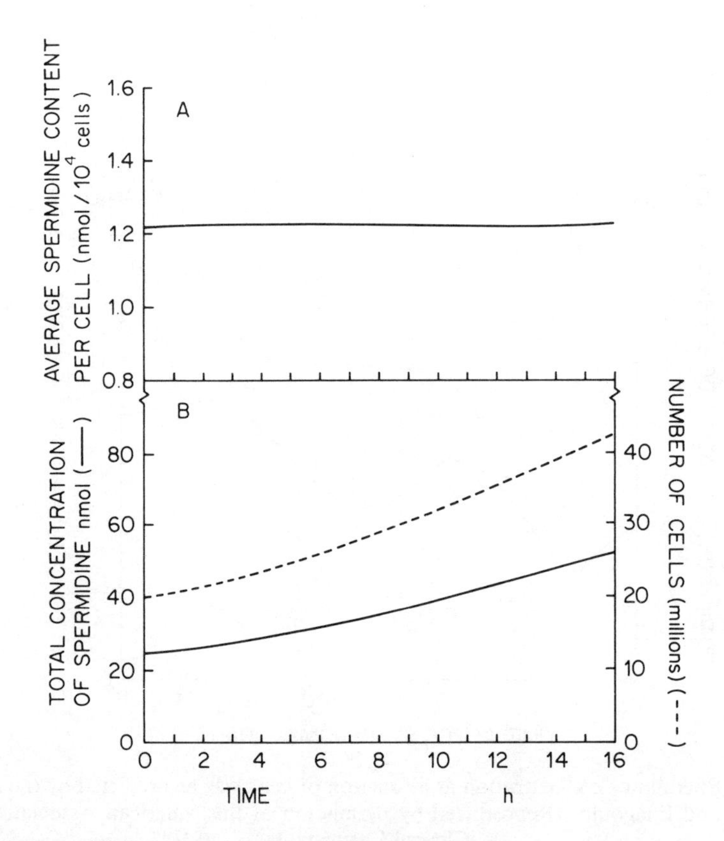

Figure 8 Spermidine concentration in proliferating tissue as predicted by model of Woo and Enagonio. (Reproduced by permission of American Association for Clinical Chemistry)

IV. DISPOSITION AND METABOLISM MODELS FOR POLYAMINES

The use of polyamines as markers of cellular growth and death has significant practical applications both for the basic study of tissues and also as a measure of efficacy of therapy of neoplastic disease. The previous models described above have all been concerned with the presence of polyamines in the tissue of interest. Often it is difficult to sample that tissue. Practical clinical use of polyamines as markers of cell growth and death depends on the ability to measure the concentration of polyamines in a fluid from which it is easy to take samples, such as blood or urine. It cannot be expected that the concentration of polyamines in either of these fluids is directly proportional to the concentration of polyamines at the tumour site at any given time. Between the tumour tissue and the sampling point lie a number of processes such as release from the tissue (Chapters 23 and 26), transfer of polyamines to other tissues by the circulatory system, metabolism of the polyamine (Chapter 18), binding of the polyamine to protein or other compounds (Chapters 23 and 25), and finally excretion processes (Chapters 25–27). All of the processes, even if linear, represent a complex network of interactions which can render the level of polyamine in fluid at any given time significantly different than the production rate of the polyamines. Thus, to be used as a marker of tissue cell growth and death, polyamine disposition must be known so that extracellular fluid levels (Chapter 27) can be related to the response of the tumour cells to therapy. Therefore, a mathematical model of spermidine distribution, metabolism, and elimination has been developed to study the response of spermidine as a marker of tumour cell death.

Recent studies by Rosenblum and Himmelstein (1978) have shown that spermidine injected intravenously in rats is not significantly distributed to tissues. Further, it has been shown that spermidine is conjugated or complexed with one or more organic species (Rosenblum and Russell, 1977). Complexed spermidine is excreted into urine while uncomplexed spermidine is not excreted. Siimes (1967) has shown that 8% of injected radioactive spermidine is metabolized per day and exhaled as $[^{14}C]CO_2$. These observations are used to develop a distribution and metabolism model for spermidine.

While the model developed is for rats, it is desirable to make the model as general as possible so that it can serve as a framework for similar models in man. Thus, as many parameters of the model as possible are associated with physiological quantities such as tissue volume and blood flow rate. It is anticipated that only intrinsic differences between the animal and man such as conjugation rate, or excretion rate, will have to be altered by measurement to generalize the model for applications to clinical situations. This modelling approach leads to a slightly more complex phrasing of the various parameters involved in the model. It serves to dissociate those parameters, however, which should be estimated from experiment or clinical study from those extrinsic properties such as blood flow rate which are species or individual dependent.

In this model we will differentiate between two sources of spermidine: an *endogenous level* which comes from the secretions of the various tissues in the mammal, and an *exogenous level* representing the presence of spermidine due to events such as massive cell kill resulting from therapy or from an exogenous source such as an injection. A schematic of the model is shown in Figure 9. Endogenous spermidine and the 'external' source of spermidine both enter the

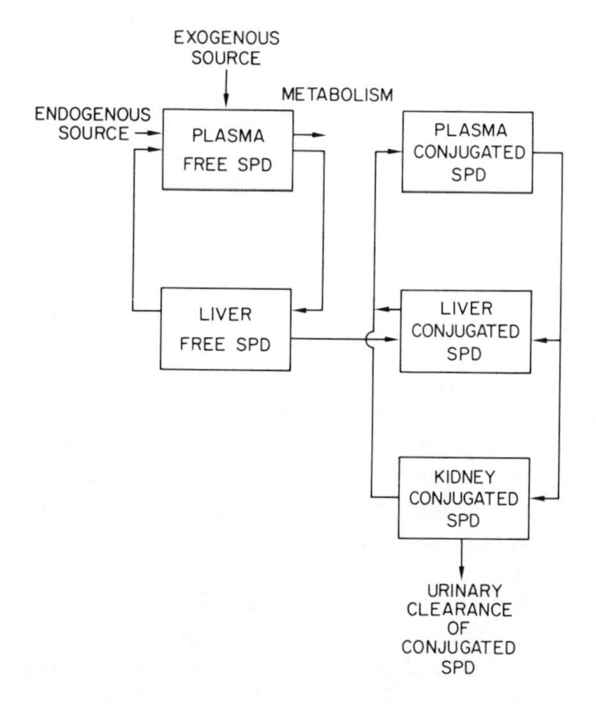

Figure 9 Schematic for spermidine pharmacokinetics model

blood compartment. Spermidine is distributed from the blood to the liver where conjugation then takes place. Conjugated spermidine is subsequently transferred to the conjugate network. The metabolism of spermidine is represented by a first order loss from the unconjugated blood pool. Conjugated spermidine is then distributed to the blood and to the kidney pool where it is excreted. Equations for this model are shown below:

$$V_B \frac{dC_{SB}}{dt} = Q_L(-C_{SB} + C_{SC}) - K_{BM}\,C_{SB}\,V_{SB} + P \tag{10}$$

$$V_C \frac{dC_{SL}}{dt} = Q_L(C_{SB} - C_{SC}) - V_L\,K_C\,C_{SC} \tag{11}$$

$$V_B \frac{dC_{CB}}{dt} = Q_L\,C_{CL} + Q_{CK} - Q_B\,C_{CB} \tag{12}$$

$$V_L \frac{dC_{CK}}{dt} = 2_L(C_{CB} - C_{CL}) + V_L\,K_C\,C_{SL} \tag{13}$$

$$V_K \frac{dC_{CK}}{dt} = Q_K (C_{CB} - C_{CK}) - K_E C_{CK} \tag{14}$$

Where V is the volume of various tissues, C is the concentration of spermidine, Q is the blood flow rate, and K_m, P, K_c are respectively the metabolism, endogenous source, and conjugation rate parameters. Subscripts represent S for spermidine, C for conjugated spermidine, L for Liver, B for blood, and K for kidney. The parameters such as the conjugation rate and the excretion rate are estimated from the experimental data while the extrinsic physiological parameters are estimated from published data (Dittmer, 1961; Hamilton and Dow, 1963; Dittmer, Altman and Grepe, 1959; Altman and Dittmer, 1962; Spector, 1956; Brookes and Murray, 1970; Brookes and Murray, 1967). The excretion rate is estimated from the slope of conjugated spermidine concentration in blood, and the conjugation rate constant is evaluated by the appearance

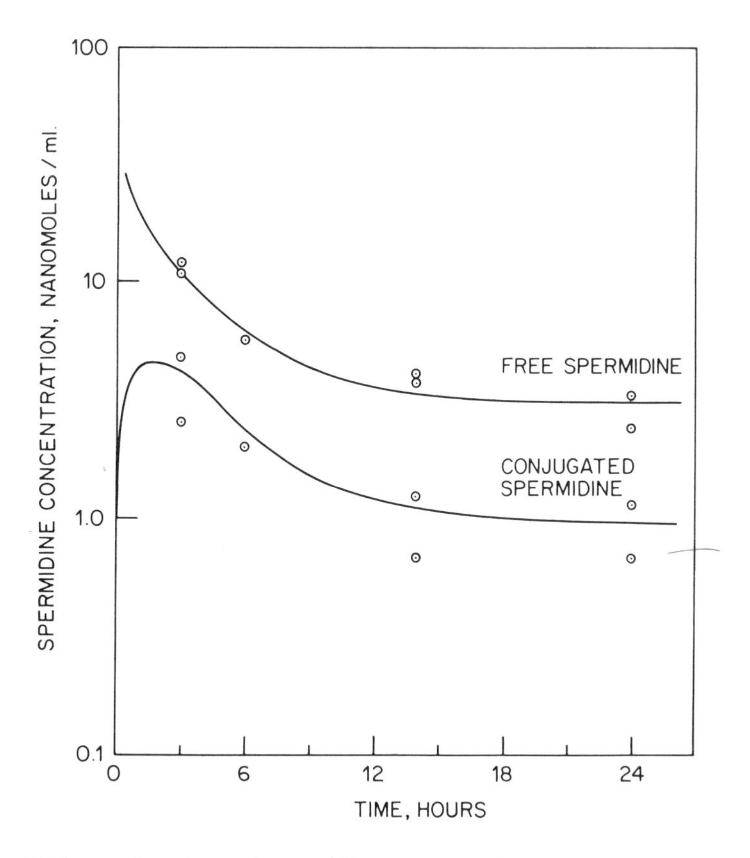

Figure 10 Free and conjugated spermidine concentration as a function of time after injection of 250 nmoles in male rats. Circles represent data, lines are model predictions

of conjugated spermidine in the blood pool following an intravenous dose of spermidine.

The model equations are then solved numerically; the results appear in Figure 10. The model represents concentration levels of both free and conjugated spermidine. The appearance of conjugated spermidine is rapid while the conjugated compound is eliminated rather slowly after conjugation is complete. The model is capable of predicting both free and conjugated concentrations of spermidine in blood and urine accurately. The application of the model would allow the estimation of the time at which one might expect peak concentrations of either free or conjugated spermidine to appear in plasma or urine after therapy. Application of the model to clinical situations would depend on experimental studies in order to quantify the production rate of spermidine from the affected tissue after therapy. It is then anticipated that these results could be extended so that one might employ the model in order to develop optimal sampling times after therapy. In addition, the model could also be extended to putrescine so that the tumour growth faction can be estimated as well as cell loss fraction.

REFERENCES

Andersson, G. K., and Agrell, I. P. (1972). *Virchows Arch. [Zellpathol].*, **11**, 1-10.

Bronk, B. V., Dienes, G. J., and Johnson, R. A. (1970). *Biophys. J.*, **10**, 487-508.

Brookes, M. and Murray (1967). *J. Anat.*, **101**, 533-541.

Brookes, M. and Murray, (1970). *J. Anat.*, **106**, 557-563.

Burns, E. R. (1969). *Growth*, **33**, 25-45.

Dittmer, D. S., Altman, P. L., and Grepe, R. M. (1959). *Handbook of Circulation*, National Academy of Sciences, Ohio.

Fuller, D. J. M., Gerner, E. W., and Russell, D. H. (1977). *Cellular Physiol.*, **93**, 81-88.

Heby, O., Marton, L. J., and Gray, S. W. (1976). *9th Coron. Nordic Soc. for Cell Biol.*, Odense Univ. Press.

Herbst, E. J. and Tangway, R. B. (1971). The interaction of polyamine with nucleic acids. In F. E. Hahan, T. T. Puck, G. F. Springes, W. Szybalski and K. Wallenfels (Eds), *Progress in Molecular and Subcellular Biology*, Springer Verlag, Berlin, pp. 166-180.

Kim, M., Bahrami, K., and Woo, K. B. (1974). *IEEE Trans. Biomed. Eng.*, **21**, 387-398.

Laird, A. K. (1964). *Brit. J. Cancer*, **18**, 490-502.

Pohjanpelto, P. and Rania, A. (1972). *Nature [New Biol]*, **235**, 247-249.

Rose, S. M. (1958). *J. Nat. Cancer Inst.*, **20**, 653-664.

Rosenblum, M., and Himmelstein, K. J., (1978) (in preparation).

Rosenblum, M., and Russell, D. H. (1977). *Cancer Res.*, **37**, 47-51.

Russell, D. H. (1973). *Life Sci.*, **13**, 1635-1647.

Russell, D. H., Durie, B. G. M., and Salmon, S. E. (1975). *Lancet*, **2**, 797-799.

Russell, D. H., Byus, C. V., and Manen, C. (1976). *Life Sci.*, **19**, 1297-1306.

Russell, D. H., and Levy, C. C. (1971). *Cancer Res.*, **31**, 248-251.

Russell, D. H., Looney, W. B., Kovacs, C. J., Hopkins, H. A., Marton, L. J., Legendre, S. M., and Morris, H. P. (1974). *Cancer Res.*, **34**, 2382-2385.

Siimes, M. (1967). *Acta. Physiol. Scand.*, Suppl., **298**, 1-66.
Spector, W. S. (1956). *Handbook of Biological Data*, Saunders, Philadelphia.
Tabor, H., and Tabor, C. W. (1964). *Pharmacol. Rev.*, **16**, 245-300.
Tabor, H., and Tabor, C. W. (1972). Adv. Enzymol., **36**, 203-268.
Weiss, P., Kavanau, J. L. (1954). *J. Gen. Physiol.*, **41**, 1-13.
Williams-Ashman, H. G., Pegg, A. E., and Lookwood, D. H. (1969). *Adv. Enzyme Regulat.*, **7**, 291-323.
Williams-Ashman, H. G., Janne, J., Coppoc, G. L., Geruch, M. E., and Schenone, A. (1972). *Adv. Enzyme Regul.*, **10**, 225-245.
Woo, K. B. and Enagonio, R. D. (1977). *Clin. Chem.*, **23**, 1409-1415.
Woo, K. B., and Simon, R. M. (1973). A quantitative model for relating tumor cell numbers to polyamine concentrations. In *Polyamines in Normal and Neoplastic Growth*, Raven Press, New York, pp. 381-393.

Chapter 25

Conjugated Polyamines in Plasma and Urine

MICHAEL G. ROSENBLUM

I. INTRODUCTION

Currently, there is a great deal of interest in developing general biochemical markers for neoplastic disease to allow objective clinical evaluation of tumour response to therapy. It has been suggested that the increased concentration of polyamines in biological fluids may provide a diagnostic tool to evaluate tumour activity (Chapter 26; Russell, 1977).

To better understand why polyamine levels are increased in cancer patients, polyamines were studied in the extracellular fluids of tumour-bearing animals. Russell and coworkers (1974b) have shown that rats bearing 3924A hepatomas demonstrated a 2-fold increase in serum spermidine levels within 36 h after 5-fluorouracil administration. This paralleled a marked decrease in the percentage of viable hepatoma cells (from 51% to 30%) in the tumour mass. The increase in serum spermidine levels was temporally associated with a depletion of polyamine pools in the tumour tissue itself.

Polyamine studies of other animals bearing tumours such as the MTW9 mammary carcinoma (Russell and coworkers, 1974a) and the L1210 leukaemia (Russell, 1972) have confirmed that a reduction in the number of viable tumour cells caused by chemotherapy, localized X-ray treatment (Russell and coworkers, 1976) or removal of the tumour hormonal support (Russell and coworkers, 1974a) resulted in a decrease in the tumour polyamine levels and an increase in extracellular polyamine levels. These investigators suggested that increased polyamine levels in extracellular fluids of tumour-bearing animals are derived mainly from tumour tissue and may be useful biochemical markers of tumour growth.

Several clinical studies of cancer patients have shown that response to therapy is well correlated with a rise in plasma and urinary polyamine levels. Conversely, patients refractory to the anti-neoplastic regimes showed no significant changes in polyamine plasma or urinary levels (Russell, 1977). Durie, Salmon and Russell, (1977) observed a close correlation between increases in the tumour [^{3}H]thymidine labelling index and increases in putrescine levels in plasma or urine. Conversely, an increase in the tumour cell loss factor caused by chemotherapy or multi-modality therapy was correlated with an increase in spermidine levels in plasma or urine.

These studies suggest that putrescine and spermidine may be useful markers of tumour cytokinetics in that changes in the tumour growth fraction and the tumour cell loss fraction closely parallel changes in putrescine and spermidine levels in extracellular fluids. Therefore, monitoring polyamine levels in cancer patients may provide a useful assessment of tumour response during treatment of patients with neoplastic disease.

In order to effectively utilize polyamine levels in plasma or urine, we must understand the factors which affect polyamine levels in extracellular fluids. Polyamine levels in plasma and urine appear to be regulated by the interplay of several dynamic events: (1) *de novo* synthesis; (2) release of intracellular pools during cell growth or cell death; (3) metabolism (conjugation); and (4) excretion. Both synthesis of polyamines and release of intracellular pools have been well-characterized in several *in vitro* and *in vivo* model systems (Fuller, Gerner and Russell, 1977; Raina and Jänne, 1968; Russell and coworkers, 1974b; Russell, 1977). Little is known about metabolism (conjugation) and excretion of polyamine, however, once they are released from tissue sites. The changes in metabolism and excretion in response to increased synthesis and release of polyamines must be studied in order to more effectively utilize polyamines as markers of neoplastic disease. Furthermore, observed changes in conjugation and excretion of polyamines in disease states such as cystic fibrosis, liver disease, or kidney disease may lead to a basic understanding of disease pathophysiology.

In this chapter, we will review recent studies on polyamine conjugation in extracellular fluids of normals and in patients with disease pathologies such as cancer or cystic fibrosis, in order to determine how changes in synthesis and release affect conjugation and excretion of polyamines. Also, we will review proposed structures for the polyamine conjugates found in human urine.

A. Early Studies of Polyamine Conjugates in Plasma and Urine

The initial clinical studies by Russell (1971), Russell and coworkers (1971) and Bachrach (1976) showed that it was not possible to detect polyamines in plasma or urine prior to acid hydrolysis of the sample. It was proposed that polyamines in plasma and urine were bound or conjugated to a carrier molecule. Several studies (Abdel-Monem and Ohno, 1977b; Nakajima, Zack and Wolfgram, 1969) have confirmed this observation and it has been suggested that polyamines are conjugated in order to be excreted into the urine.

Bachrach and Ben-Joseph (1973) showed that release of polyamines from their conjugated form requires hydrolysis in 6 N HCl at 110°C for 12 h or in 6 N KOH for 4 h. Acid hydrolysis for 4 h was shown to release only 15–18% of the conjugated polyamines. Based on the extreme conditions required for the total release of polyamines from their urinary conjugates, it was suggested that polyamines are probably covalently bound to a carrier molecule. Proposed structures for the polyamine conjugates will be discussed in Section D of this chapter.

B. Conjugation of Polyamines in the Rat

We initially chose the rat as a model system to study polyamine pharmacokinetics and conjugation. In order to assay the appearance of [^{14}C]polyamine conjugates, we developed a cation-exchange chromatography system to separate the free [^{14}C]polyamines from its [^{14}C]conjugate(s) (Rosenblum and Russell, 1977). After anaesthetizing 350–400 g rats, we placed cannulas in the left caudal vein and right caudal artery. This gave us easy access to the blood compartment for injection and blood withdrawal. At 20 min after this surgical procedure, 2.5 μCi of [^{14}C]spermidine, [^{14}C]putrescine, or [^{14}C]spermine was injected via the venous cannula. Analysis of the plasma decay curves showed that the polyamines were selectively removed from the blood in the order spermidine > putrescine > spermine. Chromatography of the plasma samples at various times after injection of the radio-label showed that both putrescine and spermidine were conjugated rapidly and extensively (Table 1). In contrast, no conjugation of the spermine label could be detected even after 60 min.

In an effort to locate the site of conjugation, we initially incubated whole blood or plasma with [^{14}C]spermidine under a variety of experimental conditions and were not able to detect appearance of [^{14}C]spermidine conjugates (Rosenblum and coworkers, 1976). This showed that blood or plasma *in vitro* is not capable of metabolizing polyamines. We then utilized our anaesthetized rat system to determine effects of various surgical procedures on conjugation. Table 1 illustrates the conjugation pattern for [^{14}C]-spermidine in nephrectomized and hepatectomized animals. After unilateral nephrectomy, the level of spermidine conjugate in plasma was higher than controls at 60 min after injection. This suggests that renal function may be an important excretory

Table 1 Appearance of [^{14}C] polyamine conjugates in plasma of rats

Treatment	Time (min)	% Conjugated
[^{14}C]putrescine	5	10
	15	22
	20	60
	60	90
[^{14}C]spermidine	5	1
	15	10
	30	50
	60	50
[^{14}]spermine	5	<0.1
	60	<0.1
[^{14}C]spermidine and unilateral nephrectomy	5	2
	15	10
	30	52
	60	75
[^{14}C]spermidine and hepatectomy	5	<0.1
	15	<0.1
	30	<0.1
	60	<0.1

Free [^{14}C] and [^{14}C]labelled conjugated polyamines were determined by Dowex 50 cation-exchange chromatography after a 2.5 μCi i.v. injection of labelled polyamines. Values are for anaesthetized male rats (300–400 g) and reflect the mean of duplicate determinations of 4 separate animals. (Data reproduced by permission of *Cancer Research*).

route of conjugated spermidine. Further, since subtotal (90%) hepatectomy totally abolished spermidine conjugation, we proposed that conjugation may occur via a liver-dependent process.

C. Conjugation of Polyamines in Humans

1. *Normal Volunteers and Cancer Patients*

Plasma pharmacokinetics and conjugation of [^{14}C]putrescine and [^{14}C]-spermidine were studied in normal volunteers and in patients with advanced cancer who had elevated (at least twice greater than normal) levels of polyamines in their plasma and urine (Rosenblum and coworkers, 1977). (All studies were approved by the Human Subjects Committee, University of

Arizona, and informed, written consent was obtained prior to each study). Blood samples (3 ml) were removed at various times after injection and were chromatographed as previously described (Rosenblum and Russell, 1977).

Both putrescine and spermidine were 95% conjugated within 5 min of injection in plasma of both normals and cancer patients (Table 2). Although the cancer patients studied had at least a 2-fold higher endogenous polyamine plasma level, this did not impede conjugation of exogenously administered radio-label. Dowex chromatography of urine samples from normal volunteers and cancer patients injected with [^{14}C]spermidine showed that greater than 90% of the excreted [^{14}C]radio-label was in the conjugated form (Figure 1, fractions 4–8). It is therefore suggested that the conjugative pathways may have an extremely high capacity for conjugating polyamines.

Table 2 Conjugation of [^{14}C] polyamines in normals and cancer patients after i.v. injection

Time after injection (min)	[^{14}C]spermidine (% of label conjugated in plasma)	
	Normals (N = 5)	Cancer patients (N = 4)
1	2	5
2	30	25
3	40	45
4	70	65
5	95	95

Time after injection (min)	[^{14}C]putrescine (% of label conjugated in plasma)	
	Normals (N = 4)	Cancer patients (N = 6)
1	10	50
2	60	60
3	90	80
4	95	90
5	95	95

In order to study the conjugation of endogenously released polyamines, we measured the changes in plasma levels of free and conjugated spermidine during chemotherapy in one patient with a large tumour mass (unpublished data). For 48 h after chemotherapy, blood samples (5 ml) were withdrawn at 4 h intervals from an indwelling intravenous line. Polyamine levels were measured in unhydrolysed (free) and hydrolysed (conjugated and free) plasma samples on a Durrum D-500 amino acid analyser (Russell and Russell, 1975).

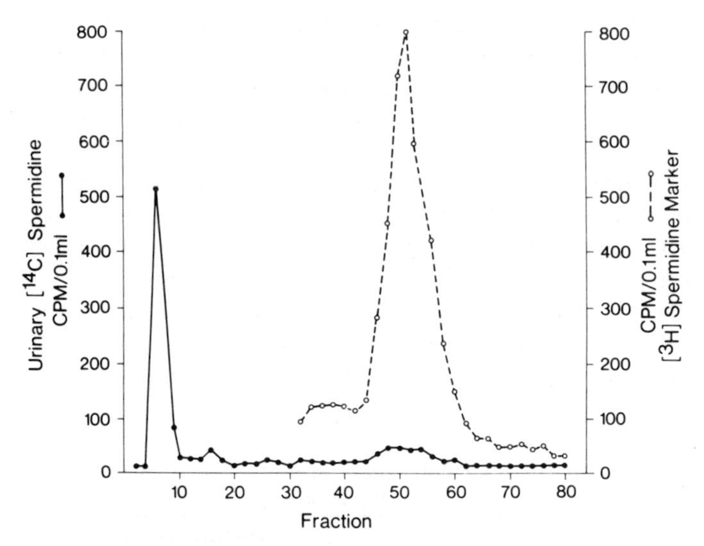

Figure 1 Dowex chromatography of [^{14}C]spermidine label in urine of cancer patients. Urine from cancer patients injected with [^{14}C]spermidine was chromatographed on Dowex 50W × 8 and eluted with a 2-5 M NaCl gradient as described previously (Rosenblum and Russell, 1977). Conjugated [^{14}C]-label appears in fractions 4–9 while the free [^{3}H]spermidine marker appears in fractions 46–60

We found that only slight fluctuations in free spermidine levels occurred during the entire 48-h period. None the less, 24 h after chemotherapy, there was a sharp 5-fold increase in plasma levels of conjugated spermidine presumably due to massive release from the tumour and rapid hepatic conjugation. These studies suggest that there are no significant differences in polyamine metabolism between normals and cancer patients. Further, the hepatic conjugation of massive levels of endogenously released polyamines (e.g. in cancer) apparently is a rapid process. Measurement of free polyamine levels in plasma or urine may be of limited value. As we have shown, increases in *de novo* synthesis and release of polyamines from tumour tissue has little effect on polyamine conjugative pathways, since they appear to have a large capacity. It therefore seems likely that increases in free levels of extracellular polyamines may only result from various disease states which affect hepatic or renal function, thus causing a change in the rate of polyamine conjugation or excretion.

2. *Cystic Fibrosis*

Cystic fibrosis (CF) is an autosomal recessive genetic disease which primarily affects the lungs and pancreas, although there are abnormalities in most exocrine glands and, possibly, in all secretory membranes (diSant'Angese and coworkers, 1953). Several studies have reported that whole blood samples from cystic fibrosis patients had altered levels of free polyamines (Lundgren, Farrell and diSant'Agnese, 1975; Rennert, Frias and Lapointe, 1973). Cohen, Lundgren and Farrell (1976) analysed various blood components to determine which cells had altered polyamine concentrations. It was found that the free spermidine concentrations in erythrocytes of patients with CF were elevated up to 150% above levels for normal controls. Whole blood samples from CF patients also were shown to have an elevated free spermidine/spermine ratio.

Recently we studied urinary polyamine levels and [^{14}C]spermidine metabolism in patients with CF (Rosenblum and coworkers, 1978a; Rosenblum and coworkers, 1978b). We found that urinary polyamine levels in CF homozygotes were 2- to 10-fold higher than in heterozygotes or normals ($P < 0.0001$). No statistically significant differences in polyamine urinary levels were found between the obligate heterozygote parents and age-matched controls. Although the [^{14}C]spermidine plasma decay curves in two CF patients with severe clinical disease were not significantly different from those of normals, urinary excretion of the radio-label by these two CF patients was only 11–13% as compared to 60–76% excreted by normals after 72 h (Table 3).

Table 3 Excretion of [^{14}C]spermidine in normals and cystic fibrosis patients

Time after injection (h)	(% of injected [^{14}C] recovered in urine)		
	Normals (N = 6)*	CF patient no. 1	CF patient no.2
4	22 ± 2	6.0	3.2
10	32 ± 3	7.0	4.0
24	47 ± 4	7.9	7.0
48	65 ± 7	10.7	9.7
72	68 ± 8	13.5	11.2

*Mean values ± s.e. mean

In another study (Rosenblum and coworkers, 1978b), plasma samples from normal volunteers and CF patients after [^{14}C]spermidine injection were chromatographed on Dowex 50 cation-exchange resin (Figure 2). This study showed that 4 min after injection of [^{14}C]spermidine, normal volunteers conjugated approximately 70% of the [^{14}C]spermidine label (Figure 2B, fractions 3–9). Cystic fibrosis patients failed to demonstrate any conjugation of the label after this interval, however. We proposed that, in CF patients, the decreased urinary excretion of the [^{14}C]sperimidine radio-label is primarily due to a delay in conjugation. Analysis of the [^{14}C]spermidine label in urine by Dowex chromatography shows that, like normals, CF patients excrete 90% of the radio-label in a conjugated form. This suggests that although the rate may be delayed, conjugation of polyamines does eventually occur in the CF patient. A posssible reason for the delay could include binding or sequestration of free polyamines in red blood cells. It is not known whether the increased free

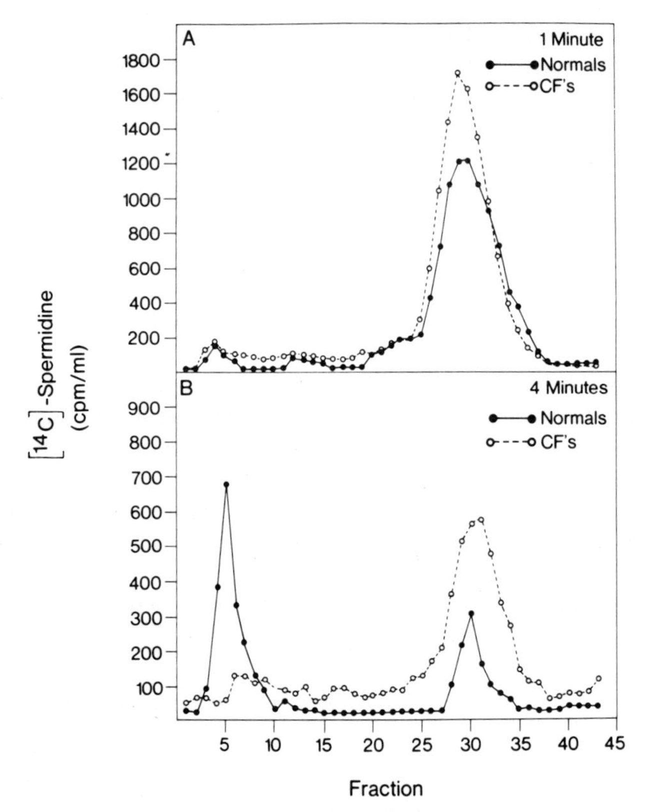

Figure 2 Dowex chromatography of plasma from normals and cystic fibrosis patients 1 min (A) and 4 min (B) after [^{14}C]spermidine injection (i.v.). The radio-label was eluted with a 2-5 M linear NaCl gradient and collected in 1 ml fractions. [^{14}C]Spermidine eluted in fractions 25–35 whereas conjugated [^{14}C]spermidine eluted in fractions 3–9 (see Rosenblum and coworkers, 1976 for further details). (Data reproduced by permission of *Science*)

spermidine levels in the red blood cells of CF patients observed by Cohen, Lundgren and Farrell (1976) and Rennert, Frias and Lapointe (1973) affect the erythrocyte membrane function. Seiler and Deckardt (1976) have, however, shown addition of exogenous spermidine can markedly alter membrane properties. An abnormal erythrocyte function causing high levels of intracellular spermidine may result in a large pool of sequestered spermidine unavailable for conjugation or excretion. An alternative theory is that, in CF, there may be a decrease in the hepatic conjugation rate for polyamines. This would result in high levels of free spermidine which may be stored in erythrocytes. Further study is needed to distinguish between these two possibilities. Abnormalities in polyamine metabolism in CF, however, may result in high levels of free spermidine in extracellular fluids. It is not clear whether these changes in polyamine metabolism contribute to the pathophysiology of cystic fibrosis.

D. Structure of the Polyamine Conjugates

Several investigators have proposed structures for the polyamine conjugates in biological fluids. Kôsaki and coworkers (1958) isolated a phospholipid containing spermine from the blood and tumour tissue of cancer patients. This lipid was named malignolipin and was isolated as a crystalline picrate. Although Kôsaki has partially characterized malignolipin (Figure 3) and has devised both colorimetric (Kôsaki, Nakagawa and Saka, 1960) and immunological techniques (Kôsaki and coworkers, 1961) for its detection in biological fluids, investigators in this country (USA) have disputed Kôsaki's

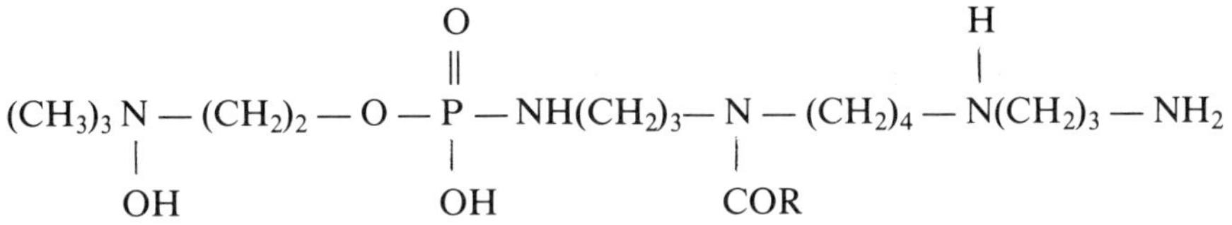

$$(CH_3)_3N(OH)-(CH_2)_2-O-P(=O)(OH)-NH(CH_2)_3-N(COR)-(CH_2)_4-N(H)(CH_2)_3-NH_2$$

Figure 3 Structure for malignolipin proposed by Kôsaki. (Data reproduced by permission of *Science*, and Kôsaki and coworkers, 1958)

findings. Several attempts to isolate malignolipin by the original procedure have failed and the histological staining techniques for malignolipin have been found to be highly non-specific. Therefore, the reliability of the malignolipin test for screening cancer was doubted (Gray, 1961; Hughes, 1960; Sax and coworkers, 1963).

Studies by Tabor (1968) and Dubin and Rosenthal (1960) have shown that *E. coli* bacteria are capable of converting spermidine to an N-acetyl derivative. In mammalian systems, *in vitro* studies by Blankenship and Walle (1977) have shown that there are enzymes in rat liver homogenates which are capable of rapidly acetylating and de-acetylating polyamines. It was proposed that these pathways are active *in vivo* to possibly control the binding of polyamines to DNA.

Several investigators have proposed that polyamines are excreted in urine of normals and cancer patients as N-acetyl derivatives. Walle (1973), Abdel-Monem and Ohno (1977a, 1977b), Nakajima, Zack and Wolfgram (1969) and Tsuji, Nakajima and Sano (1975) have characterized N^1 and N^8 acetyl spermidine and N-acetyl putrescine in human urine by a variety of infrared, dansylation, high-pressure liquid chromatographic and gas chromatography/mass spectrometry techniques.

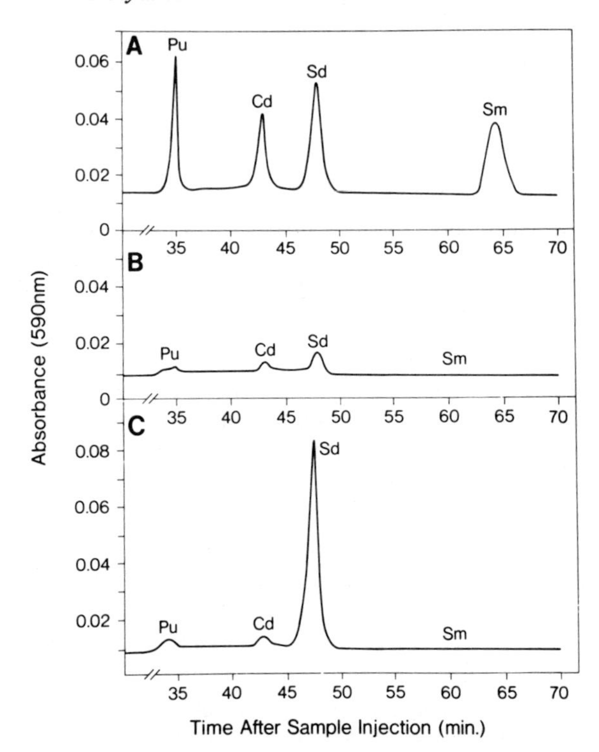

Figure 4 Analysis of purified urinary spermidine conjugate by Durrum D-500 automated amino acid analyser. (A) Standard 200 nmole mixture of putrescine (Pu), cadaverine (Cd), spermidine (Sd), and spermine (Sm). (B) 20 μl of unhydrolysed spermidine conjugate. (C) 20 μl of hydrolysed (6 N HCl at 110°C for 12 h) spermidine conjugate showing a 10-fold increase in spermidine content. (Data reproduced by permission of *Cancer Research*)

Utilizing urine isolated from healthy subjects and cancer patients injected with [^{14}C]spermidine, we have purified a [^{14}C]spermidine conjugate by organic extraction, cation-exchange chromatography, and high-pressure liquid chromatography (unpublished data). After establishing conditions for maximal recovery of the [^{14}C]label from the various techniques, we then utilized these procedures to isolate unlabelled spermidine conjugate from three independent sources: pooled urine from normal volunteers, urine from individual cancer patients, and pooled urine from several cancer patients. We found that the resulting preparations from these sources were uncontaminated by either putrescine conjugates or by free polyamines and were all able to release free spermidine upon acid hydrolysis (Figure 4). Analysis of the trimethylsilyl or trifluoracetic anhydride derivatives of unhydrolysed conjugate

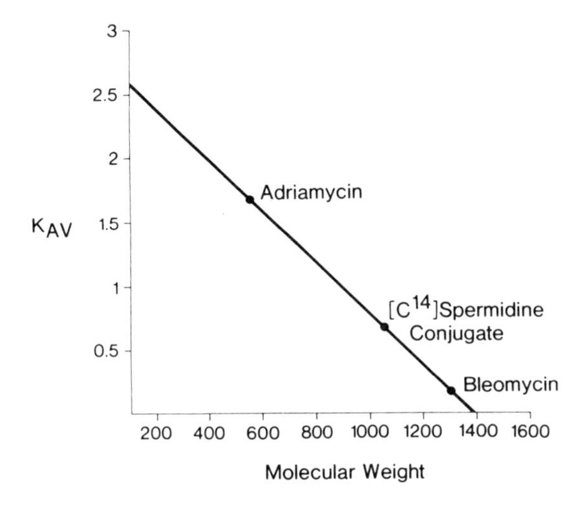

Figure 5 Sephadex G-25 molecular weight determination of purified [^{14}C]spermidine conjugate. Purified [^{14}C]spermidine conjugate isolated from urine of cancer patients was chromatographed on Sephadex G-25 in 0.05 M Na,K phosphate buffer (pH 7.2) using adriamycin (MW = 600) and bleomycin (MW = 1,300) as standards

preparations on a Finnegan gas chromatography/mass spectrometry (Model 3300 equipped with a Model 6110 data system) showed that neither N^1 nor N^8 acetyl spermidine were present. Further analysis by Sephadex G-25 chromatography showed that the purified [^{14}C]spermidine conjugate has an apparent molecular weight of approximately 1,000 daltons (Figure 5). Both plasma and urine conjugates of [^{14}C]spermidine have the same apparent molecular weight.

There may, in fact, be several urinary conjugates of the polyamines. In one recent study (Abdel-Monem and Ohno, 1977b), values for the N-acetyl spermidine content of normal urine (approximately 0.7 pmoles/24 h) appear to account for a small percentage of the total spermidine content reported for normal urine (approximately 20 pmoles/24 h, Russell, Durie and Salmon, 1975). It is possible, therefore, that small amounts of free spermidine may be converted to N-acetyl spermidine by liver enzymes *in vivo*. Alternatively, *E. coli* in the urinary tract may acetylate unconjugated spermidine present in the urine. We cannot dismiss the possibility that small amounts of spermidine conjugate lost during our isolation and purification procedures may represent the N-acetyl spermidine fraction. Recovery and [^{14}C] studies in our laboratory suggest, however, that the 1,000 MW conjugate is the major excretory route for spermidine in humans. Studies are in progress to determine the exact composition of this spermidine conjugate.

II. SUMMARY AND CONCLUSION

We have presented data that polyamine levels in plasma or urine are governed by at least four major dynamic processes: (1) intracellular synthesis, (2) release from tissue pools, (3) conjugation by an hepatic-dependent process, and (4) urinary excretion primarily of conjugated polyamines. Increased extracellular levels of polyamines in cancer patients represent increased intracellular synthesis and release, but do not appear to affect conjugation or excretion. Studies of polyamines in cystic fibrosis have shown that there is an abnormality in polyamine metabolism which affects conjugation and excretion through an unknown mechanism. Further studies may elucidate other biochemical processes and other pathological states which affect the extracellular metabolism of polyamines.

Finally, we have shown that the polyamines in urine and plasma may be present as a heterogeneous mixture of acetyl polyamines and polyamine conjugates of about 1,000 daltons. Utilization of conjugated or acetyl polyamines as the basis for immunological assays of polyamine levels in extracellular fluids may be valuable in the rapid clinical assessment of various pathological conditions and response to therapy.

ACKNOWLEDGEMENTS

The author thanks Ms Melissa Mees for typing this manuscript, and Ms Carol Kiefer, Dr. F. Meyskens, and Dr. B. G. M. Durie for their helpful suggestions.

REFERENCES

Abdel-Monem, M. M., and Ohno, K. (1977a). *J. Pharm. Sci.*, **66**, 1089–1094.
Abdel-Monem, M. M., and Ohno, K. (1977b). *J. Pharm. Sci.*, **66**, 1195–1197.
Bachrach, U. (1976). *Ital. J. Biochem.*, **25**, 77–93.
Bachrach, U., and Ben-Joseph, M. (1973). Tumor cells, polyamines, and polyamine derivatives. In D. H. Russell (Ed), *Polyamines in Normal and Neoplastic Growth*, Raven Press, New York, pp. 15–25.
Blankenship, J., and Walle, T. (1977). *Arch. Biochem. Biophys.*, **179**, 235–242.
Cohen, L. F., Lundgren, D. W., and Farrell, P. M. (1976). *Blood*, **48**, 469–475.
diSant'Agnese, P. A., Darling, R. C., Perera, G. A., and Shea, E. (1953). *Pediatrics*, **12**, 549–563.
Dubin, D. T., and Rosenthal, S. M. (1960). *J. Biol. Chem.*, **235**, 776–782.
Durie, B. G. M., Salmon, S. E., and Russell, D. H. (1977). *Cancer Res.*, **37**, 214–221.
Fuller, D. J. M., Gerner, E. W., and Russell, D. H. (1977). *J. Cell. Physiol.*, **93**, 81–88.
Gray, G. M. (1961). *Biochem. J.*, **81**, 30p–31p.
Hughes, P. E. (1960). *Stain. Technol.*, **35**, 41–42.
Kôsaki, T., Ikoda, T., Kotani, Y., Nakagawa, S., and Saka, T. (1958). *Science*, **127**, 1176–1177.
Kôsaki, T., Nakagawa, S., and Saka, T. (1960). *Mie. Med. J.*, **10**, 417–424.

Kôsaki, T., Saka, T., Nakagawa, S., and Muraki, K. (1961). *Mie. Med. J.*, **11**, 149–162.
Lundgren, D. W., Farrell, P. M., and diSant'Agnese, P. A. (1975). *Clin. Chim. Acta*, **62**, 357–362.
Nakajima, T., Zack, J. F., and Wolfgram, F. (1969). *Biochim. Biophys. Acta*, **184**, 651–652.
Raina, A., and Jänne, J. (1968). *Ann. Med. Exp. Fenn.*,. **46**, 536–540.
Rennert, O., Frias, J., and Lapointe, D. (1973). Methylation of RNA and polyamine metabolism in cystic fibrosis. In *Fundamental Problems of Cystic Fibrosis and Related Diseases*, International Book Corp., New York, pp. 41–52.
Rosenblum, M. G., Beckerman, R. C., Taussig, L. M., Durie, B. G. M., Barnett, D. and Russell, D. H. (1978a). *Pediatric Res.* (In press).
Rosenblum, M. G., Durie, B. G. M., Beckerman, R. C., Taussig, L. M., and Russell, D. H. (1978b). *Science*, **200**, 1496–1497.
Rosenblum, M. G., Durie, B. G. M., Salmon, S. E., Chang, S., and Russell, D. H. (1976). *Proc. Am. Assoc. Cancer Res.*, **17**, 15 (abs).
Rosenblum, M. G., Durie, B. G. M., Salmon, S. E., and Russell, D. H. (1977). *Proc. Am. Assoc. Cancer Res.*, **18**, 97 (abs).
Rosenblum, M. G., and Russell, D. H. (1977). *Cancer Res.*, **37**, 47–51.
Russell, D. H. (1971). *Nature, Lond.*, **233**, 144–145.
Russell, D. H. (1972). *Cancer Res.*, **32**, 2459–2462.
Russell, D. H. (1977). *Clin. Chem.*, **23**, 22–27.
Russell, D. H., Durie, B. G. M., and Salmon, S. E. (1975). *Lancet*, **(ii)**, 797–799.
Russell, D. H., Gullino, P. M., Marton, L. J., and LeGendre, S. M. (1974a). *Cancer Res.*, **34**, 2378–2381.
Russell, D. H., Levy, C. C., Schimpff, S. C., and Hawk, I. A. (1971). *Cancer Res.*, **31**, 1555–1558.
Russell, D. H., Looney, W. B., Kovacs, C. J., Hopkins, H. A., Dattilo, J. W., and Morris, H. P. (1976). *Cancer Res.*, **36**, 420–423.
Russell, D. H., Looney, W. B., Kovacs, C. J., Hopkins, H. A., Marton, L. J., Legendre, S. M., and Morris, H. P. (1974b). *Cancer Res.*, **34**, 2382–2385.
Russell, D. H., and Russell, S. D. (1975). *Clin. Chem.*, **21**, 860–863.
Sax, S. M., Harbison, P. L., Sax, M., and Baughman, R. H. (1963). *J. Biol. Chem.*, **238**, 3817–3819.
Seiler, N., and Deckardt, K. (1976). *Neurochem. Res.*, **1**, 469–499.
Tabor, C. W. (1968). *Biochem. Biophys. Res. Commun.*, **30**, 339–342.
Tsuji, M., Nakajima, R., and Sano, I. (1975). *Clin. Chim. Acta*, **59**, 161–167.
Walle, T. (1973). Gas chromatography-mass spectrometry of di- and polyamines in human urine: Identification of monoacetylspermidine as a major metabolic product of spermidine in a patient with acute myelocytic leukemia. In D. H. Russell (Ed) *Polyamines in Normal and Neoplastic Growth*, Raven Press, New York, pp. 355–365.

Chapter 26

Polyamines as Markers of Malignancy in Human Leukaemia and in other Haematological Disorders

H. DESSER

I. INTRODUCTION

The hope that the polyamines, spermidine and spermine and the diamines putrescine, cadaverine, histamine, and some of their metabolites would be useful as tumour markers has motivated some considerable amount of research on these substances during the past ten years. Many analyses of these compounds in diverse body fluids of ill and healthy humans have been performed. In the following all aliphatic di- and polyamines, histamine, and their derivatives will be referred to collectively as polyamines.

Early in the seventies D. H. Russell and her collaborators (Russell, 1971; Russell and coworkers, 1971) observed elevated diamine levels in hydrolysed urine samples of patients with malignant diseases. Succeeding investigators demonstrated the elevation of the polyamine content in the urine of patients suffering from diseases associated with highly proliferating cells (Russell, 1973a). Extension of the analyses to other physiological fluids such as whole blood, blood plasma, serum, cerebrospinal fluid, extracts from biopsy material and from isolated cell fractions, along with the variation of the methods of sample preparation and analysis has resulted in different deductions. (Raina and Jänne, 1975; Bachrach, 1976; Russell, 1977; Campbell and coworkers, 1978a; Jänne, Pösö and Raina, 1978 and Chapters 25 and 27).

Investigations on variability amongst control groups, and on the influence of *internal* (sex, age, pregnancy) and *external* (diet, hormonal influence, medication) factors have been performed, as well as studies concerning the development of specific diseases in individual untreated and treated patients. Therefore, it seems apt to summarize the recent findings in clinical polyamine research, with special reference to haematological disorders. Evaluation of data in this respect is only admissible if methods of sample preparation, analysis and the clinical status of the patients are adequately considered (Lorenz, 1975; Tabor and Tabor, 1976).

II. BIOLOGICAL MATERIAL AND ANALYTICAL PROCEDURE

A survey of the analytical methods and of sample preparation performed in studies on malignant diseases has been given by Bachrach (1976; 1978). Furthermore, comprehensive reviews exist which cover a whole range of possibilities for methods of polyamine analysis (Seiler and Deckardt, 1975; Seiler, 1976, 1977a; 1977b; Seiler, Knodgen and Eisenbeiss, 1978 and Chapter 27).

The polyamines can be of interest either as the free unbound compounds (F) or as (covalently) bound entities, which after hydrolysis, can be liberated, resulting in the detection of a 'total' polyamine content (T). Some authors (Rennert and coworkers, 1976a; Miale and coworkers, 1977) consider the determination of free polyamines advantageous since the higher quantities that are found in hydrolysed samples are compensated by a poor recovery and a tedious sample preparation. Samejima and coworkers (1976) in comparing the separations of the polyamines from hydrolysed and non-hydrolysed samples on CM-cellulose columns, favoured the hydrolysed samples since they gave better chromatograms. Both methods of sample preparation are legitimate, as concomitantly performed analysis of both total and free polyamines could lead to insights into the metabolism and physiology of these substances. For example, rapid fluctuations from unbound to bound polyamines have been observed in rat and human plasma after administration of radioactively labelled polyamines (Rosenblum and Russell, 1977; Russell, 1977; Chapter 25). How far radioimmunoassays monitor total or free polyamines depends on the free accessibility of the polyamine antigen to the specific antibody. The general consensus is that probably only the free polyamines are detected by this analytical method (Bartos and coworkers, 1975; Puri and coworkers, 1978), although some cross-reactivity with polyamine conjugates is likely (Seiler, 1977b).

Selection of sample material has to be related to the specific question posed.

Practical considerations are especially important for clinical routine investigations. Assay of polyamines in whole blood samples, despite the ease and feasibility of sample collection and preparation (Rennert and Shukla, 1978), is vulnerable to small changes in the differential and total blood cell composition which will lead to pronounced alterations in the values lastly obtained (Tabor and Tabor, 1976). Complete 24-h urine samples, although they contain large amounts of polyamines, are more difficult to collect and examine than blood samples. Further problems arise when studying out-patients. Plasma samples were suggested to be of more precise value than serum samples, since the risk of contamination by intracellular contents due to cell lysis seems to be lower in plasma (Russell, 1977). Anticoagulants can, however, have the potential for diamine oxidase liberation into the blood plasma (Maślinški, 1975) with resulting changes in polyamine levels.

III. CONTROL GROUPS AND NORMAL RANGES

Table 1 summarizes the published normal ranges of polyamines found in specimens investigated under different conditions of preparation and analysis. To give an appropriate idea of the observed findings, the data contributed from different laboratories is converted, where possible, into equivalent unit systems. In some instances when mean and standard deviations were stated the ranges were calculated with respect to the formula M $\pm$ 2 s.d. (M: mean, s.d.: standard deviation). Cadaverine is not included, since this amine was rarely found. Summaries of the histamine levels reported have been recently published (Porter and Mitchell, 1972; Maślinški, 1975; Parwaresch, 1976; Horakova, Keiser and Beavan, 1977).

In general, relatively large variations among the values obtained from normals were recorded. Therefore, consideration of age, sex, diet, hormonal status, etc., are necessary to obtain relevant reference values for the evaluation of possible aberrations in the patients under consideration. To date, only scant information on this subject is available.

In women, the putrescine excretion in the urine is higher than in men. Whereas spermine in male urine is found in higher amounts than in female urine, the amounts being 0.4 μmole putrescine/kg/24 h and 0.008 μmole spermine/kg/24 h in women, 0.2 μmole putrescine/kg/24 h and 0.18 μmole spermine/kg/24 h for men (Waalkes and coworkers, 1975). Differences between the sexes in polyamine content in the urine were also observed by Tsuji, Nakajima and Sano (1975), showing also that N-acetylspermidine was higher in female urine. Spermine again was found in larger amounts in the urine of men. Differences in the contents of free polyamines between men and women were also observed in whole blood samples (Rennert and Shukla, 1978). Increasing age of women is supposed to lead to lower putrescine levels in the urine, in contrast to the situation in men (Waalkes and coworkers,

Table 1 Normal ranges of the polyamines under different methods of preparation and analysis

References	PUT[×1]	SPD[×1]	SPM[×1]	Unit	Number of patients	Material	Method
Russell, 1971; Russell and coworkers, 1971; Schimpff and coworkers, 1973; Fair, Wehner and Brorsson, 1975; Lipton, Sheehan and Kessler, 1975; Sanford and coworkers, 1975	0.8-6.2	0.9-3.9	1.0-3.4	mg 24 h^{-1}	>200	urine (T)	HVE
Marton, Russell and Levy, 1973a; Gehrke and coworkers, 1973; Gehrke, Kuo and Zumwalt, 1974; Rodermund and Moersler, 1974; Waalkes and coworkers, 1975; Tormey and coworkers, 1975	0.2-3.7	0.3-3.2	0-0.87	mg 24 h^{-1}	>40	urine (T)	CEC
Dreyfuss, 1975	≤5	2	0.5	mg 24 h^{-1}	54 [×2]	urine (T)	TLC
Gehrke and coworkers, 1973; Marton and coworkers, 1973b; Makita, Yamamoto and Kono, 1975	1.1-2.0	0.9-2.1	0.2-0.4	mg 24 h^{-1}	>40	urine (T)	GC
Tsuji, Nakajima and Samo, 1975; Fujita and coworkers, 1976	2.9-14.3	1.4-3.7	4.3-14.5	μg mg^{-1} creatinine	77	urine (T)	HVE
Russell, Durie and Salmon, 1975; Towsend, Banda and Marton, 1976; Durie, Salmon and Russell	0.6-5.1	0.5-2.1	0.03-0.62	μg mg^{-1} creatinine	43	urine (T)	CEC
Heby and Andersson, 1978a	1.3-4.7	0.5-3.1	0-1.7	μg mg^{-1} creatinine	12	urine (T)	TLC
Desser and coworkers, 1975; Cooper, Shukla and Rennert, 1976; Cooper, Shukla and Rennert, 1978	20-160	40-120	10-160	pmole ml^{-1}	30	plasma (F)	CEC
Desser and coworkers, 1978; Samejima and coworkers, 1976	90-250	10-200	10-91	pmole ml^{-1}	22	serum (F)	CEC
Bartos and coworkers, 1975; 1977;1978		170-610	36-200	pmole ml^{-1}	17	serum (F)	RIA

Marton, Russell and Levy, 1973a; Marton and coworkers, 1973b; Nishioka and Romsdahl, 1974; Nishioka and coworkers, 1976; Samejima and coworkers, 1976; Nishioka and Romsdahl, 1977	0-490	150-510	0-80	pmole ml^{-1}	>57	serum (T)	CEC
Raina, 1962		6.2-7.0	1.1-2.3	nmole ml^{-1}	30	WB(F)	HVE
Proctor and coworkers, 1975; Lundgren and coworkers, 1976; Rennert, Frias and Shukla, 1976b; Cooper, Shukla and Rennert, 1978	0.17-0.25	0.48-7.34	0.1-5.1	nmole ml^{-1}	61	WB (F)	CEC
McEvoy and Hartley, 1975		0.21-15.9	2.8-8.1	nmole ml^{-1}	12	WB (F)	TLC
Rennert and coworkers, 1976a; Miale and coworkers, 1977	0-12.8	3.1-158	5.1-240	nmole ml^{-1}	15	BM (F)	CEC
Nishioka and coworkers, 1976	210	2120	1430	nmole ml^{-1}		BM (T)	CEC
Rennert and coworkers, 1977	70-320	0-100	0-330	pmole ml^{-1}	$7^{\times 2}$	CSF (F)	CEC
Marton and coworkers, 1974; Marton and coworkers, 1976	76-292	54-246	0	pmole ml^{-1}	$42^{\times 2}$	CSF (T)	CEC
Desser and coworkers, 1975; Chun and coworkers, 1975; Chun and coworkers, 1976; Cohen, Lundgren and Farrell, 1976; Cooper, Shukla and Rennert, 1976; Cooper, Shukla and Rennert, 1978	2.0-6.7	41-307	28-145	pmole/10^8	35	Ery (F)	CEC
	0.4-95.6	0.4-55.3	3.4-91.4	nmole/10^8	39	PMN (F)	CEC
	1.8-26.2	6.4-44.4	7.2-90.8	nmole/10^8	21	MN (F)	CEC
	0.2-27.0	0.54	3.1-4.3	pmole/10^8	23	Ptls.(F)	CEC

x1: ranges or mean ± 2 standard deviations
x2: non-malignant patients
T: Total polyamines (hydrolysed samples)
F: Free polyamines (unhydrolysed samples)
WB: Whole blood; BM: Bone marrow; CSF: Cerebrospinal fluid; Ery: Erythrocytes; PMN: Polymorphonuclear cells; MN: Mononuclear cells; Ptls: Platelets
HVE: High voltage electropheresis; CEC: Cation exchange chromatography; GC: Gas chromatography
TLC: Thin layer chromatography; RIA: Radioimmunoassay.

1975). Based on body weight children generally have been reported, by these authors, to excrete higher amounts of polyamines than adults. No numerical data, however, has been given. In whole blood samples of children higher levels of polyamines were found than observed in adults. The polyamine concentration seemed to decrease with age (Rennert and Shukla, 1978).

No influence of protein diet on the excretion of the polyamines from humans was noticed by Waalkes and coworkers (1975). Administration of arginine (a metabolic precursor of putrescine) to rats had no influence on the polyamine synthesis as determined by the amounts found in their blood (Windmueller and Spaeth, 1976). The opinion has been advanced that the polyamine synthesizing capacity of the intestinal microflora in the digestive tract can influence the amounts, especially of cadaverine, found in the urine and in the blood. A detailed investigation of this question has not yet been accomplished.

The profile of unbound polyamines in whole blood (Rennert, Frias and Shukla, 1976b) and serum (Campbell and coworkers, 1978a,b) changes during the menstrual cycle. The ratio of spermidine to spermine was found to fluctuate considerably in women, whereas in men no substantial changes were observed (Lundgren and coworkers, 1976).

The female spermidine/spermine ratio appears to rise after the last day of menstruation and subsequently falls at the end of the menstrual-cycle. Women taking oral contraceptives, and women who had been subjected to ovarectomy, had smaller individual fluctuations and no cyclic variations in their polyamine profiles. Elevated polyamine levels in the urine of women taking birth control pills have been reported, but here, too, no quantitative data was given (Lipton, Sheehan and Kessler, 1975). Therefore, in evaluating the amounts of polyamine in women, the menstrual cycle should be taken into consideration if quantitative comparisons are necessary (Lundgren and coworkers, 1976). The higher putrescine concentration found in the urine of women, in contrast to men, is not related to the menstrual cycle (Waalkes, Nakajimo and Sano, 1975). The excretion of histamine, methylhistamine and methylimidazole acetic acid during the cycle is also subject to individual fluctuations, with occasional elevations of these substances in mid-cycle (Jonassen, Granerus and Wettergrist, 1976a). Oral contraceptives do not seem to alter, however, the histamine metabolism in females (Jonassen, Granerus and Wettergrist, 1976b).

During pregnancy the putrescine level in the urine has been shown to be elevated between 1.4-fold and 3.1-fold (Russell and coworkers, 1971; Fujita and coworkers, 1976; Heby and Andersson, 1978a). Russell and coworkers (1971) found spermidine to be elevated by a factor of 2.5, while Fujita and coworkers (1976) by a factor of 2.1-fold normal. Heby and Andersson (1978a) found no difference in the spermidine values. As for spermine no conclusive data is available, although the results obtained by Heby and Andersson

(1978a) indicate no difference between pregnant and non-pregnant women. These authors and Fujita and coworkers (1976) found a decrease to normal values in the urine for putrescine *post partum*. No conclusive evidence for the same to occur in spermidine and spermine levels is available. There have been no systematic determinations done with serum or plasma of pregnant women for polyamines. But it should be kept in mind that histamine metabolism is greatly modified due to the increased activity of the diamine oxidase enzyme in the blood during pregnancy (Beaven and coworkers, 1975; Elmfors and Tryding, 1976; Dubos and coworkers, 1977). This enzyme can also use other polyamines as substrates (Maślinski, 1975). It is possible that the serum of pregnant women also contains diamine and a polyamine oxidase (Gaugas and Curzen, 1978; Chapter 22). Therefore, polyamine analysis in the serum of pregnant women has to be interpreted with caution.

Some non-malignant diseases have been described to exhibit increased amounts of the polyamines in some body fluids investigated. In particular, Waalkes and coworkers (1975) stated that 'elevations of polyamines occur in the urine of many non-cancer patients' and that 'in consequence urinary polyamine levels will have little value in the area of cancer diagnosis'. This statement was underlined by the findings of Dreyfuss and coworkers (1975) of elevations of one or more polyamines in 44% of a non-malignant group of 54 patients. Elevations of polyamine values in the urine have been found to occur in some patients with cirrhosis, liver abscess, active hepatitis, viral and bacterial infections of the lung, heart, kidney and liver; pernicious anaemia, other anaemias, rheumatoid arthritis, acromegaly, ulcerative colitis, enteritis, pancreatitis, psoriasis and in patients with inborn errors in amino acid metabolism (Russell, 1971; Russell and coworkers, 1971; Dreyfuss and coworkers, 1975; Lipton, Sheehan and Kessler, 1975; Waalkes and coworkers, 1975; Durie, Salmon and Russell, 1977; Berry and coworkers, 1978b). Serum polyamines have been found to be elevated further in *systemic lupus erythematosus* (Puri and coworkers, 1978), in renal transplant patients (Musgrave and coworkers, 1978), and in patients suffering from uraemia (Campbell and coworkers, 1978b). The polyamines have been reported to be increased in whole blood samples in patients with psoriasis, cystic fibrosis and in sickle cell anaemia (Proctor and coworkers, 1975; Chun and coworkers, 1976; Cohen, Lundgren and Farrell, 1976; Rennert, Frias and Shukla, 1976b; Berry and coworkers, 1978a; Rennert and Shukla, 1978). Histamine was elevated in the blood plasma of asthmatic patients (Narasbhat and coworkers, 1976). Children exhibiting increased cellularity in their bone marrow, as in infectious mononucleosis and in histiocytosis, were generally found to have elevations of the polyamines in bone marrow biopsy material (Rennert and coworkers, 1976a; Miale and coworkers, 1977; Rennert and Shukla, 1978).

Interpretation of results obtained from patients with severe infections or proliferative disturbances towards diagnosis of potential malignancies must therefore be considered together with the overall obtainable clinical data.

A considerable amount of data has accumulated on the incidence of elevated polyamines in malignant diseases other than haematological tumours. The scope of this chapter does not include these publications. Information can be found in recent reviews (Gingold, 1968; Russell, 1973a,b; Bachrach, 1976; Jänne, Pösö and Raina, 1978; Russell and Durie, 1978).

IV. HAEMATOLOGICAL MALIGNANCIES

Table 2 summarizes the approximate incidence of elevated values and the ratio of patient values in comparison to controls. These patients are described in the literature as having leukaemia but without a differential diagnosis being given. All polyamines were significantly elevated in the urine of untreated patients. Putrescine increased by factors of 1.2 to 4.1 spermidine by 2.4 to 9.6, spermine by 1.9 to 29.3 and acetylspermidine from 2.0 to 5.2. Differences were seen between men and women, the increased excretion of spermine in women was especially pronounced (Tsuji, Nakajuma and Sano, 1975). Chemotherapy changes the characteristic profiles. After 24–48 h, increased amounts of spermidine in patients responding to treatment were found, which was not observed in patients not responding. The putrescine concentration in the urine, however, increased in patients not responding to treatment (Russell, Durie and Salmon, 1975; Durie, Salmon and Russell, 1977; Russell, 1977). In bone marrow samples the respective increases were considerable (Nishioka and coworkers, 1976). Prolonged treatment seems to lead to a decline in the concentration of these substances, both in bone marrow and in the urine (Fujita and coworkers, 1976; Miale and coworkers, 1977). No data is available for blood samples in this grouping.

In children's leukaemia systematic studies have been performed measuring the free polyamine content in bone marrow samples (Rennert and coworkers, 1976a; Miale and coworkers, 1977) and in the cerebrospinal fluid (CSF) (Rennert and coworkers, 1977). A summary of the results of these investigations has been published (Rennert and Shukla, 1978). A high amount of the polyamines in bone marrow was correlated with increased cellularity. It has been suggested, therefore, that polyamine analysis of CSF and bone marrow aspirates is a valuable adjunct for the evaluation of the clinical status of these patients. Putrescine was elevated in the bone marrow to about 4-fold normal in cases of relapse. The levels were only slightly higher than normal in cases of remission (Rennert and coworkers, 1976a; Miale and coworkers, 1977). The CSF polyamine profile of patients with leukaemia were found to be of value in estimating meningeal involvement. Spermidine and spermine were increased considerably, 26.5-fold and 11-fold, respectively, in the CSF of patients with central nervous system (CNS) relapse, whereas putrescine was low. In contrast, putrescine was increased in the CSF by a factor of 9 in patients

Table 2 Polyamines in leukaemia patients

References	PUT		SPD		SPM		Acetyl-SPD		Number of patients/ of controls	Material remarks
	a	b	a	b	a	b	a	b		
Bachrach and Ben-Joseph, 1973	1/1: total amount of polyamines were subnormal								1/5	urine T
Russell, Durie and Salmon, 1975	n	1.9	n	2.5	n	n			6/12	urine T
Fujita and coworkers, 1976	s	4.1	s	9.6	s	4.7			4/56	urine T untreated patients
		1.1	s	2.4	s	1.9			36/56	treated patients
Durie, Salmon and Russell, 1977; Russell, 1977	s	2.1	s	3.4	s	20			68/16	urine T
Tsuji, Nakajimo and Sano, 1975	2/3	1.2	3/3	3.8	3/3	8.7	3/3	5.2	4/21	urine F men
	1/1	1.2	1/1	2.9	1/12	9.3	1/1	2.0		women
Nishioka and coworkers, 1976	s	140	s	4.8	s	12.7			n	BM

a: number of patients with values exceeding the controls *vs*. number of patients studied.
b: ratio of mean polyamine content in patients *vs*. control values.
n: no information available.
s: significant elevations
BM: Bone marrow
T: total polyamines; F: free polyamines.

with extramedullary disease and no CNS involvement (Rennert and coworkers, 1977).

In the urine of patients with acute lymphocytic leukaemia (ALL) totals of putrescine, spermidine and spermine have been found to be elevated in comparison to normal controls by factors ranging from 3 to 11 for putrescine, 11 to 12.4 for spermidine and approximately 3 for spermine (Russell and coworkers, 1971; Schimpff and coworkers, 1973; Heby and Andersson, 1978a). The data for a total of only eight adult patients have been published until now. In hydrolysed serum samples, the spermidine values of two patients were increased, but no data could be given for putrescine and spermine as the sensitivity of the method applied at that time was too low (Marton and coworkers, 1973a). In our laboratory, one case of a female patient with ALL was found to have an elevated level of free spermine in her serum, but putrescine and spermidine were in the normal range (unpublished). The situation with regard to children has been dealt with in the preceding paragraph.

Only one patient with chronic lymphocytic leukaemia (CLL) has been analysed for polyamine content in the urine so far (Russell and coworkers, 1971). This patient had an excretion of putrescine 2.6-fold normal, and of spermidine, 2.7 normal. Spermine was also elevated but the specificity of the analysis of this substance was not adequate as later observations have shown. Tsuji, Nakajimo and Sano (1975) described 4 patients with 'chronic leukemia' who had exceptionally pronounced elevations of spermine, but there too the method of high voltage electrophoresis employed brought misleading quantitative results with regard to this polyamine. Two patients with CLL were studied recently for free polyamines in their blood elements (Cooper, Shukla and Rennert, 1978). Spermine was elevated *versus* controls in whole blood by factors of 5.6 and 3.1 respectively. Polyamine analysis on the erythrocytes in one case showed no difference to normals. In the other patient the elevation was 3.3-fold normal. In mononuclear and polymorphonuclear cells some differences were encountered, and in the blood plasma in both patients, elevations 5-fold normal were observed. Spermidine and putrescine were elevated in whole blood, and in the white cells a considerable reduction of these polyamines was found. In the blood plasma no difference to normals for these two diamines were detected. We have evaluated the polyamine concentration by cation exchange chromatography in an unhydrolysed serum sample of one patient with CLL; spermine was found to be increased by a factor of approximately 15 (566 pmole/ml), putrescine by 1.9 (184 pmole/ml), and in contrast spermidine (129.5 pmole/ml) was found to be in the range of normal controls.

In adults with untreated acute myelogenous leukaemia (AML) elevations of putrescine excretion between 1.9 and 5.5-fold normal have been found in urine samples (Russell, 1971; Gehrke and coworkers, 1973; Schimpff and coworkers, 1973; Fujita and coworkers, 1976; Durie, Salmon and Russell, 1977; Heby and Andersson, 1978a). Spermidine was elevated versus normals by factors ranging from 1.4 to 21 and spermine by factors between 3.1 and 20.1. This includes instances when high voltage electrophoresis was applied for analysis. Large amounts of monoacetylspermidine and diaminopropane have also been reported in patients with AML (Walle, 1973). Altogether 23 patients have been investigated for polyamine excretion in this disease. Under successful chemotherapeutic treatment spermidine excretion increased up to 12.5-fold (Denton and coworkers, 1973a,b). Those patients not responding to treatment showed no changes. In remission putrescine and spermidine were excreted in lower amounts, while spermine remained unchanged (Russell, 1971; Schimpff and coworkers, 1973; Fujita and coworkers, 1976). In unhydrolysed bone marrow samples increased amounts (20–50%) of spermidine and spermine were detected (Rennert and coworkers, 1976a). In hydrolysed sera samples only occasional elevations of the polyamines have been reported (Marton, Russell and Levy, 1973a; Marton and coworkers, 1973b). In one patient however, a considerable increase of putrescine (440-fold normal) was reported, while spermidine was only moderately elevated and spermine was practically undetectable. In our own studies (Desser and coworkers, 1978) on untreated patients, using unhydrolysed serum, 3 out of 9 patients showed elevated putrescine values, one out of 10 showed elevated spermidine values, and 7 out of 8 showed elevated spermine values.

Chronic myelogenous leukaemia (CML) has specifically been characterized by extremely high amounts of histamine using whole blood samples (Cadiou and coworkers, 1975; Debray and coworkers, 1975; Gingold and Pecker, 1975; Zittoun and coworkers, 1975; Parwaresch, 1976; Horakova, Keiser and Beaven, 1977). No other disease studied so far has shown such high levels of histamine, reaching more than 100-fold normal control values. Other diseases, for instance polycythaemia vera, osteomyelofibrosis, acute mylogenous leukaemia and mastocytosis (Horakova, Keiser and Beaven, 1977) have been shown to have higher than normal whole blood and urine histamine contents, but never to the same extent as in CML. In blood plasma, elevations of histamine have also been reported to be specific for CML (Suzuki and coworkers, 1971; Horakova, Keiser and Beaven, 1977). Plasma levels are also elevated in mastocytosis, in patients with either cold- or exercise-induced urticarias (Horakova, Keiser and Beaven, 1977), and in asthmatic patients (Narasbhat and coworkers, 1976). Polyamine determinations have been performed in 8 cases in the hydrolysed urine of patients with CML (Russell and coworkers, 1971; Gehrke and coworkers, 1973; Fujita and coworkers, 1976; Durie, Salmon and Russell, 1977). Putrescine was found to be either in the normal range or only slightly increased. No conclusive information is available for spermine at present, although a slight increase was seen in 3 patients. Nevertheless, a considerable amount of putrescine (13.1-fold normal) was found in the urine of patients in a blastic crisis, distinguishing them from those patients in the chronic phase. In the chronic phase, total spermidine in the urine was found to be about 5.7-fold normal, but in the blast phase a factor of about 22 above normal was evaluated. No published information is available at present for polyamine levels in whole blood, serum or bone marrow samples. We were able to demonstrate elevated levels of histamine in all serum samples taken from 11 patients with CML in the chronic phase (Desser and coworkers, 1978), confirming the results obtained in whole blood and plasma samples mentioned above. It is noteworthy that in some patients with blastic crisis no histamine was detectable. Moreover, we found an elevation of putrescine values in 5 out of 6 patients during blastic transformation, but in only 4 out of 9 cases in the chronic phase of this disease. Independently of the disease stage, spermidine was elevated only occasionally. About one-half of the number of patients had elevated spermine values in the blastic and in the chronic phase.

To date, polyamine assays in malignant lymphomas have only been carried out with urine samples. More than 70 patients with lymphoma have been investigated (Schimpff and coworkers, 1973; Dreyfuss and coworkers, 1975; Russell, Durie and Salmon, 1975; Waalkes and coworkers, 1975; Durie, Salmon and Russell, 1977). Putrescine levels became elevated, with the ratio of patients' values to the mean control values ranging from 1.8 to 15. Spermidine levels were elevated by factors of 2.1 to 10.3, and after chemotherapy, they increased to over 30 times normal. An increase of spermine concentrations was also observed in these patients. Especially pronounced were the elevations found for the polyamines in the hydrolysed urine samples of patients afflicted with Burkitt's lymphoma (Waalkes and coworkers, 1975). A decline of the polyamine values was demonstrable after chemotherapy. Five patients with reticulum cell sarcoma have also been studied; putrescine elevations were between 1.1 and 4.0-fold normal, spermidine between 3.7 and 4.6-fold normal and spermine between 2.2 and 3.9-fold normal (Russell and coworkers, 1971; Fujita and coworkers, 1976; Heby and Andersson, 1978a). Lymphosarcoma patients had elevations of their putrescine values from 1.4 to 4.0-fold normal, spermidine ranging from normal to 5-fold normal, and spermine, in one case, showing a subnormal value and in the other cases exhibiting elevations between 3.9 and 8.8-fold normal. To date, urine analysis on only 18 patients with lymphosarcoma have been performed (Russell and coworkers, 1971; Bachrach and Ben-Joseph, 1973; Gehrke and coworkers, 1973; Heby and Andersson, 1978a).

The urine of 23 patients and hydrolysed serum of 5 patients with morbus Hodgkin's disease have been studied, and considerable elevations of putrescine, spermidine and spermine were shown during chemotherapeutic treatment (Russell and coworkers, 1971). The longitudinal study of one patient (stage V) showed putrescine values reaching 20-fold normal, spermidine values reaching 37-fold normal and spermine values up to 70-fold normal. In another longitudinal study on patients without therapeutic treatment, one patient showed subnormal polyamine values on admission to the hospital. Later on during his stay elevated levels in putrescine 5-fold normal, spermidine 10-fold normal and spermine from subnormal to 21-fold normal were observed. In this group of patients, putrescine values ranged between subnormal and 4-fold normal, spermidine from normal to 10-fold normal, and spermine from subnormal to 21-fold normal. Hydrolysed serum samples from morbus Hodgkin patients had increased putrescine (Marton, Russell and Levy, 1973a; Nishioka and Romsdahl, 1974), in one case reaching 150-fold normal (Marton and coworkers, 1973b). Spermidine was found in some cases to be subnormal and in other cases reaching amounts of 5-fold normal (Marton and coworkers, 1973b). Spermine was increased in only one case by a factor of 1.8 but was generally found only in small amounts (Marton, Russell and Levy, 1973a; Marton and coworkers, 1973b; Nishioka and Romsdahl, 1974).

The polyamine excretion patterns in the urine of a total of 31 patients with multiple myeloma have been recorded (Russell and coworkers, 1971; Gehrke and coworkers, 1973; Dreyfuss and coworkers, 1975; Russell, Durie and Salmon, 1975; Tsuji, Nakajima and Sano, 1975; Durie, Salmon and Russell, 1977; Heby and Andersson, 1978a). Putrescine values were between normal and 6-fold normal, spermidine was found to be between 1.4-fold and 8.6-fold normal. Interestingly, in some cases cadaverine was also detected (Dreyfuss and coworkers, 1975; Fleisher and Russell, 1975). Acetylspermidine was found to be elevated in the urine of one patient with this disease (Tsuji, Nakajimo and Sano, 1975).

Few parallel determinations on serum, urine and plasma samples of the same patients have been performed. For multiple myeloma no difference in the general picture of the polyamine pattern was observed (Russell and Russell, 1975; Russell, 1977). As a consequence of chemotherapy the putrescine and spermidine values were shown to be increased in the urine (Dreyfuss and coworkers, 1975; Russell and Russell, 1975), whereas the ratio of spermidine values post-treatment to pre-treatment increased in patients responding to therapy. The hypothesis that putrescine reflects the growth fraction of tumours was strengthened by examination of a myeloma patient (Durie, Salmon and Russell, 1977; Russell, 1977).

V. NON-MALIGNANT HAEMOPATHIES

In the urine of 11 patients with different forms of anaemia increased levels of putrescine were found to 8-fold above normal, spermidine 1.7-fold normal and spermine 1.3-fold normal. In this study 4 cases of increased cadaverine levels in the urine were detected (Dreyfuss and coworkers, 1975). In the urine of a patient with pernicious anaemia levels of putrescine 1.7-fold normal and spermidine levels 9-fold normal were found (Durie, Salmon and Russell, 1977). Histamine values in blood plasma samples of patients with this disease have also been reported to be at least 2.5-fold higher than the normally detectable level (Stopik, 1974). In the whole blood samples of 24 patients with sickle cell anaemia spermidine and spermine values 16-fold normal and 9-fold normal respectively, were found (Chun and coworkers, 1976). In addition, putrescine and to a high extent (9-fold) spermidine were also found to be elevated in the blood plasma (Cooper, Shukla and Rennert, 1978). Human red blood cells classified according to age differences, exhibited an increase of the polyamine content with advancing age (Cooper, Shukla and Rennert, 1976). Erythrocytes isolated from patients with sickle cell anaemia had higher cellular concentrations of spermidine and spermine, whereas putrescine was only moderately elevated (Chun and coworkers, 1976; Cooper, Shukla and Rennert, 1978). None the less, in this disease and in other erythropathies such as hereditary elliptocytosis, the spermine/spermidine ratio in the erythrocyte compartment was very low in contrast to the situation in malignant diseases such as CLL, lung carcinoma and psoriasis where the ratio was markedly higher than normal. In whole blood samples comparable alterations were noticed (Cooper, Shukla and Rennert, 1978). Proliferative states with high polyamine contents in the blood compartments are therefore characterized by elevated spermine/spermidine ratios in contrast to the relationship in erythropathies where this ratio is reversed.

The content of unbound polyamines in isolated leucocytes and blood plasma was the subject of one investigation with patients with *polycythaemia vera rubra* (Desser and coworkers, 1975). In this study elevations of all polyamines above the normal level were found in 7 out of 11 patients. The white blood cell compartment of these patients seemed to contain fewer polyamines than the amount found in cells isolated from normal volunteers. These results were in accordance with the findings of Chun and coworkers (1976) and Cooper, Shukla and Rennert (1976) who demonstrated that the polyamines associated with erythrocytes, white blood cells and blood plasma may be in equilibrium, and that the age and source of the erythrocytes involved are important contributing factors (Rennert and Shukla, 1978). Histamine has also been found to be moderately elevated in whole blood and urine samples of *polycythaemia vera* patients (Westin and coworkers, 1975; Horakova, Keiser and Beaven, 1977).

VI. CONCLUSION

The most pronounced deviations from normal in the content of the biogenic amines under consideration has been observed for histamine in whole blood samples of patients with CML. Spermidine and putrescine levels have been found to be considerably increased in comparison to the normal range in hydrolysed urine samples with diseases associated with high lymphoblastic cell populations, such as ALL, some blastic phases of CML, and different forms of malignant lymphomas. In particular, patients with Burkitt's lymphoma have exhibited high polyamine levels. Increases in blood plasma and serum levels for spermidine and putrescine have also been reported to be relatively high in patients with the lymphoma, but usually not to the same extent as seen in hydrolysed urine samples. Spermine has also been found to be of interest in this respect, but data is still scant and in many cases the method of determination lacked the appropriate specificity. Interestingly, in lymphoma where the highest percentage elevations have been reported, many cases with subnormal values have also been encountered.

Large increases of the polyamines in bone marrow aspirates and in the CSF have been observed associated with specific disease states, for example high spermidine and spermine values in the CSF of leukaemic children with CNS involvement, and high putrescine values in the patients without CNS involvement.

Data obtained from investigation of serum or plasma samples are rarely available. With one exception (Russell and Russell, 1975), no determinations in urine, plasma and serum, hydrolysed and non-hydrolysed have been concomitantly performed. More investigations of this kind are desirable.

Special attention is drawn to the findings of Durie, Salmon and Russell (1977) and Russell (Russell, 1977; Russell, Durie and Salmon, 1975) on chemotherapeutically treated patients. Their findings suggest that elevated levels of putrescine and spermidine reflect a function of cell growth and of cell death, respectively. Heby and Andersson (1978b), using an experimental animal model, demonstrated increased spermidine and putrescine levels mainly associated with increased cell lysis in the tumour mass.

VII. SUMMARY AND CONCLUSION

Polyamine levels in physiological fluids and biopsy material have been shown to be valuable tools in evaluating the pathogenesis of haematological diseases, as well as monitoring the protracted course of the disease with or without chemotherapeutic treatment. Many more clinical chemistry investigations will be necessary in order to demonstrate the reliability of these findings in specific disease stages.

REFERENCES

Bachrach, U. (1976). *Ital. J. Biochem.*, **25**, 77–93.

Bachrach, U. (1978). Analytical methods for polyamines. In R. A. Campbell, D. R. Morris, D. Bartos, G. D. Davis and F. Bartos (Eds), *Advances in Polyamine Research*, Vol. 2, Raven Press, New York, pp. 5–11.

Bachrach, U. and Ben-Joseph, H. (1973). Tumor cells, polyamines and polyamine derivatives. In D. H. Russell (Ed), *Polyamines in Normal and Neoplastic Growth*, Raven Press, New York, pp. 15-25.

Bartos, D., Campbell, R. A., Bartos, F. and Grettie, D. P. (1975). *Cancer Res.*, **35**, 2056–2060.

Bartos, F., Bartos, D., Dolney, A. M., Grettie, D. P. and Campbell, R. A. (1978). *Res. Commun. Chem. Path. Pharm.*, **19**, 295–309.

Bartos, F., Bartos, D., Grettie, D. P., Campbell, R. A., Marton, L. J., Smith, R. G. and Davis, G. D. (Jr.) (1977). *Biochem. Biophys. Res. Commun.*, **75**, 915–919.

Beaven, M. A., Marshall, J. R., Baylin, S. B. and Sjoerdsma, A. (1975). *Am. J. Obstr. Gynec.*, **123**, 605–609.

Berry, H. K., Glazer, H. S., Denton, M. D. and Fogesson, M. H. (1978a). Polyamine excretion by patients with defective transport of amino acids. In R. A. Campbell, D. R. Morris, D. Bartos, G. D. Davis and F. Bartos (Eds), *Advances in Polyamine Research*, Vol. 2, Raven Press, New York, pp. 313–318.

Berry, H. K., Denton, M. D., Glazer, H. S. and Kellogg, F. W. (1978b). Normal polyamines in the blood of homozygotes for cystic fibrosis. In R. A. Campbell, D. R. Morris, D. Bartos, G. D. Davis and F. Bartos (Eds), *Advances in Polyamine Research*, Vol. 2, Raven Press, New York, pp. 307–312.

Cadiou, M., Ruff, F., Meunier, F., Attalah, N., Bernadou, A., Zittoun, R., Parrot, J. L. and Bousser, J. (1975). *Nouv. Rev. Franc. Hemat.*, **15**, 261–269.

Campbell, R. A., Morris, D. R., Bartos, D., Davis, G. D. and Bartos, F. (1978a). *Advances in Polyamine Research*, Vol. 2, Raven Press, New York.

Campbell, R. A., Talwalkar, V., Bartos, D., Bartos, F., Musgrave, J., Harner, M., Puri, H., Grettie, D., Dolney, A. M. and Loggan, B. (1978b). Polyamines, uremia, and hemodialysis. In R. A. Campbell, D. R. Morris, D. Bartos, G. D. Davis and F. Bartos (Eds), *Advances in Polyamines Research*, Vol. 2, Raven Press, New York, pp. 319–343.

Chun, P. W., Rennert, O. M., Saffen, E. E. and Taylor, W. J. (1976). *Biochem. Biophys. Res. Commun.*, **69**, 1095–1101.

Cohen, L. F., Lundgren, D. W. and Farrell, P. M. (1976). *Blood*, **48**, 469–475.

Cooper, K. D., Shukla, J. B. and Rennert, O. M. (1976). *Clin. Chem. Acta*, **73**, 71–88.

Cooper, K. D., Shukla, J. B. and Rennert, O. M. (1978). *Clin. Chim. Acta*, **82**, 1–7.

Debray, J., Cheymol, G., Drulik, M., Schmitt, S. and Audebert, A. (1975). *Nouv. Rev. Franc. Hematol.*, **15**, 250–260.

Denton, M. D., Glazer, H. S., Zellner, D. C. and Smith, F. G. (1973a). *Clin. Chem.*, **19**, 904–907.

·Denton, M. D., Glazer, H. S., Walle, T., Zellner, D. C. and Smith F. G. (1973b). Clinical application of new methods of polyamine analysis. In D. H. Russell (Ed), *Polyamines in Normal and Neoplastic Growth*, Raven Press, New York, pp. 373–380.

Desser, H., Höcker, P., Weiser, M. and Böhnel, J. (1975). *Clin. Chim. Acta*, **63**, 243–247.

Desser, H., Pawlowsky, H., Stacher, A. and Partsch, G. (1978). In preparation.

Dreyfuss, F., Chayen, R., Dreyfuss, G., Dvir, R. and Ratan, J. (1975). *Israel J. Med. Sci.*, **11**, 785–795.

Dubois, A. M., Santais, M. C., Foussard, C., Dubois, F., Ruff, F., Taurelle, R. and Parrot, J. L. (1977). *Agents Actions*, **7**, 112.

Durie, B. G. M., Salmon, S. E. and Russell, D. H. (1977). *Cancer Res.*, **37**, 214–221.

Elmfors, B. and Tryding, N. (1976). *Brit. J. Obstr. Gynec.*, **83**, 6–10.

Fair, W. R., Wehner, N. and Brorsson, U. (1975). *J. Urol.*, **114**, 88–92.

Fleisher, J. H. and Russell, D. H. (1975). *J. Chromat.*, **110**, 335–340.

Fujita, K., Nagatsu, T., Maruta, K., Ito, M. and Senba, H. (1976). *Cancer Res.*, **36**, 1320-1324.
Gaugas, J. M. and Curzen, P. (1978). *Lancet*, **(i)**, 8054, 18-20.
Gehrke, C. W., Kuo, K. C. and Zumwalt, R. W. (1974). *J. Chromat.*, **89**, 231-238.
Gehrke, C. W., Kuo, K. C., Zumwalt, R. W. and Waalkes, T. P. (1973). The determination of polyamines in urine by gas-liquid chromatography. In D. H. Russell (Ed), *Polyamines in Normal and Neoplastic Growth*, Raven Press, New York, pp. 343-353.
Gingold, N. (1968). *Wiener Z. Inn. Med.*, **49**, 180-186.
Gingold, N. and Pecker, I. (1975). *Rev. Roum. Med.*, **13**, 303-308.
Heby, D. and Andersson, G. (1978a). *J. Chromat.*, **145**, 73-80.
Heby, D. and Andersson, G. (1978b). *Acta Path. Microbiol. Scand.*, **86**, 17-20.
Horakova, Z., Keiser, H. R. and Beaven, M. A. (1977). *Clin. Chim. Acta*, **79**, 447-456.
Jänne, J., Pösö, H. and Raina, A. (1978). *Biochim. Biophys. Acta*, **473**, 241-293.
Jonassen, F., Granerus, G. and Wetterqvist, H. (1976a). *Acta Obstr. Gynec. Scand.*, **55**, 297-304.
Jonassen, F., Granerus, G. and Wetterqvist, H. (1976b). *Acta Obstr. Gynec. Scand.*, **55**, 387-394.
Lipton, A., Sheehan, L. H. and Kessler, G. F. (Jr.) (1975). *Cancer*, **35**, 464-468.
Lorenz, W. (1975). *Agents and Actions*, **5**, 402-416.
Lundgren, D. W., Farrell, P. M., Cohen, L. F. and Hankins, J. (1976). *Proc. Soc. Exptl. Biol. Med.*, **152**, 81-85.
Makita, M., Yamamoto, S. and Kono, M. (1975). *Clin. Chim. Acta*, **61**, 403-405.
Marton, L. J., Russell, D. H. and Levy, C. C. (1973a). *Clin. Chem.*, **19**, 923-926.
Marton, L. J., Vaughn, J. G., Hawk, I. A., Levy, C. C. and Russell, D. H. (1973b). Elevated polyamine levels in serum and urine of cancer patients: detection by a rapid automated technique utilizing an amino acid analyzer. In D. H. Russell (Ed), *Polyamines in Normal and Neoplastic Growth*, Raven Press, New York, pp. 367-372.
Marton, L. J., Heby, O. and Wilson, C. B. (1974). *Intern. J. Cancer*, **14**, 731-735.
Marton, L. J., Heby, O., Levin, V. A., Lubbich, W. P., Crafts, D. C. and Wilson, C. B. (1976). *Cancer Res.*, **36**, 973-977.
Maslinski, C. (1975). *Agents and Actions*, **5**, 183-225.
McEvoy, F. A. and Hartley, C. B. (1975). *Ped. Res.*, **721-724.**
Miale, T. D., Rennert, O. M., Lawson, D. L., Shukla, J. B. and Frias, J. L. (1977). *Med. Ped. Oncol.*, **3**, 209-230.
Musgrave, J. E., Campbell, R. A., Bartos, D., Bartos, F., Harner, M. H., Talwalker, Y. B., Puri, H., Grettie, D. P. and Loggan, B. (1978). Serum free polyamine levels in renal transplant patients. In R. A. Campbell, D. R. Morris, D. Bartos, G. D. Davis and F. Bartos (Eds), *Advances in Polyamine Research*, Vol. 2, Raven Press, New York, pp. 351-358.
Narasbhat, K., Arroyave, C. M., Marney, S. R., Stevenson, D. D. and Tan, E. M. (1976). *J. Allerg. Clin. Immun.*, **58**, 647-656.
Nishioka, K. and Romsdahl, M. N. (1974). *Clin. Chim. Acta*, **57**, 155-161.
Nishioka, K., Romsdahl, M. N., Fritsche, H. A. (Jr.) and Hart, J. S. (1976). *Am. Cancer Congr.*, Abstract 770.
Nishioka, K. and Romsdahl, M. N. (1977). *Cancer Lett.*, **3**, 197-202.
Parwaresch, M. R. (1976). *The Human Basophil*, Springer-Verlag, Heidelberg.
Porter, J. F. and Mitchell, R. G. (1972). *Physiol. Revs.*, **52**, 361-381.
Procter, M. S., Fletcher, H. V. (Jr.), Shukla, J. B. and Rennert, O. M. (1975). *J. Invest. Dermatol.*, **65**, 409-411.

Puri, H., Campbell, R. A., Puri, V., Harner, M. H., Talwalkar, Y. B., Musgrave, J. E. Bartos, F., Bartos, D. and Loggan, B. (1978). Serum free polyamines in children with systemic lupus erythematosus. In R. A. Campbell, D. R. Morris, D. Bartos, G. D. Davis and F. Bartos (Eds), *Advances in Polyamine Research*, Raven Press, New York, pp. 359–367.

Raina, A. (1962). *Scand. J. Clin. Lab. Invest.*, **14**, 318–319.

Raina, A. and Jänne, J. (1975). *Med. Biol.*, **53**, 121–147.

Rennert, O. M. and Shukla, J. B. (1978). Polyamines in health and disease. In R. A. Campbell, D. R. Morris, D. Bartos, G. D. Davis and F. Bartos (Eds), *Advances in Polyamine Research*, Vol. 2, Raven Press, New York, pp. 195–211.

Rennert, O. M., Miale, T. D., Shukla, J. B., Lawson, D. and Frias, J. (1976a). *Blood*, **47**, 695–701.

Rennert, O. M., Frias, J. and Shukla, J. B. (1976b). *Texas Reports Biol. Med.*, **34**, 187–197.

Rennert, O. M., Lawson, D. L., Shukla, J. B. and Miale, T. D. (1977). *Clin. Chim. Acta*, **75**, 365–369.

Rodermund, O. E. and Moersler, B. (1974). *Z. Hautkrankh.*, **50**, 273–279.

Rosenblum, M. G. and Russell, D. H. (1977). *Cancer Res.*, **37**, 47–51.

Russell, D. H. (1971). *Nature (New Biol.)*, **233**, 144–145.

Russell, D. H. (1973a). *Polyamines in Normal and Neoplastic Growth*, Raven Press, New York.

Russell, D. H. (1973b). *Life Sci.*, **13**, 1635–1647.

Russell, D. H. (1977). *Clin. Chem.*, **23**, 22–27.

Russell, D. H. and Russell, S. D. (1975). *Clin. Chem.*, **21**, 860–863.

Russell, D. H., Levy, C. C., Schimpff, S. C. and Hawk, I. A. (1971). *Cancer Res.*, **31**, 1555–1558.

Russell, D. H., Durie, B. G. M. and Salmon, S. E. (1975). *Lancet*, 797–803.

Russell, D. H. and Durie, B. G. M. (1978). *Polyamines as Biochemical Markers of Normal and Malignant Growth*, Raven Press, New York.

Samejima, K., Kawase, M., Sakamoto, S., Okada, M. and Endo, Y. (1976). *Anal. Biochem.*, **76**, 392–406.

Sanford, E. J., Drago, J. R., Rohner, T. J., Kessler, G. F., Sheehan, L. and Lipton, A. (1975). *J. Urol.*, **113**, 218–221.

Schimpff, S. S., Levy, C. C., Hawk, I. A. and Russell, D. H. (1973). Polyamines — potential roles in the diagnosis, prognosis and therapy of patients with cancer. In D. H. Russell (Ed), *Polyamines in Normal and Neoplastic Growth*, Raven Press, New York, pp. 395–403.

Seiler, N. (1976). Assay procedures for polyamines and GABA in animal tissues with special reference to dansylation methods. In N. Marks and R. Rodnight (Eds), *Research Methods in Neurochemistry*, Vol. 3, Plenum Press, New York, pp. 409–441.

Seiler, N. (1977a). *J. Chromat.*, **143**, 221–246.

Seiler, N. (1977b). *Clin. Chem.*, **23**, 1519–1526.

Seiler, N. and Deckardt, K. (1975). *J. Chromat.*, **107**, 227–229.

Seiler, N., Knödgen, S. and Eisenbeiss, F. (1978). *J. Chromat.*, **145**, 29–39.

Stopik, D. (1974). *Klin. Wschrft.*, **52**, 990–992.

Suzuki, S., Ishida, F., Kono, T. and Muranaka, M. (1971). *Cancer*, **28**, 384–388.

Tabor, C. W. and Tabor, H. (1976). *Ann. Rev. Biochem.*, **45**, 285–306.

Tormey, D. C., Waalkes, T. P., Ahmann, D., Gehrke, C. W., Zumwalt, R. W., Snyder, J. and Hansen, H. (1975). *Cancer*, **35**, 1095–1100.

Townsend, R. M., Banda, P. W. and Marton, L. J. (1976). *Cancer*, **38**, 2088–2092.

Tsuji, M., Nakajina, T. and Sano, I. (1975). *Clin. Chim. Acta*, **59**, 161–167.

Waalkes, T. P., Gehrke, C. W., Tormey, D. C., Zumwalt, R. W., Hueser, J. N., Kuo, K. C., Laings, D. B., Ahmann, D. L. and Moertel, C. G. (1975). *Cancer Chemotherapy Reports*, **59**, 1103–1116.

Walle, T. (1973). Gas chromatography — mass spectrometry of di- and polyamines in human urine: identification of monoacetylspermidine as a major metabolic product of spermidine in a patient with acute myelocytic leukemia. In D. H. Russell (Ed), *Polyamines in Normal and Neoplastic Growth*, Raven Press, New York, pp. 355–365.

Westin, J., Granerus, G., Weinfeld, A. and Wetterqvist, H. (1975). *Scand. J. Haemat.*, **15**, 45–57.

Windmueller, H. G. and Spaeth, A. E. (1976). *Arch. Biochem. Biophys.*, **175**, 670–676.

Zittoun, R., Ruff, F., Mouginot, M. C., Parrot, J. L. and Bousser, J. (1975). *Nouv. Rev. Franc. Hematol.*, **15**, 298–300.

Chapter 27

Assay of Polyamines in Tissues and Body Fluids

NIKOLAUS SEILER

I. DEVELOPMENTAL PHASES OF POLYAMINE DETERMINATION

During its exploratory phase biochemical research is mainly an analytically oriented science. The development of specific, sensitive and precise methods is a prerequisite for the collection of meaningful data in this phase. Clinical application of a biochemical area requires rapid and simplified data generation and usually leads to the development of automated methods suited for screening purposes. The methods available for an analytical problem at a given time usually mirror the general status of the analytical and technical development. Usually all possible approaches to a given problem are tried out at one time or another, but soon a small number of methods emerge which best fit the practical requirements. These methods are developed further in parallel with improvements in the general technical level, until a breakthrough is made by invention of a new principle of measurement, or until an expansion of the research to new fields dictates alternate practical considerations. The history of polyamine assays faithfully follows this general picture.

Until the invention of surface chromatographic methods, separation of the polyamines from related compounds was achieved by crystallization of slightly soluble salts, such as phosphates, picrates, flavianates (Puranen, 1936; Fuchs,

1939; Hämäläinen, 1941) and chloroaurates (Wrede, 1926) and subsequent quantitative determination by gravimetry. Even in recent years complex formation with ammonium reineckate (Heby, 1972) has been used for spermine determination. These cumbersome methods were restricted in their applications: only the spermine salts were sufficiently insoluble to give high yields. Spermidine salts were too soluble for exact quantitation, and the methods were entirely inapplicable to diamines.

A first breakthrough was made by paper chromatographic methods (Bremner and Kenten, 1951; Herbst, Keister and Weaver, 1958; Blau, 1961; Ramakrishna and Adiga, 1973). Indeed the applications of paper chromatographic methods which are compiled in Table 6 of U. Bachrach's book on *Function of Naturally Occurring Polyamines* (1973) represent an impressive chapter of the biochemical work on polyamines of nearly two decades. Subsequently, the same solvents or modifications were used in combination with thin layers of cellulose (Morris and Pardee, 1966; Hammond and Herbst, 1968; White and coworkers, 1968; Smith, 1970; Kuttan and coworkers, 1971) and silica gel (Michaels and Tchen, 1968; Smith and Stevens, 1971).

The utilization of the polycationic nature of the polyamines as the basis for their separation from other cations proved to be especially useful, since it was more specific than the chromatographic methods. The first practical high voltage paper electrophoretic method was developed by Fischer and Bohn (1957). This method was modified by Herbst, Keister and Weaver (1958) and by Raina (1963) and subsequently became the most popular method of polyamine analysis until a few years ago. Electrophoretic methods are again transferable from paper sheets to thin layers of cellulose (Beer and Kosuge, 1970; Smith, 1970) and silica gel (Seiler and Al-Therib, 1974; Seiler and Knödgen 1977). For the detection of the chromatographically or electrophoretically separated amines, a number of colour forming reactions have been suggested: ninhydrin (Fischer and Bohn, 1957), diazotized *p*-nitroaniline (Lähdevirta, Raina and Heikel, 1957), amido black (Raina, 1962; 1963) and platinic iodide (Bach, 1966). Recently fluorescamine has been used as a sensitive fluorescent stain (Abe and Samejima, 1975). Among these methods reaction with ninhydrin in various modifications has gained the widest application.

Instead of separating the free amines, pre-chromatographic derivatization with one among the many reagents which are suitable to form coloured or fluorescent derivatives with primary and secondary amines (Lawrence and Frei, 1976; Seiler, 1977; Seiler and Demisch, 1977) could be used. In practice only 2,4-dinitrofluorobenzene (Holder and Bremer, 1966), 5-dimethylaminonaphthalene-1-sulfonyl chloride (Seiler and Wiechmann, 1967) and fluorescamine (Samejima, 1974; Nakamura and Pisano, 1976) have been suggested for polyamine determination. Derivatization with 5-dimethylaminonaphthalene-1-sulfonyl chloride was the first method sufficiently sensitive to allow the characterization and determination of polyamines in the picomole range and, therefore, became important for the further development of certain areas of polyamine biochemistry (Bachrach, 1978).

Among the column chromatographic methods, ion exchange column chromatography has not only the longest tradition (Rosenthal and Tabor, 1956), but is at present the most widely used separation method for polyamines. In the past, practically all kinds of commercial ion-exchange resins have been used, including cellulose phosphate (Kremzner, 1966). The amines were eluted by various buffers or acid solutions and determined in the automatically collected fractions by dinitrophenylation (Rosenthal and Tabor, 1956; Tabor and Rosenthal, 1963; Shimizu, Kakimoto and Sano, 1964; Holder and Bremer, 1966; Kremzner, Barrett and Terrano, 1970), by enzymatic methods (Unemoto and coworkers, 1963; Kremzner, Barrett and Terrano, 1970) or by condensation with *o*-phthaldialdehyde (Kremzner, 1966; Elliott and Michaelson, 1967; Kremzner, Barrett and Terrano, 1970; Hakanson and Rönnberg, 1973). With the development of commercial automated amino acid analyzers, many laboratories worked out suitable procedures for the separation of polyamines on sulfonated polystyrene resins, using with one exception (Veening, Pitt and Jones, 1974) reaction with ninhydrin and colorimetry at 570 nm for quantitation (Morris, Koffron and Okstein, 1969; Hatano and coworkers, 1970; Morris, 1971; Bremer, Kohne and Endres, 1971; Marton, Russell and Levy, 1973; Tabor, Tabor and Irreverre, 1973; Nishioka and Romsdahl, 1974; Sturman and Gaull, 1974; Gehrke, Kuo and Zumwalt, 1974; 1977; Marton and coworkers, 1974; Adler and coworkers, 1977). Polyamine determination on ion exchanger surfaces, as for instance on phosphorylated carboxymethyl cellulose thin layers (Johnson and Bach, 1968) or on cation exchange resin paper (Rinaldini, 1970), although feasible, never played a significant practical role.

A breakthrough to high sensitivity, comparable to the sensitivity of the dansylation method (Seiler, 1970; 1975), was made by the introduction of high performance liquid chromatographic equipment and the utilization of the conditions described by Roth and Hampai (1973) for the reaction of amino groups containing compounds with *o*-phthaldialdehyde (Marton and Lee, 1975; Benson and Hare, 1975). As will be discussed in detail later, the combination of modern methods of ion exchange column chromatography with one line fluorescence measurement is the most effective routine method for the determination of polyamines in tissues and body fluids.

The attempts to utilize ligand-exchange chromatography for polyamine determination (Navratil and Walton, 1975) were not adequate with regard to resolution and sensitivity. The method cannot presently be considered an alternative to ion exchange column separations, but considerable improvement is possible.

Gas-liquid chromatography was applied to polyamine separation as early as

1961 (Smith and Radford, 1961) and was subsequently used by several laboratories (Cincotta and Feinland, 1962; Johnson and Markman, 1962; Beer and Kosuge, 1970; Smith, 1970) for the detection of polyamines in various natural materials. Tailing problems, poor recoveries of the free amines, and limited sensitivity of the flame ionization detectors restricted the applicability of the method, even when modern derivatization methods were introduced (Brooks and Moore, 1969). At present, suitable derivatization methods and electron capture detection (Makita, Yamamoto and Kono, 1975) or combined gas-liquid chromatography-mass spectrometry (Walle, 1973; Smith, Daves and Grettie, 1978) allow the determination of di- and polyamines in the picomole range. These methods are representative for the developmental state of instrumental analysis.

Immunological assay methods have only recently been successfully applied to polyamine analysis (Bartos and coworkers, 1978; Bonnefoy-Roch and Quash, 1978; Campbell and coworkers, 1978). These methods are probably best suited for screening purposes and should, therefore, find wide application in problems related to disease states with alterations in polyamine metabolism.

II. EXTRACTION AND ACCUMULATION OF POLYAMINES FROM TISSUES AND FLUIDS

Although the different assay procedures require a different degree of pre-separation of the polyamines from other tissue and body fluid constituents, there are a number of procedures which are more or less generally used for polyamine extraction and/or accumulation. Most commonly the polyamines are extracted from tissues by homogenization with distilled water, addition of trichloroacetic acid (5 to 10% final concentration), followed by one h of extraction at 2–4°C. Precipitation of the macromolecular compounds is sometimes completed by heating. After centrifugation, the trichloroacetic acid is removed by repeated extraction with a total of 9 volumes of diethylether (Tabor, Tabor and Irreverre, 1973). Heating with acetic acid (Fischer and Bohn, 1957; White and coworkers, 1968) or 0.1 N HCl (Russell and Levy, 1971) has also been used. Extraction with 2 to 4 M perchloric acid (Perry, Hansen and Kloster, 1965) and deproteinization with 4% solution of 5-sulfo-salicyclic acid seem to give better results in column chromatographic procedures, than trichloroacetic acid — diethylether extraction (Marton and coworkers, 1974).

For the analysis of free polyamines in blood, the blood samples are collected in plastic syringes and subsequently transferred to plastic tubes anticoagulated with EDTA. An equal volume of blood is mixed with a 10% solution of 5-sulfosalicylic acid. After centrifugation and filtration the samples are ready for direct application to a cation exchange column (Rennert, Frias and Shukla, 1976).

The acid extracts are normally suitable for application to ion exchange columns after appropriate dilution with buffer, and they are suitable for derivative formation with dansyl chloride (Seiler, 1975). Perchloric acid can be removed from the extracts by neutralization with KOH or K_2CO_3. Addition of 3 volumes of ethanol to the neutralized solution completes $KClO_4$ precipitation. This type of extract is used in our laboratory for the separation of polyamines and their metabolites by thin-layer electrophoresis (Seiler and Al-Therib, 1974) and related methods.

Accumulation of polyamines and partial separation from inorganic salts, which may interfere with the separations can be achieved by solvent extraction from alkaline solution. Diethylether, chloroform (Bach, 1966) and *t*-butanol, but mainly *n*-butanol (McIntyre, Roth and Shaw, 1947) have been used. A version of previously published extraction procedures was described recently. The butanol extracts were suitable for direct separation of the free bases by gas-liquid chromatography (Beninati, Sartori and Argento-Cerù, 1977). The procedure is as follows: 10 ml of tissue homogenate in distilled water was treated with an equal volume of 6% perchloric acid. The precipitate was washed twice with 5 ml of 3% perchloric acid. The washings were combined with the supernatant and alkalinized (pH ± 13) with 4 N KOH. The potassium perchlorate was precipitated by centrifugation and washed with 2 N KOH. The combined supernatant and washings were extracted three times with 10 ml of *n*-butanol for 30 min The butanol extract was shaken with 10 ml of a saturated solution of sodium sulfate and was then centrifuged. The dried residue of the butanol extract was re-extracted with 20 ml of *n*-butanol and the salts were completely removed by centrifugation. The clear butanol extract was evaporated to approximately 20 μl and aliquots of this extract were injected into the gas chromatographic column.

This and similar extraction procedures have been widely applied, especially to polyamine analyses in hydrolyzates of urine (Rosenthal and Tabor, 1956; Russell, 1971; Dreyfuss and coworkers, 1973; Lipton, Sheehan and Kessler, 1975). Recoveries are reported to be 87-90% for spermine (Raina, 1963; Tabor and Tabor, 1966; Russell and coworkers, 1970). It seems, however, advisable to calculate losses by adding small amounts of labelled polyamines (Russell, Medina and Snyder, 1970; Pegg, Lockwood and Williams-Ashman, 1970), since recoveries tend to vary. Moreover, exposure of small amounts of labelled putrescine to alkaline pH produced a number of not yet identified products which were separable by thin-layer electrophoresis (N. Seiler, unpublished). The ion exchange column procedure of Inoue and Mizutani (1973) may circumvent some of the problems inherent in the extraction of di- and polyamines from alkaline solutions. In this procedure, which is similar to the sample clean-up procedure of Gehrke and coworkers (1973), the perchloric acid tissue extracts (of 0.1-1 g of tissue) are directly applied to columns (9 × 40 mm) of Dowex-50 (H^+-form). The columns are washed first with 40 ml of 0.1

M sodium phosphate buffer pH 8.0 containing 0.7 M NaCl, which elutes the amino acids (including basic amino acids), and then with 10 ml of 1 N HCl, which removes sodium. The polyamines can be recovered from the columns with 15 ml of 6 N HCl. The residue of this HCl eluate can be used directly for further analysis. The procedure was applied to the urine of cancer patients prior to paper electrophoretic determination of the polyamines (Fujita and coworkers, 1976). Since it can be carried out with many samples at the same time by using a device for multiple column chromatography (Demisch, Bochnik and Seiler, 1976), it has decisive advantages over the extraction procedure. A similar chromatographic clean-up procedure has been recently devised by Samejima and coworkers (1976), who used carboxymethyl cellulose as ion exchanger and 0.01 M and 1 M pyridine-acetic acid buffer pH 5 for the fractionated elution of amines. With a column bed volume of 1 ml it was possible to process 1 ml of hydrolyzed human urine, 1 ml of deproteinized human serum and at least 50 mg of rat liver. The method of Grettie and coworkers (1978) was devised for the purification of labelled polyamines and for the selective removal of free polyamines from serum. This is achieved with small (0.2-0.5 g) silica gel columns using 0.01 to 0.03 N HCl for the elution of the polyamines. Acid washed silica gel seems to adsorb polyamines specifically.

With few exceptions polyamines have been assayed in urine, serum and plasma samples after total hydrolysis, i.e. all polyamine conjugates were presumed to be completely split into their parent compounds. The procedure is normally as follows: 2 ml of urine are evaporated to dryness under nitrogen gas at 65–70°C and subsequently heated with 2 ml of 6 N HCl for 16 h at 100°C in pyrex screw-cap tubes tightly capped with PTFE-lined screw caps. Serum samples (5 ml) are deproteinized 5-sulfosalicyclic acid (1 ml of a 20% solution). After heating for 30 min at 70°C they are centrifuged and the supernatants are evaporated to dryness. Hydrolysis with 6 N HCl is effected as in the case of the urine samples. After hydrolysis the samples are evaporated to dryness at 70°C in a stream of nitrogen. Recovery from cancer patient urine under these conditions was for putrescine 93.8 ± 6.9, cadaverine 96.8 ± 4.3, spermidine 96.0 ± 5.1 and for spermine 97.2 ± 5.6 (Gehrke, Kuo and Zumwalt, 1974).

Hydrolysis of conjugates with 1 g barium hydroxide per 5 ml sample solution at 100°C for 4 h and subsequent removal of the excessive barium hydroxide with sulfuric acid has been recently adopted to polyamine analysis in urine samples (Adler and coworkers, 1977). In order to avoid losses by adsorption of the free amines to the precipitated barium sulfate it is necessary to store the precipitate overnight.

III. CHEMICAL METHODS

The natural di- and polyamines do not exhibit structural features which would lend themselves to sensitive direct determination of these amines, even if moderately complex mixtures are employed. The most important among the methods currently in use for their detection and quantitative measurement are therefore combinations of a physical separation method with a chemical derivatization reaction. Only the biological methods make use of specific features of the polyamines that allow assays to be performed without a separation step.

The chemical derivatization usually has the purpose of enabling sensitive detection. It may also be important, however, for the improvement of separations. This is true for all gas-liquid chromatographic methods, with the exception of the gas-liquid chromatographic separation of the free bases, which was again suggested recently (Beninati, Sartori and Argento-Cerù, 1977). It is also of importance for other pre-chromatographic derivatization methods, especially for the dansylation procedure (Seiler, 1970; 1975). Derivatization in this case improves not only detection sensitivity and separation characteristics, but allows accumulation of the polyamine derivatives simply by solvent extraction from very dilute sources and mass spectrometric characterization with small amounts of the derivatives.

For the generation of sensitively detectable molecular species the chemical reactivity of the amino groups of the polyamines is utilized, i.e. the detection reactions are limited in their specificity to primary and secondary amines. The specificity of all chemical methods of polyamine determination is therefore entirely dependent on the quality of the separation method.

A. Methods for the Separation of the Non-derivatized Amines

1. *Surface Chromatography and Electrophoresis*

Little progress has been made in thin-layer chromatography of free polyamines during the past decade. The attempt of Abe and Samejima (1975) to utilize silica-gel sintered-glass plates for separation, staining of the plates with fluorescamine (Figure 1) and subsequent direct fluorescence scanning of the plates (Seiler and Möller, 1969) is only of marginal interest. Despite a pre-cleaning step using a column with phosphorylated cellulose according to Kremzner (1966) putrescine and cadaverine could not be separated in this

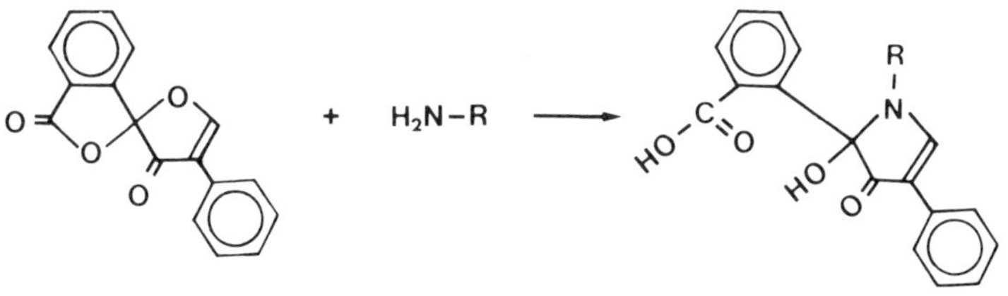

+ H_2N-R ⟶

Figure 1 Reaction of a primary amine with fluorescamine (4-phenylspiro[furan- 2(3H), 1-phthalan]-3, 3-dione)

procedure. The recent introduction of commercial high performance thin-layer plates by several companies (see Kaiser, 1976; Zlatkis and Kaiser, 1977; Zlatkis, Ettre and Dijkstra, 1977) in combination with automated sample application and high sensitivity scanning devices might successfully revive this type of approach.

High voltage paper electrophoresis has been used until recently for polyamine assay in tissues and body fluids. It was recognized, however, that urinary spermine values, as measured after staining the electrophoretograms with ninhydrin (Russell and coworkers, 1971; Lipton, Sheehan and Kessler, 1975) were considerably higher than those found with other methods (Bachrach, 1976). This method was therefore practically abandoned in favour of other methods. Paper electrophoresis and especially its more modern version, thin-layer electrophoresis, will certainly play a role in work directed at the identification of polyamine metabolites.

2. *Ion Exchange Column Chromatography*

Conventional automated amino acid analyzers, using ninhydrin as detector reagent, had proved their suitability in various practical applications of polyamine analyses. With commercial instruments, the separations achieved by multi-step gradient elution, using pH and salt gradients were mostly adequate, although time consuming and of limited sensitivity. Sample preparation was minimal, inasmuch as in most cases acidic tissue extracts could be directly separated (Morris, Koffron and Okstein, 1969; Hatano and coworkers, 1970; Morris, 1971; Bremer, Kohne and Endres, 1971; Marton, Russell and Levy, 1973; Tabor, Tabor and Irreverre, 1973; Nishioka and Romsdahl, 1974; Sturman and Gaull, 1974; Gehrke, Kuo and Zumwalt, 1974), and since automated sample applicators were available, analyses could be performed 24 h a day.

From multiple analyses of hydrolyzed urine samples Gehrke, Kuo and Zumwalt (1974) calculated relative standard deviations of ± 5.1 for putrescine, ± 4.4% for spermidine and ± 18.1% for spermine assays. These values represent the precision of the whole procedure, including sample hydrolysis with 6 N HCl. The analytical recovery of the amines usually exceeded 90%. The low precision of the spermine determinations stemmed from the fact that the concentration of total spermine in urine is near the detection sensitivity limit of the method. Indeed, the main drawback of these methods was their lack of sensitivity, which limited their applicability. Even when ninhydrin was substituted by the more sensitive fluorescamine (Figure 1) as detector reagent, the minimum detectable quantity was about 0.5 nmol for the diamines but 7 and 15 nmole for spermidine and spermine, respectively (Veening, Pitt and Jones, 1974).

Disregarding the instrumental (detector) sensitivity, detection sensitivity for column chromatographic procedures is limited by the volume of eluant necessary for complete elution of a certain compound from the column, and by the molecular features of the reaction product of the eluted compound with the detector reagent. Five or six years ago the commercial development of carefully selected ion exchange resins began to provide beads with a narrow range of diameter below 10 μm. Column diameters not exceeding 4 mm have been introduced. This requires high pressure, but both separation and sensitivity were improved considerably.

In a typical example beads of a diameter of 6–10 μm are packed to a height of 9 cm in a stainless steel column of 1.75 mm inner diameter. Columns of this type have about 7,000 theoretical plates. Elution is achieved by a three-step gradient consisting of potassium chloride-potassium citrate buffers pH 5.56 of increasing molarity. The buffer with 0.48 M potassium chloride and 0.018 M potassium citrate elutes the amino acids; the actual separations, as shown in

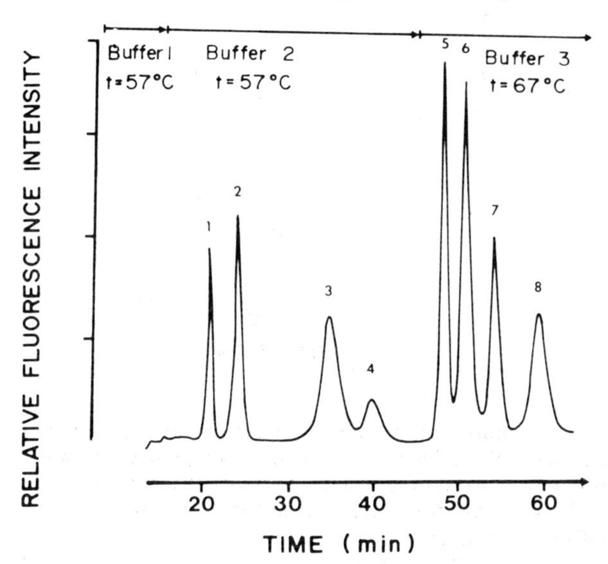

Figure 2 Ion exchange column chromatographic separation of a reference mixture of di- and polyamines. Instrumentation: Durrum amino acid analyzer combined with an Aminco Bowman fluorescence detector. Elution: Buffer 1:0.48 M potassium chloride, 0.018 M potassium citrate (pH 5.56 57°C). Buffer 2:1.44 M potassium chloride, 0.54 M potassium citrate (pH 5.56 57°C). Buffer 3: 2.4 M potassium chloride, 0.9 M potassium citrate (pH 5.56 67°C); flow rate 20 ml hour^{-1}. Detector reagent: *o*-phthaldialdehyde. Fluorescence monitoring at 455 nm (fluorescence activation at 340 nm). 1 = 1,3-diaminopropane; 2 = 1,4-diaminobutane (putrescine); 3 = 1,5-diaminopentane (cadaverine); 4 = N-3-aminopropyl-1,3-diaminopropane; 5 = spermidine (N-3-aminopropyl- 1,4-diaminobutane); 6 = N-4-aminobutyl- 1,4-diaminobutane; 7 = N,N-*bis*-(3-aminopropyl)-1,3-diaminopropane; 8 = spermine (N,N-*bis*-(3-aminopropyl)-1,4-diaminobutane. (By courtesy of Dr. J. Grove, Centre de Recherche Merrell International, Strasbourg)

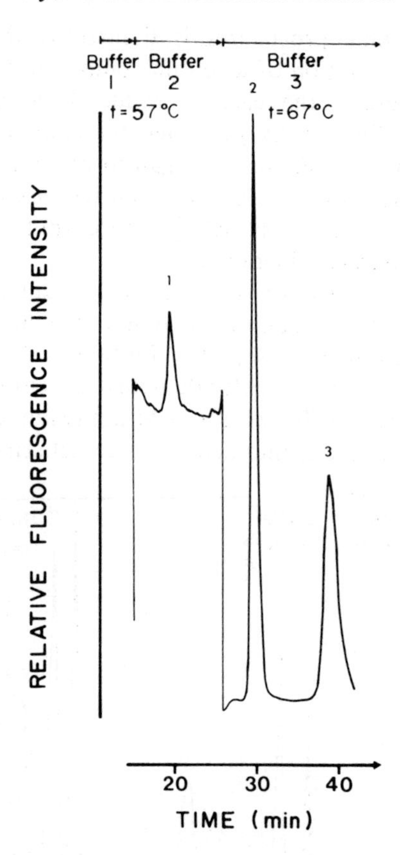

Figure 3 Polyamines in mouse brain. A 50 μl sample of the supernatant 1 + 20 mouse brain homogenate in 0.2 N perchloric acid was separated; putrescine was recorded at 10-fold higher sensitivity than spermidine and spermine. Instrumentation and separation conditions were the same as those described in the legend to Figure 2, only that the change from buffer 2 to buffer 3 was made already at 27 min. 1 = putrescine; 2 = spermidine; 3 = spermine. (By courtesy of Dr. J. Grove, Centre de Recherche Merrell International, Strasbourg)

Figures 2–4 are achieved with 1.44/0.54 M and 2.4/0.9 M buffer. At a flow rate of 20 ml h^{-1}, about 0.7 ml are needed for the elution of the putrescine zone and 1.4 ml for the spermine zone. Marton and coworkers (1974) were the first to make use of these developments and they were able to measure, with ninhydrin as detector reagent, 25–100 pmole of the polyamines. In the meantime, other modern amino acid analyzers have been adopted in a similar way to polyamine analysis (Adler and coworkers, 1977; Gehrke, Kuo and Ellis, 1977; Villanueva, Adlakha and Cantera-Soler, 1977). A further increase in sensitivity could be achieved by substitution of ninhydrin by the *o*-phthaldi-

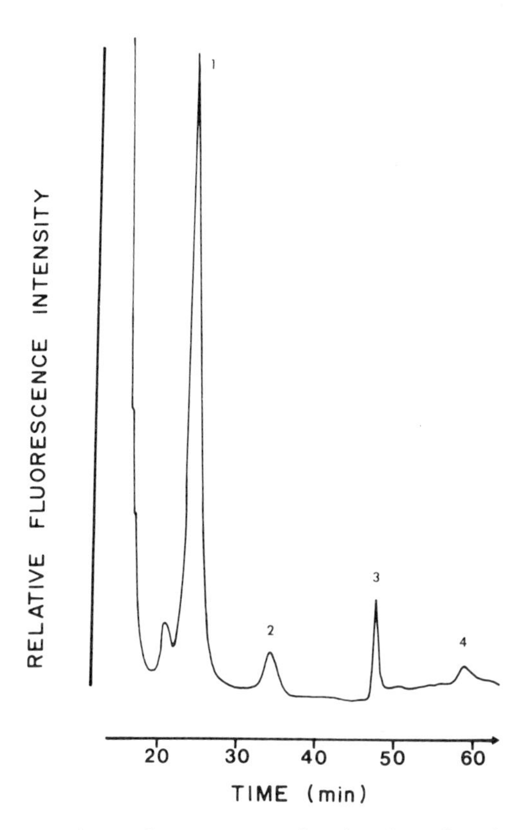

Figure 4 Ion exchange column chromatogram showing the polyamines in a hydrolyzed sample of urine of a psoriatic patient. Instrumentation and separation conditions were the same as those described in the legend to Figure 2. 1 = putrescine; 2 = cadaverine; 3 = spermidine; 4 = spermine. (By courtesy of Dr. J. Grove, Centre de Recherche Merrell International, Strasbourg)

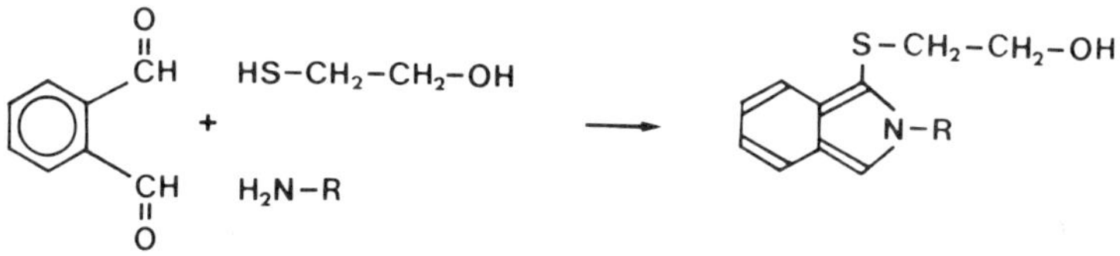

$$C_6H_4(CHO)_2 + HS{-}CH_2{-}CH_2{-}OH + H_2N{-}R \longrightarrow \text{(isoindole: } S{-}CH_2{-}CH_2{-}OH\text{, } N{-}R)$$

Figure 5 Reaction of a primary amine with *o*-phthaldialdehyde and β-mercaptoethanol

aldehyde reagent (Roth, 1971; Roth and Hampai, 1973) which consists of a 0.06 M *o*-phthaldialdehyde and 0.03 M β-mercaptoethanol in 0.4 M borate buffer pH 10.4. This reagent is continuously mixed 1 + 1 with the column effluent. It reacts with primary amino group containing compounds to give intensely fluorescent isoindoles (Figure 5) (Simons and Johnson, 1976). Most probably the di- and polyamines react with both of their primary amino groups. Fluorescence of these isoindoles is activated at 340 nm and measured at 455 mm. Three to six pmole of putrescine and spermidine and 12–15 pmole of spermine can be routinely measured after separation from biological fluids (standard deviations ± 4.6 and 4.8%, respectively). In the assay, instrumental response is linearly related to polyamine concentration over the range of 8 to 200 nmole liter^{-1} (Marton and Lee, 1975). Further improvement of sensitivity can be expected by substitution of β-mercaptoethanol by ethanethiol (Simons and Johnson, 1977) and by further improvement of the ion exchange resin columns.

Modern ion exchange column chromatographic procedures are rapid. The separation of a biological sample to an extent sufficient to allow the determination of putrescine, cadaverine, spermidine and spermine takes between 15 and 60 min. Since the equipment is fully automated, these methods are presently the most widely used routine assay procedures.

The separation programs of commercial amino acid analyzers can be extended in principle, so that certain polyamine conjugates, e.g. the monoacetyl derivatives, can be separated from each other (Tabor and Irreverre, 1973). This possibility could be of special importance for the diagnostic use of polyamine determinations. Practically applicable procedures should be worked out on this basis with modern equipment, in order to allow the collection of data on the relative excretion rates of the isomeric monoacetyl spermidines in normal human urine and by cancer patients (Abdel-Monem and Ohno, 1977).

B. Methods Separating Polyamine Derivatives

1. *Liquid-liquid Chromatographic Methods*

Among the many reagents suitable for forming derivatives with primary amines with a high molar extinction coefficient or intensely fluorescent products (Lawrence and Frei, 1976; Seiler, 1977; Seiler and Demisch, 1977), four have been suggested for prechromatographic derivatization of the polyamines; *p*-toluenesulfonyl chloride (Sugiura and coworkers, 1975), quinoline-8-sulfonyl chloride (Roeder, Pigulla nd Troschuetz, 1975), 5-dimethylaminoaphthalene-1-sulfonyl chloride (Dns-Cl) (Seiler and Wiechmann, 1967; 1970; Seiler, 1970; 1975) and fluorescamine (Samejima, 1974; Samejima and coworkers, 1976). 5-di-*n*-butylaminoaphthalene-1-sulfonyl chloride

(Figure 6) (Seiler, Schmidt-Glenewinkel and Schneider, 1973) has hitherto been used only for polyamine analyses in unpublished work. The mode of reaction of fluorescamine, which forms fluorescent derivatives only with primary amino groups containing compounds, is shown in Figure 1. In contrast to fluorescamine, the sulfonyl chlorides react with both primary and secondary amino groups.

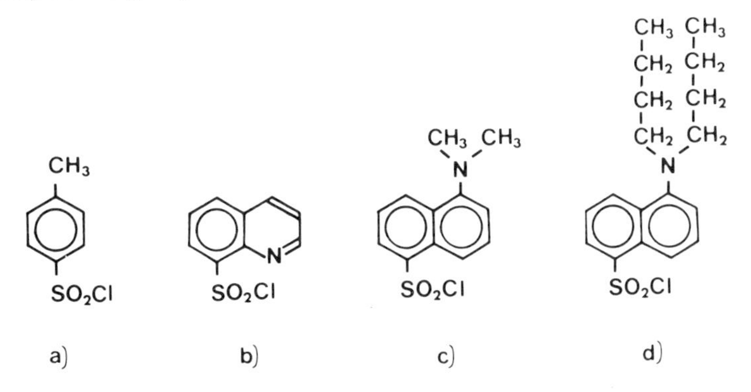

Figure 6 Aryl sulfonyl chlorides suitable to form absorbing or fluorescing derivatives with primary and secondary amines. a = *p*-toluenesulfonyl chloride; b = quinoline-8-sulfonyl chloride; c = 5-dimethylaminonaphthalene-1-sulfonyl chloride; d = 5-di-*n*-butylaminonapthalene-1-sulfonyl chloride

Since the aromatic sulfonyl chlorides do not differ greatly in their reactivity, the same principles of derivate formation are valid for all these reagents. The experience available for the derivate formation with Dns-Cl (Seiler, 1970; 1975; Seiler and Wiechmann, 1970; Seiler and Demisch, 1977) can therefore be utilized for the related reagents.

Reaction is normally achieved in alkaline (pH 8–10) buffered solutions containing an organic solvent (usually acetone) in order to ensure a homogeneous reaction medium. Similar reaction conditions are used for the derivatization with fluorescamine (Samejima, 1976; Seiler and Demisch, 1977). Reagents must be applied in excess in order to obtain quantitative reaction with all amino groups of the polyamines. One of the advantages of the pre-chromatographic derivatization methods is the solubility of the derivatives in organic solvents, which allows them to be separated from salts and other compounds that interfere with chromatographic separations and to be conveniently accumulated even from relatively large volumes of tissue extracts and body fluids.

For the separation of the di- and polyamine derivatives thin-layer and column chromatographic methods can be applied in principle. Practical use of thin-layer chromatography for the determination of polyamines in cells, tissues and body fluids has, however, been made only in the case of the Dns-derivatives (Seiler and Wiechmann, 1967; Seiler and coworkers, 1969; Cohen, Morgan and Streibel, 1969; Dion and Herbst, 1970; Seiler and Askar, 1971; Bachrach and Ben Joseph, 1973; Dreyfuss and coworkers, 1973; 1975; Seiler and Lamberty, 1975a,b; Fleischer and Russell, 1975; Abdel-Monem and Ohno, 1975; Heby and Andersson, 1978). The thin-layer chromatographic method of Nakamura and Pisano (1976) in which the compounds are derivatized at the origin of the thin-layer plates with fluorescamine, seems not to have found application.

For the separation of the Dns-derivatives of the di- and polyamines mostly silica gel thin layers have been used. The developing solvents have been adapted to the specific needs. Since the Dns-derivatives are intensely fluorescent and at the same time easily extractable from silica gel, they are estimated quantitatively either by *in situ* scanning of fluorescence or, after elution, by conventional fluorometry (Seiler and Wiechmann, 1966; 1967; 1970; Seiler, 1970). The method is highly sensitive. With suitable equipment, including an 8 μl flow cell, 2–10 pmole of putrescine, spermidine and spermine can be measured with a s.d. of $\pm$ 3–5%, after separation on high performance thin-layer plates (Seiler and Knödgen, 1977). With a sensitive scanning device it should be possible to increase sensitivity even further.

High sensitivity and improvement of specificity is obtained by mass spectrometric evaluation of thin-layer chromatographically separated Dns-derivatives. Especially the mass-spectrometric measurement of *bis*-Dns-putrescine proved to be of great value (Seiler and Knödgen, 1973; Seiler and Schmidt-Glenewinkel, 1975). The appropriate application of chromatographic methods, however, allows one to obtain correct results as well. The specificity of the method is documented by the fact that the first correct tissue levels of putrescine were obtained by thin-layer chromatography of its Dns-derivative (Seiler and Askar, 1971). These values were subsequently confirmed by quantitative mass spectrometry (Seiler and Schmidt-Gelenwinkel, 1975; Seiler and Lamberty, 1975a) and by ion exchange column chromatography (Marton and coworkers, 1974).

Besides the advantages of sensitivity and versatility, the dansylation method and in general the derivatization methods in combination with thin-layer chromatography have the disadvantage that they cannot be automated. Furthermore, side products of the reaction and especially reaction products which are formed in the presence of high concentrations of sugars (as for instance in sucrose density gradients) and high concentrations of ammonia and/or amino acids in hydrolyzed urine and blood samples can cause problems. The thin-layer chromatographic separations can be considerably improved by a purification step. The dansylated sample is dissolved in toluene and applied to a small silica gel column (5 $\times$ 30 mm). The non-polar side products of the dansylation reaction are eluted with toluene-triethylamine (10

+ 1) and the Dns-di- and polyamines (together with other amine derivatives) with ethyl acetate. The amino acid derivatives are not eluted under these conditions, so that the origin of the thin-layer plate is not overloaded (Seiler, Knödgen and Eisenbeiss, 1978).

The recently increased interest in the polyamine field obviously encouraged a number of groups to design separation methods for polyamines by HPLC. The advantages of this method are mainly good separations and the possibility for automation. It is not the aim of this review to discuss these methods in detail, because little practical experience presently exists with most of them. It should be mentioned, however, that in our hands adsorption chromatographic methods were not suitable for routine determinations because of the lack of stability of the columns. The separation characteristics of the columns changed after repeated runs or, at least, very long equilibration times were necessary for the establishment of reproducible separations, if solvent gradients were used. Very good separations could be obtained by the application of appropriate solvent gradients. (Isocratic methods led to considerable broadening of the spermine peak, if the separations were otherwise adequate for the complete separation of the homologous diamines.)

Reversed phase systems proved to be reliable in routine quantitative determinations. The tendency of all researchers to develop rapid separation systems is obvious, but in some cases adequate separations of the homologous diamines have not been shown. The application of these methods is restricted. Normally 20–50 min are needed for the separation of the derivatives of the di- and polyamines. Figure 7 shows the separation of some dansylated di- and polyamines on a reversed phase column, using a water methanol gradient. It should be pointed out that even a fully automated system cannot compete with a thin-layer chromatographic method in the analysis rate.

Toluene sulfonamides (Sugiura and coworkers, 1975; Hayashi and coworkers, 1978) and quinoline-8-sulfonamides (Roeder, Pigulla and Troschuetz, 1977) are quantitated in the column effluents by UV absorption measurement at 254 nm and 230 nm, respectively. The linear range of detector response, as given by the authors of these methods, is 2–50 μg for the toluene sulfonamides and 20–200 ng for the quinoline-8-sulfonamides. The high molar absorption coefficient of the Dns-derivatives (ϵ_{338} = 8,850; ϵ_{251} = 27,800 for *bis*-Dns-ethylenediamine in methanol as solvent (Seiler and Wiechmann, 1966; Seiler, 1970)) allows the measurement of 10 ng tri-Dns-spermidine by UV absorptimetry (Abdel-Monem and Ohno, 1975; Macek and coworkers, 1978; Abdel-Monem and coworkers, 1978). The great advantage of the Dns-derivatives is their sensitive detectability by fluorescence monitoring of the column effluents (Newton, Ohno and Abdel-Monem, 1976; Seiler, Knödgen and Eisenbeiss, 1978; Saeki, Uehara and Shirakawa, 1978). Using the separation system described in Figure 7, in combination with a Perkin-Elmer fluorescence spectrophotometer 204, equipped with an 8 μl flow cell and a

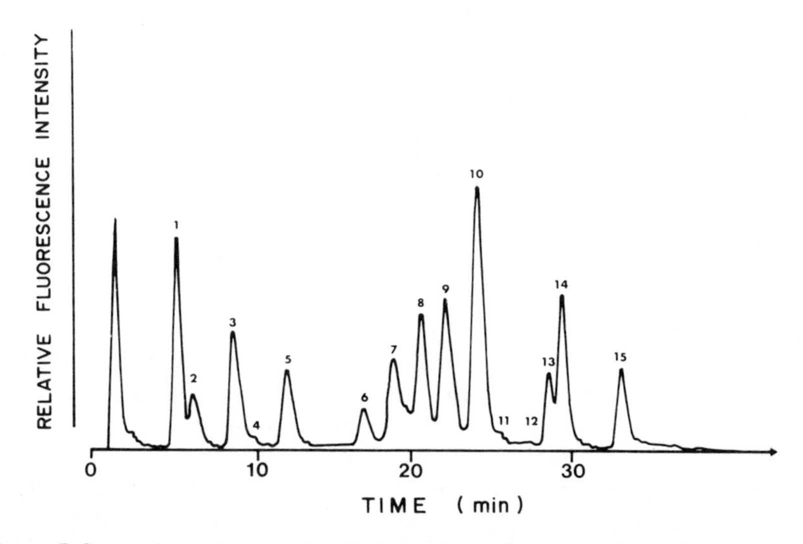

Figure 7 Separation of some dansyl-derivatives of mono-, di- and polyamines by reversed phase high-performance-liquid chromatography. Column: Hibar prepacked column (250 × 3 mm); Lichrosorb RP-8 (7 μm). Chromatograph: Varian 8500. Detector: Perkin-Elmer fluorescence spectrophotometer Model 204 A with 8 μl flow cell. Fluorescence activation at 360 nm; fluorescence measurement at 510 nm. Elution: three step water-methanol gradient (60 ml $hour^{-1}$). 1 = Dns-ammonia; 2 = Dns-ethanolamine; 3 = Dns-methylamine; 4 = Dns-2-oxopyrrolidine (reaction product of γ-aminobutyrate); 5 = Dns-dimethylamine; 6 = *bis*-Dns-agmatine: 7 = Dns-β-phenylethylamine; 8 = *bis*-Dns-putrescine; 9 = *bis*-Dns-cadaverine; 10 = *bis*-Dns-diaminohexane (suitable as internal standard for polyamine determinations); 11 = *bis*-Dns-histamine; 12 = O,N-*bis*-Dns-serotonin; 13 = O,N-*bis*-Dns-*p*-tyramine; 14 = *tris*-Dns-spermidine; 15 = tetrakis-Dns-spermine (For details, see Seiler, Knödgen and Eisenbeiss, 1978)

mercury arc lamp for excitation at 360 nm, (fluorescence measurement at 510 nm) it was possible to measure routinely less than 20 pmoles of *bis*-Dns-putrescine in tissue samples, with a linear recorder response up to at least 2 nmoles (linear regression coefficient ± 0.988). Peak evaluation by peak height measurements gave a reproducibility (standard deviation) of less than ± 7% of measurements performed on three different days. With more sensitive detectors and electronic integration of peak areas it is possible to improve the precision of the method considerably. For the separation of fluorescamine derivatives by HPLC, a comparable sensitivity is reported (Samejima and coworkers, 1976). 5-Di-*n*-butylaminonaphthalene-1-sulfonyl derivatives have a somewhat higher fluorescence quantum yield than the Dns-derivatives (Seiler, Schmidt-Glenewinkel and Schneider, 1973). Since they have a higher affinity for the lipophilic stationary phase of the reversed phase column, they are eluted at higher methanol concentrations than the Dns-derivatives. The detection sensitivity of these derivations is therefore about three times higher than for the corresponding Dns-derivatives. 3-5 pmoles of the putrescine derivative could be routinely measured with the above mentioned equipment.

2. *Gas-liquid Chromatography*

Despite the high standards in the technical development of gas-liquid chromatography, this method is not very widely used for polyamine analyses at present. The reason for this is presumably the convenience of the ion exchange column chromatographic procedures, which allow the direct application of tissue and body fluid samples on the separating columns. Gas-liquid chromatographic procedures normally require a clean-up step which consists of a relatively simple ion exchange column chromatography (or a related method). The clean-up step may result in low analytical recoveries of amines (Gehrke and coworkers, 1973).

Flame ionization detectors allow the measurement of 200-600 pmole amounts. Since the tailing problems of the separations of free amines seem to have been solved, by using Corning glass beads coated with 1% KOH and 4 permill Carbowax 20 M, free polyamines may now be successfully separated at column temperatures between 100 and 200°C (Beninati, Sartori and Argento-Cerù, 1977). The method has been applied to polyamine analysis of rat tissues.

A simple derivatization reaction which forms strongly basic reaction products and therefore requires similar separation conditions as the free polyamine bases was described by Giumanini, Chiavari and Scarpini (1976). Using formaldehyde-sodium borohydride, the polyamines are permethylated. The methylated products are recommended especially for quantitative evaluation by mass fragmentography, but flame ionization detection is also suitable.

Most commonly the trifluoroacetates of the polyamines are prepared to avoid peak tailing and to allow the application of both flame ionization (Gehrke and coworkers, 1973; Denton and coworkers, 1973 a, b; Walle, 1973) and electron capture detectors.

In one procedure, derivative formation is carried out by adding equal portions of acetonitrile and trifluoroacetic anhydride to the dried samples and heating to 100°C for 5 min. Acetonitrile and trifluoroacetic anhydride are removed by evaporation with N_2 gas at room temperature and the samples are redissolved in 0.5 ml diethylether containing the internal standard, e.g. phenanthrene (Gehrke and coworkers, 1973). This procedure, however, leads to partial hydrolysis of monoacetylated polyamines and the formation of the corresponding per-trifluoro-acetyl-derivatives. In the case where free polyamines are measured in samples containing large amounts of polyamine conjugates, e.g. non-hydrolyzed urine, the values obtained with this derivate forming reaction may be too high (Smith, Daves and Grettie, 1978). The method of Walle (1973) avoids splitting of acetyl-polyamines. Unfortunately the details of this procedure seem not to have been published. A further disadvantage of trifluoracetylation is that diperfluoroacetylation of the primary amino groups may occur as an unwanted side reaction (Ehrsson and Brötell, 1971). A more quantitative formation of the peracylated polyamines may be obtained using pentafluorobenzoyl chloride. Makita, Yamamoto and Kono (1975) have devised a rapid method using this reagent, which allows them to measure 1-5 pmole of the polyamines in urine samples. The usual liquid phase for the separation the trifluoroacetamides and related derivatives of the polyamines is OV 17 on Chromosorb. The temperature programming typically starts around 120°C and runs to 300°C at a rate of 10–15°C min^{-1}.

3. *Mass Spectrometry*

Coupling of gas-liquid chromatography with mass spectrometry is a powerful tool for the identification of unknown compounds. It is also important for the establishment of the uniformity of gas chromatographically separated peaks before a certain species of samples is routinely measured, e.g. using an electron capture detector. Walle (1973) was the first to use gas-liquid chromatography-mass spectrometry for the identification of polyamine derivatives in urine. He was able to identify monoacetyl-spermidine in the form of its di-trifluoroacetamide as a major metabolic product of spermidine. More recently, Kneifel, Rolle and Paschold (1977) applied the same method combination to the identification of homologues of spermidine in green algae, and Smith and Daves (1977); Smith, Daves and Grettie (1978) showed its usefulness as a quantitative tool for polyamine analysis in normal human serum. In addition to the trifluoroacetamides, the trimethylsilyl derivatives have been used, but they had no advantage over the trifluoroacetamides. As is common in quantitative gas-liquid-chromatography coupled with mass spectrometry, deuterated analogues of the polyamines were used as internal standards. These standards considerably simplify the procedure and increase the reliability of the method, because losses during sample preparation and sensitivity changes in the instrument are of little consequence for the quantitative result. In a systematic study, the gas-liquid chromatography-mass spectrometry of the trifluoroacetamides was compared with ion exchange column chromatography and with an immunological method (Bartos and coworkers, 1977). The results obtained with these three methods were well correlated.

From the standpoint of specificity, coupling of gas-liquid chromatography with mass spectrometry is superior to any other existing method. Since the method is cumbersome, as are all gas-liquid-chromatoraphic methods, and since expensive and complicated instrumentation is used with a corresponding need for high-quality technical personnel, it cannot be the routine method of choice. However, the method is precise (s.d. ± 4.4% for spermidine and ± 7.7% for putrescine determinations in human serum) and sensitive (it works efficiently with a few picomoles). It is therefore the method of choice for checking less sophisticated techniques, as was done in the above mentioned comparative study. As a tool for unambiguous identification of unknown compounds, mass spectrometry either coupled with gas-liquid chromatography, or in combination with another separation method, as for instance thin-layer chromatography of the Dns-derivatives (Seiler, 1975), is indispensable. In view of the already existing separation methods for polyamine derivatives, suitable for mass spectrometric analyses, the recent trend to develop devices for the direct coupling of a liquid chromatograph to a mass spectrometer, may also prove useful in the future for the study of certain aspects of polyamine biochemistry. In this respect the 5-di-*n*-butylaminonaphthalene-1-sulfonyl derivatives (Seiler, Schmidt, 1973) could be especially useful.

Mass spectrometry has not been applied to the quantitative determination of the non-derivatized di- and polyamines. During the last decade ionization techniques have been developed that produce high yields of molecular ions, even of slightly volatile polar compounds. The field desorption method (Beckey and Schulten, 1975) in combination with the so-called integrated ion current technique (Majer and Boulton, 1973) should be a promising approach to rapid, sensitive and specific polyamine analysis of body fluids.

IV. BIOLOGICAL METHODS

The early methods using biological detectors (Herbst, Keister and Weaver, 1958; Guirard and Snell, 1964) did not distinguish between the different polyamines and putrescine. Therefore, they never found much application. The enzymatic assays for polyamines, although also not widely used, have a number of characteristics, which are worth bearing in mind, the most important being that they do not require a separatory step. The most recent development in this regard, the radio-immunological assays, are not only sensitive but at the same time sufficiently specific and rapid to allow routine assays in large numbers. These methods suggest themselves for various clinical problems.

A. Enzymatic Assay Methods

1. *Putrescine*

An assay for putrescine has been published (Harik, Pasternak and Snyder, 1973) which utilizes the activating effect of putrescine on S-adenosyl-L-methionine-1-carboxylyase (Pegg and Williams-Ashman, 1969; Zappia, Carteni-Farina and Della Pietra, 1972). In practice the production of $[^{14}C]O_2$ from $[1\text{-}^{14}C]$S-adenosylmethionine is determined under standardized conditions. Since the enzyme preparations from yeast are fairly specific, putrescine can be measured in 10 pmole quantities directly in 40 μl tissue extracts. Only large amounts of 1,3-diaminopropane and 1,5-diaminopentane (cadaverine) would interfere with this assay.

2. *Spermidine*

Spermidine is selectively oxidized to Δ^1-pyrroline by lyophilized cells of *Serratia marcescens* (Bachrach and Oser, 1963). Δ^1-pyrroline is reacted with *o*-aminobenzaldehyde to form 2,3-trimethylene-1,2-dihydroquinazolinium (Figure 8), which can be measured by spectrophotometry at 435 nm (ϵ_{435} = 1,860 l. mole^{-1}. cm^{-1}). The method is sensitive enough for the determination of 5-10 nmole spermidine. An alternative is to use dichlorophenolindophenol as electron acceptor in the enzymic oxidation of spermidine (Campello, Tabor and Tabor, 1965).

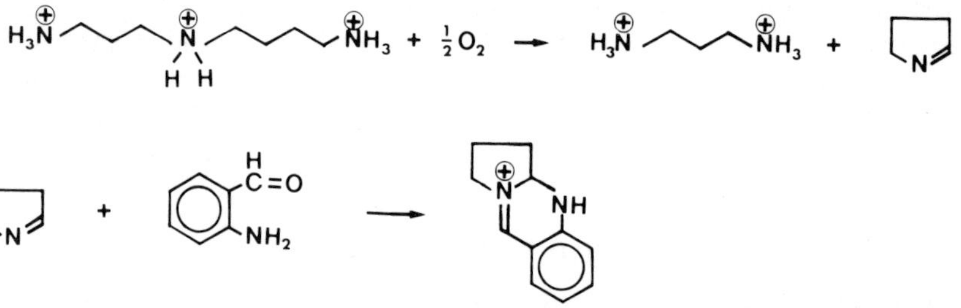

Figure 8 Enzymic oxidation by *Serratia marcescens* of spermidine to Δ^1-pyrroline, and its subsequent transformation by *o*-aminobenzaldehyde to 2,3-trimethylene-1,2-dihydroquinazolinium

3. *Spermidine Plus Spermine*

No specific enzymatic method is available for spermine. Spermidine and spermine are both stoichiometrically oxidized by plasma amine oxidase to the corresponding aminoaldehydes (Figure 9; and see Chapters 18 and 22). If the oxidation is carried out in the presence of resorcinol, a fluorescent product is obtained which can be quantitatively determined at 520 nm (excitation of fluorescence at 366 nm). The method allows the determination of 200 nmole amounts (Unemoto and coworkers, 1963). Another possibility is to use N-methyl-2-benzothiazolone hydrazone as a reagent, which forms with the aminoaldehydes a blue cation as is illustrated in Figure 9 (Bachrach and

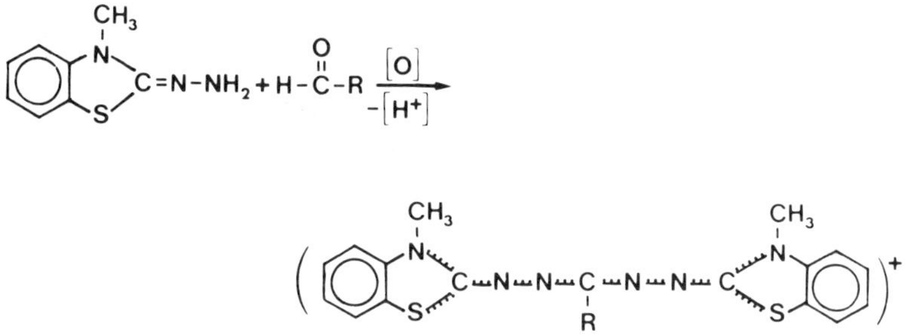

Figure 9 Reaction of aldehydes with N-methyl-2-benzothiazolone hydrazone

Reches, 1966). The molar extinction coefficient at 660 nm of the reaction products of spermidine and spermine are 6,250 and 12,500 $1.mole^{-1}.cm^{-1}$, respectively. The method allows the determination of 3-5 nmole of spermidine and spermine. It may be possible to improve the sensitivity of the method by application of purified enzymes and thus lower the blank values.

B. Immunological Methods

Antibodies against spermidine, putrescine and cadaverine were produced more than ten years ago and used in immunological and oncological experiments (Quash and coworkers, 1973; Quash, 1967). Methods have now been developed which allow the measurement of bound polyamines, using immunolatex and tritiated antipolyamine antibodies (Bonnefoy-Roch and Quash, 1978; Quash and coworkers, 1978).

Of more general use are the radioimmunological assays for spermine (Bartos and coworkers, 1975) and spermidine (Bartos and coworkers, 1978). Thanks to adequate purification of the antibodies (Bartos and coworkers, 1978; Bartos and Bartos, 1978), these methods have now overcome the initial problems of limited specificity. This was shown by the above mentioned systematic comparison of the radio-immunoassay with ion exchange column chromatography and coupled gas-liquid chromatography-mass spectrometry (Bartos and coworkers, 1977). The methods are now available for routine applications and have in fact already been used in clinical trials (Bagdade and coworkers, 1978; Musgrave and coworkers, 1978; Puri and coworkers, 1978; Campbell and coworkers, 1978).

The procedure is briefly as follows: the assay mixture contains [^{3}H] spermine (or [^{3}H] spermidine), antiserum, and 0.01-0.02 ml of serum sample. After 16 h incubation at 4°C, labelled bound and free spermine (or spermidine) are separated by adsorbing the labelled bound polyamine to charcoal. The sensitivity of the method allows measurement of 1 pmole of spermidine and spermine per 0.002 ml of serum. The intra-assay precision was 12.8% at 0.3 nmole of spermidine per ml of serum (Bartos and coworkers, 1977).

Radioimmunological methods open a new dimension of polyamine assay, because of the superior sensitivity of the method and the large number of samples analyzable with sufficient precision.

V. SUMMARY AND CONCLUSION

The arsenal of analytical methods now available for the measurement of putrescine, spermidine and spermine allows the solution of almost all problems for which sensitive and specific determination of these amines is crucial. The still increasing rate with which scientific papers related to polyamines appear obviously mirrors the ease with which polyamine determinations can now be carried out routinely and automatically. A shortcoming of most methods is their limited versatility which restricts their application to the above mentioned amines. Even extension of the determinations to the homologous diamines may cause problems, and unfortunately some of the procedures do not even take into account the possibility of the presence of cadaverine in biological samples.

There is presently no method available which allows the routine measurement of even simple polyamine conjugates, such as their monoacetyl derivatives. The dansylation method in combination with suitable separation techniques might be a basis for a practical solution (Abdel-Monem and Ohno, 1977a,b; Abdel-Monem and coworkers, 1978). Improvement of ion exchange column chromatographic separations could be another alternative to the solution of this problem. Coupling of gas-liquid chromatography and mass spectrometry (Walle, 1973) has already been shown suitable in this respect. The methods presently in use to detect polyamine metabolites in urine and tissues (Perry, Hansen and Kloster, 1972; Tsuji, Nakajima and Sano, 1975; Noto, Tanaka and Nakajima, 1978) are only of limited practical value. Future systematic work in this area may prove especially important for the improvement of our knowledge of polyamine functions. A still unsolved problem of great significance is the cellular and subcellular localization of the polyamines. It can be hoped that immunological methods will fill out this gap in the near future.

REFERENCES

Abe, F. and Samejima, K. (1975). *Anal. Biochem.*, **67**, 298–308.
Abdel-Monem, M. M. and Ohno, K. (1975). *J. Chromat.*, **107**, 416–419.
Abdel-Monem, M. M. and Ohno, K. (1977). *J. Pharm. Sci.*, **66**, 1089–1094.
Abdel-Monem, M. M. and Ohno, K. (1977). *J. Pharm. Sci.*, **66**, 1195–1197.

Abdel-Monem, M. M., Ohno, K., Newton, N. E. and Weeks, C. E. (1978). Thin-layer chromatography and high-pressure liquid chromatography of dansyl polyamines. In R. A. Campbell, D. R. Morris, D. Bartos, G. D. Daves, Jr. and F. Bartos (Eds), *Advances in Polyamine Research*, Vol. 2, Raven Press, New York, pp. 37–49.
Adler, H., Margoshes, M., Snyder, L. R. and Spitzer, C. (1977). *J. Chromat.*, **143**, 125–136.
Bach, M. K. (1966). *Anal. Biochem.*, **17**, 183–191.
Bachrach, U. (1973). *Function of Naturally Occurring Polyamines*. Academic Press, New York and London.
Bachrach, U. (1976). *Ital. J. Biochem.*, **25**, 77–93.
Bachrach, U. (1978). Analytic methods for polyamines. In R. A. Campbell, D. R. Morris, D. Bartos, G. D. Daves, Jr. and F. Bartos (Eds), *Advances in Polyamine Research* Vol. 2, Raven Press, New York, pp. 5–11.
Bachrach, U. and Ben Joseph, M. (1973). Tumor cells, polyamines and polyamine derivatives. In D. H. Russell (Ed), *Polyamines in Normal and Neoplastic Growth*, Raven Press, New York, pp. 15–26.
Bachrach, U. and Oser, I. S.(1963). *J. Biol. Chem.*, **238**, 2098–2101.
Bachrach, U. and Reches, B. (1966). *Anal. Biochem.*, **17**, 38–48.
Bagdade, J., Campbell, R. A., Grettie, D., Bartos, D. and Bartos, F. (1978). Effects of polyamines on human arterial smooth muscle cells in tissue culture. In R. A. Campbell, D. R. Morris, D. Bartos, G. D. Daves, Jr. and F. Bartos (Eds), *Advances in Polyamine Research*, Vol. 2, Raven Press, New York, pp. 345–349.
Bartos, F. and Bartos, A. (1978). Antipolyamine antibodies. In R. A. Campbell, D. E. Morris, D. Bartos, G. D. Daves, Jr. and F. Bartos (Eds), *Advances in Polyamine Research*, Vol. 2, Raven Press, New York, pp. 65–70.
Bartos, F., Bartos, D., Dolney, A. M., Grettie, D. P. and Campbell, R. A. (1978). *Res. Commun. Chem. Pathol.*, **19**, 295–309.
Bartos, F., Bartos, D., Grettie, D. P., Campbell, R. A., Marton, L. J., Smith, R. G. and Daves, Jr., G. D. (1977). *Biochem. Biophys. Res. Commun.*, **75**, 915–919.
Bartos, D., Campbell, R. A., Bartos, F. and Grettie, D. P. (1975). *Cancer Res.* **35**, 2056–2060.
Beckey, H. D. and Schulten, H. R. (1975). *Angew, Chem. Int. Ed.*, **14**, 403–415.
Beer, S. R. and Kosuge, T. (1970). *Virology*, **40**, 930–938.
Beninati, S., Sartori, C. and Argento-Cerù, M. P. (1977). *Anal. Biochem.*, **80**, 101–107.
Benson, J. R. and Hare, P. E. (1975). *Proc. Natl. Acad. Sci. USA*, **72**, 619–622.
Blau, K. (1961). *Biochem. J.*, **80**, 193–200.
Bonnefoy-Roch, A. M. and Quash, G. A. (1978). Development of immunolatex procedures for measuring bound polyamines. In R. A. Campbell, D. R. Morris, D. Bartos, G. D. Daves,Jr. and F. Bartos (Eds), *Advances in Polyamine Research*, Vol. 2, Raven Press, New York, pp. 55–63.
Bremer, H. J., Kohne, E. and Endres, E. (1971). *Clin. Chim. Acta*, **32**, 407–418.
Bremner, J. M. and Kenten, R. H. (1951). *Biochem. J.*, **49**, 651–655.
Brooks, J. B. and Moore, W. E. (1969). *J. Microbiol.*, **15**, 1433–1447.
Campbell, R. A., Talwalker, Y. B., Harner, M. H., Bartos, D., Bartos, F., Musgrave, J. E., Puri, H., Grettie, D. P. Dolney, A. M. and Loggan, B. (1978). Polyamines, Uremia and Hemodialysis. In R. A. Campbell, D. R. Morris, D. Bartos, G. D. Daves, Jr. and F. Bartos (Eds), *Advances in Polyamine Research*, Vol. 2, Raven Press, New York, pp. 319–343.
Campello, A. P., Tabor, C. W. and Tabor, H. (1965). *Biochem. Biophys. Res. Commun.*, **19**, 6–9.
Cincotta, J. J. and Feinland, R. (1962). *Anal. Chem.*, **34**, 774–776.

Cohen, S. S., Morgan, S. and Streibel, E. (1969). *Proc. Natl. Acad. Sci. USA*, **64**, 669–676.

Demisch, L., Bochnik, H. J. and Seiler, N. (1976). *Clin. Chim. Acta*, **70**, 357–369.

Denton, M. D., Glazer, H. S., Walle, T., Zellner, D. C. and Smith, F. G. (1973a). Clinical application of new methods of polyamine analysis. In D. H. Russell (Ed). *Polyamines in Normal and Neoplastic Growth*, Raven Press, New York, pp. 373–380.

Denton, M. D., Glazer, H. S., Zellner, D. C. and Smith, F. G. (1973b). *Clin. Chem.*, **19**, 904–907.

Dion, A. S. and Herbst, E. J. (1970). *Ann. N.Y. Acad. Sci.*, **171**, 723–734.

Dreyfuss, F., Chayen, R., Dreyfuss, G., Dvir, R. and Ratan, J. (1975). *Israel J. Med. Sci.*, **11**, 785–795.

Dreyfuss, G., Dvir, R., Harrel, A. and Chayen, R. (1973). *Clin. Chim. Acta*, **49**, 65–72.

Ehrsson, H. and Brötell, H. (1971). *Acta Pharm. Suecica*, **8**, 591–598.

Elliott, B. and Michaelson, I. A. (1967). *Anal. Biochem.*, **19**, 184–187.

Fischer, F. G. and Bohn, H. (1957). *Hoppe-Seyler's Z. Physiol. Chem.*, **308**, 108–115.

Fleisher, J. H. and Russell, D. H. (1975). *J. Chromat.*, **110**, 335–340.

Fuchs, H. (1939). *Hoppe-Seyler's Z. Physiol. Chem.*, **257**, 149–150.

Fujita, K., Nagatsu, T., Maruta, K., Ito, M., Senba, H. and Miki, K. (1976). *Cancer Res.*, **36**, 1320–1324.

Gehrke, C. W., Kuo, K. C. and Ellis, R. L. (1977). *J. Chromat.*, **143**, 345–362.

Gehrke, C. W., Kuo, K. C., Zumwalt, R. W. and Waalkes, T. P. (1973). The determination of polyamines in urine by gas-liquid chromatography. In D. H. Russell (Ed) *Polyamines in Normal and Neoplastic Growth*, Raven Press, New York, pp. 343–353.

Gehrke, C. W., Kuo, K. C. and Zumwalt, R. W. (1974). *J. Chromatog.*, **89**, 231–238.

Giumanini, A. G., Chiavari, G. and Scarpini, F. L. (1976). *Anal. Chem.*, **48**, 484–489.

Grettie, D. P., Bartos, D., Bartos, F., Smith, R. G. and Campbell, R. A. (1978). Purification of radiolabelled polyamines and isolation of polyamines from serum by silica gel chromatography. In R. A. Campbell, D. R. Morris, D. Bartos, G. D. Daves, Jr. and F. Bartos (Eds), *Advances in Polyamine Research*, Vol. 2, Raven Press, New York, pp. 13–21.

Guirard, B. H. and Snell, E. E. (1964). *J. Bacteriol.*, **88**, 72–80.

Hämäläinen, R. (1941). *Acta Soc. Med. 'Duodecim', Ser. A.*, **23**, 97–165.

Hakanson, R. and Rönnberg, A. L. (1973). *Anal. Biochem.*, **54**, 353–361.

Hammond, J. E. and Herbst, E. J. (1968). *Anal. Biochem.*, **22**, 474–484.

Harik, S. I., Pasternak, G. W. and Snyder, S. H. (1973). *Biochim. Biophys. Acta*, **304**, 753–764.

Hatano, H., Shimizu, K., Rokushika, S. and Murakami, F. (1970). *Anal. Biochem.*, **35**, 377–383.

Hayashi, T., Sugiura, T., Kawai, S. and Ohno, T. (1978). *J. Chromat. Biomed. Appl.*, **145**, 141–146.

Heby, O. (1972). *Insect Biochem.*, **2**, 13–22.

Heby, O. and Andersson, G. (1978). *J. Chromat. Biomed. Appl.*, **145**, 73–80.

Herbst, E. J., Keister, D. L. and Weaver, R. H. (1958). *Arch. Biochem. Biophys.*, **75**, 178–185.

Holder, S. and Bremer, H. J. (1966). *J. Chromat.*, **25**, 48–57.

Inoue, H. and Mizutani, A. (1973). *Anal. Biochem.*, **56**, 408–416.

Johnson, H. G. and Bach, M. K. (1968). *Arch. Biochem. Biophys.*, **128**, 113–123.

Johnson, M. W. and Harkman, R. (1962). *Virology*, **17**, 276–281.

Kaiser, R. E. (Ed) (1976). *Einführung in die Hochleistungs-Dünnschicht-Chromatographie*, Eigenverlag, Institut für Chromatographie, Bad Dürkheim.

Kneifel, H., Rolle, I. and Paschold, B. (1977). *Z. Naturforsch.*, **32**, 190–192.
Kremzner, L. T. (1966). *Anal. Biochem.*, **15**, 270–277.
Kremzner, L. T., Barrett, R. E. and Terrano, M. J. (1970). *Ann. N. Y. Acad. Sci.*, **171**, 735–748.
Kuttan, R., Radhakrishnan, A. N., Spande, T. and Witkop, B. (1971). *Biochemistry*, **10**, 361–365.
Lähdevirta, J., Raina, A. and Heikel, T. (1957). *Scand. J. Clin. Lab. Invest.*, **9**, 345–348.
Lawrence, J. F. and Frei, R. W. (1976). Chemical derivatization in liquid chromatography. In *J. Chromat. Library Vol. 7*, Elsevier, Amsterdam, Oxford, New York.
Lipton, A., Sheehan, L. M. and Kessler, G. F. (1975). *Cancer (Brussels)*, **35**, 464–468.
Macek, K., Deyl, Z., Jiránek, J. and Smrz, M. (1978). *Sbornik-lékarsky*, **80**, 44–48.
Majer, J. R. and Boulton, A. A. (1973). *Meth. Biochem. Anal.*, **21**, 467–514.
Makita, M., Yamamoto, S. and Kono, M. (1975). *Clin. Chim. Acta*, **61**, 403–405.
Marton, L. J., Heby, O., Wilson, C. B. and Lee, P. L. Y. (1974). *FEBS Lett.*, **41**, 99–103.
Marton, L. J. and Lee, P. L. Y. (1975). *Clin. Chem.*, **21**, 1721–1724.
Marton, L. J., Russell, D. H. and Levy, C. C. (1973). *Clin. Chem.*, **19**, 923–926.
McIntyre, F. C., Roth, L. W. and Shaw, J. L. (1947). *J. Biol. Chem.*, **170**, 537–544.
Michaels, R. and Tchen, T. T. (1968). *J. Bacteriol.*, **95**, 1966–1967.
Morris, D. R. (1971). Automated determination of polyamines. In H. Tabor and C. W. Tabor (Eds), *Methods in Enzymology* Vol. 17B, Academic Press, New York and London, pp. 850–853.
Morris, D. R., Koffron, K. L. and Okstein, C. H. (1969). *Anal. Biochem.*, **30**, 449–453.
Morris, D. R. and Pardee, A. B. (1966). *J. Biol. Chem.*, **241**, 3129–3135.
Musgrave, J. E., Campbell, R. A., Bartos, D., Bartos, F., Harner, M. H., Talwalker, Y. B., Puri, H., Grettie, D. P. and Loggan, B. (1978). Serum-free polyamine levels in renal transplant patients. In R. A. Campbell, D. R. Morris, D. Bartos, G. D. Daves, Jr. and F. Bartos (Eds), *Advances in Polyamine Research*, Vol. 2, Raven Press, New York, pp. 351–358.
Nakamura, H. and Pisano, J. J. (1976). *J. Chromat.*, **121**, 33–40.
Navratil, J. D. and Walton, H. F. (1975). *Anal. Chem.*, **47**, 2443–2446.
Newton, N. E., Ohno, K., Abdel-Monem, M. M. (1976). *J. Chromat.*, **124**, 277–285.
Nishioka, K. and Romsdahl, M. M. (1974). *Clin. Chim. Acta*, **57**, 155–161.
Noto, T., Tanaka, T. and Nakajima, T., (1978). *J. Biochem.*, **83**, 543–552.
Pegg, A. E., Lockwood, D. H. and Williams-Ashman, H. G. (1970). *Biochem. J.*, **117**, 17–31.
Pegg, A. E. and Williams-Ashman, H. G. (1969). *J. Biol. Chem.*, **244**, 682–693.
Perry, T. L., Hansen, S., Foulks, J. G. and Lang, G. M. (1965). *J. Neurochem.*, **12**, 397–405.
Perry, T. L., Hansen, S. and Kloster, M. (1972). *J. Neurochem.*, **19**, 1395–1396.
Puranen, U. H. (1936). *Dtsch. Z. Ges. Gerichtl. Med.*, **26**, 366–381.
Puri, H., Campbell, R. A., Puri, V., Harner, M. H., Talwalker, Y. B., Musgrave, J. E., Bartos, F., Bartos, D. and Loggan, B. (1978). Serum-free polyamines in children with systemic Lupus erythematosus. In R. A. Campbell, D. R. Morris, D. Bartos, G. D. Daves, Jr. and F. Bartos (Eds), *Advances in Polyamine Research*, Vol. 2, Raven Press, New York, pp. 13–21.
Quash, G. A. (1967). *C. R. Acad. Sci. Ser. D.*, **265**, 934–936.
Quash, G., Bonnefoy-Roch, A. M., Gazzolo, L. and Niveleau, A. (1978). Cell membrane polyamines: A qualitative and quantitative study using immunolatex and

tritiated anti-polyamine antibodies. In R. A. Campbell, D. R. Morris, D. Bartos, G. D. Daves, Jr. and F. Bartos (Eds) *Advances in Polyamine Research*, Vol. 2 Raven Press, New York, pp. 85–92.

Quash, G. A., Fresland, L., Delain, E. and Huppert, J. (1973). Antipolyamine antibodies and growth. In D. H. Russell (Ed) *Polyamines in Normal and Neoplastic Growth*, Raven Press, New York, pp. 157–165.

Raina, A. (1962). *Scand. J. Clin. Lab. Invest.*, **14**, 318–319.

Raina, A. (1963). *Acta Physiol. Scand.*, **60**, (Suppl. 218), 7-81.

Ramakrishna, S. and Adiga, P. R. (1973). *J. Chromat.*, **86**, 214–218.

Rennert, O. M., Frias, J. and Shukla, J. B. (1976). *Tex. Rep. Biol. Med.*, **34**, 187–196.

Rinaldini, L. M. (1970). *Anal. Biochem.*, **36**, 352–367.

Roeder, E., Pigulla, I. and Troschuetz, J. (1977). *Z. Anal. Chem.*, **288**, 56–58.

Rosenthal, S. M. and Tabor, C. W. (1956). *J. Pharmacol. Exp. Ther.*, **116**, 131–138.

Roth, M. (1971). *Anal. Chem.*, **43**, 880–882.

Roth, M. and Hampai, A. (1973). *J. Chromat.*, **83**, 353–356.

Russell, D. H. (1971). *Nature (New Biol.)*, **233**, 144–145.

Russell, D. H. and Levy, C. C. (1971). *Cancer Res.*, **31**, 248–251.

Russell, D. H., Levy, C. C., Schimpf, S. C. and Hawk, I. A. (1971). *Cancer Res.*, **31**, 1555–1558.

Russell, D. H., Medina, V. J. and Snyder, S. H. (1970). *J. Biol. Chem.*, **245**, 6732–6738.

Saeki, Y., Uehara, N. and Shirakawa, S. (1978). *J. Chromat. Biomed. Appl.*, **145**, 221–229.

Samejima, K. (1974). *J. Chromat.*, **96**, 250–254.

Samejima, K., Kawase, M., Sakamoto, S., Okada, M. and Endo, Y. (1976). *Anal. Biochem.*, **76**, 392–406.

Seiler, N. (1970). *Meth. Biochem. Anal.*, **18**, 259–337.

Seiler, N. (1975). Assay procedures for polyamines and GABA in animal tissues with special reference to dansylation methods. In N. Marks and R. Rodnight (Eds), *Research Methods in Neurochemistry* Vol. 3, Plenum Publishing Corp. New York, pp. 409–441.

Seiler, N. (1977). *J. Chromat.*, **143**, 221–246.

Seiler, N. and Al-Therib, M. J. (1974). *Biochem. J.*, **144**, 29–35.

Seiler, N. and Askar, A. (1971). *J. Chromat.*, **62**, 121–127.

Seiler, N. and Demisch, L. (1977). Fluorescent derivatives. In K. Blau and G. S. King (Eds) *Handbook of Derivatives for Chromatography*, Heyden, London; Bellmawr, Rheine, pp. 346–390.

Seiler, N. and Knödgen, B. (1973). *Org. Mass Spectrom.*, **7**, 97–105.

Seiler, N. and Knödgen, B. (1977). *J. Chromat.*, **131**, 109–119.

Seiler, N., Knödgen, B. and Eisenbeiss, F. (1978). *J. Chromat. Biomed. Appl.*, **145**, 29–39.

Seiler, N. and Lamberty, U. (1975a). *J. Neurochem.*, **24**, 5–13.

Seiler, N. and Lamberty, U. (1975b). *Compt. Biochem. Physiol.*, **52B**, 419–425.

Seiler, N. and Möller, H. (1969). *Chromatographia*, **2**, 470–475.

Seiler, N., Schmidt-Glenewinkel, T. and Schneider, H. H. (1973). *J. Chromat.*, **84**, 95–107.

Seiler, N. and Schmidt-Glenewinkel, T. (1975). *J. Neurochem.*, **24**, 791–795.

Seiler, N., Werner, G., Fischer, H. A., Knödgen, B. and Hinz, H. (1969). *Hoppe-Seyler's Z. Physiol. Chem.*, **350**, 676–682.

Seiler, N. and Wiechmann, M. (1966). *Z. Anal. Chem.*, **220**, 109–127.

Seiler, N. and Wiechmann, M. (1967). *Hoppe-Seyler's Z. Physiol. Chem.*, **348**, 1285–1290.

Seiler, N. and Wiechmann, M. (1970). TLC analysis of amines as their DANS-derivatives. In A. Niederwieser and G. Pataki (Eds), *Progress in Thin-Layer Chromatography and Related Methods*, Vol. 1, Ann Arbor-Humphrey Humphrey Science publishers, Ann Arbor, pp. 95–144.

Shimizu, H., Kakimoto, J. and Sano, I. (1964). *J. Pharmacol. Exp. Ther.*, **143**, 199–204.

Simons, S. S. Jr. and Johnson, D. F. (1976). *J. Amer. Chem. Soc.*, **98**, 7098–7099.

Simons, S. S. Jr. and Johnson, D. F. (1977). *Anal. Biochem.*, **82**, 250–254.

Smith, E. D. and Radford, R. D. (1961). *Anal. Chem.*, **33**, 1160–1162.

Smith, R. G. and Daves, G. D. (1977). *Biomed. Mass Spectrometry*, **4**, 146–151.

Smith, R. G., Daves, Jr. G. D. and Grettie, D. P. (1978). Gas-chromatography — mass spectrometry analysis of polyamines in serum using deuterium-labelled analogs as internal standards. In R. A. Campbell, D. R. Morris, D. Bartos, G. D. Daves, Jr. F. Bartos (Eds), *Advances in Polyamine Research*, Vol. 2, Raven Press, New York, pp. 23–35.

Smith, T. A. (1970). *Anal. Biochem.*, **33**, 10–15.

Smith, T. A. (1970). *Phytochem.*, **9**, 1479–1486.

Smith, T. A. and Stevens, R. A. J. (1971). *Biochem. J.*, **122**, 17.

Sturman, J. A. and Gaull, G. E. (1974). *Pediat. Res.*, **8**, 231–237.

Sugiura, T., Hayashi, T., Kawai, S. and Ohno, T. (1975). *J. Chromat.*, **110**, 385–388.

Tabor, C. W. and Rosenthal, S. M. (1963). Determination of spermine, spermidine, putrescine and related compounds. In S. P. Colowick and N. O. Kaplan (Eds) *Methods in Enzymology*, Vol. 6, Academic Press, New York and London, pp. 615–622.

Tabor, C. W. and Tabor, H. (1966). *J. Biol. Chem.*, **241**, 3714–3723.

Tabor, H., Tabor, C. W. and Irreverre, F. (1973). *Anal. Biochem.*, **55**, 457–467.

Tsuji, M., Nakajima, T. and Sano, I. (1975). *Clin. Chim. Acta*, **59**, 161–167.

Unemoto, T., Ikeda, K., Hayashi, M. and Miyaki, K. (1963). *Chem. Pharm. Bull. (Tokyo)*, **11**, 148–151.

Veening, H., Pitt, W. W. Jr. and Jones, G. R. Jr. (1974). *J. Chromat.*, **90**, 129–139.

Villanueva, V. R., Adlakha, R. C. and Cantera-Soler, A. M. (1977). *J. Chromat.*, **139**, 381–385.

Walle, T. (1973). Gas-chromatography-mass spectrometry of di- and polyamines in human urine: Identification of monoacetyl spermidine as a major metabolic product of spermidine in a patient with acute myelocytic leukemia. In D. H. Russell (Ed) *Polyamines in Normal and Neoplastic Growth*, Raven Press, New York, pp. 355–366.

White, W. F., Cohen, A. I., Rippel, R. H., Story, J. C. and Schally, A. V. (1968). *Endocrinology*, **82**, 742–752.

Wrede, F. (1926). *Hoppe-Seyler's Z. Physiol. Chem.*, **153**, 291–313.

Zappia, V., Carteni-Farina, M. and Della Pietra, G. (1972). *Biochem. J.*, **129**, 703–709.

Zlatkis, A., Ettre, L. S. and Dijkstra, G. (Eds) (1977). In *Advances in Chromatography* (Proceedings of the Twelfth International Symposium, Amsterdam Nov. 7-10, 1977). Chromatography Symposium Department of Chemistry, University of Houston, Houston, Texas.

Zlatkis, A. and Kaiser, R. E. (Eds) (1977) High performance thin-layer chromatography. *Journal of Chromat. Library Vol. 9*, Elsevier, Amsterdam, Oxford, New York.

Index

ABBREVIATIONS

cAMP	cyclic adenosine 3′:5′monophosphate
DNA	deoxyribonucleic acid
methyl-GAG	methylglyoxal *bis*(guanylhydrazone)
ODC	ornithine decarboxylase
RNA	ribonucleic acid
SAMD	S-adenosyl-L-methionine decarboxylase
TPA	12-*0*-tetradecanoylphorbol-13-acetate